COMPLEX AND DISTRIBUTED SYSTEMS
Analysis, Simulation and Control

IMACS TRANSACTIONS ON SCIENTIFIC COMPUTATION - 85
11th IMACS World Congress
on Scientific Computation
Oslo, Norway, 5-9 August 1985

Series Editors:

R. VICHNEVETSKY
Rutgers University
New Brunswick, N.J., U.S.A.

R. HENRIKSEN
Institute for Technical Cybernetics
Trondheim, Norway

B. WAHLSTRÖM
Technical Research Center of Finland
Espoo, Finland

VOLUME IV

INTERNATIONAL ASSOCIATION FOR MATHEMATICS AND COMPUTERS IN SIMULATION
(Association Internationale pour les Mathématiques et Calculateurs en Simulation)

COMPLEX AND DISTRIBUTED SYSTEMS
Analysis, Simulation and Control

edited by

SPYROS TZAFESTAS
Computer Science Division
National Technical University
Athens, Greece

and

PIERRE BORNE
Laboratoire d'Automatique et d'Informatique Industrielle
Institut Industriel du Nord
Villeneuve d'Ascq, France

N·H
P&C
1986

NORTH-HOLLAND
AMSTERDAM · NEW YORK · OXFORD · TOKYO

ISBN: 0 444 70097 8
ISBN set: 0 444 70082 x

Published by:
ELSEVIER SCIENCE PUBLISHERS B.V.
P.O. Box 1991
1000 BZ Amsterdam
The Netherlands

Sole distributors for the U.S.A. and Canada:
ELSEVIER SCIENCE PUBLISHING COMPANY, INC.
52 Vanderbilt Avenue
New York, N.Y. 10017
U.S.A.

Library of Congress Cataloging-in-Publication Data

Complex and distributed systems.

(IMACS transactions on scientific computation ; v. 4)
Includes bibliographies and index.
1. Distributed parameter systems. 2. System analysis. 3. Control theory. I. Tzafestas, S. G., 1939- . II. Borne, Pierre. III. Series: IMACS World Congress on Systems Simulation and Scientific Computation (11th : 1985 : Oslo, Norway). IMACS transactions on scientific computation ; v. 4.
QA402.C55 1986 003 86-16218
ISBN 0-444-70097-8 (U.S.)

PRINTED IN THE NETHERLANDS

PREFACE

This fourth volume of the *1985 IMACS Transactions on Scientific Computation* contains selected papers on the analysis, simulation and control of complex and distributed systems, from those presented at the *11th IMACS Congress* (Oslo, Norway, August 1985). The authors were invited to provide extended and/or improved versions of their papers.

The volume is divided into the following five sections:

1. Complex system modelling, simulation and identification,
2. Bond graph analysis and modelling,
3. Nonlinear oscillators and chaotic systems,
4. Distributed parameter systems,
5. Control of complex systems,

which represent five major areas of system theory, and show the modern trend in the growing field of complex and distributed parameter systems theory.

In view of the extensive coverage of many timely and important aspects of complex and distributed systems theory and the rich set of new results derived by internationally known experts, it is hoped that this volume, together with the other volumes of the present IMACS Transactions, will be a useful addition to the technical literature used for research and teaching.

Athens, April 1986

Spyros G. Tzafestas

CONTENTS

Section V
CONTROL OF COMPLEX SYSTEMS

Section I

COMPLEX SYSTEM MODELLING, SIMULATION AND IDENTIFICATION

Complex and Distributed Systems: Analysis, Simulation and Control
S.G. Tzafestas and P. Borne (editors)
Elsevier Science Publishers B.V. (North-Holland)

A CONCEPTUAL FRAMEWORK FOR MODELLING THE DYNAMICS OF ENVIRONMENTAL SYSTEMS

DR. I. MOFFATT

Department of Environmental Science
University of Stirling
STIRLING
Scotland

A conceptual framework for modelling the dynamics of environmental systems is presented. It is argued that apparently stable systems can evolve via bifurcation when critical thresholds are exceeded. When a system is forced further away from equilibrium dissipative structures emerge. These dissipative structures are characterised by stochastic, non-linear feedback mechanisms which have the capacity to transform an apparently stable environmental system into a relatively more complex one which evolves. Some examples of these structures are simulated using system dynamics and the implications for further research are discussed.

INTRODUCTION

One of the major difficulties in building dynamic models of environmental systems resides in resolving the paradox that these systems are both stable yet evolve. Generally, model builders have concentrated their efforts on understanding the dynamics of stable systems. Whilst this research is well established it is clear that by focussing attention on stable systems model builders have, by and large, ignored the evolution of such apparently stable systems. This paper outlines a conceptual framework which can simultaneously accommodate both the dynamics of stable and developing environmental systems.

In the following section a brief definition and discussion of environmental systems is offered. It will be argued that the study of environmental systems transcends conventional disciplinary boundaries and, through necessity, has to embrace both hard and soft systems simultaneously. This discussion is then followed, in section three, by presenting a conceptual framework for modelling the dynamics of environmental systems. By drawing upon and extending the work of the Brussels school (1) it is argued that an apparently stable system can pass through a chaotic mode of behaviour which, if driven further from a previous position of equilibrium, can then undergo a radical transformation which has the potential for a new qualitatively different system to emerge. These dissipative structures are characterised by stochastic, non-linear feedback mechanisms and are present, but latent, in many environmental systems. The fourth section uses system dynamics and DYNAMO to simulate the various modes of behaviour described in the conceptual framework for modelling environmental systems. Finally, some of the implications of this conceptual framework for further research into the dynamics of environmental systems are discussed.

ENVIRONMENTAL SYSTEMS ARE HARD AND SOFT

According to Bennett and Chorley an environmental system can be defined very broadly as an interdisciplinary study embracing 'physical, biological, man-made, social and economic reality' (2). Obviously, such a broad definition covers a whole host of disciplines and it is, perhaps, useful to consider environmental systems as the intersection of three sets namely the ecological; the economic and socio-political systems. The study of ecological systems is primarily concerned with the explicit elucidation of the structure and function of a plant or animal community and its natural habitat. The habitat can consist of both organic and inorganic material. Several texts have shown that the structure and functioning of ecological systems can be understood by use of computer simulation (3, 4). Whilst ecological studies are one important facet of environmental science it would be misleading to suggest that all environmental scientists are concerned solely with ecological problems. Increasingly, society's economic activities are having a major impact upon ecological systems. The misuse of ecological systems for short term economic gain can have a major, if not catastrophic, impact on the life support systems of this planet. If economic and ecological systems are not integrated in a holistic manner then serious repercussions may result from our short sighted negligence (5).

The study of the inter-relationships between economic and ecological systems do not, however, take place in a socio-political vacuum. Increasingly, decisions made by socio-political institutions can have a major impact on the environment. It is, therefore, essential that environmental scientists consider the way in which material aspects of our culture support a particular set of political ideas as opposed to more ecologically sensible political philosophies and practices. Pepper, for example, notes that 'the British Conservative Government in 1980-1 put so much research money into nuclear power rather than 'soft' energy sources perhaps because of the power of the pro-nuclear lobby and also because it wanted to break the political power of the coalminers'(6). In an attempt to explain the way in which vested interests manipulate environmental decision making environmental scientists need to consider critically the ethical principles and political practices of these groups.

Clearly, an environmental system is a complex phenomena and it cannot be studied in its entirety by adopting a purely ecological or economic or socio-political perspective. To try and explain the dynamics of environmental systems from any one or two perspectives is myopic. Yet, to try and develop a conceptual framework which can accommodate all three sets in an integrated, holistic dynamic framework is exceedingly difficult. One of the reasons for this difficulty resides in the fact that environmental systems are simultaneously hard and soft systems.

In a recent reappraisal of systems analysis Checkland has argued that hard systems are a special case of the so-called soft-systems methodology (7). A hard system can be characterised as the search for an efficient means of reaching a clearly defined objective or goal, once the goal or objective is clearly defined, then systematic appraisal of alternative solutions to the problem, helped by various techniques, enables the problem to be solved. A classic example of this approach was the successful attempt by the American nation to land a man on the moon. A soft systems approach provides a way of tackling ill-structured problems without imposing on them the means-end dichotomy which is characteristic of the hard systems approach. In many cases the use of verbal models helps to clarify the major interactions in a system without degenerating into arid polemic. Alternatively, some simulation modelling of soft systems can 'degenerate into science fiction, in which fragmentary data are pushed far beyond any limit of credibility (8).

The distinction between hard and soft systems is clearly illustrated below (see Figure 1). Despite the differences between these two approaches to using a systems approach to either optimize or to learn about a specific system of interest several researchers have failed to grasp the significance of the soft methodology (9). Often they have attempted to apply the hard systems approach to problems which are soft. One of the results of this major methodological mistake has been to ignore the interaction of clashing value systems found in soft systems by assuming that the system model builders implicit values are the only ones which are important. Witness, for example, the heated exchange over the value systems embedded in the World and Urban Dynamic models (10, 11, 12). If hard and soft systems are to be integrated in a coherent way then this problem of incorporating dialectical changes within our models must be tackled. Fortunately, the conceptual framework outlined in this paper is capable of achieving this synthesis of hard and soft systems.

A CONCEPTUAL FRAMEWORK

In order to make progress in the dynamic modelling of environmental systems it is essential that a framework is developed which can incorporate both the hard and soft elements of these complex systems. Furthermore, it is fundamentally important that the conceptual framework can resolve the paradox that environmental systems are both stable yet evolve. Generally, model builders have concentrated their efforts on understanding the dynamics of stable systems. Whilst this research is well established it is clear that by focussing attention on stable systems model builders have, by and large, ignored the evolution of such apparently stable systems. This section outlines a conceptual framework which can simultaneously accommodate both the dynamics of stable and developing hard and soft environmental systems.

Any dynamic model may be defined as a simplification of a real world system which changes through time. This apparently straightforward definition of dynamic models hides a bewildering richness of dynamic behaviour (13). This dynamic behaviour can be described as synchronic or diachronic change (14). Synchronic change

	Paradigm 1 Hard systems thinking	Paradigm 2 Soft systems thinking
C: Customers	Decision-makers who command real world systems	Participants who debate the differences between the models and the expression of the problem situation
A: Actors	External analysts and engineers	Those who choose to take part: analaysts and/or problem owners
T: Transformation	Information into advice to decision-makers	Information into specific learning for the "actors"
W: Worldview	Real world problem is systemic. Methodology is systematic Optimization is possible	Real world problem is problematical. Methodology is systemic Learning is possible
O: Owners	Decision-makers/ clients	"Actors" as defined above, or the analyst
E: Ethics	Power structures and value systems of the decision-maker clients	As little as possible compatible with achieving change in the problem situation

Figure 1 Hard and Soft Systems (After Checkland 1984)

describes the way in which the elements of a system alter through space-time within a fixed structure. Diachronic change, however, describes the processes whereby the structure of a system is transformed into another form. Most dynamic models of environmental systems have examined synchronic structures but it is becoming increasingly obvious that diachronic change must be examined if we are to understand the complexities of the dynamics of environmental systems.

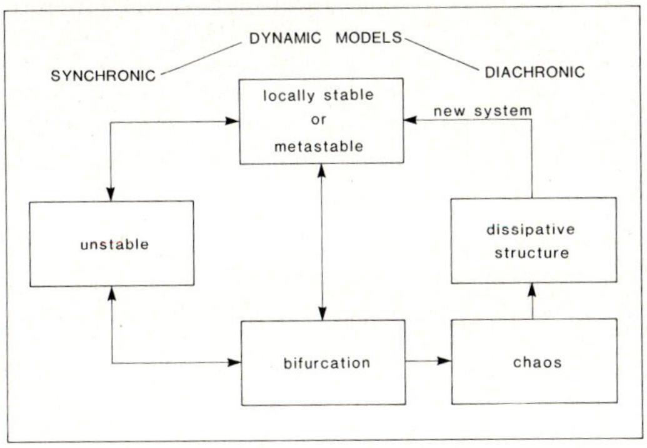

Fig. 2. Various modes of behaviour in dynamic models

In figure 2 a conceptual framework which attempts to integrate both synchronic and diachronic structures is illustrated. Beginning with systems which are locally stable, emphasis is placed on those systems which are in dynamic equilibrium (i.e. whose macroscopic state variables do not fluctuate through time although the microscopic elements may, in fact, change). Included in this class of local stability is the concept of stable cyclical oscillations in the behaviour of a system. As an open environmental system is driven further away from locally stable locations by either exogenous or endogenous change then unstable behaviour is exhibited. An obvious manifestation of such change is a demographic system collapsing to extinction or increasing exponentially.

Apart from the obvious forms of instability a more interesting case is that of bifurcation. Several workers have noted that primary bifurcation or the hysteresis phenomenon can be exhibited in a variety of environmental, biological and chemical systems (15). By forcing a system beyond a threshold of stability the system can achieve a new locally stable state. Alternatively, these primary bifurcations can be developed so that the trajectory of the system changes in a more complex manner. As the system moves away from one position of dynamic equilibrium further bifurcations are possible until chaotic behaviour is observed even in systems with deterministic equations (16).

Chaotic behaviour is observed in many dynamic system models which have one, or more, feedback loops. These non-equilibrium systems, especially when interacting with the outside world may also form the genetic phase for the formation of new structures. These new structures are termed self-organising or dissipative structures. They may appear locally stable but are, as Prigogine and Stengers write 'essentially a reflection of the global situation of non-equilibrium (processes) producing them' (17). Unlike bifurcation and chaotic behaviour in dynamic models of environmental systems these dissipative structures are generated by a mix of deterministic and stochastic elements. It is important to realise that such stochastic perturbations may be very small in any environmental system but can alter fundamentally the entire system of interest. Furthermore, these structures are created and maintained in open systems by a continuous influx of energy or matter (18). In this way the dynamics of environmental systems can exhibit diachronic as well as synchronic change.

SOME EXAMPLES

Numerous examples could be given to illustrate the use of this new conceptual approach to modelling the dynamics of environmental systems. Three examples (population growth using a logistic equation; a dynamic version of Christaller's theory of central places; and the environmental management implications of dynamic models) are described briefly below. In each example, there is a wide range of dynamic behaviour implicit in these relatively simple models.

When a population is composed of single generations with no overlap between successive generations then the population growth occurs in discrete steps. In such circumstances it is convenient to model these systems of interest as difference equations (19). The logistic equation (S-shaped curve) is probably the simplest form of non-linear equation used in ecological studies. There are various ways of writing the logistic equations but a discrete version could be written as follows

$$\Delta p_t = rp_t \left(1 - \frac{p_t}{L}\right) \qquad \text{(equation 1)}$$

where p is the number of people, L is the upper limit of carrying capacity on this number and r(1-P/L) is the rate at which new people are recruited into the system. This difference equation can be modelled using system dynamics to produce a variety of dynamic behaviour. When $0 < r < 1$ then the system converges monotonically towards L. Oscillating convergence towards L is observed when $1 < r < 2$. If, however, the condition $2 < r < 2.57$ a series of stable limit cycles of period 2^n can develop. When $r > 2.57$ then the logistic equation model enters into chaotic behaviour. As the name implies, chaotic behaviour is unpredictable. Any value of r which falls into this regime can generate trajectories that may settle into a stable limit cycle of any integral period or may never settle into a finite cycle (20) in Figure 3.

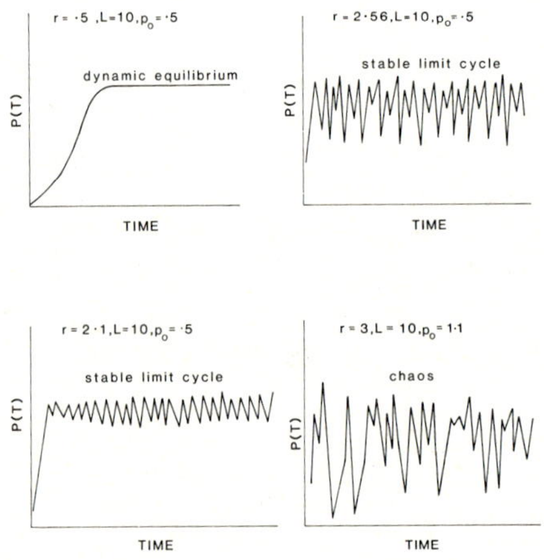

Fig. 3.
Various modes of behaviour in a simple difference equation
see text for explanation

As is well known, Christaller's (1933) theory of central places showed three different hexagonal lattices depending on the marketing, transport or administrative principles (k = 3, 4 and 7 respectively). Each of these sets of hexagonal patterns appear to be timeless, optimised spatial configurations. It was, however, clear to Christaller that these static patterns represented 'only a snapshot of the existing world in continuous change; the stationary state is only fiction whereas motion is reality' (21). In the past decade several researchers have attempted to provide a dynamic version of Christaller's pioneering work (22, 23, 24).

In the 1970s a dynamic model of interacting urban centres, combining both stochastic and deterministic elements, was developed in an attempt to describe the evolution of a central place system (25-28) 1978, 1979a, 1979b, 1981). Using a modified logistic equation in which the natural carrying capacity L (equation 1) of a particular place can be increased by its potential employment capacity as used in the familiar Lowry model (29). Unlike the previous example, however, each population centre is in competition with other centres of activity located elsewhere. Furthermore, each central location is able to act as a focus of production and consumption for the inhabitants of the central place and those in the immediate hinterland. By incorporating non-linearities and stochastic processes into this dynamic model a qualitatively more realistic evolution of central place patterns has been produced. Preliminary empirical work indicates that this model is sufficient to describe correctly the evolution of tertiary employment and residential structure in the Bastogne region of Belgium, 1947-70 (30).

In the previous two examples it is clear that even in the case of some simple, non-linear dynamic models there is a very sharp transition from stable to unstable or chaotic behaviour as a parameter exceeds a critical threshold value. Within the range of a critical threshold value it is possible to optimise some aspect of the behaviour of the modelled system. In this way it may be possible to develop simple models which may be used for environmental management and planning (31). It is, however, very clear that if simple dynamic models are to be used in environmental management and planning then there is an urgent need to develop rigorous methods to determine sensitive parameters before policies emanating from these models are put into practice.

Apart from the technical problems involved in using dynamic models of environmental systems for management or planning purposes there are also deeper ethical considerations to be taken into account. O'Riordan (32) notes that the environmental ethic exists to change people's outlook on the world, their values and behaviour and not just to shift public policies and redirect particular decisions. One way of addressing ethical issues involved in environmental management and planning is to use a simple dynamic model of structural conflict in the environment. In the case of advanced industrial societies three different attitudes towards the environment can be discerned. The conservative approach suggests that the market mechanism will solve environmental problems when they arise. The liberals, however, suggest that such problems can be eradicated only if further funding is given to environmental management. Finally, the radicals argue that a fundamental shift in attitudes to the environment is required to resolve the problems. These three different attitudes have been built into a dynamic model which incorporates a dissipative structure. This dissipative structure is triggered by stochastic perturbations in a non-linear feedback loop. The result of triggering this dissipative structure can reveal the ways in which conflicting values can lead to different forms of social evolution which are either antagonistic to, or in harmony with, the environment (33).

CONCLUSION

This paper has described the nature of environmental systems as the interaction of ecological, economic and socio-political sets. One of the problems in studying environmental systems resides in the fact that they are both hard and soft. Furthermore, in attempting to model the dynamics of these systems it is clear that both synchronic and diachronic change must be considered. The conceptual framework outlined in the paper incorporates hard and soft systems as well as synchronic and diachronic change.

By drawing upon the notion of dissipative structures it is possible to portray the dynamics of environmental systems as cyclic phenomena moving from stability, into instability, bifurcations, chaos and into dissipative structures. These latter structures can radically transform the behaviour and structure of the entire system of interest. These revolutionary changes are embedded in complex systems and are triggered by low probability stochastic changes which cause fundamental shifts in the structure and function of hitherto apparently stable systems.

Whilst this conceptual framework for modelling the dynamics of environmental systems is tentative it is clear that several environmental systems do in fact exhibit these modes of behaviour. Several examples of the dynamics of environmental systems have been discussed using the method of system dynamics simulation. These examples include stable but oscillating predator-prey relationships; the chaotic behaviour of urban dynamics and dissipative structures illustrating the emergence of a more ecologically sane society as a result of dialectical conflict in a model of an environmental system. Obviously, much more detailed empirical and theoretical research needs to be undertaken in order to comprehend the dynamics of environmental systems. Nevertheless, the conceptual framework described above offers environmental scientists a new and useful way of understanding and changing environmental systems as they unfold around us.

REFERENCES

(1) Prigogine, I. and Stengers, I., Order out of chaos (Heinemann, London, 1982).

(2) Bennett, R.J. and Chorley, R.J., Environmental systems: philosophy, analysis and control (Methuen, London, 1978).

(3) Hall, A.S. and Day, J.W. (eds) Ecosystem modelling in theory and practice (John Wiley, New York, 1977).

(4) Jeffers, J.N.R. An introduction to systems analysis: with ecological applications (Edward Arnold, London, 1978).
(5) Wilson, A.G. and Kirkby, M.J., Mathematics for Geographers and Planners(Oxford University Press, Oxford, 1980).
(6) Pepper, D., The roots of modern environmentalism (Croom Helm, London, 1984).
(7) Checkland, P., Rethinking a systems approach, in: Tomlinson, R. and Kiss,I. (eds), Rethinking the process of operational research and system analysis (Pergamon, Oxford, 1984).
(8) Coyle, R.G., Futures (1984), 16, 6, 594-609.
(9) Morgan, R.K., Area (1981) 13, 3, 219-23.
(10) Forrester, J.W., Urban dynamics (M.I.T., Massachusetts, 1969).
(11) Forrester, J.W., World dynamics (M.I.T., Massachusetts, 1971).
(12) Moffatt, I., Some methodological and epistemological problems involved in system dynamics modelling, in: Morecroft, J.D.W., Anderson, D.F. and Sterman, J.D., (eds) The 1983 International System Dynamics Conferance, (Massachusetts, 1983) pp. 339-57.
(13) May, R.M. (ed) Theoretical ecology: principles and applications (Blackwell, London, 1976).
(14) Huckfeldt, R.R., Kohfeld, E.W. and Likens, T.W., Dynamic modelling an Introduction (Sage, Beverley Hills, 1982).
(15) Oldfield, F., Geography (1983), 68, 3, 245-56.
(16) May, R.M., Nature (1976), 261, 459-67.
(17) Nicolis, G. and Prigogine, I., Self-organisation in non-equilibrium systems from dissipative structures to order through fluctuations (Wiley, New York, 1977).
(18) Peacocke,A.R., An introduction to the physical chemistry of biological organisation (Clarendon Press, Oxford, 1983).
(19) May, R.M., Science (1976), 186, 645-47.
(20) Kohfeld, C.W. and Salert, B., Political Methodology (1978) 8, 1, 1-45.
(21) Christaller, W., Central places in Southern Germany (Prentice Hall, New Jersey, 1966).
(22) Moffatt, I. A theory of settlement at the global scale (unpublished M.Sc. thesis, University of Newcastle Upon Tyne, Dept. of Geography, 1974).
(23) White, D.W. Geographical Analysis (1977), 9, 226-43.
(24) White, D.W. Geographical Analysis (1979), 1, 201-8.
(25) Allen, P.M. and Sanglier, M., Journal of Social and Biological Structures (1978), 1, 3, 265-80.
(26) Allen, P.M. and Sanglier, M., Journal of Social and Biological Structures (1979), 2, 4, 269-78.
(27) Allen, P.M. and Sanglier, M., Geographical Analysis, (1979), 11, 3, 256-72.
(28) Allen, P.M. and Sanglier, M., Environmental Planning A, (1981), 13, 167-83.
(29) Lowry, I.S. A model of metropolis, (Rand Corporation, R.M. 203 R.C.,1964).
(30) Allen, P.M., Sanglier, M., Boon, F., Deneubourg, J.L. and DePalma, A., Models of urban settlement and structures as dynamic self-organising systems (U.S. Dept. of Transportation, DPB, 1981).
(31) Moffatt, I. Environmental management, models and policies, in: Murray-Smith, D.F., (ed), Proceedings of the 1984 U.K.S.C. Conference in Computer Simulation (Butterworths, London, 1984).
(32) O'Riordan, T., Journal of Geography in Higher Education, (1981), 5, 1, 3-17.
(33) Moffatt, I., European Journal of Operational Research, (1986), 25, (in print).

Complex and Distributed Systems: Analysis, Simulation and Control
S.G. Tzafestas and P. Borne (editors)
Elsevier Science Publishers B.V. (North-Holland)

ON THE STOCHASTIC SIMULATION AND OPTIMIZATION IN PRODUCTION ENGINEERING

G. GIORLEO
Dip. Progettazione e Produzione Industriale, Facoltà di Ingegneria - Università di Bari

R. PASQUINO
Istituto di Ingegneria Meccanica, Facoltà di Ingegneria - Università di Salerno

ABSTRACT - The problem of the stochastic simulation and optimization of machining processes in production engineering is here considered with particular attention to optimization analysis. An application to grinding processes is also developed.

1. INTRODUCTION

The construction of reliable and rapid simulation models holds basic importance in the design and inspection of the production systems. Moreover it is well understood, see for instance refs. /1÷3/, that the simulation of a machining operation in production engineering can be defined by input-output models with random parameters or noise. The stochasticity of such simulation models links up, in general, with the uncertainly in which works the whole production system (for the continual changes of the various costs, for the increasing bonds in environment, for the tumultuous technological development, etc.) and, in particular, with the fact that each machining operation is not deterministically repeated (e.g., when the production is realized by several operating machines working in different operation conditions).

More in details on the content of this paper, the second section contains the description of two classes of input-output models which can define a general framework suitable to include consistent simulations of several machining processes, such as the ones considered in refs. /4÷7/, and which will be afterwords analysed in the application. The third section defines the structure and the formulation of the optimization process; in this section the main lines to be followed in the computer aided optimization are also indicated. The final section contains an application to the analysis of grinding process, a problem which has been already considered by the Authors in the paper /4/, where a stochastic formulation of the Mayne and Malkin model /5/ was proposed.

2. A FRAMEWORK FOR MODELLING AND SIMULATION

A framework for two simple models towards the simulation of machining processes is here considered in a general form, namely without any particularization. The application realized in the last section will then show how the simulation of the grinding process can be cast into the above announced framework.

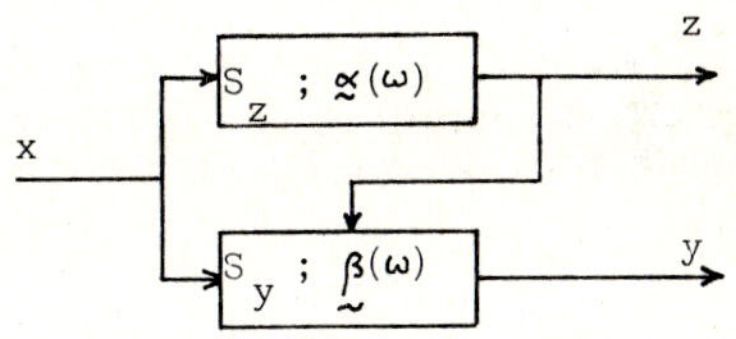

Fig. 1 - Input-output scheme (model 1)

With reference to the input-output scheme sketched in fig.1, the first simulation model (model 1) is characterized by the following items:

a) x is the input, variable technological parameters characterizing the machining process (cutting velocity, machining dept, etc);

b) z is the output, quantitative production obtained as a stochastic algebric map from x into z:

$$z = S_z(x; \underset{\sim}{\alpha}(\omega)) \tag{1}$$

where $\underset{\sim}{\alpha}$ is a set of random parameters characterizing the simulation model;

c) y is the energy spent to obtain z and

is defined as a stochastic algebric map from x and z into y:

$$y = S_y(x,z;\underset{\sim}{\beta}(\omega)) \qquad (2)$$

where $\underset{\sim}{\beta}(\omega)$ is a set of random parameters characterizing the simulation model.

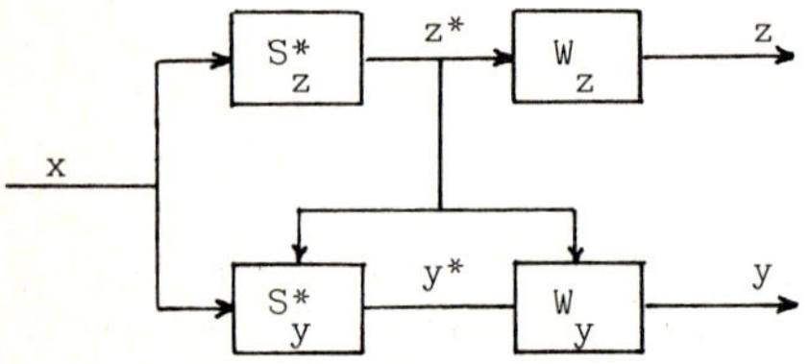

Fig. 2 - Input-output scheme (model 2)

A simulation model alternative to the above model 1 is now considered. Such a model (model 2), sketched in fig.2, is characterized by following items:

a*) x, y and z have the same meaning as in model 1;

b*) z is obtained by the superposition of a deterministic map $S^*_z(x)$ and a random variable $W_z(\omega; z^*)$, with zero mean value and whose probability density involves z* as a parameter:

$$z = z^* + W_z(\omega; z^*) \quad , \quad z^* = S^*_z(x) \qquad (3)$$

c*) y is obtained by the superposition of a deterministic map $S^*_y(x,z^*)$ and a random variable $W_y(\omega; y^*, z^*)$, with zero mean value and whose probability density involves y* and z* as parameters:

$$y = y^* + W_y(\omega; y^*, z^*) \quad , \quad y^* = S^*_y(x, z^*) \qquad (4)$$

d*) the identification of the random variables W_z and W_y is realized as follows:

$$W_z = \varepsilon(z^*)\,\gamma(\omega) \quad ; \quad \gamma \in [-\tfrac{1}{2}, \tfrac{1}{2}] \quad , \quad P(\gamma) : |-\tfrac{1}{2}, \tfrac{1}{2}| \rightarrow \mathbb{R}_+ \qquad (5.1)$$

$$W_y = \eta(y^*, z^*)\,\delta(\omega) \quad ; \quad \delta \in |-\tfrac{1}{2}, \tfrac{1}{2}| \quad , \quad P(\delta) : |-\tfrac{1}{2}, \tfrac{1}{2}| \rightarrow \mathbb{R}_+ \qquad (5.2)$$

where ε and η are scaling operators, with $\varepsilon, \eta \geq 0$, and γ and δ are random variables characterizing the simulation model; moreover

$$E(\gamma) = E(\delta) = 0 \qquad (6)$$

$$\text{ord}(z^*)\uparrow \, , \; \text{ord}(y^*)\uparrow \; \Rightarrow$$
$$\Rightarrow \; \varepsilon\uparrow, \; \eta\uparrow, \; V(\gamma)\uparrow, \; V(\delta)\uparrow \qquad (7)$$

where the terms E and V are, as usual, the functions "mean value" and "variance".

One can note that in the above identification (model 2) ε and η have the meaning of scaling operators so that the implication (7) states, according to the phenomenological behaviour of the real system, that as y and z increase also all the uncertainesses of the simulation increase. The actual quantitative identification can be realized parametrizing the probability density and supplying an estimation of these parameters after moment calculations of experimental data. This method is well known in probability theory and consequently it is not repeated here.

3. OPTIMIZATION

For both models proposed in the preceding section, the calculation of the moments both for y and for z can be realized with the known /6,7/ techniques. In particular the following equations can be stated:

Model 1

$$E(z^p) = \varphi^p_z(x) = \int_A S^p_z(x; \underset{\sim}{\alpha}(\omega))\; P(\underset{\sim}{\alpha})\; d\underset{\sim}{\alpha} \qquad (8)$$

$$E(y^p) = \varphi^p_y(x) =$$
$$= \int_{AB} S^p_y(x, z = S_z(x; \underset{\sim}{\alpha}(\omega)); \underset{\sim}{\beta}(\omega))\; P(\underset{\sim}{\alpha}, \underset{\sim}{\beta})\; d\underset{\sim}{\alpha}\; d\underset{\sim}{\beta} \qquad (9)$$

where A and B are the domains of $\underset{\sim}{\alpha}$ and $\underset{\sim}{\beta}$.

Model 2

$$E(z^p) = \psi^p_z(x) =$$
$$= \int_C (S^*_z(x) + W_z(\omega; z^*))^p\; P(\gamma)\; d(\gamma) \qquad (10)$$

$$E(y^p) = \psi^p_y(x) =$$
$$= \int_D (S^*_y(x, z^*) + W_y(\omega; y^*, z^*))^p\; P(\delta)\; d\delta \qquad (11)$$

where C and D are the domains of γ and δ, respectively. In the above calculation the knowledge of the means of the variables γ and δ is usefull:

$$E(\gamma^p) = m^p_\gamma(x) = \int_{-1/2}^{1/2} \gamma^p\; P(\gamma; x)\; d\gamma \qquad (12)$$

$$E(\delta^p) = m^p_\delta(x) = \int_{-1/2}^{1/2} \delta^p\; P(\delta; x)\; d\delta \qquad (13)$$

These prelimanies are the basis of the optimization problem which, in general lines, following /3/, can be formulated in a fashion that:

finding the optimum value $x=x_{ott}$ such that the output z is maximized in mean with y minized in mean; moreover the dispersion both for z and y should be minimized, with a distortion for z towards the large values of z and for y towards the small values of y.

Before any formalization of the above conceptual line one has to establish if y, namely the required energy, is a free variable or a constrained one (for instance, with values below a given value). In general the constrained can be formulated as follows:

$$p = 1,2,\dots : E(y^p) \leq y_p \tag{14}$$

The unconstrained problem will be here firstly considered and afterwards some indication will be given for the constrained one. Accordingly let us propose the following objective function:

$$J=J(x)=k_1E(z) - k_2E(z^2) + k_3E(z^3) + -h_1E(y) - h_2E(y^2) - h_3E(y^3) \tag{15}$$

where $k_{1,2,3}$ and $h_{1,2,3}$ are positive constants to be constructed, on the basis of experimental information, giving the right weights to the "costs" of the production process; the sign preceding the constants k_i and h_i is indicated in the previously formulated statements on the optimization problem.

Note that the function J has different expression for each of the simulation models, formally:

Model 1

$$J = k_1\varphi_z^1 - k_2\varphi_z^2 + k_3\varphi_z^3 + - h_1\varphi_y^1 - h_2\varphi_y^2 - h_3\varphi_y^3 \tag{16}$$

Model 2

$$J = k_1\Psi_z^1 - k_2\Psi_z^2 + k_3\Psi_z^3 + - h_1\Psi_y^1 - h_2\Psi_y^2 - h_3\Psi_y^3 \tag{17}$$

At this end the "optimization problem" can be formulated as follows:

finding the optimum value $x=x_{ott}$, with $0 < x \leq x_M$ (where x_M is the maximum value technologically attainable for x), such that J=J(x) attains a maximum for $x=x_{ott}$.

Note that since J is a continuous function of x certainly at least a maximum exists, which can be searched with the known techniques /8/ of function analysis.

In the constrained case the condition (14) has to be considered. In this case the simultaneous computation of y and J has to be realized and the maximum of J has to be searched by a "univariate search method" /8/ over a discrete set of values of x belonging to the range $[0,x_M]$.

4. APPLICATION AND DISCUSSION

In order to apply and test the consistency of the proposed simulation-optimization technique, the problem of the steel grinding process simulation is here considered. As already mentioned in the introduction, the relevant feature of such a process can be simulated by the Mayne and Malkin model /5/ and developed, through the analysis of its characteristics, as briefly discussed in ref. /4/. Since an analysis of such a model was already realized by the Authors following the line of "model 1", the application considered in this section analyses the grinding process, namely its simulation and optimization, in the framework of "model 2".

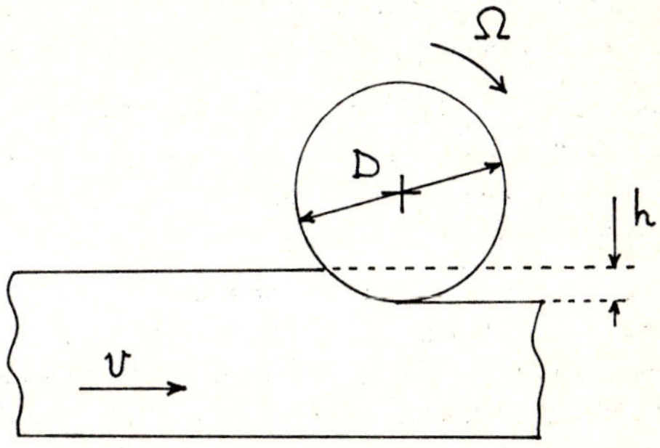

Fig. 3 - Draft of the grinding process

Consider then the system drafted in fig.3 and recall that the metal removual rate per unit workpiece width "Y" and the power "W" spent to obtain such a removual, according to the previously mentioned model /5/, are given by the following equations:

$$Y=a_1 hv \tag{18}$$

$$W=a_2 Y+a_3 \Omega D+a_4 A\Omega D^{3/2} h^{1/2}+a_5 AD^{-1/2} h^{1/2} v \tag{19}$$

where h = wheel depth of cut;
v = workpiece velocity;
A = wheel wear flat area (as fraction of the wheel surface area);
D = grinding wheel diameter;
Ω = wheel angular velocity.

According to the formalism of "model 2", we can set:

$$\underset{\sim}{x} = \{x_1=h,\ x_2=v,\ x_3=\Omega,\ x_4=D,\ x_5=A\}$$
$$y = W$$
$$z = Y$$

Moreover the experimentation on "3 G 71 M Grinding Machine Tool" already realized in ref./4/ supplies the following further indications:

a) the range of the conceivable values of z* and y* is the following

$$0\leq z^*\leq 100\ \text{mm}^2/\text{s} \quad , \quad 0\leq y^*\leq 1\ \text{kW/mm}$$

b) the limit values of the ε and η operators are

$$\varepsilon(z=0) = 0 \quad , \quad \varepsilon(z^*=100) = 0.12$$

$$\eta(y=z=0) = 0 \quad , \quad \eta(y^*=1,z^*=100) = 0.16$$

and, in this range of z* and y*, a sufficiently accurate simulation can be realized assuming a linear influence of z* upon ε and a quadratic influence of (y*,z*) upon η;

c) analogous influence is verified for the variances of the random variables $\gamma(\omega)$ and $\delta(\omega)$ that range up to a maximum value of 0.01 and 0.22, respectively.

Of course the above indications are consistent with the experiments realized in /4/ and can hopefully either improved or adapted to particular working conditions. An insight into this problem deeper than the one of /4/ is not here realized.

After these preliminaries, the mathematical formulation of the simulation model, as far as the input-output process is concerned, can now have the following form:

$$z=z^*+\varepsilon(z^*)\ \gamma(\omega;z^*) \quad , \quad z^*=S^*_z(\underset{\sim}{x}) \tag{20}$$

$$y=y^*+\eta(y^*,z^*)\ \delta(\omega;y^*,z^*) \quad , \quad y^*=S^*_y(\underset{\sim}{x},z^*) \tag{21}$$

where

$$S^*_z(\underset{\sim}{x})=a_1 x_1 x_2 \tag{22.a}$$

$$S^*_y(\underset{\sim}{x},z^*)=a_1 a_2 x_1 x_2+a_3 x_3 x_4+a_4 x_1^{1/2} x_3 x_4^{3/2} x_5+ a_5 x_1^{1/2} x_2 x_4^{-1/2} x_5 \tag{22.b}$$

$$\varepsilon(z^*)= 1.2\cdot 10^{-3}\ z^* \tag{23.a}$$

$$\eta(y^*,z^*)=1.6\cdot 10^{-3}\ y^* z^* \tag{23.b}$$

At this end the particularization of the objective function J, characterized simply by first two moments, can be realized. The calculation of the first and second order moments of z and y, realized in order to compute J, gives the following result:

$$E(z)=S^*_z(\underset{\sim}{x}) \tag{24}$$

$$E(y)=S^*_y(\underset{\sim}{x}) \tag{25}$$

$$E(z^2)=(S^*_z(\underset{\sim}{x}))^2+0.01\ \varepsilon^2(\underset{\sim}{x})\ S^*_z(\underset{\sim}{x}) \tag{26}$$

$$E(y^2)=(S^*_y(\underset{\sim}{x}))^2+0.22\ \eta^2(\underset{\sim}{x})\ S^*_z(\underset{\sim}{x})\ S^*_y(\underset{\sim}{x}) \tag{27}$$

Consequently

$$J(x)=k_1 S^*_z(\underset{\sim}{x})-h_1 S^*_y(\underset{\sim}{x})-k_2((S^*_z(\underset{\sim}{x}))^2+ +0.01\ \varepsilon^2(\underset{\sim}{x})S^*_z(\underset{\sim}{x}))-h_2((S^*_y(\underset{\sim}{x}))^2+ +0.22\ \eta^2(\underset{\sim}{x})\ S^*_z(\underset{\sim}{x})\ S^*_y(\underset{\sim}{x}) \tag{28}$$

and replacing the actual expression of ε and η as defined by eqs. 23 and dropping the argument $\underset{\sim}{x}$ in all functions:

$$J=S^*_z(k_1-k_2 S^*_z-1.44\cdot 10^{-8} k_2 S^{*2}_z)+ -S^*_y(h_1+h_2 S^*_y)-0.56\cdot 10^{-6} h_2 S^{*3}_z S^{*3}_y \tag{29}$$

In the study of the optimization of the steel grinding process it is reasonable to refer initially to the working conditions at assigned characteristics of the workpiece and machine. In these terms, for the search of the optimum value of x, is here investigated the influence of x_1 and x_2, namely of the depth of the cut and of the workpiece velocity , at constant values of the remaing components of $x^{(*)}$.

To this end, typical ranges and values for the examined technological parameters are fixed:

$$0 \le h \le 0.1 \text{ mm} \quad , \quad 0 \le v \le 600 \text{ mm/s}$$
$$A = 0.01 \; , \; \Omega = 300 \text{ s}^{-1} \; , \; D = 200 \text{ mm}$$

and the values obtained by experiments and already reported in tab. 1 of ref. /4/ are taken for the constants a_i of the eqs. 22, namely

$$a_1=0.95 \; , \; a_2=0.0125 \; , \; a_3=9.6\cdot10^{-7}$$
$$a_4=8.5\cdot10^{-6} \; , \; a_5=2$$

In order to visualize the results of the above concept, then, two sets of exemplifying values for the "weights" k_i and h_i are choosed ($k_1=1$, $k_2=0.03$, $h_1=h_2=0.5$ and $k_1=1$ $k_2=h_1=h_2=0.05$); the behaviour of J versus x_2, for different values of x_1, is plotted in the figures 4 and 5, respectively. This curves show in the previously fixed technological range, among other things,

- a typical "bell-shape";
- as the systems appears to be affected especially by variation of the h and v;
- as optimum high workpiece velocity must not be used with high wheel depth of cut;
- the sensitivity of the system performance, at high values of x_1, to nonoptimum choises of x_2;
- a first indication on the influence of the "weights" over the optimum values of x_2.

Extensive and deep investigations on the productive aspects of the grinding process (which however are not included in the aims of the present work) can allow an actual evaluation of the weights both of the production rate and cost and of the respective fluctuations, for a righter construction of the pertinent objective function. In any case the model here proposed is consistent with grinding experimental tests and supplies a useful contribution to the optimum definition of the economic machining rate, through a stochastic approach, and, in these terms, is a interesting "tool" both for the design and for the production control in steel grinding process.

(*) In general the wheel wear flat area (parameter A) grows during the machining for attritious wear of grinding wheel; however, in many production grinding, this growth is minimal because the wheel is redressed frequently and/or right cutting oils are used.

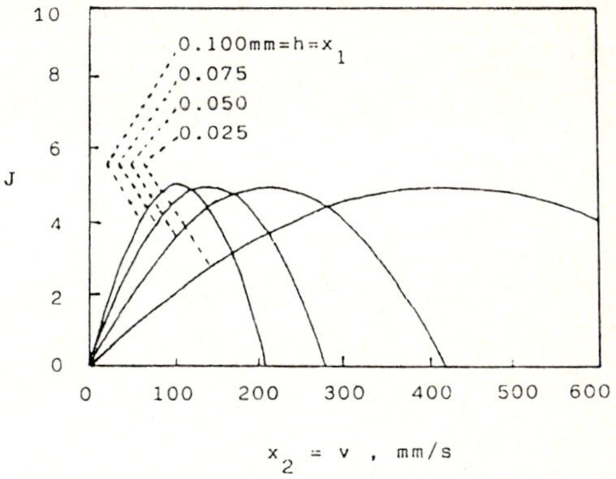

Fig. 5 - J versus x_2, for different x_1 values (with "weights" $k_1=1$, $k_2=h_1=h_2=0.05$)

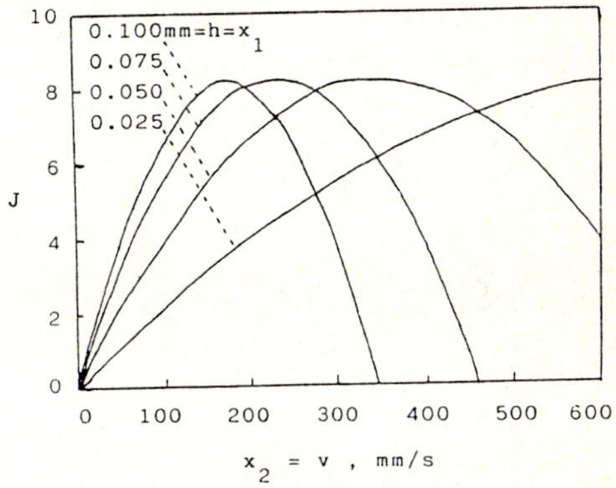

Fig. 4 - J versus x_2, for different x_1 values (with "weights" $k_1=1$, $k_2=0.03$, $h_1=h_2=0.5$)

REFERENCES

1 - BURNEY, F.A., et al., "Stochastic Approach to Characterization of Machine Tool Systems Dynamics under Actual Working Conditions", J. Eng. Ind., ASME Trans., May(1976), 614-619.

2 - AMITAY, G., "Adaptive Control Optimization of Grinding", J. Eng. Ind., ASME Trans., Feb. (1981), 103-108.

3 - GIORLEO, G., and PASQUINO, R., "Metodologie

Probabilistiche nel Campo dei Sistemi Industriali", Progettare, 35, (1983), 47-50.

4 - GIORLEO, G., and PASQUINO, R., "On the Steel Grinding Process: Stochastic Modelling and Computer Aided Optimization", Proc. 1st Int. Symp. on Design and Synthesis, Tokyo, July (1984), 80-85.

5 - MAYNE, R.W., and MALKIN, S., "Parameter Optimization of the Steel Grinding Process", J. Eng. Ind., ASME Trans., August (1976), 1040-1052.

6 - SOONG, T., "Random Differential Equations", Ac. Press., New York, (1974).

7 - PAPOULIS, "Calcolo delle Probabilità e Processi Stocastici", Boringhieri, Torino (1976).

8 - DORNY, N., "Vector State Approach to Optimization", Wiley, New York, (1974).

Complex and Distributed Systems: Analysis, Simulation and Control
S.G. Tzafestas and P. Borne (editors)
Elsevier Science Publishers B.V. (North-Holland)

UNICUS - A NEW APPROACH TO THE SIMULATION AND DESIGN OF CONTROL SYSTEMS

D.Matko, M.Šega*, B.Zupančič

Faculty of Electrical Engineering, Tržaška 25, 61000 Ljubljana
*Institute J. Stefan, Jamova 39, 61000 Ljubljana
Yugoslavia

The paper describes a new approach to systems simulation and design which is included in the interactive program package for analysis and design of systems ANA. It enables an universal computer simulation (UNICUS) which is block oriented and where each block can be either a source of the signal, a signal joining block a linear transfer block, an analog computer, or a general block, written by the user in the form of a subroutine. By means of general blocks and analog computer the concept of ANA can be extended to nonlinear and time variable systems. Analog computer included in the simulation require the real time simulation. Due to the concept of UNICUS the simulation is in spite of its generality as fast as a simulation in FORTRAN can be. Also some special blocks for on line implementation are available: parameter estimator, state estimator and Kalman filter. For off line applications an optimization is available. An example of optimal open loop response design and closed loop PID controller optimization is included to show the applicability of UNICUS. UNICUS is written in a standard FORTRAN and can be run on every computer while interpreter (written in FORTRAN too) is implemented on PDP 11/34 with RT 11 operating system.

1. Introduction

The paper describes a compiler for control systems simulation, which is included in the interactive program package for analysis and design of control systems ANA. The compiler enables an universal computer simulation (UNICUS), which is block oriented and where each block can be either a source of the signal, a signal joining block, a linear transfer block, an analog computer block or a general block, written by the user in the form of a subroutine, an analog computer or real process connected to the digital compu-ter via A-D and D-A converters. All blocks are multivariable. As signal sources some standard signals (step, pulse, ramp, Gaussian noise), signals generated (computed or recorded) by ANA and signals generated by an user written subroutine are available. Among signal joining blocks summing and subtracting are implemented. Linear transfer blocks are transfer blocks of the ANA (discrete transfer function matrices and state space description). By means of general blocks and analog computer the concept of ANA can be extended to nonlinear and time variable systems. Analog computer and real process included in the simulation require the real time simulation. Due to the concept of UNICUS the simulation is in spite of its generality as fast as simulation in FORTRAN can be. Also some special blocks for on line implementation are available: parameter estimator, state estimator and Kalman filter.

2. The language structure

The block in the control structure can be either a source of the signal, a signal joining block, a linear transfer block, an analog computer or a general block written by the user in the form of a subroutine. As signal sources some standard signals (step, pulse, ramp, Gaussian noise), signals generated (computer or recorded) by ANA and signals generated by an user writtern subroutine are available. Among signal joining blocks summing and subtracting are implemented. Linear transfer block are transfer blocks of the ANA (discrete transfer functions matrices and state space description). The general block written by the user represents a significant extension of the simulation possibilities because the nonlinear and time varying blocks can be included to the simulation scheme. The general block is written by the user in FORTRAN prior to entering ANA and describes the desired input output relation. If the block called "process" is written by the user in the form of a general block, the simulated delay of the process must be 1 less as the real one. Also two vectors - "parameters" and "states" of the general block are available. All other variables (parameters and states of other blocks are available through their names). An analog computer or a real process can be included to the UNICUS via A-D and D-A converters. This of course requires the real time simulation and due to the concept of UNICUS the simulation is in spite of its generality as fast as a simulation in FORTRAN can be. The number of "real time" blocks is limited to 1. It is forseen that a digital simulation language will be included in UNICUS. As parameter estimation blocks three standard procedures are included (recursive least squares, recursive extended least squares and recursive maximum likelihood methods), it is forseen to include also instrumental variables and output error methods. The state estimator and Kalman filter are generated by user in ANA and included into the simulation.

3. Communication with the user

The interpreter of ANA helps the non skilled user to generate the data base for UNICUS. It is possible to choose the standard control structure, which includes one process, controller in the direct and feedback path, prefilter, two noise filters on the input and the output of the process, feed forward controller, process parameter estimation and state observer or Kalman filter. If a more complicated scheme is required, it is possible to extend the standard structure to the structure required by the user. The interactive definition of concrete scheme is based on semigraphics capabilities of videoterminal VT 100 which is one of the input-output units of the package. The screen is divided into three parts: the last 9 rows are reserved for interactive communication with the user (SCROLL area). The first 17 rows are devided in two areas: the left part is the area of the standard structure and the right part is reserved for additional ele-ments and connections, which enable more general simulation scheme applica-tions. During the interacvtive defini-tion of the scheme first the maximal standard structure is shown in the left upper part of the screen. This maximal structure, shown in Fig. 1 is then interactively adjusted to the structure designed by the user.

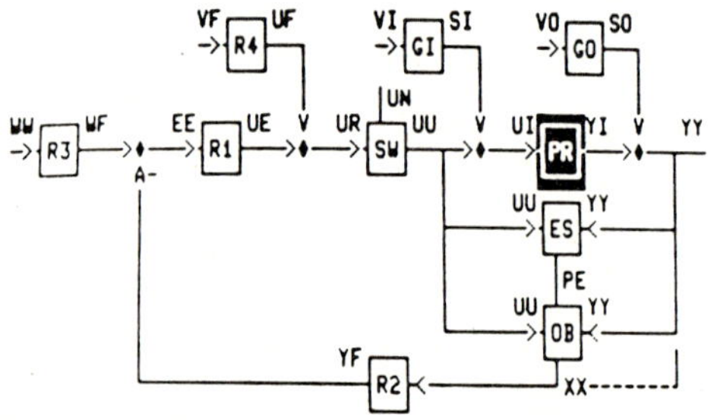

Fig.1. The maximal standard structure of UNICUS.

The block being determined is shown in reverse mode and if it exists its type must be given. The type of each block implies the control structure and so the determination of the remaining scheme. When every block of the standard structure is determined, the existing scheme, if necessary can be extended through additional elements shown on the right upper part of the screen. The connections of the additional elements with the subset of the standard scheme are shown by means of markers. The data base contains also the data of the timing of the simulation: initial time, final time, maximal number of steps and the data for output, which is in the format of ANA. Some additional conditions for the termination of the simulation run can be added in the form of a logical expression. The data for optimization option (discussed later) are also generated interactively and are included into the data base.

4 The optimization option

For off line applications the optimization is available. In this case in each simulation run the criterion function is evaluated and optimization gives the optimal values of the parameters. The criterion function can be chosen from a menu of criterion functions or written by user in form of a subroutine (ex. $\Sigma(\underline{x}^T\underline{\underline{Q}}\underline{x} + \underline{u}^T\underline{\underline{R}}\underline{u})$). $\underline{x}$ and $\underline{u}$ are two vectors of variables of the control scheme and $\underline{\underline{Q}}$ and $\underline{\underline{R}}$ corresponding weighting matrices. For the first step, for the last step and for other steps of the simulation three different values of the weighting matrices $\underline{\underline{Q}}$ and $\underline{\underline{R}}$ can be specified. As optimization method a method EXTREM (1) is included. It is forseen to include some other methods. The limitations of the optimization is included in the form of logical expressions (maximum 10) or a user written subroutine. By means of this off line optimization the PID controllers, optimal open loop responses of batch processes etc . can be designed.

5. The compilation and run procedures

UNICUS compiles the input data base into a program and a subroutine in FORTRAN. The compilation is done in several steps where various tests of the simulation scheme validity and sorting procedure are performed. The interpreter calls first the FORTRAN compiler to compile the program and subroutine generated by UNICUS, then a linker to link the UNICUS program and subroutine with user written subroutines and subroutines of the UNICUS library and finally runs the simulation program. The results of the simulation are in the format of ANA and can be dealt with the interactive package output routines which are accessible directly from the run time interpreter. Also some modification of the simulation scheme can be done without repeated running of the compiler: the parameters of linear blocks and output variable list. So different control possibilities can be tested by UNICUS and evaluated by ANA.

6. The NaOH dissolution example

The sodium hydroxide solution NaOH, used in processing technology of ironoxide pigment production however in exactly defined concentration is prepared in iron dissolving basin. A suitable mathematical model of reactor was prepared in order to simulate the exothermic chemical reaction of dissolving /3/. Fig.2 shows the corresponding mathematical model.

In Fig. 2 Eqns 6,7,8 and 10 represent the reactor, Eqns 1,2 and 5 the cooling system and Eq. 9 the heat exchange with the enviroment. The NaOH dissolution is a typical endpoint problem. The objective of the control procedure is that the temperature in the reactor after 40 hours is 25 ^{o}C and that the quantity of undissolved NaOH is minimal. As the high temperature accelerates the dissolution according to the Arrhenius equation it is not recommended to keep the temperature of the reactor on 25° C during the initial phase of the dissolution. On the other hand the capacity of the cooling system is limited not only through the maximal flow (0.67 l/s) but mainly through the

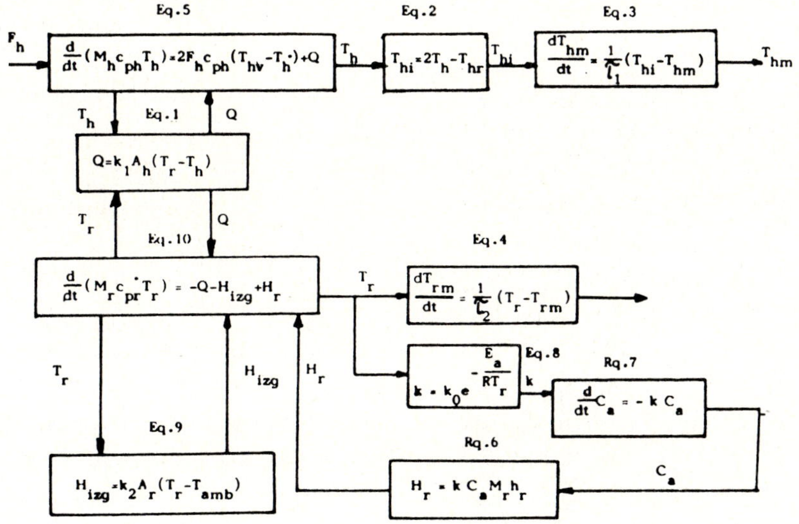

Fig.2. The mathematical model of a chemical reactor with exothermic reaction

limited temperature (12° C) of the mixed cooling water returning to the cooler.

According to these requirements we choose the following optimization criterion:

$$I = \int_{40h}^{50h} ((T_r - 25)^2 + C_a^2)dt \qquad (1)$$

where T_r is the reactor temperature and C_a is the quantity of the undissolved NaOH. Also the constraint

$$T_{hii} \leq 12^oC \qquad (2)$$

is taken into account, where T_{hii} is the temperature of the mixed water returning to the cooler.

First the open loop responses were obtained by the optimization option of the UNICUS. The simulation scheme for this case shown in Fig. 3 consists of two blocks: the proces shown in Fig.2 and the block which generates the input signal in the form of third order Tschebyshev polynomial sum. The optimizing parameters were the weighting factors for particular polynomials.

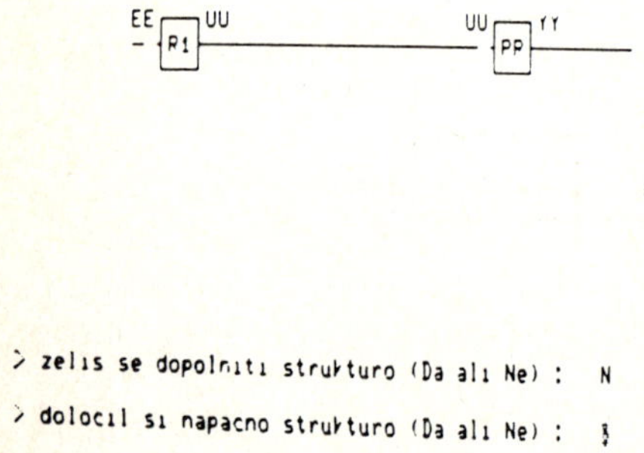

Fig.3. The simulation scheme for optimal open loop responses

The optimal open loop responses are shown in Fig. 4 where T_r^*, F_h^* and T_{hii} mean temperature of reactor, three way valve position and temperature of water returning to cooler respectively. This is naturally a suboptimal solution, since only 3 terms of the Tschebyshev polynomial sum were used as the input function. So we decided to modify the optimal reactor temperature response by forcing the temperature after 40h to be 25° C.

In this case the simulation scheme consists of four blocks: the process, the PID controller and two blocks generating optimal three way valve position and optimal temperature response. The simulation scheme is shown in Fig.5. The optimizing parameters are the parameters of the PID controller.

Fig.6 shows the open loop response (no PID) of the reactor temperature (T_r^-) at the +10% change in termal conductivity of the cooling system.

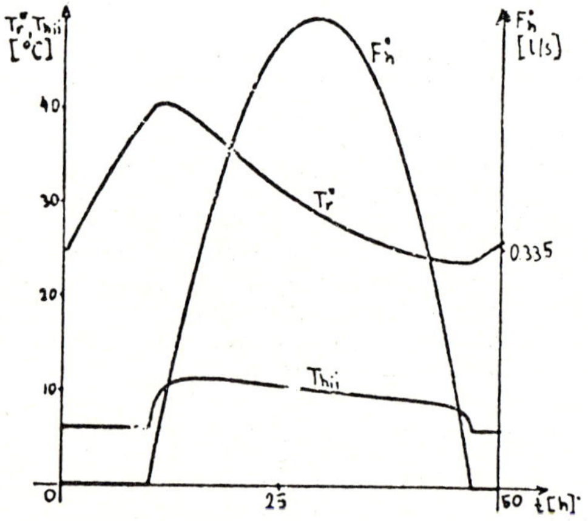

Fig.4. Optimal open loop responses of the reactor

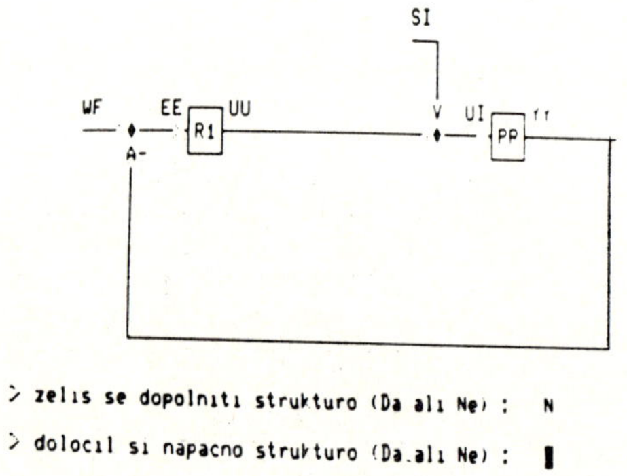

Fig.5. The PID controller optimization scheme

After the PID controller was applicated into the control loop the response (T_r^+) became very similar to the optimal one. Also the three way valve position (F_h) is shown in the same figure. In Fig.7 the optimal responses and valve position at the -10% change in thermal conductivity are shown.

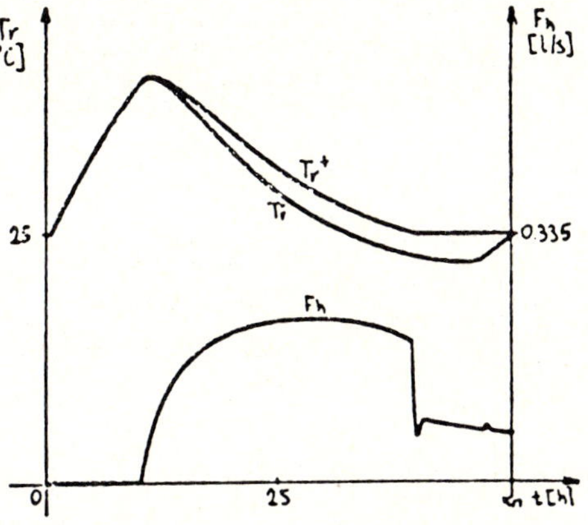

Fig.6. Responses at the +10% change in thermal conductivity

6. CONCLUSION

A new approach to simulation and optimization of control systems is described. It is included in the interactive program package for analysis and design of control systems ANA. We believe that it might be of great help to the engineers designing all kinds of systems.

7. References

1. ANA - interactive CAD package - manual, Faculty of electrical engineering, Ljubljana, 1984 (in slovene).
2. Franks, R.G.E.(1972) Modelling and simulation in chemical engineering, Wiley Interscience, New York.
3. Matko. D. et all., Hierarchical optimization of NaOH dissolution, Preprints of the first European Workshop on the "Real Time Control of Large Scale Systems", Patras, 1984.

Complex and Distributed Systems: Analysis, Simulation and Control
S.G. Tzafestas and P. Borne (editors)
Elsevier Science Publishers B.V. (North-Holland)

MODELLING AND SIMULATION IN DESIGN, OPTIMIZATION AND DIAGNOSTIC OF TURBOSET CONTROL SYSTEM

Zygfryd DOMACHOWSKI
Institute of Fluid-Flow Machinery of Polish Academy of Sciences
Fiszera 14, 80-952 Gdańsk, Poland
Tlx 0512042 Imp pl

Examples of modelling and simulation technique applied to the design, optimization and diagnostics of turboset control system have been presented. The turboset control system models were tested on analogue as well as digital computers according to the investigation particularities and the preferences of every one.

1. INTRODUCTION

When experiments are expected to be dangerous or harmful to the equipment or if they are difficult to be carried out, then the simulation technique is of great importance.
This is the case for the automatic control systems of turboset. The simulation technique is a very appropiate support to the design, optimization and diagnostics of rotational speed and load automatic control of turbosets.
On the one hand the experience in simulation of turboset control system confirms a general rule concerning the need of some compromise between the complexity of process model (the great accuracy of process model) and the computing possibility and costs.
On the other hand a great accuracy of simulating models is often not needed in the simulation investigations of automatic control of turbosets. An approximate model or a model of similar system to be controlled is then taken under consideration.
Several examples of modelling and simulation technique applied to the design, optimization and diagnostics of turboset control systems have been presented.

2. DESIGN AND COMPARATIVE INVESTIGATIONS

The accurate system model is unknown in such studies because the simulation comes before the realization of the system. Therefore an approximate model is elaborated on the basis of already existing systems and the knowledge of the process. Such a model is usually enough accurate for choosing the governor structure. Namely, the simulation methods allow to compare some properties of different kinds of governor using the same model of the system to be controlled and using the same mathematical criterion.
For some cases a control system under design is in a few parts the same as the other already known ones. Thanks to these similarities the simulation model can be easily elaborated. In such a way it is possible to compare different control systems from the same point of view.

2.1 Optimum Turboset Feedback in Generating Unit Control System

The simulation technique allows among others to compare different kinds of governor's structure and parameters.
The mode of feedback in turboset (a set of turbine and electric energy generator) power control system determines significantly the behaviour of the whole generating unit (a set of steam generator, turbine and electric energy generator) control system, i.e. the behaviour of turboset as well as of steam generator control systems.
The so called power output (PO) feedback, which up to now is mostly applied in turboset control system, has been compared to the so called turbine control valve position (TCVP) feedback [4]. Fig. 1 represents a block diagram of a generating unit control system under consideration.
The comparative analysis, by means of simulation techniques, has shown some important advantages of the turbine control valve position feedback. They are:

i/ the angular velocity overshoot after turboset shutdown is significantly less,
ii/ there is practically no overcontrol of turbine control valve positioning and

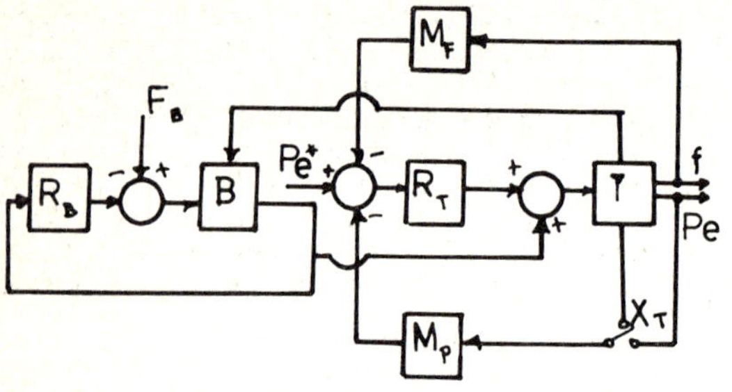

Figure 1

Block diagram of generating unit control system, B - steam generator, T - turboset, R_B - controller of steam generator, R_T - controller of turboset, M_f - frequency meter, M_P - power or control valve position meter, F_B - steam generator command signal, P_e^* - turboset power output (P_e - set value), f - frequency, P_T - iive steam pressure, X_T -control valve position

iii/ the live steam pressure drop is also considerably less in response to a step increase of turboset power set value, Fig. 2a.

iv/ the transient processes in generating unit control system are much insensitive to deviations of turboset controller parameters.

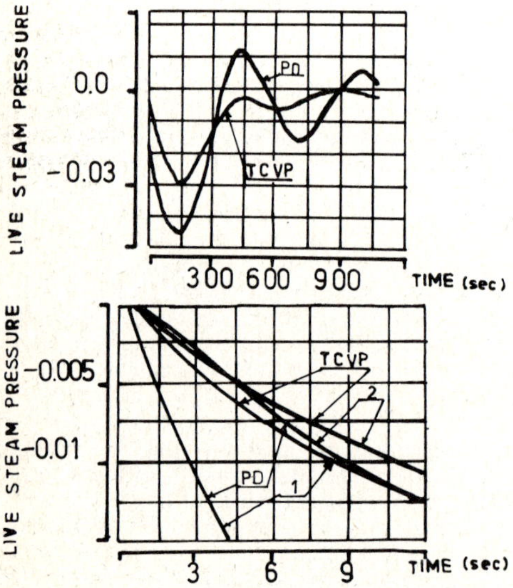

Figure 2

Live steam pressure in response to turboset power set value step increase, 1 - optimum, 2 - sub-optimun parameters of turboset controller

2.2 Turboset Frequency and Voltage Coupled Control System Design

An interaction of frequency and voltage controls appears during changes, in particular fast and rapid, of active and reactive power in an electric power grid. Such an interaction can be utilized to improve the turboset control characteristics. For this purpose a stabilizing device using the turboset slip signal on its input has been proposed, Fig. 3. Modelling investigations of the frequency and voltage coupled control system of turboset have been carried out [1.2]. They have included:

i/ concept verification of frequency and voltage coupled control loops of turboset, under several operational modes,

ii/ exploration of interdependence between turboset frequency and voltage,

iii/ consideration of the possibility of turboset frequency and voltage control interloop coupling to improve control system transients,

iv/ optimization of turboset control system,

v/ consideration of the sensitivity of turboset control system characteristics with regard to some parameters.

Disturbances of electric system frequency as well as of electric system voltage have been applied. The simulation research has demostrated that the influence of a signal proportional to the turboset slip acting onto the turboset voltage command improves transient processes in turboset control system, resulting from electric grid frequency or/and voltage disturbances. There are such improvements in a large margin of control system model parameters. Moreover the sensitivity of transient processes to the control system model parameters is significantly low. The above mentioned improvements occur without inconveniences related to some other devices. The results of modelling investigations of the control system transients have been compared to those ones corresponding to applying, for the same purpose, a device supplied on its input by the turboset active power signal (Electricité de France, Brown Boveri). This comparison gives the presented proposition more profitable.

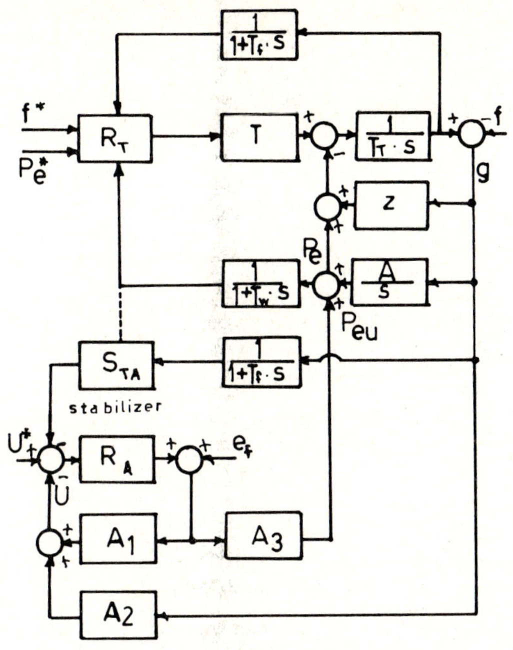

Figure 3

Simplified block diagram of turboset frequency and voltage coupled control loops, --- Electricité de France or Brown Boveri concept, A_1, A_2, A_3 - electric energy generator, $\bar{U}$ - voltage, P_{eu} - power output consecutive to terminal voltage, g - frequency slip (other main notations see Fig. 1).

2.3. Nuclear Generating Unit Control System Design

At present the nuclear power plants generally supply a constant load. However when the proportion of such plants in electric power grid increases, then they will be expected to take an increasing part in the frequency and power control of electric power system. There are some specific regulation problems in nuclear power plants because of structural differences between nuclear and fossil fuel plants. The comparison between drum boiler plant and reactor plant control system, in open loop and in closed loop, as well as some investigations concerning the reactor plant controller design have been carried out by means of simulation [3].

Reactor - and boiler-turboset plants differ in the source of their primary energy; this leads to structural differences which are essential in the case of steam generators (boilers). The nuclear reactor (PWR, BWR, breed) corresponds to a conventional boiler while the steam generator (a set of heat exchangers) corresponds to an evaporator with a primary reheater.

However, despite of the above mentioned constructional differences, it can be assumed analogous structure of reactor- and boiler-turboset plant models for the needs of a comparative analysis. The differences between models will concern parameters of some of their components.

2.3.1. Comparison between Boiler and Reactor-Turboset Plant Control Systems. Comparative investigations of regulation characteristics of reactor- and boiler-turboset plants illustrate the faster power increase response, to a step signal of fuel supply rate, in a reactor -turboset. The fossil fuel boiler-turboset plant achieves the steady state after a 6 times longer period than a reactor-turboset plant. This is caused mainly by the differences in the accumulation time constant of the named above two types of turboset plants.

In response to a unit step signal acting onto control valves of turbine, the boiler-turboset power follows much more slowly the control signal due to the time constant of the secondary reheater in this turboset, being much higher than that of vapour separator in a reactor-turboset.

The model of an automatic frequency and power control system of a reactor- and a fossil fuel drum boiler-turboset plant was investigated with respect to a power balance jump in the electric power grid, Fig. 4 continuos line. The resulting electric network frequency change has affected then the turbogenerator and the turboset power set value.

The turboset power transient, P_T, of reactor-turboset plant and of fossil fuel drum boiler plant one are quite similar. The live steam pressure, p_T and the fuel supply rate, F_B, are accompanied by slightly higher amplitude and c.a. five times higher oscillation frequency in a reactor- than in a boiler -turboset plant.

2.3.2. Reactor-Turboset Plant System Design. Up to now the boiler- and the reactor-turboset plant were assumed with so called "driving turbine" control. This implies that a control signal (corresponding to a change in the electric power system frequency) acts basicly on the turbine control valve; the boiler adjusts its power to the turboset power. This type control signal action has been applied in the comparative investigations discussed above. However, the "driving turbine" control of a reactor-turboset plant appears to be undesirable due to

excessive overshoot and oscillations of the fuel supply rate, Fig. 5, and live steam pressure as well as those of the turbine control valves and turboset angular velocity. Both the above mentioned signals (fuel supply rate and live steam pressure) should be limited because of the reactor and piping security.
Application of an additional feed forward signal acting at the reactor input (see Fig. 4, line RF) allows to improve, but slightly, as live steam pressure as fuel supply rate transients, Fig. 5.
"Driving boiler" control (slide pressure regulation, see Fig. 4, line RS) of a reactor-turboset plant makes that live steam pressure increases in the response to the electric network power balance aperiodically. The overshoot of the fuel supply rate transient is small as well as the oscillations in this transient process are strongly damped see Fig. 5, curve RS, which is significantly important from the point of view of the mentioned above security of the reactor's materials.

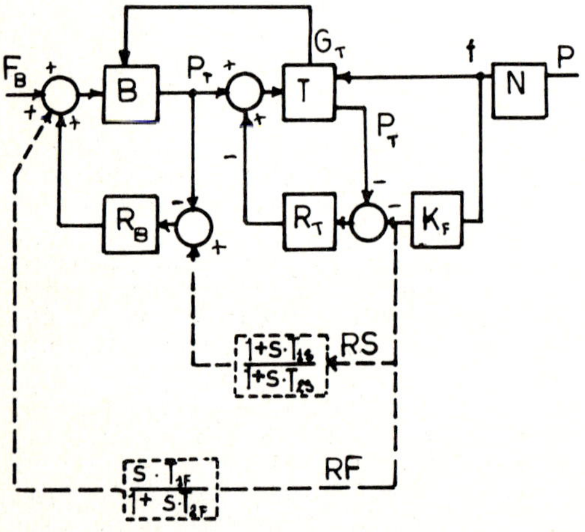

Figure 4

Block diagram of conventional or nuclear power plant control system, B - boiler or evaporator, T - turboset, N - electric power network, R_B - boiler or evaporator controller, R_T - turboset controller, K_F - frequency regulation coefficient of governor, f - power system, P_T - turboset power, P_N - electric network power balance, F_B -input signal to the boiler or reactor, G - flow of steam, P_T - live steam pressure, RF - feed forward variant, RS - slide pressure regulation.

2.4. Design of Automatic Electrohydraulic Control System for Industrial Power Network

The plants of synthetic fibre, paper and chemical industry require , among others, reliable power supply. Furthermore, due to technological reasons, the frequency and voltage of supplied electric current should fulfil certain conditions, they should be kept within prescribed limits.
Under usual conditions the requirements concerning the reliability and the quality of the electric power supply can be fulfilled most satisfactory by connecting on industrial power network to a large external power grid, e.g. the state grid. However, the disturbances and failures in the external power grid (local atmospheric disturbances, short-circuits, overvoltages, breakdowns, connection errors) force disconnections of the industrial network from the external one.
Frequency and voltage control of an industrial network requieres a special system or systems of automatic control. The basic task of this system is

i/ to reduce the frequency and voltage rise after isolating an industrial network from the external one.
ii/ to control the frequency and voltage in the industrial network.
iii/ to synchronize the industrial network with the external one.

A concept of electrohydraulic frequency and power control system of an industrial network has been worked out and patented in Poland [5]. The block diagram of such a system is shown in Fig 6.
Simulation investigations, before in a heat and power station application of the considered control system, have included:

i/ concept verification of industrial network control system,
ii/ exploration of control system characteristics under different operation circumstances (a sudden disconnection of the industrial network from the external grid, industrial network load changing, turboset shut-down),
iii/ test on the effect of steam pressure regulation in turbine bleeds on frequency control in an industrial network.

The most severe operation conditions for the described above control system occur after a sudden disconnection of the industrial network from the external grid. This results from the excess of the total power generated in the industrial network over the total power absorbed in this network. Due to this reason during the modelling as well as in field investigations the controllers were set in such a way as to limit properly the maximum frequency rise during a transient following a sudden disconnection of the industrial network from the external one and to reduce the duration of this transient.

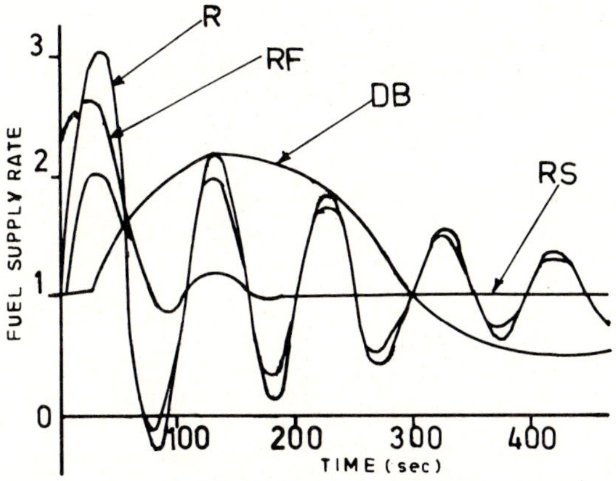

Figure 5

Unit step response of fuel supply rate to electric network power balance (notations see Fig. 4).

The presented above concept of an industrial network control system has been applied in a heat and power station. All the demands set before it have been met. The correctness of operation of the above mentioned automatic control system has been verified under various operation conditions. Its dynamic characteristics are very advantageous both at the industrial system load disturbances and at the disturbances in turbine bleeds.

3. OPTIMIZATION

During optimization by means of simulation no model of the whole control system in known. A few parts of it are usually under design supported by simulation studies, as that has been ilustraded in the precedent sections. The estimation of its parameters is then based on the data from similar systems. Most of these parameters are fixed and next the main parameters characterizing the performances of control system are determined as the value ranges.
Model investigations neglect as usual some particularities of controlled systems, e.g. in the industrial network control system modelling the nonlinearities in the controlled system were neglected which have been particularly significant in servomotors of turbine control valve [5]. Therefore the accurate values of parameters have to be adjusted after the control realization.
The optimization investigations of the turboset automatic control system have to be carried out in at least two kinds of operation conditions, i.e. after a sudden discharge of turboset (angular velocity control) and during the turboset loading (load control). However, the optimum controller parameters differ for these two cases. For this reason it is needed to find a compromise between them and to determine some sub-optimum parameters.

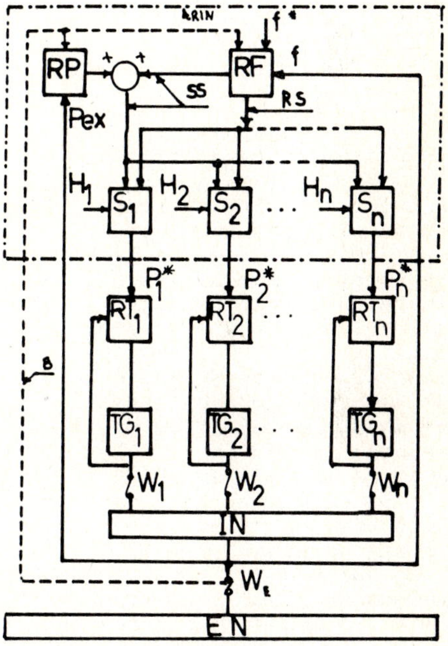

Figure 6

The concept of frequency and power control of an industrial network, IN - industrial network, RIN - industrial system frequency and power controller, B - disconnection signal, W - switch, f - frequency (f* - set value), P_{ex} - exchanged power, RF - frequency controller RP - exchanged power controller, RS - rapid signal, SS - slow signal, RT - turbine power controller, S - turboset controller, H - hand signal, P_e* - set value of turboset controller.

4. DIAGNOSTICS

The causes of improper operation of the power plant control system equipment are usually very difficult to be detected. Solving this problem requires to set up and check certain hypothesis (usually a hypothetical model of improperly operating equipment). Such a model cannot be identified in a usual way because the experiments are quite impossible. Thus the comparison between the model characteristics and those of the actual equipment ones is not possible. The model is verified by means of some auxiliary signals recorded from the process. These signals are compared with the respective ones recorded from the model under investigation.
Thanks to such simulation investigations some causes of damaging oscillations of the control valve actuator of the 200 MW condensing turbine as well as of the 30 MW condensing extraction turbine were found.

4.1 Security of Installation

The angular velocity overshoot after a shut-down of turboset should be limited from the point of view of turboset security. Its maximum admissible value should be less than 10 % (e.g. 8 %) of nominal value of turboset angular velocity.
The angular velocity transients after a shut-down of turboset have been compared in consideration of the kind of feedback signal in turboset power control system [4], see Fig. 1. In the case of power output (PO) feedback the turboset angular velocity overshoot has been equal to about 12.4 %. It has been equal to about 7.2 % in the case of turbine control valve position (TCVP) feedback. For these two cases that value is practically independent on the parameters of PL controller.
The value of 12.4 of angular velocity overshoot is unacceptable. In other words such a situation requires a certain additional device to limit the turboset angular velocity overshoot.

4.2 Component Longevity

As a response to the command as well as disturbance signals some oscillations occur within a turboset control system. From this point of view limitations are imposed in particular on turbine control valve and servomotor behaviour to avoid stress fatigue of material.
Both the overcontrol and oscillations of turbine control valves (of high pressure turbine and of low pressure turbine) are greater in the case of PO feedback than of TCVP one. This fact appears under command signal as well as disturbance signal changes.

5. CLOSING REMARKS

The turboset control system models were tested on analogue and digital computers according to the investigation particularities and the preferences of every one.
The analogue computers MEDA 42TA and MEDA 41TC, and digital computer R-32 (using simulation language CSMP) were applied.

REFERENCES

[1] Domachowski Z.: Sur l'utilisation du signal de glissement dans la régulation de tension et de fréquence des groupes turbo-alternateurs. Archiwum Energetyki, 1984, No. 1 p. 35-49.
[2] Domachowski Z.: Simulation as a tool in design frequency and voltage coupled control of turboset. Proceedings of the International Conference on Power Plant Simulation, November 19-21, 1984, Cuernavaca, Morelos, México, p. 108-115.
[3] Domachowski Z. and Bugala G.: Simulation as a Tool of Comparison between Nuclear and Fossil Fuel Generating Unit Control Systems. Preprints of European Simulation Meeting "Simulation in Research and Development", Eger, Hungary, 1984, p. 267-276, and Proceedings of IMACS European Simulation Meeting on Simulation in Research and Development Eger, Hungary, 27-30 August, 1984, Edited by A. Javor, North Holland.
[4] Domachowski Z. and Bugata G.: Influence of the Feedback Kind in Turboset Power Control System on Control Characteristics of Generating Unit. Archiwum Energetyki, 1984, No. 3/in Polish/
[5] Domachowski Z. and Kilmacki Z.: Automatic electrohydraulic control system for industrial power network. Proceedings of the First Third-World Electric Power Conference "Problems and Local Solutions", October 22-25, 1984, Teheran-Iran vol. 1, p. 174-183.

Complex and Distributed Systems: Analysis, Simulation and Control
S.G. Tzafestas and P. Borne (editors)
Elsevier Science Publishers B.V. (North-Holland)

A METHOD FOR ON-LINE STRUCTURAL IDENTIFICATION OF MULTIVARIABLE SYSTEMS

M. D'Apuzzo (*) L. Sansone (**)

(*) Department of Electrical Engineering - University of Naples
(**) Department of Computer Science - University of Naples

A method for on-line structural identification of linear multivariable systems from noisy input-output signals is proposed. The method starts by determining an over-parametrized structure using a Model Reference Adaptive System algorithm. Then the true structure of the system is determined using the 'Product Moment Matrix' obtained upon convergence of the previous algorithm. This matrix allows us to obtain the invariant observability indices of the multivariable system. Therefore the proposed method gives an 'a priori' prediction of the model structure without a previous parameters evaluation. The results obtained from the application of this procedure to a simulated system with different noise statistical properties are also reported.

1. INTRODUCTION

The identification of multivariable systems from input-output sequences involves two successive steps: (i) the structural identification (ii) the parametric identification. For such a system the minimum number of parameters to be identified depends not only on the system order, which is a global structural information, but also on the observability indices. They are more precise structural informations and define canonical models of the Multi-Input, Multi-Output system (MIMO). Therefore these indices must be evaluated before the parametric identification in order to avoid trial and error procedures in the determination of a suitable structure for the system to be identified.

For the structural identification, Guidorzi [1] utilizes a sequence of symmetrical increasing-dimension matrices obtained from the matrix of input-output data and select in this sequence the non-singular matrices. Then the parametric identification is carried out in one shot by means of the least-square formula. This method can be extended to the noisy case if the noise covariance matrix is known. In order to avoid the knowledge of the statistical properties of the disturbing noise, Guidorzi [2] introduces the 'range error test' which gives a prediction on the reconstruction error of models with different structures without the actual estimation of these models. This test is effective up to 30 percent additive noise.

The 'range error test' can be considered as a development of structural tests used for Single-Input, Single-Output (SISO) system. For instance, Wellstead [3] performs a singularity test on the instrumental Product Moment Matrix Λ. Young, Jakeman and Mc Murtries [4] utilize a test based on some statistical properties of the inverse of the matrix Λ. The Authors [5], instead, perform a singularity test using directly the inverse of matrix Λ.

In this paper a method for structural identification of linear MIMO systems in presence of noise is proposed. The method is based on the following steps: (i) determining a suitable over-parametrized auxiliary system (ii) estimating the observability indices of the MIMO system. In (i) the auxiliary system is identified by means of a Model Reference Adaptive System (MRAS) algorithm. This auxiliary system allows us to determine a matrix Λ uncorrupted by noise. In (ii) a minimal sequence of symmetrical increasing-dimension matrices directly obtained from the matrix Λ is constructed. This sequence allows us to apply successively singularity tests in order to determine the structural realization of the MIMO system. Note that the proposed method uses the same recursive algorithm both in structural and in parametric identification. As consequence the procedure can be nicely implemented in an on-line environment using a dedicated microprocessor.

In the following the method is described, then the computation procedure is indicated and the fundamental conditions concerning the implementation of the algorithm are examined. Finally, experimental results concerning a two-input, two-output system with different noise to signal ratio will be reported.

2. STRUCTURAL IDENTIFICATION

The input-output description of a linear multivariable system in terms of difference equations can be put in the form

$$y_s(k) = \sum_{i=1}^{m} \sum_{j=1}^{\nu_{si}} a_{si,j}\, y_i(k-\nu_{si}+j-1) + \sum_{i=1}^{r} \sum_{j=1}^{\nu_s} b_{si,j}\, u_i(k-\nu_s+j-1) \quad s=1,\dots,m \tag{1}$$

where:

- m is the number of outputs
- r is the number of inputs
- ν_{si}, ν_s are the observability indices
- $a_{si,j}, b_{si,j}$ are the system parameters
- u_i, y_i are observed sequences of input-output signals.

The observability indices are defined as

$$\begin{aligned} \nu_{si} &= \nu_s && \text{for } s = i \\ \nu_{si} &= \min(\nu_s+1, \nu_i) && \text{for } s > i \\ \nu_{si} &= \min(\nu_s, \nu_i) && \text{for } s < i \end{aligned} \tag{2}$$

Equation (1) can be written in a compact form as

$$y_s(k) = \underline{\theta}_s^T\, \underline{x}_s\,(k-1) \tag{3}$$

where:

$$\underline{\theta}_s^T = \left[a_{s1,1},\dots,a_{s1,\nu_s},\dots,a_{sm,1},\dots a_{sm,\nu_{sm}};\; b_{s1,1},\dots,b_{s1,\nu_s},\dots,b_{sr,1},\dots,b_{sr,\nu_s}\right] \tag{4}$$

and

$$\left[x\,(k-1)\right]^T = \left[y_1(k-\nu_{s1}),\dots,y_1(k-1),\dots,y_m(k-\nu_{sm}),\dots,y_m(k-1);\; u_1(k-\nu_s),\dots,u_1(k-1),\dots,u_r(k-\nu_s),\dots,u_r(k-1)\right] \tag{5}$$

Each equation (3) defines a subsystem of the whole MIMO system to be identified.

a) Determination of the auxiliary system

Let

$$\nu_1 = \nu_2 = \dots = \nu_m = \nu \tag{6}$$

being ν an observability index higher than that one presumed on each subsystem. Because of the given definitions (2) the structure of the auxiliary system model has been completely defined, therefore referring to (1) or (3) it remains to determine the parameters.

To identify these parameters one can choose an output error configuration ('parallel' estimation model). In this paper a Parallel Least Square Method (PLSM) is employed [6],[7]. This algorithm is suitable in noisy environment because it allows to determine unbiased values of auxiliary system model parameters.The recursive identification algorithm has the general form

$$\hat{\underline{\theta}}_s(k) = \hat{\underline{\theta}}_s(k-1) + \underline{\gamma}_s(k-1)\delta_s(k) \quad s=1,\dots,m \tag{7}$$

where $\hat{\underline{\theta}}_s(k)$ is the vector of the parameters estimated at k-step. In the case of PLSM algorithm:

$$\underline{\gamma}_s(k-1) = \frac{P_s(k-1)\underline{\omega}_s(k-1)}{1+\underline{\omega}_s^T(k-1)P_s(k-1)\underline{\omega}_s(k-1)} \tag{8}$$

$$\delta_s(k)=y_s(k)-\hat{\underline{\theta}}_s^T(k-1)\underline{\omega}_s(k-1)+\sum_{i=1}^{\nu_s} \hat{c}_{si}(k-1)\varepsilon_{s(k-i)} \tag{9}$$

being

$$P_s(k) = \left[I-\underline{\gamma}_s(k-1)\underline{\omega}_s^T(k-1)\right] P_s(k-1) \tag{10}$$

$$\left[\underline{\omega}_s(k-1)\right]^T = \left[z_1(k-\nu_{s1}),\dots,z_1(k-1),\dots,z_m(k-\nu_{sm}),\dots,z_m(k-1);\; u_1(k-\nu_s),\dots,u_1(k-1),\dots,u_r(k-\nu_s),\dots,u_r(k-1)\right] \tag{11}$$

$$z_s(k) = \hat{\underline{\theta}}_s^T(k-1)\underline{\omega}_s(k-1) \tag{12}$$

$$\hat{c}_{si}(k-1) = -\hat{a}_{ss,i}(k-1) \qquad i=1,\dots,\nu_s \tag{13}$$

$$\varepsilon_{sk} = y_s(k)-\hat{\underline{\theta}}_s^T(k)\underline{\omega}_s(k-1) \tag{14}$$

It may be noted that $z_1(k), z_2(k), \dots, z_m(k)$ are the instrumental variables generated by the parallel adjustable model chosen as the auxiliary system.

Utilizing the recursive algorithm PLSM, by application of the inversion lemma [8], one obtains too

$$\left[P_s(k)\right]^{-1}=\left[P_s(k-1)\right]^{-1}+\underline{\omega}_s(k-1)\underline{\omega}_s^T(k-1) \tag{15}$$

As the following is true [4]

$$\Lambda_s = \left[P_s \right]^{-1} \tag{16}$$

upon convergence the PLSM algorithm provides directly by (15) the instrumental product moment matrices Λ_s.

Moreover, in consequence of (6), the vectors (11) are equal to each other and, through (10) (16), it follows that

$$\Lambda_1 = \Lambda_2 = \ldots = \Lambda_m = \Lambda \tag{17}$$

Λ is a $[\nu(m+r)]$ symmetrical square matrix and consists of m+r submatrices each having ν vectors. When the convergence of the PLSM algorithm has been reached the extraneous disturbances are removed from the elements of the matrices P_s and by (16) also the instrumental product matrices Λ_s are independent on the noise dynamics. In consequence of this fact, for the structural identification, a deterministic technique, such as a singularity test, can be used. Moreover, owing to (17), the test procedure can use the elements of only one small matrix.

b) Estimation of the observability indices

Define the matrix $\Lambda_s(1,1,\ldots,1)$ as 'kernel' of each s-th subsystem to be identified. This 'kernel' is extracted from the matrix obtained in a).

Construct now the sequence of symmetrical increasing-dimension matrices $\Lambda_s(\mu_1,\mu_2,\ldots,\mu_{m+r})$ where μ_1 is the number of vectors in the first submatrix, μ_2 is the number of vectors in the second, etc..

Determine then the observability indices in the following way. Let

$$F = \{i_{f1}, i_{f2}, \ldots, i_{fn1}\} \tag{18}$$

the set of subsystems structurally identified,

$$G = \{i_{g1}, i_{g2}, \ldots, i_{gn2}\} \tag{19}$$

the set of subsystems to be identified.

The determinant ratio defined as

$$LDR_s(\hat{\nu}_s) = \frac{\det \Lambda_s(\mu_1,\mu_2,\ldots,\hat{\nu}_s,\ldots,\mu_{m+r})}{\det \Lambda_s(\mu_1,\mu_2,\ldots,\hat{\nu}_s+1,\ldots,\mu_{m+r})} \tag{20}$$

where $\mu_i=\nu_i$ $i\in F$ and $\mu_i=\hat{\nu}_i$ $i\in G$ and $i\neq s$, increases rapidly when $\hat{\nu}_s=\nu_s$ being ν_s the true observability index of the s-th subsystem. In fact, if $\hat{\nu}_s=\nu_s$ then $LDR_s(\hat{\nu}_s) >> LDR_s(\hat{\nu}_s-1)$ and $LDR_s(\hat{\nu}_s+1) \simeq LDR_s(\hat{\nu}_s)$. This obviously means that the $\Lambda_s(\mu_1,\mu_2,\ldots,\hat{\nu}_s+1,\ldots,\mu_{m+r})$ is a singular matrix and in presence of noise a quasi-singular matrix. Therefore ν_s can be estimated from the evolution of the LDR_s quantity.

It may be noted that the matrices Λ_s are similar to the 'information matrices' proposed in [1]. But these 'information matrices' depend on the sequences of the input-output signals being the output data corrupted by noise. The matrices Λ_s, instead, depend on the sequences of the input signals and the sequences of the instrumental variables z_s being last ones statistically orthogonal to the noise.

3. COMPUTATION METHOD

For the structural identification the procedure is carried out according the steps a) and b).

In the first step the observability index ν must be chosen so as to ensure that it is the structure of the unknown system which is detected by the instrumental product moment matrix, and not that of the instrument generator. Then, the model of the auxiliary system should be over-parametrized in comparison with the model of the system to be tested. In others words, it ensures that the matrix Λ, obtained upon convergence of PLSM algorithm, consists only of submatrices having a number of vectors surely higher than the one presumed on the actual submatrices.

The procedure relating to the parametric identification does not give problems requiring the typical conditions to be satisfied with reference to the chosen method [6],[7].

In the second step b) it is necessary to pay attention to choosing the minimal sequences of the matrices Λ_s. Once the 'kernel' common to all subsystem has been obtained, the sequence of symmetrical increasing-dimension matrices is given by

$$\begin{array}{lll}
\Lambda_1(2,1,1,\ldots,1) & \Lambda_2(2,2,1,\ldots,1) & \ldots \\
\Lambda_1(3,2,2,\ldots,2) & \Lambda_2(3,3,2,\ldots,2) & \ldots \\
\Lambda_1(4,3,3,\ldots,3) & \Lambda_2(4,4,3,\ldots,3) & \ldots \\
\vdots & &
\end{array} \tag{21}$$

Let $\Lambda_s(\mu_1,\mu_2,\ldots,\mu_{m+r})$ be the first singular matrix in the considered sequence and μ_s, $s=1,2,\ldots,m$ the index incremented by one with respect to the last non-singular matrix that has been found. Then is follows that $\nu_s=\mu_s-1$. The index μ_s takes now the value ν_s and is not further incremented, while remaining ones are cyclically incremented until a second singular matrix is found and a second scalar ν_s is determined. The procedure ends when all the m scalars ν_s have been determined. It may be noted that if two adjacent matrices in sequence (21) are considered, all the elements of the first are present in the subsequent one that can thus be obtained by adding only a limited number of terms already computed in the matrix Λ.

For the singularity test, the sequences of the LDR quantities are given by

first subsystem

$$\mathrm{LDR}\ (1) = \frac{\det\ (1,1,1,\ldots,1)}{\det\ (2,1,1,\ldots,1)}\ ,$$
$$\mathrm{LDR}\ (2) = \frac{\det\ (2,1,1,\ldots,1)}{\det\ (3,2,2,\ldots,2)}\ , \tag{22}$$
$$\vdots$$

second subsystem

$$\mathrm{LDR}\ (1) = \frac{\det\ (1,1,1,\ldots,1)}{\det\ (2,2,1,\ldots,1)}\ ,$$
$$\mathrm{LDR}\ (2) = \frac{\det\ (2,2,1,\ldots,1)}{\det\ (3,3,2,\ldots,2)}\ , \tag{23}$$
$$\vdots$$

From the evolution of the quantities LDR_1, LDR_2, etc. the observability indices $\nu_1, \nu_2, \ldots$ and so on are estimated and thus the whole structure of MIMO system is identified.

In order to evaluate the MIMO system parameters, the same algorithm PLSM, used in a), is employed to assure that the estimated parameters are unbiased.

The proper application of this structural identification method requires the same conditions be satisfied, already reported in [5] and here, for sake of clarity, repeated:

- the noise signals must be a stationary stochastic type with expected value equal to zero. This last condition leads to the identification of the unbiased parameters;
- the noise signals must be statistically orthogonal to the instrumental variables z_s in order to filter the extraneous disturbances;
- the input signals should test the system under different dynamic conditions as long as the amplitudes of the signals lie in the system linear range. For instance, a binary pseudo-random signal (PRBS) can be properly used.

4. RESULTS OF SIMULATION

The proposed method has been verified for ascertaining the effectiveness of the identification procedure for determining the observability indices using the matrix Λ obtained by means of the PLSM algorithm. For such a reason, several simulations have been performed with different models whose structure had been estimated by means of the LDR test.

The performance of the procedure employed with various noise to signal ratio (σ) has been examined; for instance, PRBS noise has been used.

In order to illustrate the effectiveness of the LDR test, here are reported the results obtained by considering an two-input, two-output linear system whose input-output description in terms of normalized polynomial matrices is [9]:

$$\begin{bmatrix} z-0{,}75z+0{,}125 & 0{,}5z-0{,}25 \\ 0{,}5z-0{,}6 & z-z+0{,}24 \end{bmatrix} \underline{y}(k) = \begin{bmatrix} z-0{,}2 & 0{,}1 \\ 0{,}3 & -0{,}5 \end{bmatrix} \underline{u}(k) \tag{24}$$

Fig. 1 shows the evolution of the LDR quantities. The matrix Λ has been obtained upon convergence of PLSM algorithm with ν assumed

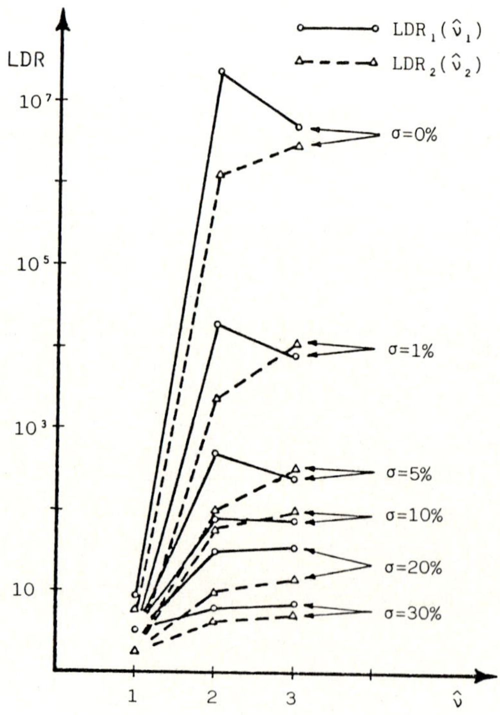

Fig. 1 - Evolution of LDR_1, LDR_2 quantities for determining the observability indices ν_1, ν_2.

equal to 4. The identification of the auxiliary system has been made on a number of samples ranging from 1763 to 2774 depending on the noise to signal ratio.
Fig. 1 outlines the rise of the LDR_1 and LDR_2 quantities when $\hat{\nu}=\nu_1=\nu_2=2$. The sharpness of this rise is dulled as the extraneous noise level is increased. The test is effective up to 30 per cent noise to signal ratio.

Finally it should be noted that the simulation results always lead to the minimum values of observability indices assuring the minimal state space realization.

5. CONCLUSIONS

In this paper a method for the on line structural identification of a multi-input, multi-output system has been proposed.

The procedure is based on the evaluation of determinant ratios and utilizes input-output data of an auxiliary system and of the actual system.

This test allows to perform the structural identification as an 'a priori' step and does not require the previous estimation of the parameters of tentative models with different structures which, in the case of multivariable systems, can easily reach large numbers also for limited number of outputs and for modest order.

Moreover this test offers the advantage of a direct applicability to the noisy data without any need to know or estimate the additive noise statistics.

Finally, this test operates on small matrices directly obtained during the identification recursive algorithm. Therefore the procedure is inexpensive from the computational standpoint and is suitable for implementation on dedicated microprocessor systems.

AKNOWLEDGEMENTS

The Authors wish to express their gratitude to Prof. G. Savastano for the many stimulating discussions which gave a significant contribution to the development of this paper.

REFERENCES

[1] R.P. Guidorzi: Invariants and Canonical Forms for Systems Structural and Parametric Identification. Automatica, Vol. 17, (1981) 117.

[2] R.P. Guidorzi, M.P. Losito, T. Muratori: The Range Error Test in the Structural Identification of Linear Multivariable Systems. IEEE Trans. on Aut. Contr., Vol. AC-27, (1982) 1044.

[3] P.E. Wellstead: An Instrumental Product Moment Test for Model Order Estimation. Automatica, Vol. 14, (1978) 89.
P.E. Wellstead, R.A. Rojas: Instrumental Product Moment Model Order Testing: Extensions and Applications. Int. J. Control, Vol. 35, (1982) 1013.

[4] P. Young, A. Jakeman, R. Mc Murtries: An Instrumental Variable Method for Model Order Identification. Automatica Vol. 16, (1980) 281.

[5] M. D'Apuzzo, L. Sansone: Consistent Structure Determination of Noisy Measurement System Model. Proc. of IV Int. IMEKO-Symposium on Measurement and Estimation, (1984) 67.

[6] I.D. Landau: Unbiased Recursive Identification Using Model Reference Adaptive Techniques. IEEE Trans. on Aut. Contr., Vol. AC-21, (1976) 194.
A. Gauthier, I.D. Landau: On the Recursive Identification of Multi-Input, Multi-Output Systems. Automatica, Vol. 14, (1978) 609.
L. Dugond, I.D. Landau: Recursive Output Error Identification Algorithm. Theory and Evaluation. Automatica, Vol. 16, (1980) 443.

[7] M. D'Apuzzo, L. Sansone: Simulation of Noisy Measurement System. Simulation of System '79, North-Holland, (1980) 357.

[8] J.M. Mendel: Discrete Techniques of Parameter Estimation. The Equation Error Formulation, M.Dekker (1973).

[9] N.K. Sinha, Y.H. Kwong: Recursive Estimation of the Parameters of Linear Multivariable Systems. Automatica, Vol. 15, (1979) 471.

Complex and Distributed Systems: Analysis, Simulation and Control
S.G. Tzafestas and P. Borne (editors)
Elsevier Science Publishers B.V. (North-Holland)

PARAMETER IDENTIFICATION IN VOLTERRA TYPE INTEGRAL SYSTEMS

N.K. Patel
Kalinga Iron Works
MATKAMBEDA-758036
INDIA

P.C. Das
Indian Institute of Technology
KANPUR-208016
INDIA

S.S. Prabhu
Indian Institute of Technology
KANPUR-208016
INDIA

Numerical algorithms, based on the quasilinearization technique, are developed for the determination of parameters in a system of Volterra type integral equations, minimizing a seminorm. The convergence proofs are given for the two algorithms developed, under different restrictions on the system equations. It is seen that the first algorithm is quadratically convergent and much less computational effort is required in each iteration as compared to the second but it imposes more constraints in the system equations and the convergence is local in nature. Methods of solution, for the linearized problem appearing in the algorithms, are given for few special cases.

1. INTRODUCTION

The usual problem of approximation theory can be stated as follows:

Let (X,d) be a metric space and $V \subset X$ be a subset of X. Let $x \in X$ be a given point in X. It is required to find an element $v^* \in V$ such that

$$d(x_o, v^*) \leq d(x_o, v) \;\forall\; v \in V$$

The work here is devoted to the development of numerical algorithms in the case X is restricted to the space of all continuous functions in $[0,1]$ with range in R^n and with the metric a seminorm. Further the set V is restricted to the set of solutions of a system of Volterra type integral equations (VTIE) with parameters.

The algorithms developed are based on continuous linerarization of the integral equations and the solution of the linear approximation problem. The basic idea in this is the quasilinearization technique due to Bellman and Kalaba [1].

Section -2 contains the problem formulation and the linearized problem. In Section -3 and Section -4 two numerical algorithms are presented and their convergence established under certain restrictions on various system quantities. The proofs for the auxilary results, used in establishing the convergence of the algorithms, are contained in Appendix -I. Methods of solution of the linearized problem appearing in the algorithms are given for few special cases in Appendix -2.

2. PROBLEM FORMULATION AND LINEARIZED PROBLEM

Let $\tilde{h}: [0,1] \times R^n \times R^m \times R^p \to R^n$, $q \in C_1[0,1]$ be given. Further let $T: C_n[0,1] \to C_1[0,1]$ be a linear operator. Here $C_r[0,1]$ denotes the space of all continuous functions on $[0,1]$ with range in R^r.

A solution $x^*(\cdot \mid \mu^*, \beta^*)$ of the integral equation,

$$x(t) = \tilde{h}(t, x(t), \beta, \mu) = \tilde{f}(t, \mu) + \int_0^t \tilde{g}(t, s, x(s), \beta)\, ds \qquad (2.1)$$

is sought, so that $P(q - Tx^*)$ is a minimum, under all the solutions $x(\cdot \mid \mu, \beta)$ of the equation (2.1). Here P is a seminorm in $C_1[0,1]$.

In future the above problem will be assumed to have the following form :

$$x(t) := F(x, \mu)(t) = f(t, \mu) + \int_0^t g(t, s, x(s))\, ds$$

$F: C_n[0,1] \times R^p \to C_n[0,1]$ be a Frechet differentiable mapping and the derivative of F at (x, μ) will be denoted by $F'(x, \mu)$.

Let $A \subset R^p$ be convex. Define,

$B \subset C_n[0,1] \times R^p$ as $B := C_n[0,1] \times A$

Problem (GP) :

Minimize $P(q-Tx)$

$(x,\mu) \in B \wedge x = F(x,\mu)$

Let $(x,\mu) \in B$. Linearize F at $(\tilde{x},\tilde{\mu})$ and set,

$$F_{(\tilde{x},\tilde{\mu})}(x,\mu) = f(t,\tilde{\mu}) + \frac{\partial f}{\partial \mu}(t,\tilde{\mu})(\mu-\tilde{\mu}) + \int_0^t g(t,s,\tilde{x}(s))\,ds + \int_0^t \frac{\partial g}{\partial x}(t,s,\tilde{x}(s))\,(x(s)-\tilde{x}(s))ds$$

The linearized problem at $(\tilde{x},\tilde{\mu})$ denoted by $(LP_{(\tilde{x},\tilde{\mu})})$ can be written as follows:

$(LP_{(\tilde{x},\tilde{\mu})})$: Minimize $P(q-Tx)$

$(x,\mu) \in B \wedge x = F_{(\tilde{x},\tilde{\mu})}(x,\mu)$.

3. FORMULATION OF THE FIRST ALGORITHM AND CONVERGENCE PROOF

For the formulation of the first algorithm (without λ-strategy) and proof of its convergence the following assumptions and auxilary results will be used:

Assumptions :

(V1.a): f is continuous and defined $\forall t \in R$

(V1.b): For each constant $\bar{K} > 0$ and each bounded $S \subset R^n$ $\exists$ a measurable function $K(t,s) \ni$

$|g(t,s,x) - g(t,s,y)| \leq K(t,s)\,|x-y|$,

wherever $0 \leq s \leq t \leq \bar{K}$ and both x and y are in S. For each $t \in [0,\bar{K}]$ the function $K(t,s)$ is in $L^1(0,t)$ as a function of s and $\lim_{r \to 0} \int_t^{t+r} K(t+r,s)ds = 0$.

(Thi is obviously the case if $K(t,s)$ is integrable in the product space, by Fubini's Theorem). (Here $|\cdot|$ denotes the Euclidean norm).

Further $\int_0^1 \sup_{t\in[0,1]} K(t,s)ds$ exists.

(V2) : There exist a local minimal solution (x^*,μ^*) of (GP) with the following properties;

(x^*,μ^*) is a strongly unique local minimal solution of (GP) if the following holds :

Let $(x^*,\mu^*) \in B \wedge x^* = F(x^*,\mu^*)$ then $\exists \delta_1 > 0, K_1 > 0 \ni (x,\mu) \in B \wedge \|x-x^*\| \leq \delta_1 \wedge x = F(x,\mu) \Rightarrow P(q-Tx)-P(q-Tx^*) \geq K_1 \|x-x^*\|$

(V3.a) : f is linear in μ

(V3.b) : Let $\exists \delta_2 > 0$, a measurable function $\hat{K}(t,s) \ni |\frac{\partial g}{\partial x}(t,s,\bar{x}(s)) - \frac{\partial g}{\partial x}(t,s,x(s))| \leq \hat{K}(t,s)\,|\bar{x}-x|\,s$ $\forall(\bar{x},\bar{\mu}) \in B \wedge (x,\mu) \in B$, wherever $\|\bar{x}-x^*\| \leq \delta_2 \wedge \|x-x^*\| \leq \delta_2$.

Lemma -1:

(i) $\forall (x,\mu) \in B\ \exists (\bar{x},\bar{\mu}) \in B \ni \bar{x} = F(\bar{x},\bar{\mu})$ and $\|\bar{x}-x\| \leq e^L \|x-F(x,\mu)\|$, where $L = \int_0^1 \sup_{t\in[0,1]} K(t,s)\,ds$

(ii) $\forall (\tilde{x},\tilde{\mu}) \in B, \forall(\bar{x},\bar{\mu}) \in B\ \exists (x,\mu) \in B \ni \bar{x} = F_{(\tilde{x},\tilde{\mu})}(\bar{x},\bar{\mu})$ and $\|\bar{x}-x\| \leq e^L \|x-F_{(\tilde{x},\tilde{\mu})}(x,\mu)\|$

Corollary -1:

(i) $\forall K > e^L C$

$\inf.\{P(q-Tx) + K\|x-F(x,\mu)\| : (x,\mu) \in B\}$

$= \inf.\{P(q-Tx) : (x,\mu) \in B,\ x = F(x,\mu)\}$

(ii) $\forall K > e^L C\ \forall (\bar{x},\bar{\mu}) \in B$

$P(q-T\bar{x}) + K\|\bar{x}-F(\bar{x},\bar{\mu})\| = \inf.\{P(q-Tx) + K\|x-F(x,\mu)\| : (x,\mu) \in B\}$

$\Longleftrightarrow (\bar{x},\bar{\mu})$ minimal solution of (GP).

(iii) Similar statements are valid for problem $(LP_{(\tilde{x},\tilde{\mu})})$, $(\tilde{x},\tilde{\mu}) \in B$.

Here, $C := \sup.\{P(Tx) : x \in C_n[0,1], \|x\| = 1\}$

Lemma - 2:

Assumption (V2) $\Longrightarrow \forall (x,\mu) \in B \wedge \|x-x^*\| \leq \delta_1/2$

$P(q-Tx) - P(q-Tx^*) + K_2\|x-F(x,\mu)\| \geq K_1\|x-x^*\|$, where $K_2 = (C+K_1)\,e^L$

Lemma -3:

Assumption (V2) and convexity of $B \Rightarrow \forall (x,\mu) \in B$,

$P(q-Tx) - P(q-Tx^*) + K_2\|x-F_{(x^*,\mu^*)}(x,\mu)\| \geq K_1\|x-x^*\|$

Lemma -4:

From Assumptions (V3) and convexity of

B it follows that $\forall\, (\bar{x},\bar{\mu}) \in B, (x,\mu) \in B;$

$\|\bar{x}-x^*\| \leq \delta_2 \wedge \|x-x^*\| \leq \delta_2 \Longrightarrow$

$\|F(x,\mu)-F_{(\bar{x},\bar{\mu})}(x,\mu)\| \leq \frac{1}{2} L_1 \|x-\bar{x}\|^2$

where $L_1 = \sup_{t\in[0,1]} \int_0^1 \hat{K}(t,s)\, ds$

ALGORITHM (A1) : (Iteration method without λ-strategy)

Start : Choose $(x^{(1)}, \mu^{(1)}) \in B$

Iteration step : $(x^{(i)}, \mu^{(i)})$ is already constructed.

Construct $(x^{(i+1)}, \mu^{(i+1)})$ as the minimal solution of the linear approximation problem ($LP_{(x^{(i)},\, \mu^{(i)})}$).

Theorem 1. (First convergence of Alg-A1) Let A be convex and (x^*, μ^*) be a local minimal solution of (GP). Let assumptions (V1), (V2) and (V3) be valid. Then the algorithm (A1) converges locally quadritically to (x^*, μ^*), i.e.

$\exists \delta > 0,\ C_1 > 0 \ni (x^{(1)},\mu^{(1)}) \in B \wedge$
$\|x^{(i)}-x^*\| \leq \delta \Longrightarrow$

$\forall\, i \in N,\ \|x^{(i+1)}-x^*\| \leq C_1 \|x^{(i)}-x^*\|^2$

where $\{(x^{(i)},\ \mu^{(i)})\}_{i\in N}$ is a sequence obtained through algorithm (A1) with $(x^{(1)},\ \mu^{(1)})$ as its initial member. Here, N is the set of natural numbers.

Proof :

Choose δ_0 with $0 < \delta_0 < \delta_2$ and $L_1 K_2 \delta_0 < K_1$. Let $(x,\mu) \in B$ with $\|x-x^*\| \leq \delta_0$ and $(\bar{x},\bar{\mu})$ be minimal solution of ($LP_{(x,\mu)}$). Then by Lemma -3,

$K_1 \|\bar{x}-x^*\| \leq P(q-T\bar{x}) - P(q-Tx^*) +$
$K_2 \|\bar{x}-F_{(x^*,\mu^*)}(\bar{x},\bar{\mu})\|$
$\leq P(q-T\bar{x}) - P(q-Tx^*) + K_2 \|\bar{x}-F_{(x,\mu)}(\bar{x},\bar{\mu})\|$
$+ K_2 \|F_{(x,\mu)}(\bar{x},\bar{\mu}) - F_{(x^*,\mu^*)}(\bar{x},\bar{\mu})\|$ (3.i)

Consider,

$\|F_{(x,\mu)}(\bar{x},\bar{\mu}) - F_{(x^*,\mu^*)}(\bar{x},\bar{\mu})\| =$

$\|F(x,\mu) + \frac{\partial F}{\partial \mu}(x,\mu)(\bar{\mu}-\mu)$
$+ \frac{\partial F}{\partial x}(x,\mu)(\bar{x}-x) - F(x^*,\mu^*) - \frac{\partial F}{\partial \mu}(x^*,\mu^*)$
$(\bar{\mu}-\mu^*) - \frac{\partial F}{\partial x}(x^*,\mu^*)(\bar{x}-x^*)\|$

$= \|[F(x,\mu) + \frac{\partial F}{\partial \mu}(x,\mu)(\mu^*-\mu) +$
$\frac{\partial F}{\partial x}(x,\mu)(x^*-x) - F(x^*,\mu^*)]$
$- \frac{\partial F}{\partial x}(x,\mu)x^* + \frac{\partial F}{\partial x}(x^*,\mu^*)\, x^* +$
$(\frac{\partial F}{\partial x}(x,\mu) - \frac{\partial F}{\partial x}(x^*,\mu^*))\ \bar{x}\|$
$\leq \|F(x^*,\mu^*) - F_{(x,\mu)}(x^*,\mu^*)\| +$
$\|(\frac{\partial F}{\partial x}(x,\mu) - \frac{\partial F}{\partial x}(x^*,\mu^*))\ (\bar{x}-x^*)\|$
(3.ii)

By Corollary - 1 (iii.b)

$P(q-T\bar{x}) + K_2\|\bar{x}-F_{(x,\mu)}(\bar{x},\bar{\mu})\| \leq P(q-Tx^*)$
$+ K_2\|x^*-F_{(x,\mu)}(x^*,\mu^*)\| \Longrightarrow$
$P(q-T\bar{x}) - P(q-Tx^*) + K_2\|\bar{x}-F_{(x,\mu)}(\bar{x},\bar{\mu})\|$
$\leq K_2\|x^*-F_{(x,\mu)}(x^*,\mu^*)\|$ (3.iii)

Substituting in (3.i) from (3.ii) and (3.iii) $K_1\|\bar{x}-x^*\| \leq 2K_2\|F(x^*,\mu^*) - F_{(x,\mu)}(x^*,\mu^*)\| +$
$K_2\|(\frac{\partial F}{\partial x}(x,\mu) - \frac{\partial F}{\partial x}(x^*,\mu^*)).\ (\bar{x}-x^*)\|$
$\leq K_2 L_1 \|x-x^*\|^2 + K_2 L_1 \|x-x^*\| \cdot \|\bar{x}-x^*\|$
$\leq K_2 L_1 \|x-x^*\|^2 + K_2 L_1 \delta_0 \|\bar{x}-x^*\|$
$\Rightarrow \|\bar{x}-x^*\| \leq \frac{K_2 L_1}{K1 - L_1 K2 \delta_0} \|x-x^*\|^2 =$
$C_1 \|x-x^*\|^2$

So the assertion follows for every δ with $0 < \delta < \delta_0$ and $C_1 \delta < 1$.

3. 1. Extension of Convergence of Algorithm (A1) :

In Theorem-1, the function $f(t,\mu)$ was restricted to be linear in μ. The uniqueness in the local minimal solution of (GP) was assumed with respect to the space C_n [0,1] . In this section the case where $f(t,\mu)$ is not necessarily linear in μ has been studied. For the convergence proof of Algorithm (A1) the uniqueness of the solution is assumed in the set $B(:= C_n[0,1] \times A)$. In this case the theorem possesses a local character with respect to μ as well.

For this purpose the modifications in assumptions and auxilary results are given below:

Assumptions :

(V2.m) : (x^*,μ^*) is a strongly unique local minimal solution of (GP)

if the following holds;

$(x^*,\mu^*) \in B$, $x^* = F(x^*,\mu^*)$ and $\exists\, \delta_1 > 0$, $K_1 > 0 \ni (x,\mu) \in B \wedge$ $(\|x-x^*\| + \|\mu-\mu^*\|) \leq \delta_1 \wedge x = F(x,\mu) \Longrightarrow$ $P(q-Tx) - P(q-Tx^*) \geq K_1(\|x-x^*\| + \|\mu-\mu^*\|)$

(V3.m) (a): $\frac{\partial f}{\partial \mu}$ is local Lipschitz continuous in a neighbourhood of μ^*, i.e.,

$\exists\, \delta_2' > 0$, $L_1' > 0$ $\forall$ $\mu_1, \mu_2 \in R^p \ni$

$\|\mu_1-\mu^*\| < \delta_2' \wedge \|\mu_2-\mu^*\| \leq \delta_2' \Longrightarrow$

$\|\frac{\partial f}{\partial \mu}(t,\mu_1) - \frac{\partial f}{\partial \mu}(t,\mu_2)\| \leq L_1' \|\mu_1-\mu_2\|$

(b) : Same as (V3.b).

Lemma -5:

Assumption (V2.m) $\Longrightarrow$

$\{\forall (x,\mu) \in B \wedge (\|x-x^*\| + \|\mu-\mu^*\|) \leq \delta_1/2 \Rightarrow$ $P(q-Tx) - P(q-Tx^*) + K_2\|x-F(x,\mu)\| \geq$ $K_1 (\|x-x^*\| + \|\mu-\mu^*\|)\}$, where $K_2 = (C+K_1)\, e^L$.

Lemma -6:

Assumption (V2.m) and convexity of $B \Longrightarrow$ $\forall\, (x,\mu) \in B$

$P(q-Tx) - P(q-Tx^*) + K_2\|x - F_{(x^*,\mu^*)}(x,\mu)\|$ $\geq K_1(\|x-x^*\| + \|\mu-\mu^*\|)$

Lemma -7:

From (V3.m) and convexity of B it follows that,

$(\bar{x},\bar{\mu}) \in B, (x,\mu) \in B$, $\|\bar{x}-x^*\| \leq \delta_2, \|\bar{\mu}-\mu^*\| \leq \delta_2'$, $\|x-x^*\| \leq \delta_2$,

$\|\mu-\mu^*\| \leq \delta_2' \Longrightarrow \|F_{(\bar{x},\bar{\mu})}(x,\mu) - F(x,\mu)\| \leq \frac{1}{2}\bar{L}(\|x-\bar{x}\|^2 + \|\mu-\bar{\mu}\|^2)$

where $\bar{L} = \mathrm{Max.}\,(L_1', L1)$.

Theorem -2 :(Second convergence proof of Algorithm (A1))

Let A be convex and (x^*,μ^*) be a local minimal solution of (GP). Let assumptions (V1),(V2.m) and (V3.m) be valid. Then the algorithm (A1) converges locally quadratically to (x^*,μ^*), i.e.,

$\exists\, \delta > 0$, $C_1 > 0 \ni (x^{(1)},\mu^{(1)}) \in$ $B \wedge (\|x^{(i)}-x^*\| + \|\mu^{(i)}-\mu^*\|) \leq \delta$ $\Longrightarrow (\|x^{(i+1)}-x^*\| + \|\mu^{(i+1)}-\mu^*\|) \leq$ $C_1(\|x^{(i)}-x^*\|^2 + \|\mu^{(i)}-\mu^*\|^2)$

$\forall\, i \in N$, where $\{(x^{(i)},\mu^{(i)})\}_{i \in N}$ is the sequence obtained through algorithm (A1) with $(x^{(1)},\mu^{(1)})$ as starting point; N is the set of natural numbers.

Proof :

Choose δ_0 with $0 < \delta_0 \leq \min(\delta_2', \delta_2)$ and $\bar{L} K_2 \delta_0 < K_1$. Let $(x,\mu) \in B$ with $(\|x-x^*\| + \|\mu-\mu^*\|) \leq \delta_0$ and $(\bar{x},\bar{\mu})$ the minimal solution of $(LP_{(x,\mu)})$.

From Lemma -6,

$K_1(\|\bar{x}-x^*\| + \|\bar{\mu}-\mu^*\|) \leq P(q-T\bar{x}) - P(q-Tx^*) +$ $K_2\|\bar{x} - F_{(x^*,\mu^*)}(\bar{x},\bar{\mu})\|$

$\leq P(q-T\bar{x}) - P(q-Tx^*) + K_2\|\bar{x} - F_{(x,\mu)}(\bar{x},\bar{\mu})\|$ $+ K_2\|F_{(x,\mu)}(\bar{x},\bar{\mu}) - F_{(x^*,\mu^*)}(\bar{x},\bar{\mu})\|$ (3.1.i)

Consider, $F_{(x,\mu)}(\bar{x},\bar{\mu}) - F_{(x^*,\mu^*)}(\bar{x},\bar{\mu}) =$ $F(x,\mu) + F'(x,\mu)((\bar{x},\bar{\mu}) - (x,\mu)) - F(x^*,\mu^*) - F'(x^*,\mu^*)((\bar{x},\bar{\mu}) - (x^*,\mu^*))$
$= F(x,\mu) + F'(x,\mu)((x^*,\mu^*) - (x,\mu)) - F(x^*,\mu^*)$
$- F'(x,\mu)(x^*,\mu^*)$

$+ F'(x,\mu)(\bar{x},\bar{\mu}) + F'(x^*,\mu^*)((x^*,\mu^*) - (\bar{x},\bar{\mu}))$
$= -(F(x^*,\mu^*) - F_{(x,\mu)}(x^*,\mu^*)) - (F'(x,\mu)$
$- F'(x^*,\mu^*)).((x^*,\mu^*) - (\bar{x},\bar{\mu})) \Longrightarrow$

$\|F_{(x,\mu)}(\bar{x},\bar{\mu}) - F_{(x^*,\mu^*)}(\bar{x},\bar{\mu})\| \leq$
$\|F(x^*,\mu^*) - F_{(x,\mu)}(x^*,\mu^*)\| +$
$\|(F'(x,\mu) - F'(x^*,\mu^*))((x^*,\mu^*) - (\bar{x},\bar{\mu}))$

(3.1.ii)

By Corollary-1 (iii.b)

$P(q-T\bar{x}) + K_2\|\bar{x} - F_{(x,\mu)}(\bar{x},\bar{\mu})\| \leq \inf.$ $\{P(q-Ty) + K_2\|y - F_{(x,\mu)}(x,\tilde{\mu})\|$: $(x,\mu) \in B \leq P(q-Tx^*) + K_2\|x^* - F_{(x,\mu)}(x^*,\mu^*)$

(3.1.iii)

Substituting in (3.1.i) from (3.1.ii) and (3.1.iii)

$K_1(\|\bar{x}-x^*\| + \|\bar{\mu}-\mu^*\|) \leqslant 2K_2\|F(x^*,\mu^*)$
$-F_{(x,\mu)}(x^*,\mu^*)\| + K_2\|(F'(x,\mu)-F'(x^*,\mu^*))$
$((x^*,\mu^*)-(\bar{x},\bar{\mu}))\| \leqslant K_2\bar{L}\ (\|x-x^*\|^2+\|\mu-\mu^*\|^2)$
$+K_2\bar{L}\,\delta_0(\|\bar{x}-x^*\|+\|\bar{\mu}-\mu^*\|) \Rightarrow \|\bar{x}-x^*\| + \|\bar{\mu}-\mu^*\|$
$\leqslant \frac{K_2\bar{L}}{K_1-K_2\bar{L}\,\delta_0}(\|x-x^*\|^2+\|\mu-\mu^*\|^2)$
$=C_1(\|x-x^*\|^2+\|\mu-\mu^*\|^2)$

Thus the assertion follows for every δ, with $0<\delta<\delta_0$ and $C_1\delta < 1$.

4. FORMULATION OF THE SECOND ALGORITHM AND CONVERGENCE PROOF

This section presents an algorithm with λ-strategy; that converges with essentially weaker assumptions.

For this purpose the necessary assumptions, definitions and auxilary results are given below;

Assumptions:

$(\bar{V}1)$: Same as assumption (V1)

$(\bar{V}2)$: $K \in R^+$ with $K > e^L C$, $0<\beta<1$
$(\lambda_j)_{j\in N} \subset R^+$, $\lambda_1 = 1$, $\alpha := \inf_{j\in N} \frac{\lambda_{j+1}}{\lambda_j} > 0$

$\lim_j \lambda_j = 0$. The set A is convex and compact.

$(\bar{V}3)$: f is linear in μ.

Definitions:

1. $(\hat{x},\hat{\mu}) \in B$ is called a stationary pt. of (GP) iff $(\hat{x},\hat{\mu})$ is a minimal solution of $(LP_{(\hat{x},\hat{\mu})})$.

2. The penalty function $\varphi : B \to R$ is defined as, $\varphi(x,\mu) := P(q-Tx) + K\|x-F(x,\mu)\|$

3. The linearized penalty function $\varphi(\hat{x},\hat{\mu}) : B \to R$, $(\hat{x},\hat{\mu}) \in B$ is defined as,

$$\varphi_{(\hat{x},\hat{\mu})}(x,\mu) := P(q-Tx)+K\|x-F_{(\hat{x},\hat{\mu})}(x,\mu)\|$$

Lemma -8:

Let $(x,\mu) \in B$ and $(\bar{x},\bar{\mu})$ be the solution of $(LP_{(x,\mu)})$. Then, $P(q-T\bar{x}) =$

$P(q-T\bar{x}) + K\|\bar{x}-F_{(x,\mu)}(\bar{x},\bar{\mu})\|$

$\leqslant P(q-Tx)+K\|x-F(x,\mu)\| = \varphi(x,)$.

Corollary-2:

In case min $P(q-Tx)$ over all $(x,\mu)\in B$ such that $x = F(x,\mu)$ is unique and min. $P(q-T\bar{x})$ over all $(x,\mu) \in B$ with $\bar{x} = F_{(x,\mu)}(\bar{x},\bar{\mu})$ is unique, then both the minima must coincide.

Lemma -9:

Let $(x,\mu) \in B$. $(\bar{x},\bar{\mu}) \in B$ be the solution of $(LP_{(x,\mu)})$ with $\varphi(x,\mu) > P(q-T\bar{x})$.
Then $\exists\ \delta > 0 \ni$ for $0<\lambda<\delta$,
$\varphi(x+\lambda(\bar{x}-x), \mu+\lambda(\bar{\mu}-\mu)) \leqslant \varphi(x,\mu) - \beta\lambda(\varphi(x,\mu)-P(q-T\bar{x}))$

Remark :

From Lemma -9 one concludes;

(x^*,μ^*) is a local minimal solution of (GP) $\Longrightarrow$ (x^*,μ^*) is a stationary point of (GP) $\Longrightarrow x^* = F(x^*,\mu^*)$.

ALGORITHM (A2). (iteration method with λ-strategy)

Start : Choose $(x^{(1)}, \mu^{(1)}) \in B$.

Iteration Step :

Suppose $(x^{(i)}, \mu^{(i)})$ is already constructed. Determine $(\bar{x}^{(i)}, \bar{\mu}^{(i)})$ as the minimal solution of the linerazied approximation problem $(LP_{(x^{(i)},\mu^{(i)})})$.

In case $P(q-T\bar{x}^{(i)}) = \varphi(x^{(i)},\mu^{(i)})$; $(x^{(i)},\mu^{(i)})$ is a stationary point of (GP) and the iteration will be stopped. Otherwise, $\varphi(x^{(i)},\mu^{(i)}) > P(q-T\bar{x}^{(i)})$

Take; $h^{(i)} = \bar{x}^{(i)} - x^{(i)}$

$1^{(i)} = \bar{\mu}^{(i)} - \mu^{(i)}$

Determine the smallest index $j := j^{(i)} \in N$ with $\varphi(x^{(i)} + \lambda_j h^{(i)}, \mu^{(i)} + \lambda_j 1^{(i)}) \leqslant \varphi(x^{(i)}, \mu^{(i)}) - \beta\lambda_j(\varphi(x^{(i)},\mu^{(i)}) - P(q-T\bar{x}^{i}))$

Set, $x^{(i+1)} = x^{(i)} + \lambda_j h^{(i)}$

$\mu^{(i+1)} = \mu^{(i)} + \lambda_j 1^{(i)}$.

Theorem-3 (Convergence proof of Algorithm (A2))

Let A be convex, compact and assumptions $(\bar{V}1)$, $(\bar{V}2)$ and $(\bar{V}3)$ be satisfied. Starting with $(x^{(1)}, \mu^{(1)}) \in B$ the iteration method (A2) is carried out. In case the method terminates at the ith iteration $(x^{(i)}, \mu^{(i)})$ must be a stationary point of (GP). Otherwise the method provides a bounded sequence $\{(x^{(i)}, \mu^{(i)})\}_{i \in N}$ in B, which possesses a limit point. Every limit point is a stationary point of (GP).

Proof :

(i) The sequence possesses a limit point in B :

By construction $\{\varphi(x^{(i)}, \mu^{(i)})\}_{i \in N}$ is a monotone decreasing sequence and therefore for some constant $\xi > 0, \forall i \in N$

$$\|x^{(i)} - F(x^{(i)}, \mu^{(i)})\| \leq \xi$$

It follows from here, $\forall i \in N$, $\forall t \in [0,1]$;

$$|x^{(i)}(t)| \leq \xi + |F(x^{(i)}, \mu^{(i)})(t)| \leq \xi + \|F(o, \mu^{(i)})\| + \int_0^t K(t,s) \cdot |x^{(i)}(s)|\, ds$$

By Gronwall's Lemma one obtains a bound for $x^{(i)}$ and hence an uniform bound from boundedness of A.

Obviously $\{x^{(i)}\}_{i \in N}$ is uniformly bounded and equicontinuous. So by Ascoli's theorem, $\{(x^{(i)}, \mu^{(i)})\}_{i \in N}$ possesses a limit point in B, as A is compact and f is continuous.

(ii) Every limit point of $\{(x^{(i)}, \mu^{(i)})\}_{i \in N}$ in B is a stationary point of (GP) :

Let $(x^*, \mu^*) \in B$ be a limit point of $\{(x^{(i)}, \mu^{(i)})\}_{i \in N}$

we prove by contradiction.

Let (x^*, μ^*) be a non-stationary point. Then,

$$\varphi(x^*, \mu^*) > m := \inf.\{P(q - Tx) : (x, \mu) \in B, x = F_{(x^*, \mu^*)}(x, \mu)\}$$

Since φ is continuous and $\{\varphi(x^{(i)}, \mu^{(i)})\}_{i \in N}$ is a monotone decreasing sequence, the following holds;

$$\varphi(x^*, \mu^*) = \inf.\{\varphi(x^{(i)}, \mu^{(i)}) : i \in N\}$$

Now we shall show that $\exists\, i \in N \ni$

$$\varphi(x^{(i+1)}, \mu^{(i+1)}) < \varphi(x^*, \mu^*) \qquad (4.i)$$

Let $\eta := \varphi(x^*, \mu^*) - m > 0$ and $(\bar{x}, \bar{\mu}) \in B$ with $\varphi_{(x^*, \mu^*)}(\bar{x}, \bar{\mu}) = m$.

Let $(\bar{x}^{(i)}, \bar{\mu}^{(i)})$ be a minimal solution of $(LP_{(x^{(i)}, \mu^{(i)})})$, consequently,

$$\varphi(x^{(i)}, \mu^{(i)}) - P(q - T\bar{x}^{(i)}) \geq \varphi(x^*, \mu^*) - \varphi_{(x^{(i)}, \mu^{(i)})}(\bar{x}, \bar{\mu}).$$

Choosing $(x^{(i)}, \mu^{(i)})$ sufficiently close to (x^*, μ^*); $|\varphi_{(x^{(i)}, \mu^{(i)})}(\bar{x}, \bar{\mu}) - \varphi_{(x^*, \mu^*)}(\bar{x}, \bar{\mu})| \leq$

$$K \|F_{(x^{(i)}, \mu^{(i)})}(\bar{x}, \bar{\mu}) - F_{(x^*, \mu^*)}(\bar{x}, \bar{\mu})\|$$

$$\leq K \|F(x^{(i)}, \mu^{(i)}) - F_{(x^*, \mu^*)}(x(i), \mu(i))\| + K \|\int_0^t (\partial g(t,s,x^{(i)}(s)) - \partial g(t,s,x))) / \partial x \cdot (\bar{x} - x^{(i)})_s\, ds \leq \eta/2$$

From the above estimates, $\varphi(x^{(i)}, \mu^{(i)}) - P(q - T\bar{x}^{(i)}) \geq \varphi(x^*, \mu^*) - (\varphi_{(x^*, \mu^*)}(\bar{x}, \bar{\mu}) + \eta/2) \geq \eta/2$

Let $R > 0$ be $\ni \forall i \in N$, $\|x^{(i)}\| \leq R \wedge \|\bar{x}^{(i)}\| \leq R$.

We have $\forall (x, \mu) \in B\, \exists \delta \ni 0 < \delta < 4R \wedge \|x - x^*\| \leq \delta \Longrightarrow$

$$\|F(x, \mu) - F_{(x^*, \mu^*)}(x, \mu)\| \leq \frac{(1-\beta)\eta}{8R} \|x - x^*\|$$

Now we give an estimate for $\lambda_j (j = j^{(i)})$ below :

For $0 < \lambda < \frac{\delta}{4R} \leq 1$ because of $\|x^{(i)} + \lambda h^{(i)} - x^*\| \leq \|x^{(i)} - x^*\| + \lambda \|h^{(i)}\| \leq \frac{\delta}{2} + \frac{\delta \cdot 2R}{4R} = \delta$

The following holds;

$$\varphi(x^{(i)} + \lambda h^{(i)}, \mu^{(i)} + \lambda 1^{(i)}) \leq \varphi_{(x^*, \mu^*)}(x^{(i)} + \lambda h^{(i)}, \mu^{(i)} + \lambda 1^{(i)}) + K \|F(x^{(i)} + \lambda h^{(i)}, \mu^{(i)} + \lambda 1^{(i)}) - F_{(x^*, \mu^*)}(x^{(i)} + \lambda h^{(i)}, \mu(i) + \lambda 1^{(i)})\|$$

$$\leq \varphi_{(x^*, \mu^*)}(x^{(i)} + \lambda h^{(i)}, \mu(i) + \lambda 1(i)) + \frac{(1-\beta)\eta}{8R} \cdot \left(\frac{\alpha\delta}{2} + 2R\lambda\right) \quad \left(\text{taking } \|x^{(i)} - x^*\| \leq \frac{\alpha\delta}{2}\right)$$

$$\leq \varphi_{(x^{(i)}, \mu^{(i)})}(x^{(i)} + \lambda h^{(i)}, \mu^{(i)} + \lambda 1^{(i)}) + \frac{(1-\beta)\eta}{4}\left(\frac{\alpha\delta}{4R} + \lambda\right) \leq (1 - \lambda)\varphi_{(x^{(i)}, \mu^{(i)})}(x^{(i)}, \mu^{(i)}) + \lambda\varphi_{(x^{(i)}, \mu^{(i)})}(\bar{x}^{(i)}, \bar{\mu}^{(i)}) +$$

$$\frac{(1-\beta)\eta}{4}\left(\frac{\alpha\delta}{4R} + \lambda\right) \leq \varphi(x^{(i)}, \mu^{(i)}) - \lambda\beta(\varphi(x^{(i)}, \mu^{(i)}) - P(q - T\bar{x}^{(i)})) + \eta\frac{(1-\beta)}{4}\left(\frac{\alpha\delta}{4R} + \lambda\right)$$

$$- \frac{(1-\beta)}{2}\eta\lambda \leq \varphi(x^{(i)}, \mu^{(i)}) - \lambda\beta(\varphi(x^{(i)}, \mu^{(i)}) - P(q - T\bar{x}^{(i)})) + \frac{(1-\beta)\eta}{4}\left(\frac{\alpha\delta}{4R} - \lambda\right)$$

i.e.

$$\varphi(x^{(i)}+\lambda h(i),\mu^{(i)}+\lambda_1^{(i)}) \leq \varphi(x^{(i)},\mu^{(i)}) - \lambda\beta(\varphi(x^{(i)},\mu^{(i)}) - P\ (q-T\bar{x}^{(i)})\ )$$

In case $\frac{\alpha\delta}{4R} \leq \lambda < \frac{\delta}{4R}$

Since, $\inf\{\frac{\lambda_{j+1}}{\lambda_j} : j \in N\} = \alpha > 0$,

there is a $j' \in N$ with $\frac{\alpha\delta}{4R} \leq \lambda_{j'} < \frac{\delta}{4R}$

By the definition of $j^{(i)}$ in the algorithm (A2) it follows,

$j = j^{(i)} \leq j'$ and $\lambda_j \geqslant \lambda_{j'} > \frac{\alpha\delta}{4R}$

The inequality (4.i) is obtainable as,

$$\varphi(x^{(i+1)},\mu^{(i+1)}) = \varphi(x^{(i)}+\lambda_j h\ (i),\mu^{(i)}+\lambda_j 1(i)\ ) \leq \varphi(x^{(i)},\mu^{(i)}) - \lambda_j\beta(\varphi(x^{(i)},\mu^{(i)}) - P\ (q-T\bar{x}^{(i)})\ ) \leq \varphi(x^*,\mu^*) + \frac{\alpha\delta\beta\eta}{16R} - \frac{\alpha\delta}{4R}\ \frac{\beta\eta}{2} < \varphi(x^*,\mu^*\ )$$

Remark :

In the above convergence proof $\tilde{f}$ is assumed to be linear in μ. The proof can be extended to the case where $\hat{f}$ is not necessarily linear in μ but is differentiable with respect to μ.

For the convergence proof in this general case it is only required to replace $\|x\|$ by $(\|x\| + \|\mu\|)$ and $\|x_1 - x_2\|$ by $(\|x_1-x_2\| + \|\mu_1-\mu_2\|)$ in the above proof.

5- CONCLUSION

The convergence proof of the two algorithms developed for the numerical determination of parameters in a system of VTIE to minimize a given seminorm have been given. Each iteration of the algorithm without λ-strategy (Alg.A1) requires much less computational efforts as compared to that with λ-strategy (Alg. A2) and the convergence is quadratic. However, the system constraints are strigent and the convergence is only local.

The first convergence proof of algorithm A1 (with f linear in μ) and the convergence proof of algorithm A2, includes the corresponding results of Hoffman and Klostermair [2] as a special case. Besides, the condition (V4) in their lemma 5.3 has been relaxed.

The second convergence proof of algorithm A1 (with f linear or nonlinear in μ) is of different nature than the results of Hoffman and Klostermair [2] on account of (V2.m).

APPENDIX - 1

Proofs of the auxilary results in Sec.3 and Sec. 4:

Lemma - 1(Proof):

(i) Consider any $(x,\mu) \in B$. $\exists(\bar{x},\mu) \in B \ni$

$$\|\bar{x}-x\|_t \leq \|f(t,\mu) + \int_0^t g(t,s,\bar{x}(s))ds - x(t)\|_t$$

(Here $\|\cdot\|_t$ denotes norm in $C_r[o,t]$)

$$= \|f(t,\mu) + \int_0^t g(t,s,x(s))ds - x(t) + \int_0^t [g(t,s,\bar{x}(s)) - g(t,s,x(s))]\ ds\ \|_t$$

$$\leq \|x-F(x,\mu)\| + \int_0^t K(t,s)\ \|\bar{x}-x\|_s\ ds$$

Applying Gronwell's Lemma;

$$\|\bar{x}-x\| \leq e^L\ \|x-F(x,\mu)\|$$

(ii) The proof follows by replacing $F(\bar{x},\mu)$ by $F_{(\tilde{x},\tilde{\mu})}(\bar{x},\mu)$ in the above proof.

Corollary -1(Proof)

Proofs for (i) and (ii) are given here and (iii) follows **is** a similar manner.

(i) First consider the proof for '$\leq$'

$$\text{Inf.}\{P(q-Tx) + K\|x-F(x,\mu)\| : (x,\mu) \in B\}$$
$$\leq \text{inf.}\{P(q-Tx) + K\|x-F(x,\mu)\| : (x,\mu) \in B, x = F(x,\mu)\}$$
$$\leq \text{inf.}\{P(q-Tx) : (x,\mu) \in B, x = F(x,\mu)\}$$

Next, we prove '$\geqslant$'

$$\text{inf.}\{P(q-Tx) : (x,\mu) \in B, x = F(x,\mu)\}$$
$$\leq \text{inf.}\{P(q-T\bar{x}) : (x,\mu) \in B,\ (\bar{x},\bar{\mu}) \in B,\ \bar{x} = F(\bar{x},\bar{\mu}),\ \|\bar{x}-x\| \leq e^L\ \|x-F(x,\mu)\|\}$$
$$\leq \text{inf.}\{P(q-T\bar{x}) - P(T(x-\bar{x})) + C\|\bar{x}-x\| : (x,\mu) \in B,\ (\bar{x},\bar{\mu}) \in B,\ \bar{x} = F(\bar{x},\bar{\mu}), \|\bar{x}-x\| \leq e^L\ \|x-F(x,\mu)\|\}$$
$$\leq \text{Inf.}\{P(q-Tx) + C\|\bar{x}-x\| : (x,\mu) \in B,\ (\bar{x},\bar{\mu}) \in B,\ \bar{x} = F(\bar{x},\bar{\mu}),\ \|\bar{x}-x\| \leq e^L\ \|x-F(x,\mu)\|\}$$
$$\leq \text{inf.}\{P(q-Tx) + C\ e^L\ \|x-F(x,\mu)\| : (x,\mu) \in B\}$$
$$\leq \text{inf.}\{P(q-Tx) + K\|x-F(x,\mu)\| : (x,\mu) \in B\}$$

(ii) '$\Longrightarrow$'

$$\text{inf.}\{P(q-Tx) + K\|x-F(x,\mu)\| : (x,\mu) \in B\}$$
$$= P(q-T\bar{x}\ ) + K\|\bar{x}-F(\bar{x},\bar{\mu})\|$$
$$= P(q-T\bar{x}) + \bar{K}\|\bar{x}-F(\bar{x},\bar{\mu})\| + (K-\bar{K}) \cdot \|\bar{x}-F(\bar{x},\bar{\mu})\|$$

(Here $\bar{K}$ is defined as $e^L C < \bar{K} < K$)

$\geq \inf.\{P(q-Tx) : (x,\mu) \in B,\ x=F(x,\mu)\} + (K-\bar{K}) \| \bar{x}-F(\bar{x},\bar{\mu})\|$

(by (i))

$= \inf. \{P(q-Tx)+K \| x-F(x,\mu)\| : (x,\mu) \in B\} + (K-\bar{K}) \|\bar{x}-F(\bar{x},\bar{\mu})\|$

$\Longrightarrow \bar{x} = F(\bar{x},\bar{\mu})$

$\Longrightarrow P(q-T\bar{x}) = \inf. \{P(q-Tx) : (x,\mu) \in B,\ x = F(x,\mu)\}$

$\Longrightarrow (\bar{x},\bar{\mu})$ minimal solution of (GP).

Lemma -2: (Proof)

By Lemma (1.1)

$\exists\ (\bar{x},\bar{\mu}) \in B$ with $\bar{x} = F(\bar{x},\bar{\mu})$ and $\|\bar{x}-x\| \leq e^{L} \| x-F(x,\mu)\|$

Let $\|\bar{x}-x^*\| \leq \delta_1$. Then by assumption (V2),

$K_1 \| x-x^*\| \leq K_1 (\|x-\bar{x}\| + \|\bar{x}-x^*\|)$

$\leq K_1 \| x-\bar{x}\| + P(q-T\bar{x}) - P(q-Tx^*)$

$\leq K_1 e^{L} \| x-F(x,\mu)\| + P(q-Tx) + P(T(x-\bar{x})) - P(q-Tx^*)$

$\leq K_1 e^{L} \| x-F(x,\mu)\| + P(q-Tx)-P(q-Tx^*) + C \| x-\bar{x}\|$

$\leq (K_1+C)e^{L} \| x-F(x,\mu)\| + P(q-Tx)-P(q-Tx^*)$

$= K_2 \| x-F(x,\mu)\| + P(q-Tx) -P(q-Tx^*)$

Suppose $\|\bar{x}-x^*\| > \delta_1$

$K_1 \| x-x^*\| \leq K_1 \| x-\bar{x}\| = K_1 \| x-\bar{x}\| + P(q-Tx^*)-P(q-Tx^*)$

$\leq K_1 \| x-\bar{x}\| + P(q-T(x^*-x+x))-P(q-Tx^*)$

$\leq K_1 \| x-\bar{x}\| + P(q-Tx) -P(q-Tx^*) + P(T(x-x^*))$

$\leq K_1 \| x-\bar{x}\| + P(q-Tx)-P(q-Tx^*)+C \| x-x^*\|$

$\leq (K_1+C)e^{L} \| x-F(x,\mu)\| + P(q-Tx)-P(q-Tx^*)$

$= K_2 \| x-F(x,\mu)\| + P(q-Tx)-P(q-Tx^*)$

Lemma -3: (Proof)

Let $x_\lambda := \lambda x + (1-\lambda) x^*$

and $\mu_\lambda := \mu^* + \lambda(\mu-\mu^*)$

for sufficiently small $\lambda > 0$ Lemma -2 and convexity of B $\Longrightarrow$

$P(q-Tx_\lambda) - P(q-Tx^*) + K_2 \| x_\lambda - F(x_\lambda,\mu_\lambda)\| \geq K_1 \| x_\lambda - x^*\| \Longrightarrow \lambda P(q-Tx) - \lambda P(q-Tx^*) + K_2 \| x_\lambda - F(x_\lambda,\mu_\lambda)\| \geq \lambda K_1 \| x-x^*\| \Longrightarrow P(q-Tx)-P(q-Tx^*) + K_2 \| x-F(x^*,\mu^*) - \frac{F(x_\lambda,\mu_\lambda)-F(x^*,\mu^*)}{\lambda} \| \geq K_1 \| x-x^*\|$

$\Longrightarrow P(q-Tx)- P(q-Tx^*) + K_2 \| x-F(x^*,\mu^*) - \frac{\partial f(\mu^*)}{\partial \mu} (\mu-\mu^*) - \int_0^t \frac{\partial g}{\partial x} (t,s,x^*(s)) (x-x^*)_s\, ds \| \geq K_1 \| x-x^*\|$

$\Longrightarrow P(q-Tx)- P(q-Tx^*) + K_2 \| x-F'_{(x^*,\mu^*)}(x,\mu)\| \geq K_1 \| x-x^*\|$

Lemma -4(Proof):

Define $x_\lambda := \bar{x} + \lambda(x-\bar{x})$

$(F(x,\mu)-F'_{(\bar{x},\bar{\mu})}(x,\mu))(t) = f(t,\mu) + \int_0^t g(t,s,x(s))\, ds - f(t,\bar{\mu}) - \frac{\partial f}{\partial \mu} (t,\bar{\mu}) (\mu - \bar{\mu}) - \int_0^t g(t,s,\bar{x}(s))ds - \int_0^t \frac{\partial g}{\partial x}(t,s,\bar{x}(s))(x-\bar{x})_s ds$

$= \int_0^t [g(t,s,x(s))-g(t,s,\bar{x}(s))]\, ds - \int_0^t \frac{\partial g}{\partial x} (t,s,\bar{x}(s))(x-\bar{x})_s\, ds = \int_0^1 \int_0^t \frac{\partial g}{\partial x}(t,s,x_\lambda(s)) (x-\bar{x})_s\, ds\, d\lambda - \int_0^t \frac{\partial g}{\partial x}(t,s,\bar{x}(s))(x-\bar{x})_s\, ds = \int_0^1 \int_0^t [\frac{\partial g}{\partial x} (t,s,x(s)) - \frac{\partial g}{\partial x}(t,s,\bar{x}(s))] (x-\bar{x})_s\, ds\, d\lambda \Longrightarrow$

$\| F(x,\mu)-F'_{(\bar{x},\bar{\mu})}(x,\mu)\| \leq \sup_{t\in[0,1]} \int_0^1 \int_0^t |\frac{\partial g}{\partial x}(t,s,x_\lambda(s)) - \frac{\partial g}{\partial x}(t,s,\bar{x}(s))|\, ds.d\lambda \|x-\bar{x}\|$

$\leq \sup_{t\in[0,1]} \int_0^1 \int_0^t \hat{K}(t,s) \| x_\lambda-\bar{x}\|_s\, ds\, d\lambda \cdot \| x-\bar{x}\| \leq \sup_{t\in[0,1]} \int_0^1 \int_0^t \hat{K}(t,s)ds\, d\lambda \| x-\bar{x}\|^2$

$\leq \frac{1}{2} L_1 \|x-\bar{x}\|^2$

Lemma -5:(Proof)

By Lemma -1 (i)

$\exists\ (\bar{x},\bar{\mu}) \in B$ with $\bar{x} = F(\bar{x},\bar{\mu})$ and $\|\bar{x}-x\| \leq e^{L} \| x-F(x,\mu)\|$

In case $(\|\bar{x}-x^*\| + \|\bar{\mu}-\mu^*\|) \leq \delta_1$

By assumption (V2.m) $K_1 (\|x-x^*\| + \|\mu-\mu^*\|) \leq K_1 (\| x-\bar{x}\| + \|\bar{x}-x^*\| + \|\bar{\mu}-\mu^*\| + \|\mu-\bar{\mu}\|) \leq K_1 e^{L} \| x-F(x,\mu)\| + P(q-T\bar{x}) - P(q-Tx^*)$

(By construction in Lemma -1(i), $\mu = \bar{\mu}$)
$\leq K_1 e^L \|x - F(x,\mu)\| + C\|x - \bar{x}\| + P(q - Tx) - P(q - Tx^*) \leq K_2 \|x - F(x,\mu)\| + P(q - Tx) - P(q - Tx^*)$

Now consider the case $(\|\bar{x} - x^*\| + \|\bar{\mu} - \mu^*\|) > \delta_1$
$K_1(\|x - x^*\| + \|\mu - \mu^*\|) \leq K_1(\|\bar{x} - x^*\| + \|\bar{\mu} - \mu^*\| - \|x - x^*\| - \|\mu - \mu^*\|) \leq K_1(\|\bar{x} - x\| + \|\bar{\mu} - \mu\|) = K_1\|\bar{x} - x\|$, since $\bar{\mu} = \mu$
$= K_1\|\bar{x} - x\| + P(q - Tx^*) - P(q - Tx^*)$
$\leq K_1\|\bar{x} - x\| + P(q - Tx) + C\|x - x^*\| - P(q - Tx^*)$
$\leq K_1\|\bar{x} - x\| + C\|\bar{x} - x\| + P(q - Tx) - P(q - Tx^*)$
$\leq (K_1 + C)e^L \|x - F(x,\mu)\| + P(q - Tx) - P(q - Tx^*)$
$= K_2 \|x - F(x,\mu)\| + P(q - Tx) - P(q - Tx^*)$

Lemma -6 (Proof)

Define $x_\lambda := \lambda x + (1 - \lambda)x^*$
$\mu_\lambda := \mu^* + \lambda(\mu - \mu^*)$

For sufficiently small $\lambda > 0$, Lemma-5 and convexity of $B \Longrightarrow \lambda P(q - Tx) + P(q - Tx^*) + K_2 \|\lambda x - \lambda F(x^*,\mu^*) + F(x^*,\mu^*) - F(x_\lambda, \mu_\lambda)\| \geq K_1 \lambda(\|x - x^*\| + \|\mu - \mu^*\|) \Longrightarrow P(q - Tx) - P(q - Tx^*) + K_2 \|x - F(x^*,\mu^*) - \frac{F(x_\lambda,\mu_\lambda) - F(x^*,\mu^*)}{\lambda}\| \geq K_1(\|x - x^*\| + \|\mu - \mu^*\|) \Longrightarrow \hat{P}(q - Tx) - P(q - Tx^*) + K_2 \|x - F_{(x^*,\mu^*)}(x,\mu)\| \geq K_1(\|x - x^*\| + \|\mu - \mu^*\|)$; as $\lambda \to 0$

Lemma -7: (Proof)

Define $x_{\lambda_1} = \bar{x} + \lambda_1 (x - \bar{x})$ and
$x_{\lambda_2} = \mu + \lambda_2 (\mu - \bar{\mu}) \cdot \|F_{(\bar{x},\bar{\mu})}(x,\mu) - F(x,\mu)\|_t$
$= \|\int_0^1 f'(\mu_{\lambda_2})(\mu - \bar{\mu})\, d\lambda_2 - f'(\bar{\mu})(\mu - \bar{\mu})$
$+ \int_0^1 G'(x_{\lambda_1})(x - \bar{x})d\lambda_1 - G'(\bar{x})(x - \bar{x})\|_t$
Here, $f'(\mu_{\lambda_2})$ represents $\frac{\partial f}{\partial \mu}\Big|_{\mu = \mu_{\lambda_2}}$ and $G'(x_{\lambda_1})(x - \bar{x})(t)$ represents $\int_0^t \frac{\partial g}{\partial x}(t,s,x_{\lambda_1}(s))(x - \bar{x})_s\, ds$. $\|F(\bar{x},\bar{\mu})(x,\mu) - F(x,\mu)\|_t \leq$
$\int_0^1 \|f'(\mu_{\lambda_2}) - f'(\bar{\mu})\|_t\, d\lambda_2 \|\mu - \bar{\mu}\|$
$+ \int_0^1 \|G'(x_{\lambda_1}) - G'(\bar{x})\|_t\, d\lambda_1 \|x - \bar{x}\|$
$\leq \int_0^1 \lambda_2 L_1'\, d\lambda_2 \|\mu - \bar{\mu}\|^2 + \int_0^1 \lambda_1 L_1\, d\lambda_1 \|x - \bar{x}\|^2$
$\leq \frac{1}{2} L_1 \|x - \bar{x}\|^2 + \frac{1}{2} L_1' \|\mu - \bar{\mu}\|^2 \Longrightarrow$

$\|F_{(\bar{x},\bar{\mu})}(x,\mu) - F(x,\mu)\| \leq \frac{1}{2}\bar{L}(\|x - \bar{x}\|^2 + \|\mu - \bar{\mu}\|^2)$

Lemma-8: (Proof)

Suppose the assertion is not true. Then,
$P(q - T\bar{x}) > P(q - Tx) + K\|x - F(x,\mu)\|$
$> P(q - Tx) - C\|x - \check{x}\| + K\|x - F(x,\mu)\|$
$> P(q - T\check{x}) - Ce^L\|x - F_{(x,\mu)}(x,\mu)\| + K\|x - F(x,\mu)\|$
(taking $(\check{x},\check{\mu}) \ni \|x - \check{x}\| \leq e^L \|x - F_{(x,\mu)}(x,\mu)\|$)
$> P(q - T\check{x}) + (K - Ce^L)\|x - F_{(x,\mu)}(x,\mu)\|$
$\Longrightarrow P(q - T\bar{x}) > P(q - T\check{x}) \wedge \check{x} = F_{(x,\mu)}(\check{x},\mu)$

$\Longrightarrow P(q - T\bar{x})$ is not a minimal solution of $(LP_{(x,\mu)})$. $P(q - T\bar{x}) \leq \varphi(x,\mu)$.

Lemma -9: (Proof)

φ possesses in $(x,\mu) \in B$ a right derivative $\varphi'(x,\mu)$ and the following holds; $\varphi'(x,\mu)((\bar{x},\bar{\mu}) - (x,\mu)) = \varphi'_{(x,\mu)}(x,\mu)((\bar{x},\bar{\mu}) - (x,\mu)) \leq \varphi'_{(x,\mu)}(\bar{x},\bar{\mu}) - \varphi(x,\mu) = P(q - T\bar{x}) - \varphi(x,\mu) < 0$.
From here the assertion follows as
$0 < \beta < 1$.

APPENDIX-2

Methods of solving the linear problem :

The linearized problems arising in the iteration schemes can be stated as ;
Minimize $\underset{t}{P}(q - Tx)$; $(x,\mu) \in B \wedge x(t) = \underline{a}(t) + D(t)\mu + \int_0^t K(t,s)\, x(s)\, ds$, $t \in [0,1]$
Here, $B = C_n^0[0,1] \times A$, $A \subset R^p$, A is a convex subset.
We consider below the solution methods for few special cases of this problem :

L-1 : Problem with known resolvent kernel.

If the resolvent kernel for the system equation is known apriori, x can be explicitely expressed as a function of μ. Let R be the resolvent Kernel.
Then, $x(t) = \underline{a}(t) + D(t)\mu + \int_0^t R(t,s)(\underline{a}(s) + D(s)\mu)ds = \underline{b}(t) +$

$H(t)\mu$, where $\underline{b}$ and H are known function;

$\underline{b}: [0,1] \to R^n$, $H:[0,1] \to R^n \times R^p$.

The problem reduces to,

$$\underset{\mu \in A}{\text{Minimize}}\ P(q - T(\underline{b} + H\mu))$$

This problem can be solved using a suitable technique depending on the seminorm P.

L-2 : Problem with Degenerate Kernel

In this case the problem is easily reduced to one with linear differential equation constraints and thus can be solved using standard techniques.

To do this let us consider the system equation,

$$x(t) = \underline{a}(t) + D(t)\mu + \int_0^t K(t,s)\, x(s)\, ds$$

$x(t)$ can be written as,

$$x(t) = \underline{a}(t) + D(t)\mu + \sum_{m=1}^{M} G_m(t) Y_m(t) \qquad (L\text{-}2.i)$$

where,

$$K(t,s) = \sum_{m=1}^{M} G_m(t)\, H_m(s) \qquad (L\text{-}2.ii)$$

$$Y_m(t) = \int_0^t H_m(s)\, x(s)\, ds \qquad (L\text{-}2.iii)$$

Now differentiating (L-2.iii) and substituting from (L-2.i),

$$\dot{Y}_m(t) = H_m(t).\underline{a}(t) + H_m(t)D(t)\mu + H_m(t)\cdot \left(\sum_{i=1}^{M} G_i(t)\, Y_i(t)\right) \qquad (L\text{-}2.iv)$$

With this the problem of minimization turns out to be equivalent to,

$$\underset{\mu \in A}{\text{Minimize}}\ P\left(q - T\left(\underline{a}(t) + D(t)\mu + \sum_{m=1}^{M} G_m(t)\cdot Y_m(t,\mu)\right)\right)$$

subject to

$$\dot{Y}_i(t,\mu) = H_i(t)\left[\underline{a}(t) + D(t)\mu + \sum_{m=1}^{M} G_m(t)\, Y_m(t,\mu)\right]$$

$i = 1,2, \ldots, M$.

L-3 : Problem with Integral Cost functional :

A set of sufficiency conditions are given for the following problem :

Problem (P) :Minimize

$$J = \int_0^1 f_0(s, x(s))\, ds \qquad (L\text{-}3.i)$$

subject to the constraints,

$$x(t) = f(t,\mu) + \int_0^t K(t,s)\, x(s)\, ds \qquad (L\text{-}3.ii)$$

where f and f_0 are convex and differentiable with respect to μ and x respectively.

Theorem :

The pair (x^*, μ^*) is a solution of the optimization problem (P) if it satisfies the following conditions,

(i) $\exists$ a row vector $\psi : [0,1] \to R^n \ni$

$$\psi(t) = f_{0_x}(x^*(t), t) + \int_t^1 \psi(s) K(s,t)\, ds, \quad t \in [0,1]$$

and

$$\int_0^1 \psi(t)\, f_\mu(t, \mu^*)\, dt = 0$$

(ii) x^* is a solution of the constraint equation (L-3.ii) corresponding to the parameter value μ^*, i.e., $x^*(t) = f(t,\mu^*) + \int_0^t K(t,s)\, x^*(s)\, ds$, $t \in [0,1]$.

Proof:

For any (x, μ) satisfying (L-3.ii) and (x^*, μ^*) as above, we have, $\Delta J = J(x,\mu) - J(x^*,\mu^*) =$

$$\int_0^1 [f_0(x(s),s) - f_0(x^*(s),s)]\, ds$$

$$\geq \int_0^1 f_{0_x}(x^*(s),s)\, (x^*(s) - s(x))\, ds$$

(by convexity assumption)

$$= \int_0^1 \Big[\psi(u)x^*(u) - \int_0^1 \psi(s)K(s,u)\,ds\; x^*(u) - \psi(u)\, x(u) + \int_u^1 \psi(s)K(s,u)\,ds\; x(u)\Big] du$$

$$= \int_0^1 \psi(u)[f(u,\mu^*) - f(u,\mu)]\, du \geq \int_0^1 \psi(u)\, f_\mu(u,\mu^*)\, (\mu^* - \mu)\, du$$

(by convexity)

$$= \int_0^1 \psi(u)\, f_\mu(u,\mu^*)\, du\, (\mu^* - \mu) = 0$$

$\Rightarrow (x^*, \mu)$ is an optimal pair.

REFERENCES

[1] Bellman, R.E., Kalaba, R.E., 'Quasilinearization and Nonlinear Boundary-value Problems'. The RAND Corporation, 1700 Main St., Santa Monica, California (1965).

[2] Hoffman, K.H., Klostermair, A., 'Approximation with Solutions of Differential Equations ' (in German), 556 Lecture Notes in Mathematics, Springer Verlag.

Complex and Distributed Systems: Analysis, Simulation and Control
S.G. Tzafestas and P. Borne (editors)
Elsevier Science Publishers B.V. (North-Holland)

IDENTIFICATION OF LINEAR PARAMETER-UNCERTAIN MODELS FOR MECHANICAL VIBRATION TEST BENCHES

V. Tews
Institute for Control Engineering
Section Control Systems Theory
Technical University of Darmstadt
Schloßgraben 1
D-6100 Darmstadt, FRG

Summary: Treated is a class of servohydraulic vibration test benches. After description of applications of such test benches the problems in connection with controlling these systems are stated. It is shown that parameter-uncertain frequency response models are best suited for a general description of a broad class of such test benches. Methods for an experimental identification are described. Based on a-priori-knowledge and on these measurements a general model for the whole class of test-benches under consideration here is given. Determination of the specific parameters of a single exponent of this class is shown by the example of a car axis test bench.

1. INTRODUCTION

In many areas life-span tests or strength tests on mechanical structures are more and more carried out by automatic test benches reproducing predescribed time-signals of forces or accelerations in the laboratory. This paper is dealing with a certain class of servohydraulic vibration test benches. These test benches are used as "road simulators" for life-span tests of parts of cars or whole carbodies (see fig. 1) on one hand or as "earth-quake simulators" for strength tests on models or parts of buildings or other mechanical structures on the other hand. Slightly modified, mainly as purely electro-mechanical test benches, they are also used for vibration tests of electronic components or devices.

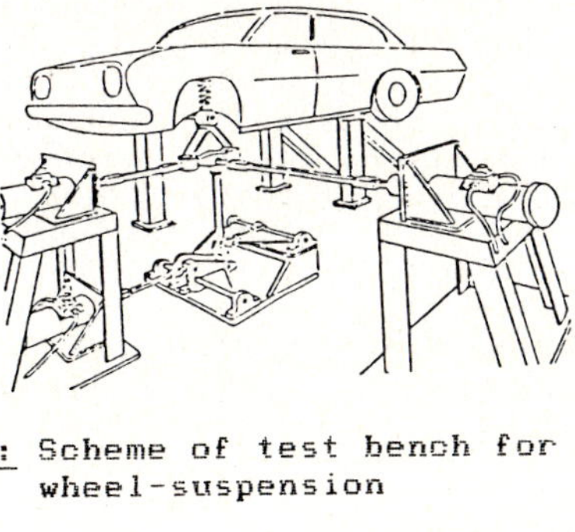

Fig. 1: Scheme of test bench for one wheel-suspension

Despite of different size and application the basic structure of such test-benches is similar: a passive mechanical specimen is driven by hydraulic actuators in several directions. The problem of controlling these test benches is basicly the same with all these devices except for the different values of the actual parameters. From a paper of Zachmann, Hillberry and Kettelkamp [1] it can be seen that even for such applications as load simulators for prosthetic knee joints the structure and the main characteristics are the same. Therefore the aim of this paper is to give a general model for this class of test benches. In the next chapter the main problems in connection with controlling these devices are described and the requirements on a model are stated. Chapter 3 shows that parameter-uncertain frequency-domain models are best suited for these purposes. In chapter 4 methods are described to identify these models. Chapter 5 finally gives a general frequency response model for the regarded class of servohydraulic vibration test benches. The specific parameters of a single exponent of this class are given by the example of a car axis test bench.

2. REQUIREMENTS ON THE MODEL

The main problems in controlling these test benches will be summed up with the example of a street simulator. The aim of street simulators is to reproduce in the laboratory those loads on a test specimen that can be measured by test drives on special test roads. Input signals are therefore force- or acceleration courses that have been measured on real test drives. The actuators of the test bench have to be controlled in a way, that at the measuring points on the test specimen these forces or accelerations are reproduced. Because forces are considered in different directions and at several excitation points, we have to deal with a multi-input-multi-output system, e.g. with excitation in all 3 coordinate directions on all four wheels with a 12x12 system. Some of these control loops are strongly coupled by the test specimen or by special geometric effects. The single control loops can carry nonlinearities, like nonlinear characteristics of springs and anti-shock-pads, bounds on the possible displacements, hysteresis through rubber elements in the spring suspensions, friction in several joints, nonlinear behaviour of the hydraulic cylinders etc.. Transport effects in the hydraulic system cause dead-times. Longterm tests or heavy loads on the test-specimen cause fatigue, i.e. the mechanical behaviour and the system parameters change. Other effects to be mentioned here are disturbances from the outside, like measurement disturbances, cross-coupling on electric lines, pressure deviations in the hydraulic system etc.

During life-span tests lasting for weeks or months one test sequence of a few minutes length is repeated permanently. The control problem reduces to reproduce this test sequence. Real-time computing is not neccessary, so more complicated and therefore time-consuming procedures are possible to reach the desired accuracy. The adaptation to long-term parameter changes can also be done more slowly in parallel to the actual test. One important requirement is the possibility to perform the whole controller design nearly automatically on a digital computer. For the behaviour of the whole system will be influenced drastically by the test-specimen, so for each new specimen or even for each constructive modification an entirely new calculation of the drive signals is necessary. During the development of a new car this will very often be the case. So we demand the controller design to be as much automated as possible. Furtheron the design should possibly be done by testsite engineers without the necessity to be control engineering specialists.

3. CHOICE OF MODEL TYPE

The design of a control is i.g. based on a model of the plant to be controlled. The real plant which is nonlinear with possibly dead time can be described by a nonlinear model or approximately by a linear model. To establish a nonlinear model one starts i.g. with theoretical modelling based on physical equations. A completely experimental determination is normally not possible. A linear model in contrast can be identified with only litte a-priori-information about the system. There is no general theory for the design of nonlinear control loops. The procedure relies very much on the actual system and demands a lot of experience from the designer. For linear systems there are a number of design procedures, that can easily be implemented on a digital computer and work nearly completely automatic. Surely a carefully worked out nonlinear model will give a better description of the systems behaviour, and a nonlinear control based thereon will show a better performance. But the expenses and the demands on the design engineer for such a design are very high. The requirement of easy handling cannot be met, an automation of the drive signal calculation is hardly possible. That is why nonlinear models are assumed to be not appropriate for our control problem. Instead we will try to develop approximate linear models which give a sufficient performance. First experiences with the control of these test-benches showed that in general "classical" design procedures based on linear models with constant parameters, like dead-beat design or state feedback, did not produce satisfactory results. So we shall concentrate on linear systems, however with uncertain or varying parameters.

In the frequency domain there exist several design procedures which take into account parameter-variations which can have their origins also in nonlinearities or disturbances.

There are basicly two approaches to the controller design. With the "robust" approach under certain circumstances fixed controllers can be constructed based on a predefined deviation-range of the parameter-variations so that the transfer function of the closed loop does not leave a presdescribed area in the frequency domain (fig. 2.).
The design of such MIMO-systems can be done either by sequential closing of the single control loops (according to Mayne [2]), with the single control loops being shaped by a "robust" procedure like the one described by Horowitz/Sidi [3] or they can be done by direct "robust" procedures with loop

couplings considered as disturbances as proposed by Horowitz [4] (see also Horowitz/Löcher [5]). CAD implementation of the "robust" design of the first one of the procedures is described in [6], an application of it to the control of vibration test benches can be found in Gräser [7]. But all these procedures are relatively complicated and need a skilled design-engineer.

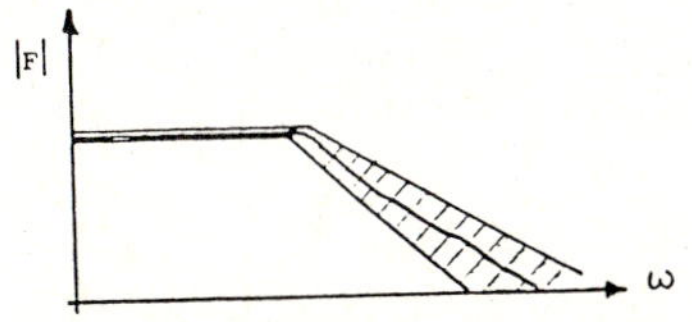

Fig. 2: Range of permitted amplitude variation of the closed loop frequency response

Another approach to our control problem was taken by one of the manufacturers of such test-benches, the Carl Schenck AG, Darmstadt [8]. The analog controllers in the control loops are not optimized as to bound the variations in the frequency response, but instead an iterative procedure on a higher level is used to control these variations. The drive signal to the test-bench is prefiltered by the inverse of a linear system model. Due to the inaccuracy of the linear model with respect to the nonlinear plant the prefiltered input will not exactly give the desired output signal. By an iterative modification of the system model and/or the drive signals the output error can be brought near to zero, however.

Since the inverse of a physical system cannot be represented by a causal system (number of zeroes > number of poles) this modification is performed not in a closed loop but iteratively on a sequence of sampled data points. Since we are interested only in the shapes of the signals and not in the absolute time scale, this modification can be performed in the frequency-domain.

For both these approaches, robust design and prefiltering with the inverse system, we need a parameter varying representation of a linear system of the form

$$F(j\omega) = \hat{F}(j\omega) + \Delta F(j\omega) \qquad (1)$$

or equivalently

$$F(j\omega) = (I + \delta F(j\omega))\,\hat{F}(j\omega) \qquad (2)$$

where $\hat{F}(j\omega)$ is a "nominal" system with constant parameters and $\Delta F(j\omega)$ or $\delta F(j\omega)$ resp. a (possibly) time dependent parameter variation (I denotes the unity matrix).

If certain bounds on $\Delta F(j\omega)$ can be determined, the system $F(j\omega)$ can be represented by an area in the frequency domain. Fig. 3 shows the special case where $|\Delta F(j\omega)|$ is assumend to be bounded by $|\Delta F(j\omega)| \le \Delta F_{max}(j\omega)$.

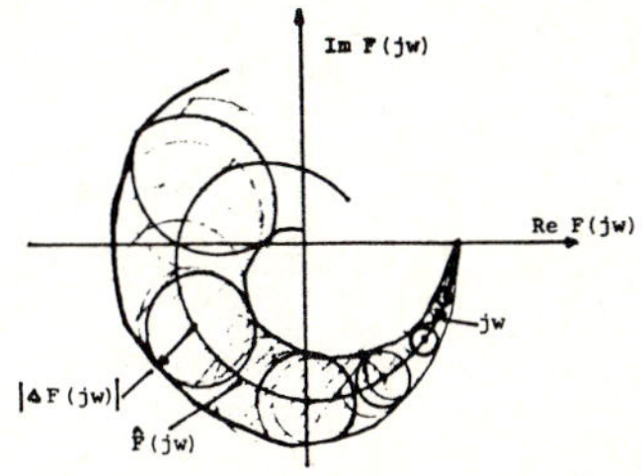

Fig. 3: Frequency response $\hat{F}$ with variation area

For each frequency $j\omega$ a circle with radius $\Delta F_{max}(j\omega)$ can be drawn around $\hat{F}(j\omega)$. The envelope of all these circles bounds the possible values of the system's frequency response $F(j\omega)$. In the next chapter a method will be derived to estimate the possible variation area of a nonlinear system for a certain kind of input signal. Once $\Delta F_{max}(j\omega) = |\Delta F(j\omega)|_{max}$ is known bounds on amplitude and phase of $F(j\omega)$ can be calculated (see also fig. 4).

$$|\hat{F}| - \Delta F_{max} \le |F| \le |\hat{F}| + \Delta F_{max} \qquad (3)$$

and

$$\arg(\hat{F}) - \arcsin\left(\frac{\Delta F_{max}}{|\hat{F}|}\right) \le \arg(F)$$
$$\le \arg(\hat{F}) + \arcsin\left(\frac{\Delta F_{max}}{|\hat{F}|}\right) \qquad (4)$$

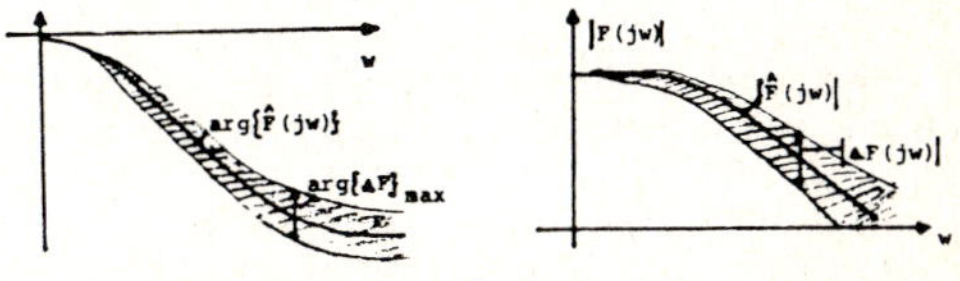

Fig. 4: Bode-plot of frequency response of fig. 3

For a design following robust approaches or the prefilter approach only these bounds are of interest.

4. EXPERIMENTAL IDENTIFICATION

Consider a frequency response like the one in (1) represented on a digital computer by a number of spectral lines with frequencies

$$\omega_k = k \cdot \omega_s \qquad k \in [1,N]$$

$$\omega_s = \frac{2\pi}{N} \cdot f_s \qquad (5)$$

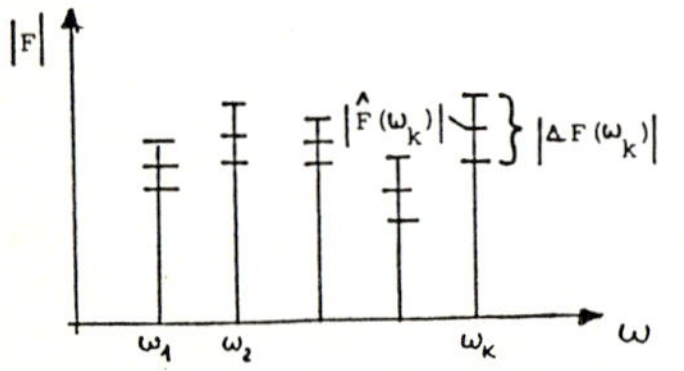

Fig. 5: Computer representation of $|F(\omega)|$

We assume the system input to be a sinusoidal signal with frequency ω_k. If our system was linear with constant parameters the frequency response would be one spectral line $\hat{F}(\omega_k)$ only, $\Delta F(\omega_k)$ would be zero. Long-term parameter variations cause $\Delta F(\omega_k)$ varying dependent on t, $\Delta F(\omega_k, t)$. Noise causes $\Delta F(\omega_k)$ to be a random number. Deadtime (as a linear effect) results in a phase shift of $\Delta F(\omega_k)$.

If our system is nonlinear the response to a sinusoidal signal with only one frequency ω_k will cover several spectral lines, besides the frequency ω_k the harmonics $\omega_i = i \cdot \omega_k$, $i = 2, 3, \ldots \infty$. If the input contains two frequencies ω_k and ω_1 there will be additional modulation products $\omega_k \pm \omega_1$.

If the number of frequencies in the input signal increases, the number of modulation products increases significantly.

Now let our input signal be a random signal consisting of a large number of spectral lines with arbitrary amplitudes and random phase shifts. Each line of the output spectrum $Y(\omega_k)$ then consists of three parts:

$Y_1(\omega_k)$: part of the output at ω_k directly correlated to the input with the same frequency ω_k

$Y_2(\omega_k)$: parts of the output at ω_k correlated with other input frequencies ω_j or modulation products $(p\omega_i \pm q\omega_j)$

$Y_3(\omega_k)$: part of the output at ω_k independent of the input (e.g. measurement noise).

If the phase shifts of $X(\omega_k)$ are chosen randomly and independently from one another, $Y(\omega_k)$ as a random variable is a sum of a great number of independent random variables ($Y_1(\omega_k) + Y_3(\omega_k)$ + the various terms in $Y_2(\omega_k)$). If the system under consideration is at least BIBO-stable, all these random variables have finite means and variances. Thus the central limit theorem can be applied, stating $Y(\omega_k)$ to be a normally distributed random variable. As $X(\omega_k)$ is to be considered as fixed for one experiment, also the frequency response

$$F(\omega_k) = \frac{Y(\omega_k)}{X(\omega_k)}$$

is a normally distributed random-variable. If several $F_j(\omega_k)$ are measured for several different $X_j(\omega_k)$, mean $\hat{F}(\omega_k)$ and variance σ_F^2 can be estimated. $\hat{F}(\omega_k)$ can be treated as the "nominal" frequency response in (1). σ_F^2 can be used to calculate confidence intervals for $F(\omega_k)$, a 3-σ-interval e.g., giving estimates for $|\Delta F(\omega_k)|$.

If, as a special case, the distribution density function of the phase shift of the input signal is symmetric with a mean value of $k*\pi$, $k = 0, \pm 1, \pm 2, \ldots$, it can be shown, that the mean value of $F(\omega_k)$, $\hat{F}(\omega_k)$, is given by

$$\hat{F}(\omega_k) = \frac{Y_{1i}(\omega_k)}{X_i(\omega_k)}$$

which is constant over all i, and is an optimal linearisation (in the mean-square sense) of the nonlinear system $F(\omega_k)$. $E\{Y_2(\omega_k)\}$ can be shown to be zero, while $E\{Y_3(\omega_k)\}$ is assumed to be zero, thus $E\{\Delta F\} = 0$.

Being the optimal linearization, $\hat{F}(\omega_k)$

can also be computed following the well-known identification procedure involving spectral density functions

$$\hat{F}(\omega) = \frac{S_{xy}(\omega)}{S_{xx}(\omega)} = \frac{E\{Y(\omega)\cdot X^*(\omega)\}}{E\{X(\omega)\cdot X^*(\omega)\}} \quad (5)$$

where S_{xy} denotes the cross-power spectral density between input and output calculated as the expected value of the Fourier-transformed output signal $Y(\omega_k)$ times the conjugate complex Fourier-transformed input signal $X^*(\omega_k)$, and S_{xx} denotes the autopower spectral density of the input signal $X(\omega_k)$. The accuracy of this linear model can be estimated by the coherence function

$$\gamma^2 = \frac{|S_{xy}|^2}{S_{xx}\cdot S_{yy}} = 1 - \frac{S_{nn}}{S_{yy}},$$

$$0 \le \gamma^2 \le 1 \quad (6)$$

where S_{nn} denotes the power spectral density caused by nonlinearities and disturbances.

If we define our variation ΔF to be a linear system giving at its output the same power spectral density S_{nn} we obtain from (6)

$$S_{nn} = E\{|\Delta F|^2\} = |\hat{F}|^2 \frac{1-\gamma^2}{\gamma^2} \quad (7).$$

Keeping in mind that the variance of a complex random variable F is defined by

$$\sigma_F^2 = E\{|F-\hat{F}|^2\} = E\{|\Delta F|^2\} \quad (8)$$

we obtain

$$\sigma_F^2 = |\hat{F}|^2 \frac{1-\gamma^2}{\gamma^2} \quad (9)$$

which means that we can calculate the variance of the frequency response from the measured coherence.

So it is possible to get the optimal linear system $\hat{F}$ and an estimation of the variations thereof, $|\Delta F|$, from well-known identification procedures, the spectral analysis, and using conventional measurement equipment with only little postprocessing.
Another well-known procedure for an experimental identification is the orthogonal correlation. This method excites the system with a harmonic signal

$$x = x_0 \sin \omega_0 t \quad (10)$$

and gives as a result amplitude y_0 and phase φ_0 of an output signal

$$y = y_0 \sin(\omega_0 t + \varphi_0) \quad (11)$$

correlated to the input. This method allows a very accurate measurement of the frequency response even in presence of big noise. With nonlinear systems the input signal may contain additional frequency components to bring the system near to its set point. Correlated are only the signals with frequency ω_0.

A linearisation is performed in the sense of the describing function.

If we use a random signal with random phase shifts (as described earlier), the linearisation in the sense of the describing function can be shown to give the same results as the linearisation in the least-squares-sense described earlier.

So both methods, spectral analysis and orthogonal correlation, give the same "nominal" frequency response and can be combined.

For the identification of the axis test bench described in the next chapter we used a spectrum analyzer to get a quick overview of the systems behaviour and to estimate the variations of the frequency response. Where more accurate measurements were required (e.g. in the neighbourhood of resonance frequencies) orthogonal correlation was performed.

The results of this identification are the numerical values of the frequency response and its variation bounds. For our analysis the knowledge of these numerical values is sufficient. If a parametric form is required, the frequency response can be matched by a computer program following procedures suggested by Strobel [9], Levi [10] and Sanathanahan/Koerner [11].

5. GENERAL FORM OF THE MODEL

Let us first examine to how many inputs and outputs our MIMO-system can be reduced. As an example we consider a car test-bench with tied car-body (fig. 6). The couplings between the horizontal actuators can be neglected because the wheel suspensions are very stiff

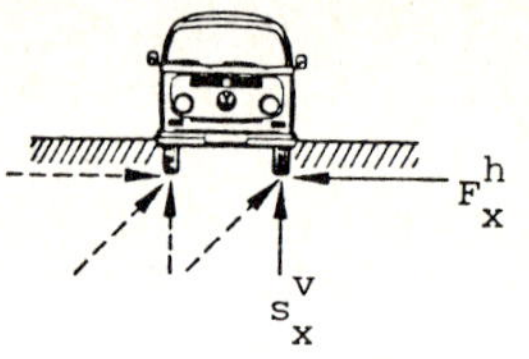

Fig. 6: Scheme of tied-body test bench

against horizontal loads and the actuators are deviated only very little. The couplings from the horizontal to the vertical actuators can be neglected because of the small horizontal elasticity. The dynamic behaviour of the horizontal loops is very similar, so i.g. it will suffice to perform the controller design for the vertical loop and one horizontal loop. This structure remains the same when we look at test benches with excitation of one wheel only (see fig. 1) or with the axis test bench described later.

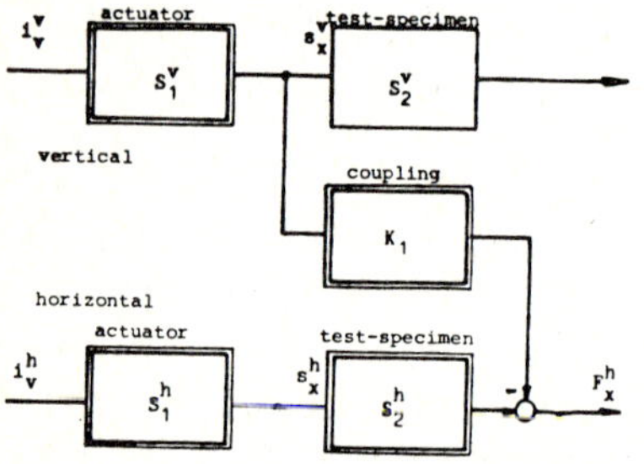

Fig. 7: Block-diagram of test bench + test specimen

We obtain a block-diagram according to fig. 7. In the vertical loop the controlled signal is the displacement s_x^v at the actuated point. In the horizontal loop the force F_x^h is to be controlled. Between the two control loops there is a geometrical coupling (fig.8)

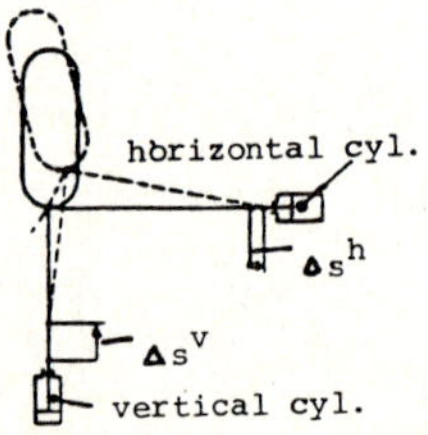

Fig. 8: Geometrical coupling between vertical and horizontal loop

If the vertical actuator is deviated by Δs^v the horizontal actuator has to follow by the displacement Δs^h. The characteristic of this correction is approximately quadratic. In the following the single blocks of the block diagram in fig. 7 will be discussed. As a specific example we treat an axis test bench where lifespan tests on a passenger car rear axis are performed with control of horizontal force and vertical displacement (fig. 9).

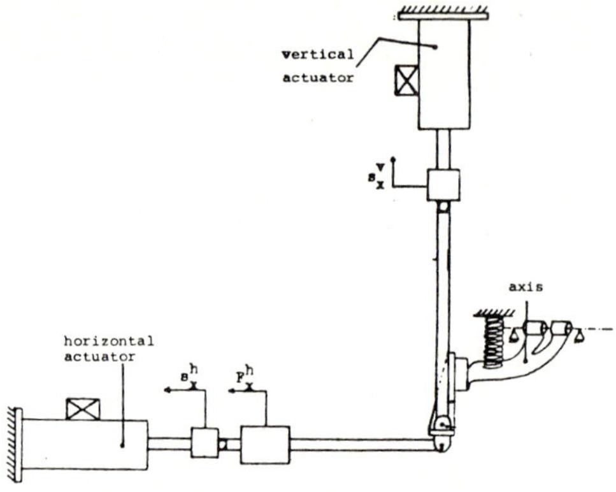

Fig. 9: Scheme of axis test bench

The frequency range in the actual test is from 1-50 Hz, some identification measurements were made up to 200 Hz. In general identification below 1Hz was found critical because of a low signal / noise ratio in this range.

In [7] Gräser derived a nonlinear model for the hydraulic actuators. His measurements on a carbody test bench showed, however, that the influence of the nonlinear characteristic of the valve is quite small. The frequency response we identified from the horizontal actuator of the axis test bench (fig. 10) showed the same characteristics as stated in [7]. For lower frequencies there is integral behaviour, in the frequency range from 25 Hz to about 50 Hz we have reactions from the axis back to the hydraulic actuators. Due to the nonlinear behaviour of the axis the resonances of these reactions vary slightly when the excitation level is changed. These nonlinear effects are however too small to create a significant variation $\Delta F(\omega)$. For higher frequencies the actuator behaves like several lagterms. Larger actuators for heavier loads may show deadtime due to the transport of the hydraulic oil. On the relatively small axis test-bench, however, no such effects were detected. A time-delay would cause an additional phase-shift but as long as only the numerical values of the frequency response are taken into consideration no additional error arises, because, after an experimental identification as discussed in chaer 4, this phase-shift is included in the model. The vertical actuator shows IT_n behaviour (fig. 11).

For higher frequencies the output signal becomes very small, so $|\Delta F(\omega)|$ is greater than the nominal frequency response $|\hat{F}|$.

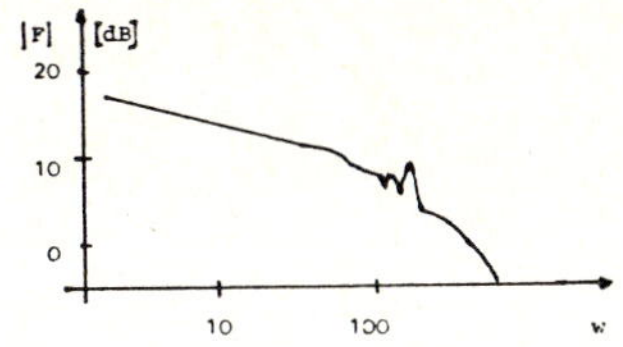

Fig. 10: Amplitude diag. of horizontal actuator

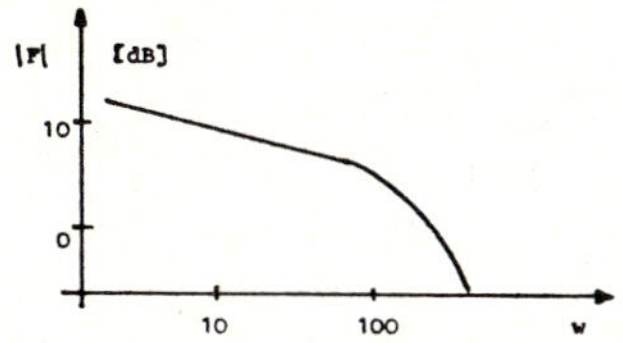

Fig. 11: Amplitude diag. of vertical actuator

The test specimen is a relatively complicated mechanical system showing several resonances and natural frequencies. But as long as car test benches are concerned the frequency response is similar, because even if whole car bodies are tested, the dynamic behaviour is mainly conducted by the behaviour of the wheel suspension.

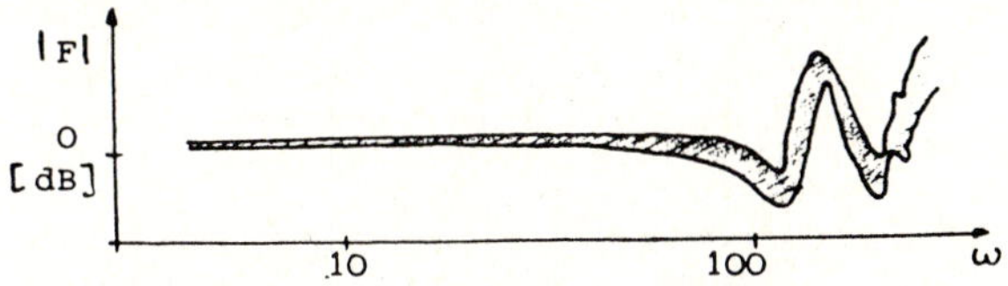

Fig. 12: Axis frequency response

Fig. 12 shows the amplitude of the axis frequency response. Changes in the excitation level cause changes in the resonance frequencies, so a relatively broad deviation $|\Delta F(\omega)|$ can be estimated. If these frequency responses are to be matched to parametric models, 6th order equations have been found to be satisfactory in all cases. So the test specimen can be modelled by the mechanical system shown in fig. 13. For a structural analysis or if the parameter uncertainty $|\Delta F(\omega)|$ is quite large, it may even be sufficient to use an analogue system with one mass, two springs and one damper only (fig. 14).

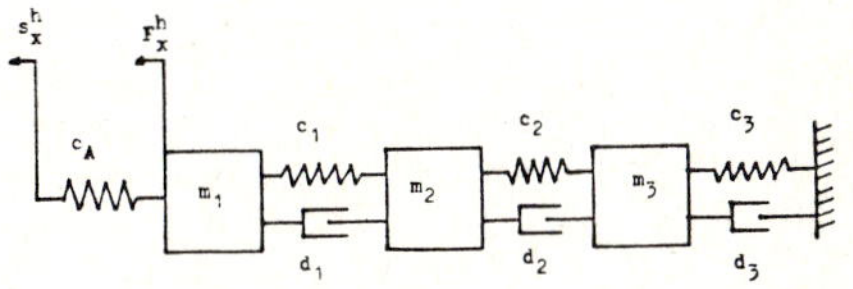

Fig. 13: Analogous mechanical system for axis S_2^h

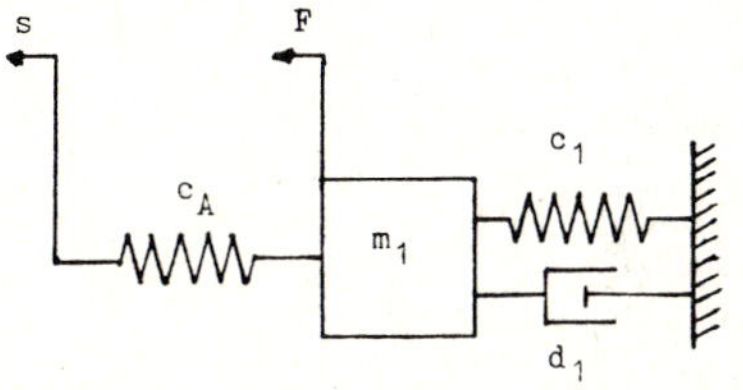

Fig. 14: Simplified model for axis S_2^h

In our block-diagram fig. 7 the output of the coupling K is connected to the output of the horizontal loop F_x^h. The geometric coupling described above, however, affects S_x^h in front of the axis. So, if we measure the frequency response between F_x^h and S_x^v this will be affected by the frequency response of the axis. The amplitude diagram of the coupling K clearly shows the characteristic influence of the axis (fig. 15).

Fig. 16 shows the block diagram of this axis test bench including controllers.

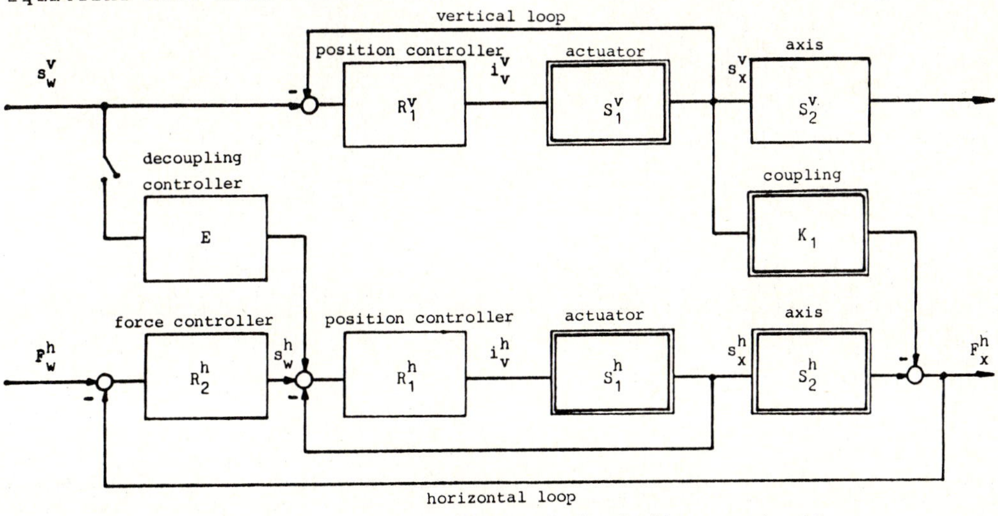

Fig. 16: Block-diagram of whole test bench including controllers

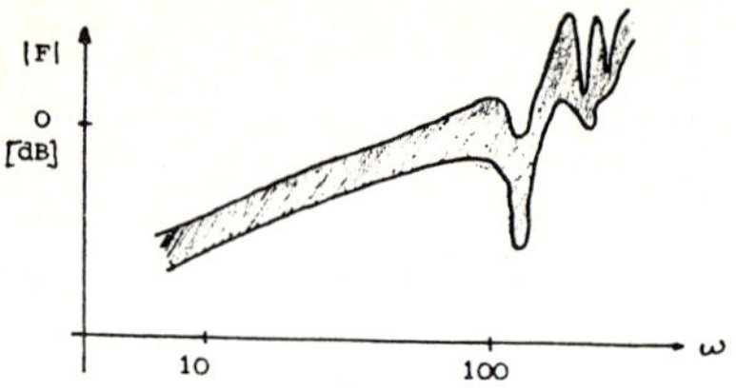

Fig. 15: Coupling amplitude diagram

For the controller-design no "robust" design procedures were employed, instead on general experience based simple PD-controllers for position control and an I-controller for the force-control are used. To obtain a sufficient accuracy in the horizontal loop a cascade of two controllers, position control and force control, is normally necessary. A decoupling compensator is used to approximately compensate the geometrical coupling. As a linear block this compensation can only be exact for one frequency. In this example the decoupling was optimized for 3.0 Hz.

At last we shall show the frequency response of the closed-loops (fig. 17). The horizontal loop shows a good proportional behaviour for lower frequencies. For higher frequencies the mechanical resonances of the axis come into effect. The vertical loop shows a nearly proportional behaviour over a broad frequency range. For higher frequencies the lag-terms of the actuator affect the frequency response.

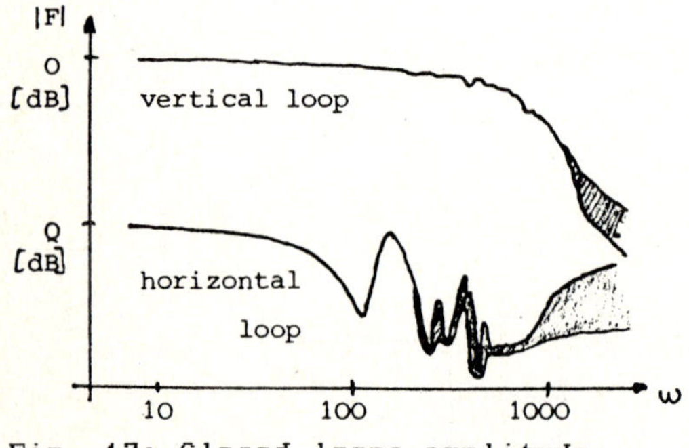

Fig. 17: Closed loops amplitude diagram

If - as in this case - in the controller design no effort is made to assure that the closed loop output meets the specifications on correspondence between test stand and road loading of the specimen to be tested, this has to be achieved by shaping the input signals which can be done by the ITFC-method described earlier [8].

6. CONCLUSIONS

A general frequency domain model was derived for a certain class of vibration test benches with uncertain or varying parameters. This model consisting of a "nominal" frequency response $\hat{F}(\omega)$ and variations $\Delta F(\omega)$ thereof, can be used for the design of "robust" controllers for these test benches or for precompensation and additional special iterative drivesignal calculation procedures. The coherence function, which can be measured during experimental identification, is used to estimate bounds on the variations $|\Delta F(\omega)|$. Further research will be focussed on the topic how specific nonlinearities affect the nominal frequency response and variations of the mechanical systems under consideration.

7. LITERATURE

[1] Zachmann, N.J., Hillberry, B.M. and Kettelkamp, D.B.: Design of a load simulator for the dynamic evaluation of prosthetic knee joints, ASME-publication 78-DET-59, Minneapolis 1978.

[2] Mayne, D.Q.: The design of linear multivariable systems, Automatica 9 (1973), pp. 201-207.

[3] Horowitz, I. and Sidi, M.: Synthesis of feedback systems with large plant ignorance for prescribed time domain tolerances. Int. J. Control, 16 (1972), pp. 287-309.

[4] Horowitz, I.: Quantitative synthesis of uncertain multiple input/output feedback systems. Int. J. Control, 30 (1979), pp. 81-106.

[5] Horowitz, I. and Löcher, C.: Design of a 3x3 multivariable feedback system with large plant uncertainty. Int. J. Control, 33 (1981), pp. 677-699.

[6] Gräser, A., Neddermeyer, W. and Tolle, H.: CAD of the Horowitz / Sidi-design for feedback systems with large plant parameter uncertainty. IFAC Symposium, Purdue University, West Lafayette, Indiana (USA), Sept. 15-17, 1981.

[7] Gräser, A.: Zur Modellbildung und Regelung mehraxialer Kraftfahrzeugprüfstände. Fortschrittsber. VDI-Z., Reihe 8, Nr. 50.

[8] ITFC-Iterative Transfer Function Compensation, Manuals of the Carl Schenck AG, Darmstadt.

[9] Strobel, H.: Systemanalyse mit determinierten Testsignalen, VEB-Verlag Technik, Berlin 1968.

[10] Levi, C.: Complex curve fitting. IEEE Trans. Autom. Contr., 4 (1959), pp. 37-43.

[11] Sanathanahan, C.K., Koerner, I.: Transfer function synthesis as a ratio of two complex polynomials IEEE Trans. Autom. Contr., 9 (1963), pp. 56-58

Section II

BOND GRAPH ANALYSIS AND MODELLING

Complex and Distributed Systems: Analysis, Simulation and Control
S.G. Tzafestas and P. Borne (editors)
Elsevier Science Publishers B.V. (North-Holland)

BOND GRAPHS REVISITED : A SYSTEMIC VIEW

Francis LORENZ

Fabrique Nationale - R&D CFAO
Voie de Liège, 33
B 4400 HERSTAL, BELGIUM *

The General System Theory defines a model (of some object) as any representation related by homomorphism with the object under study and by isomorphism with the so called General System. The path from the object to the model actually flows through a Representation System which should itself be isomorphic with the General System. That implies that the Bond Graph Theory, as a particular Representation System applying to some class of objects, should exhibit all the characteristics of the General System : it should be able to describe the structure, the activity and the evolution of an object involved in an environment to achieve some goals.

This idea leads to some new concepts to be included in the Bond Graph Theory. The present paper is intended to present these concepts more as a challenge to Bond Graph theorists, than as a finite addition to the method.

1. INTRODUCTION

Despite its already impressive interest, the Bond Graph Theory, when faced with real-world systems, quickly reveals its limits. The purpose of this paper is to present these limits as well as some proposals to go beyond them. The ideas developed here came from a confrontation of the Bond Graph Theory with the General System Theory which is supposed to define what modelisation is.
However, no formal solution will be proposed. On the contrary, it is thought that the question is still open so other ways of reflexion are possible.

2. SYSTEMIC BOND GRAPHS

We will shortly review the main ideas of the General System Theory and then see how they apply to the Bond Graph Theory to finally identify what is lacking in the latter theory.

2.1 The General System Theory

The modelisation of an object, following the GST, consists in designing a new object, called the model, related by homomorphism with the object under study and by isomorphism with the General System (i.e. having the same properties). Now, this General System is just an artefact. It is by no way the model of some real-world system. So, it does not make sense to ask if the systemic paradigm corresponds to reality (systems exist in human mind, not in nature); one may only ask whether it is useful or not [1]. I will here suppose it is.

The modelisation process implies the existence of some Representation System. This Representation System is itself isomorphic with the General System. That is the point on which the whole paper is based, the Bond Graph Theory beeing just a Representation System applying to some class of objects.

The General System consists in three aspects : structure, activity, evolution. Moreover, it is generally involved in some environment to achieve some goal.

* detached from CIG South, Avenue de l'Informatique, 9, B, 4430 ALLEUR, BELGIUM

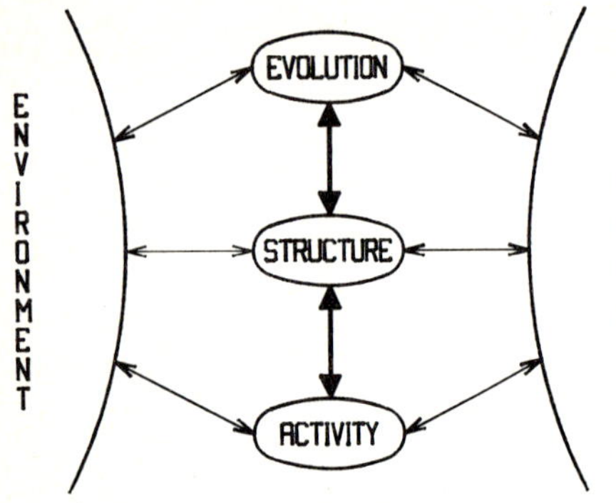

The basic notion is the notion of <u>process</u>. A distinction must be made between processors and processed objects. A processor will move the processed object in a TSF (time, space, form) reference. This leads to a typology of processors : T-processors, S-processors and F-processors. There is also a typology of processed objects. These objects are constituted of matter, energy or information. This typology is however hierarchical. One needs energy to process matter, and one needs information to process energy. Notice that this hierarchy corresponds to an increasing degree of abstraction.

The universe is represented as a set of interconnected processors, the links beeing oriented, neutral and acausal. The structure of the system can be represented by a matrix relating the inputs of the set of processors to their outputs. We can also define the state equations which relate the outputs of each processor to its inputs (activity of the system). Both are obviously required to describe completely the system. However, a large number of systems will exhibit, for some values of their state vector, a modification of their structure and/or activity. This is the evolution aspect of the system.

Lastly, we can observe that there exists some processors which are outside the system. These are sources or sinks and represent the environment of the system. No formal name exists in the General System Theory for these processors; I would propose gates or G-processors. The goals aspect will not be considered here. It is only important for goal-seeking or purposeful systems.

2.2. The Bond Graph Theory

Beyond the formalism, I would define a Bond Graph as a topological representation of energy flows between discrete elements of a system.

This representation is already coherent with the GST, the elements beeing the processors and the bonds beeing the oriented, neutral and acausal links.
In classical bond graphs, only a few S-processors exists. It is essentially the Ø and 1 junctions, which represent a distribution, not a transportation. The TF and GY elements, although they generally also imply some transportation, are functionnally F-processors. The storage elements I and C are clearly T-processors. Lastly, the sources and sinks (S, SE, SF, R) are G-processors, representing the environment.

We can then redraw the field representation of Rosenberg [2] :

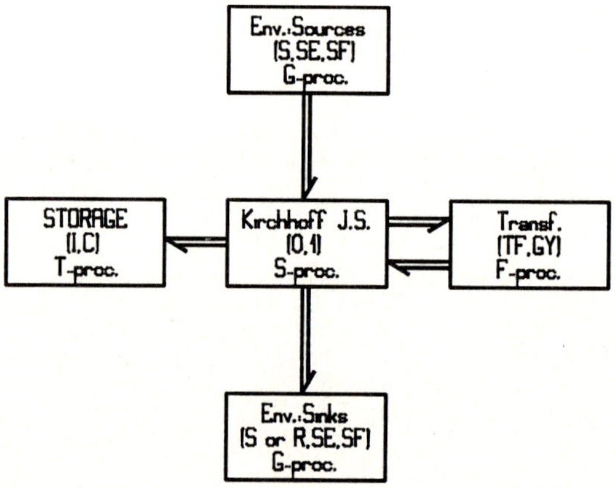

The apparent lack of an output link to the storage field is easily understood according to Breedveld [3].

In classical bond graphs, the processed object is energy. The degree of abstraction of the notion of energy is probably, together with the fact that the representation is only topological, the key of the bond graph's universality. It seems difficult and it is eventually not desirable to include processing of matter in the theory. At the opposite, the hierarchy information-energy-matter practically imposes considering the representation of information. Here come the "activated bonds" of Karnopp and Rosenberg [4].

2.3. Requested New Features of the Bond Graph Theory

A first requirement of the Bond Graph Theory is an improved use of the information bonds. All authors use them, but always seem more or less embarrassed to do so. I would take as a proof the fact that, although the bond graph language has been almost fixed for a long time [5], no (normalized) formalisation of its information part ever appeared. I claim that the use of information bonds is not a shame but a necessity. I consequently call for such a formalisation. A block diagram would then appear just as a special bond graph which is linear and which does not have any power bond.

Notice that a formal inclusion of information bonds in the Bond Graph theory leads to unexpected conclusions. According to Zeigler [6], the models can be classified as continuous and discrete time and state. Energy phenomena generally suppose continuous time and continuous state so they are governed by differential equations. Informationnal control, however, can present any of the four combinations (examples : analog feed-back, on/off switches, sampling systems, digital control) so they can be governed by discrete events as well. Now, the question is how does this interfer with the Bond Graph formalism ?

A second requirement is an improved representation of the system environment. A closed system, according to the first law of thermodynamics, can include neither sources nor sinks. Only a few systems can be modelized that way. So the classical bond graph sources and sinks (resistors) already represent the environment. However, I would call a classical bond graph a closed model of an open system because that model cannot be included as a submodel in a bigger one. I call for some formalism enabling a designer to conceive open models.

A third requirement is the formal inclusion of evolution. If we consider two masses coming into contact, we need to represent a shock. If the shock is partially or completely elastic, it implies a local modification of the laws governing the movement (activity aspect), with or without energy loss. If the shock is completely dissipative, it also implies a modification of the structure, as a new bond did appear, involving moreover a local modification of the causality.

That kind of problem is very common in mechanical systems, but can also be frequently found in other types of systems (example : ideal diode or non-return valve). A means to rigorously deal with evolutionnal systems is required here (a resistor with zero or infinite resistance, beyond the numerical problems, is not equivalent to no resistor, at least from a sructural point of view).

3. SYSTEMIC BOND GRAPH FEATURES

We can now imagine an augmented Bond Graph Theory including the abovementioned features. No strict formalism will be given, as numerous ways of applying the ideas can be found.

3.1. Basic Elements of a Bond Graph Model

The typology of bonds is easy to define. I would propose the following :

scalar bonds   vector bonds

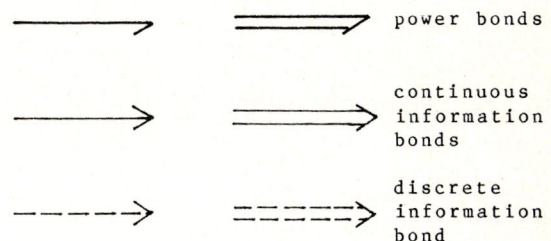

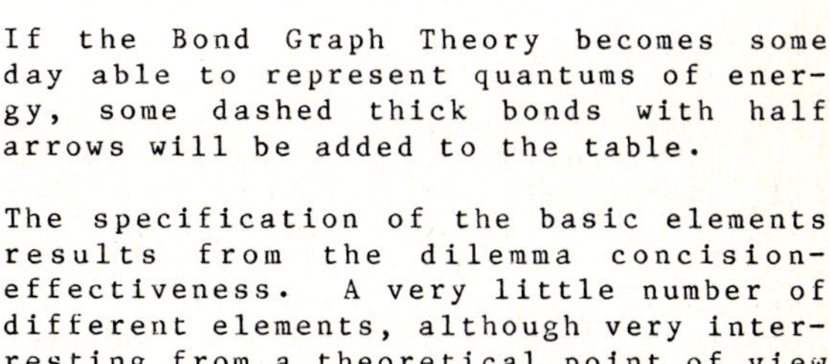

If the Bond Graph Theory becomes some day able to represent quantums of energy, some dashed thick bonds with half arrows will be added to the table.

The specification of the basic elements results from the dilemma concision-effectiveness. A very little number of different elements, although very interesting from a theoretical point of view [3,7], leads to almost non legible bond graphs. At the opposite, a very large number of different elements leads to an almost non manageable representation, due to the high degree of redundancy of the language. A compromise should be found.

The classical elements should be included as well as a set of new normalized elements dealing with both types of information. These can probably be inspired by the block diagram formalism, extended to allow non linearities. The matricial version of all these elements are also very useful (see for instance the works of Tiernego and Bos [8]). Some other elements may be useful or strictly necessary : direct sum [9], gyristor [10], sympletic gyrator [11], dissipating transducer [12], transmission line [13],...
The mine is the subsystem. A subsystem is a single element representing another bond graph. The latter must of course be what I called an open model. To achieve this, I simply replace all (or may be some) of the souces and sinks by explicit G-processors, which I name gates.

3.2. Alternative Descriptions

The bond graph only describes the structure of the object. We must also add the laws governing the elements (at least the non linear ones) and the parameters. At that point, some alternative description can be desired. For instance, a subsystem can be described with one structure and one set of laws but more than one set of parameters. This corresponds to different sizes of the same device. If more then one set of laws are given, it corresponds to different assumptions about the activity (linear and non linear sets, for instance). If there are more than one structure, it corresponds to different assumptions about the structure (with or without some leakage for instance). Multi-structured models will also be useful to represent the evolution of the system.

3.3. Catastrophy Management

The evolution aspect corresponds to a modification of the structure or the activity of the model. Although it can reveal some differences, the process is quite similar to what is studied in the Catastrophy Theory [14,15]. It implies several structures and laws for a single system and a means to switch among them. That means is the catastrophy management. The catastrophy management monitors the activity of the system for some condition. When the condition is realised, it takes the control to switch to a modified model and reinitializes this model. It also updates its condition set (the condition(s) to be monitored in one structure is (are) different from the condition(s) to be monitored in the modified structure).

4. SPECIFICATIONS OF A FORMALIZED LANGUAGE

We will not propose a language here, but we will only attempt to identify its specifications to include the foregoing ideas.

4.1. Structure of the Language

A system is composed of three blocks of description : structure, laws and catastrophy management, plus a set of parameters. More than one version of each of these blocks can be described. Multiple choice switches are required to enable the selection of a specific block at a specific time. Inside a block, the description is not procedural. Because of potential naming problems when more than one designer work on a single project, a nesting of the systems and subsystems, similar to the nesting of procedures in PASCAL, should be implemented. It further has the advantage of allowing local and global definition of parameters. However, that organisation is not always sufficient and a true parameter transmission mechanism should be provided.

4.2. Use of the Language

The language should accept partial description to be completed in the future. So, some laws must be accepted for gate elements, to allow testing a submodel without incorporating it in a bigger model. I would call these laws limit conditions. At the opposite, it should be possible to test a system even if its subsystems are only roughly known. This can be achieved if some laws are also accepted for subsystem elements, which is a black box representation of the subsystem. So, top-down as well as bottom-up design will be feasable.

Moreover, the notion of submodel has the advantage to allow the constitution of a library of often used subsystems. Indeed, the bond graph supplies a very unique way to build such a library, as acausal submodels make no assumption about the way they are connected to the rest of the system, which is not the case, for instance, with a library of FORTRAN subroutines or even block diagrams. Of course, a means of saving the (non- linear) laws with all causalities restorable is also required.

4.3. Open Language

Lastly, it should be highly desirable to allow a user to define its own new elements and bonds. The black box use of subsystem only partially covers this feature. It allows, for instance, neither a redefinition of a Ø or 1 junction nor the definition of a new bond type. The only way to completely achieve the goal is to describe the basic elements and bonds in some metalanguage rather than embedding them in the program.

5. CONCLUSIONS

The paper presented some new ideas to augment the power of the Bond Graph Theory. To conclude, I will once again refer to the General System Theory. It is now rather well known that, although the entropy of a closed system can only increase, it can actually decrease if the system is open to its enrironment and receives enough information from it. The author wishes to be such an open system and consequently welcomes all the criticisms and comments about his paper. Indeed, as an object, the paper itself should be submitted to evolution.

ACKNOLEDGEMENTS

I wish to thank Professor Wolper of the University of Liège and Pierre Levert of the Fabrique Nationale who both did so much that without their help this paper would not exist.

REFERENCES

[1] Le Moigne J.L., la Théorie du Système Général - Théorie de la Modélisation, Presses Universitaires de France, Paris, 1977.

[2] Rosenberg R.C., "State-Space Formulation for Bond Graph Models of Multiport Systems",J. Dyn. Syst., Meas. & Control, trans. ASME, Series G, Vol. 93, N.1, May 1971, pp. 35-40

[3] Breedveld P.C., Physical Systems Theory in Terms of Bond Graphs, Ph. D. Thesis, Twente University of Technology, Enschede, The Netherlands, 1984.

[4] Karnopp D.C., Rosenberg R.C., Systems Dynamics : A Unified Approach, Wiley, New York, 1975.

[5] Rosenberg R.C., Karnopp D.C., "Definition of the Bond Graph Language", J. Dyn. Syst., Meas. & Control, Trans. ASME, Series G, Vol. 94, N.3, September 1972, pp. 179-182.

[6] Zeigler B.P., Theory of Modelling and Simulation, Wiley, New York, 1976.

[7] Rosenberg R.C., "On Gyrobondgraphs and Their Uses", J. Dyn. Syst., Meas. & Control, Trans. ASME, Series G, Vol. 100, N.1, March 1978, pp. 76-82.

[8] Tiernego M.J., Bos A.M., "Modelling the Dynamics and Kinematics of Mechanical Systems with Multibond Graphs", J. Franklin Inst., Vol. 319, N. 1/2, January/February 1985, pp. 37-50.

[9] Breedveld P.C., "Proposition for an Unambiguous vector bond graph notation", J. Dyn. Syst., Meas. & Control, Trans. ASME, Series G, Vol. 104, N.3, September 1982, pp. 267- 270.

[10] Allen R.R., "Multiport Representation of Inertia Properties of Kinematic Mecanisms", J. Franklin Inst., Vol. 308, N.3, 1979, pp. 235-254.

[11] Breedveld P.C., "Thermodynamic Bond Graphs : A New Synthesis", International J. Modelling & Simulation , Vol. 1, N.1, Acta Press, Anaheim, 1980.

[12] Thoma J.U., Atlan H., "Network Thermodynamics with Entropy Stripping", J. Franklin Inst., Vol. 303, N.4, 1977, pp. 319-328.

[13] Tsai N.T., Wang S.M., "Delay-Bond Graph Models for Geared Tortional Systems", J. Applied Mechanics, Trans. ASME, Series E, Vol. 41, N.2, June 1974, pp. 366-370.

[14] Stewart I., "Oh ! Catastrophe", Les chroniques de Rose Polymath, Belin, Paris, 1982.

[15] Thom R., Modèles Mathématiques de la Morphogenèse, ed. Christian Bourgeois, Paris, 1980.

Complex and Distributed Systems: Analysis, Simulation and Control
S.G. Tzafestas and P. Borne (editors)
Elsevier Science Publishers B.V. (North-Holland)

JUSTIFIED USE OF ANALOGIES IN SYSTEMS SCIENCE

P.M.A.L. HEZEMANS

Eindhoven University of Technology
Dept. of Mech. Eng.
P.O. Box 513
5600 MB Eindhoven
Netherlands

L. van GEFFEN

Metal Institute TNO
APA Section
P.O. Box 541
7300 AM Apeldoorn
Netherlands

Research in the use of analogies in Science and Techology of the last two centuries reveals that besides striking successes of appliations of analogies there are also dramatic failures, not to speak of sophistical missuses in reasoning of logic. Searching for the reason for success and failure forces us to lose ourself in a mechanism of analogy thinking governed by logic, resulting in a sharp division to be created between analogy-reasoning and analogy-application.
By realistically starting from the idea that an analogy besides the similarities, also supposes the differences it is possible to make use of a justified use of analogies in Systems Science. In order to yield this in this article some conditions are formulated. From some examples one can see that the use of analogies does good service for modelling. With the help of bond graphs the analogy can thus be critically applied, that the analogy can be seen visually, physically and mathematically at the same time, in a recognizable way. The application of algebraic topology in the use of analogies makes it possible to design a dual system from the original system. Finally some rigorous conclusions are drawn with respect to justified use of analogies of bond graphs in the modelling process.

1. INTRODUCTION

1.1. Definition

Coming from the Greek word ana-logon, literally "in proportion", one can shortly define it as concurrence in some aspects between phenomena further essentially different. It is not quite impossible that analogy sometimes primarily points to a very big difference and secondly to some resemblance.

1.2. Analogy-use and -reasoning

One can make use of analogies in a correct and justified way in order to elucidate and comprehend e.g. a complicated problem, an abstract representation, an obscure relation between certain facts or a system, difficult to approach. In Systems Science it is required that analogy use is only correct and justified if a proof of this analogy can be given.

1.2.1. Analogy-reasoning

Analogy-reasoning is a reasoning, whereby from a correspondence between two things in a certain aspect it is decided upon similar correspondence in an other connected respect. We try to prove something with analogies. The analogy prove is an argumentation starting from partial similarity between two cases, in order to conclude complete similarity. Generalizing like this by analogies could lead to dangerous generalizations. Many analogies failed because of the unallowed use of the analogy prove. The attribution of properties to an entity by means of analogy may accidentally reveal a new phenomenon and thus advance knowledge, but it is always accidental and as a systematic approach to knowledge it is an extraordinarily clumsy technique. The price of analogy is, indeed, eternal vigilance.

1.2.2. Critical view on the use of analogies

Historical examples prove that analogies have turned to good advantage in the first phase of certain sciences, but failed in later research due to their imperfections. Analogies certainly form a reliable quide in the phenomena we can expect, but they never give definitive insight in what we will discover. Create a hypothesis from an analogy and test it critically.

1.2.3. Criterion for analogy use in Systems Science

If one wishes to come to a critical use of analogies, one has to restrict oneself to things between which analogies have been determined or proved. From here interpretations and conclusions can be drawn that only have validity with regards to these aspects and no other aspects.

1.2.4. Requirements for analogy application

Before one decides to use analogy application three requirements have to be fulfilled:

1. Mathematical analogy implies that two models obey identical sets of equations; 2. Physical analogy implies that generalized components obey identical physical laws; 3. Visual analogy implies visual association of shape between elements or systems.

1.3. Homology

Homology is defined as morphological resemblance between systems. Apart from the component function homology only determines the geometrical resemblance in the system structure. Its function is here completely insignificant. Analogous models that have identical mathematical functions need by no means be homologous. We can see this in fig. 1.

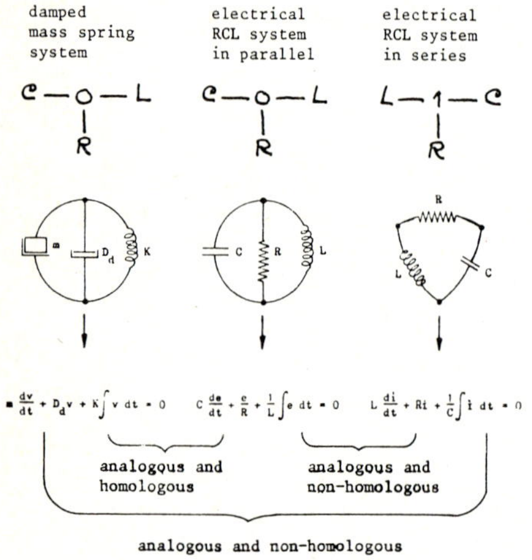

fig. 1 Analogous models with their differences

1.4. Duality

This strangely developed relation between analogy and homology can be explained with 'duality'-considerations. Duality means ambiguity. On one side one tries to formalize two different principles, conceptions, equations or systems into one higher principle, one broader conception, one more universal equation or one more generalized system. On the other side, if a system model is built strictly formal, one often finds different possible conclusions, from which seemingly contradictory (dual) interpretations can be drawn. In systems theory duality results in a formalized system model having two analogous, but non-homologous system models. In terms of the preceding example the "series-model" is a dual o the "parallel-model".

2. ANALOGY BETWEEN SYSTEMS

In order to quickly set up a system model a clear arrangement in thinking is required: "analogy thinking". It is therefore necessary to identify analogy between systems and analogy between treatment methods. Analogous systems are systems obeying identical mathematical models and analogous treatment methods are identical methods to treat, even if the systems in question are not analogous. Analogy-"seeing" has to take place systematically observing the already mentioned requirements. Before deciding on analogy between systems one has to check if each of the three partial analogies are fulfilled: Analogy between 1. variables, 2. system components, 3. system structures.

2.1. Analogy between variables

Firestone [1] introduced the distinction between two types of physical variables: across variables and through variables: across-variables, like velocity, voltage, pressure, temperature and concentration (chemical potential) are expressed as differences and are measured across two spatially different points. The corresponding through variables, like force, current, fluid flow, heat flow (entropy flow) and mass flow are measurable at a single point. The product of each variable pair has the dimension of power. The work of Trent [2] provided a more rigorous basis for Firestone's "complete" analogies as well as algorithmic rules for their construction, by the use of the linear graph and its associated matrix algebra. In table 1 α and τ variables are displayed for some physically different systems.

System	Across variable α	Through variable τ
Electrical	Voltage difference U	Current I
Fluid	Pressure difference p	Fluid flow Q
Mechanical translational	velocity difference v	Force F

table 1 Across and through variables for some physical systems

2.2. Analogy between system components

In physically different energy domains one comes across certain components with analogous functions. In table 2 a functional classification is seen of analogous components with matching function descriptions defined with α and τ variables.

generalized resistor	generalized capacitor	generalized inductor
$\alpha = R.\tau$	$\tau = C.\frac{d\alpha}{dt}$	$\alpha = L.\frac{d\tau}{dt}$

R = generalized ideal resistance: 1/b = reciprocal translational damping, R = electrical resistance, R_f = fluid resistance, R_t = thermal resistance

C = generalized ideal capacitance: m = mass, C = electrical capacitance, C_f = fluid capacitance, C_t = thermal capacitance

L = generalized ideal inductance: 1/k = reciprocal translational stiffness, L = electrical inductance, L_f = fluid inductance

table 2

2.3 Analogy between system structures

If two physically different systems display identical structures they are called structurally analogous. It means they possess identical structure equations (B-matrix).

2.4. Definition of analogous system

If all three partial analogies are fulfilled we speak of analogous systems. In the fig. 2 analogous electrical, hydraulical and mechanical systems are drawn together with their three partial analogies.

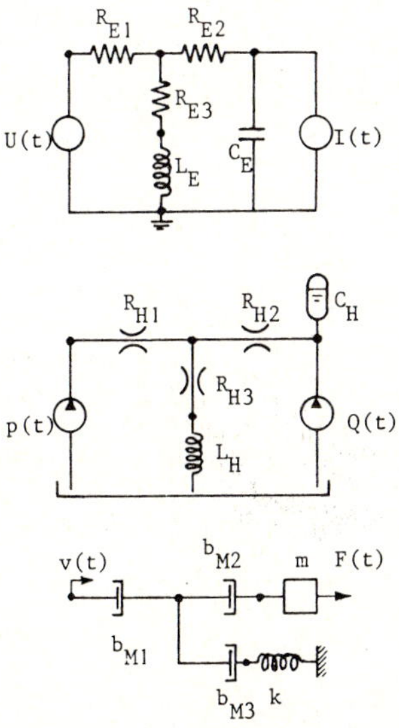

1. Variable analogy:

general	α	τ
electrical	U	I
hydraulical	p	Q
trans. mech.	v	F

2. Component analogy:

$R_{E1} = R_{H1} = 1/b_{M1}$
$R_{E2} = R_{H2} = 1/b_{M2}$
$R_{E3} = R_{H3} = 1/b_{M3}$
$L_E = L_H = 1/k$
$C_E = C_H = m$
$SA_E = SA_H = SA_M$
$ST_E = ST_H = ST_M$

3. Structure analogy:

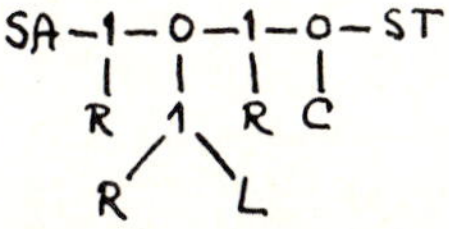

fig. 2 Partial analogies

3. ANALOGY BETWEEN DERIVING METHODS

In literature on Systems Science one comes across two different methods for setting up so called structure equations, linear graphs and bond graphs. If a bond graph is set up with the power direction arrows pointing towards the elements, including sources, TF's, GY's and multiports and if it is agreed that the power absorbed in and produced by the energy port is respectively positive and negative, a generalized method can be constructed for deducting the structure equations. The following example is taken:

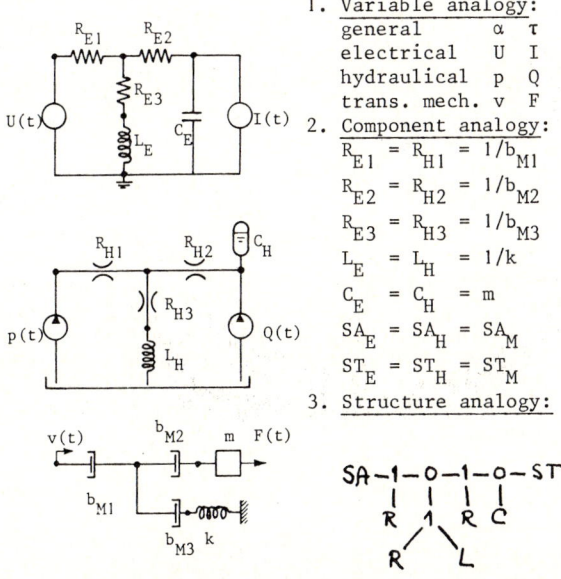

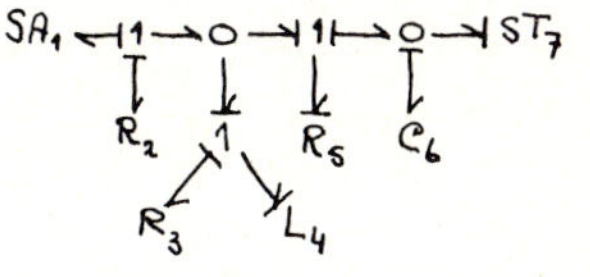

fig. 3 Example

Only the energy ports are numbred, not the bonds. Two groups of components are formed:

Group	System components	Port numbers
$\vdash\!\rightharpoonup$	SA_1, R_2, R_3, C_6	1, 2, 3, 6
$\mid\!\leftharpoonup$	L_4, R_5, ST_7	4, 5, 7

table 3

In the left-hand column vector of the matrix structure equation one first places the through variables of $\vdash\!\rightharpoonup$ and after that the across variables of $\mid\!\leftharpoonup$ and in the right-hand vector first the across variables of $\vdash\!\rightharpoonup$ and after that the through variables of $\mid\!\leftharpoonup$, all of them in rank-order of index numbers. Substitution of the partitioned matrices B and C from the bond graph structure completes the matrix equation:

$$\begin{bmatrix}\tau_1\\ \tau_2\\ \tau_3\\ \tau_6\\ \alpha_4\\ \alpha_5\\ \alpha_7\end{bmatrix} = \begin{bmatrix} & & & & 1 & 1 & 0\\ & \underline{0} & & & 1 & 1 & 0\\ & & & & 1 & 0 & 0\\ & & & & 0 & 1 & -1\\ -1 & -1 & -1 & 0 & & & \\ -1 & -1 & 0 & -1 & & \underline{0} & \\ 0 & 0 & 0 & 1 & & & \end{bmatrix} \cdot \begin{bmatrix}\alpha_1\\ \alpha_2\\ \alpha_3\\ \alpha_6\\ \tau_4\\ \tau_5\\ \tau_7\end{bmatrix}$$

with

$$\begin{bmatrix}1 & 1 & 0\\ 1 & 1 & 0\\ 1 & 0 & 0\\ 0 & 1 & -1\end{bmatrix} = [B] \quad \text{and} \quad \begin{bmatrix}-1 & -1 & -1 & 0\\ -1 & -1 & 0 & -1\\ 0 & 0 & 0 & 1\end{bmatrix} = [C]$$

A known curiosity is revealed: $B = -C^T$ (T = transposed).

Immediately the association with linear graphs comes up. Indeed one recognises $\vdash\!\rightharpoonup$ as lines of the "tree" of a linear graph and $\mid\!\leftharpoonup$ as the lines of the "cotree". Obviously for bond graph theory one can now reap the fruits of network theory for linear graphs, i.e. network synthesis, network optimization, interpretation possibilities by means of reciprocity, passivity and symmetry theorems and network analysis of three- and more-dimensional networks. Famous works as Gabriel Gabriel Kron's "Diakoptics" [3] can then be studied with less effort. (and more α and τ).

4. ANALOGY BETWEEN PHYSICAL THEORIES

As it appeared to be possible to set up analogous systems in physical different domains intuitively one feels that there has to exist a common analogy for the physical theories as well, in order to generalize electrical, hydraulical, mechanical, thermodynamical, etc. theories. Tonti [4] proved this was true. The instigator of it was Gabriel Kron who continually emphasized that there must be an underlying justification for the proliferation of electrical and mechanical analogs used in modelling a diverse range of physical phenomena. He introduced the use of homology theory (topology) in this context as the basis of his network and systems theory. By extending Kron's and more recent works into a unified theory beginning with a topological analysis of "System Invariants" this leads to systematic procedures for deriving and solving matrix (tensor) models and efficient solution algorithms and a philosophy which should contribute towards the understanding and teaching methodology in general [5]. Branin [6] presents a treatment of the network concept not only in its usual context related to linear graph theory (or "1-networks") but also in relation to a higher dimensional homolog called the 3-network. Then, by introducing the concept of a dual 3-network, he demonstrates that the resulting algebraic structure is the progenitor of an isomorphic operational structure that characterizes the vector calculus. Branin's paper is a prelude to and companion of Tonti's very remarkable work [4] on a formal treatment of physical theories of an impressive variety. Using algebraic-topological concepts Tonti classifies numerous physical phenomena as 1-, 2-, 3-, and even 4-network problems. Tonti shows how a major portion of theoretical physics comes within the scope of this algebraic-topological approach, or what may in its broadest sense be called homological network theory. This theory uses cell-complexes, homolog and cohomolog sequences, chains and cochains, boundary- and coboundary operators. The physical interest of the process of constructing sequential chains comes from the fact that the equations of structure of every physical theory states that one chain is the coboundary of another.

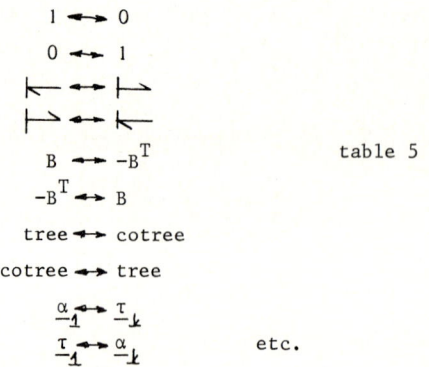

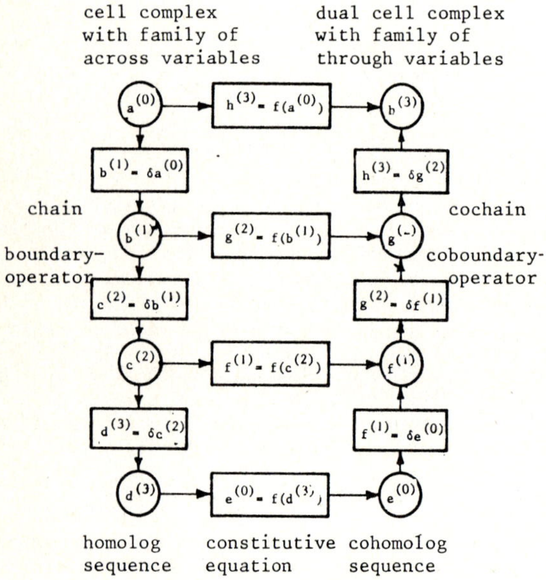

fig. 4 Coboundary process

In fig. 4 this coboundary process for "4-complexes" is displayed. In [4] Tonti works out examples from different energy domains; he combined all physical theories in one homology theory. As regards the ατ-classification Kron and several different authors [7], [8] confirmed that one cannot dissociate the topology from the "through"- and "across" -concepts.

5. DUALITY BETWEEN SYSTEMS

In order to determine duality between systems it is also here necessary to identify three partial dualities: Duality between 1. variables, 2. system components, 3. system structures.

5.1. Duality between variables

As through and across variables can be formalized into the higher concepts of "physical quantities" the α variables of a certain system can dualy be understood as τ variables of another system and vice versa. The energy nature remains unchanged.

5.2. Duality between system components

It is striking that with regard to a generalized inductor it holds: α equals a constant times the time derivative of τ, and for a generalized capacitor: τ equals a constant times the time derivative of α. By interchanging the α's and τ's one can obtain the dual system components, like:

original			dual	
inductor	L	↔	capacitor	C
capacitor	C	↔	inductor	L
resistor	R	↔	conductor	G=1/R
α-source	A	↔	τ-source	T
τ-source	T	↔	α-source	A
transformer	n	↔	dual transformer	1/n
gyrator	r	↔	dual gyrator	1/r

table 4

5.3. Duality between system structures

In the dual case interchanging α's and τ's leads to dual structure equations according to the next table:

1	↔	0	
0	↔	1	
$\leftarrow\!\mid$	↔	$\mid\!\rightharpoonup$	
$\mid\!\rightharpoonup$	↔	$\leftarrow\!\mid$	
B	↔	$-B^T$	table 5
$-B^T$	↔	B	
tree	↔	cotree	
cotree	↔	tree	
$\underline{\alpha}_1$	↔	$\underline{\tau}_k$	
$\underline{\tau}_1$	↔	$\underline{\alpha}_k$	etc.

5.4. Definition of dual system

If two given systems contain all these three partial dualities one system is by definition dual to the other.

6. EXAMPLES

6.1. Illustrative example

In fig. 5 the dualization process in its essence is pictured:

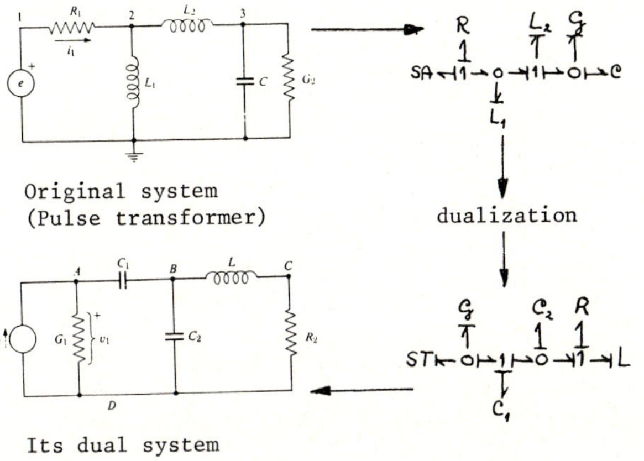

fig. 5 Dualization process

Further on it is possible to get a dual system component from the graphical characteristic of the original system component by constructing the dual characteristic (induced orientation [9]).

6.2. Worked example

Controlled frictional clutch and its dual, see fig. 6 and 7

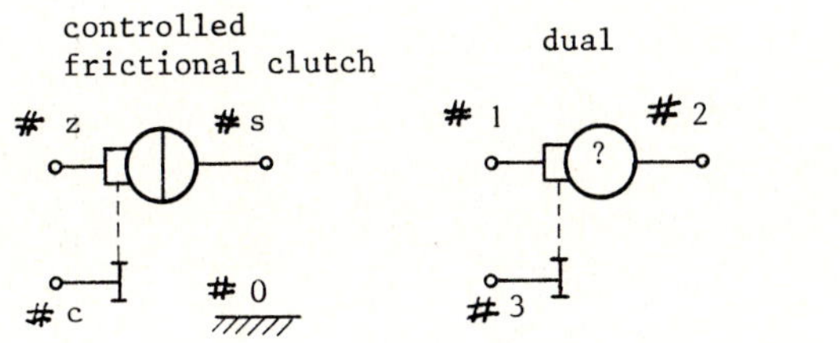

fig. 6 Controlled frictional clutch and its dual

We want to find a dual for the controlled frictional clutch. By reason of duality considerations in turns out to be a controlled speed regulator. How can it be constructed? From an abstracted scheme in figure 7 a bond graph is set up (fig. 8) that is transformed in a dual bond graph (fig. 10). From here "reality" is being re-transformed (fig. 12).

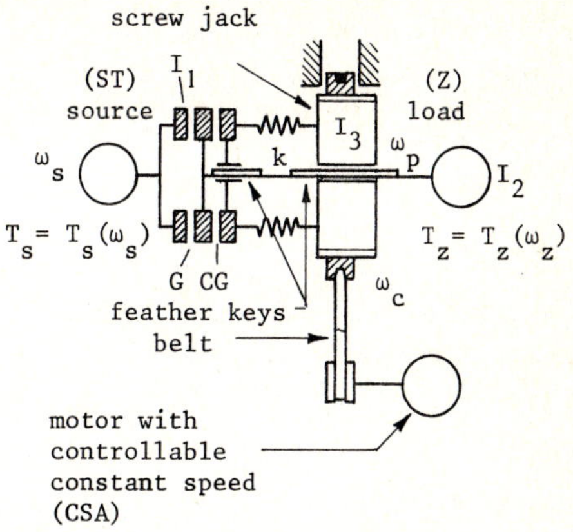

fig. 7 Controlled frictional clutch

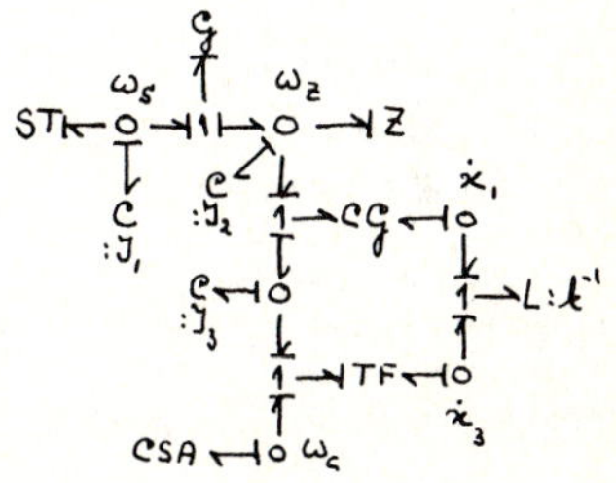

fig. 8 Bondgraph of fig. 7

In fig. 8 CG is defined by: 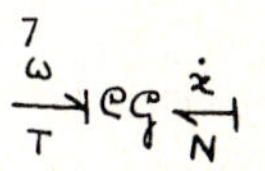 with coulomb friction:

$$T = T(N,\omega) = f.N.R$$

f = friction coefficient.

After reduction the bondgraph looks like

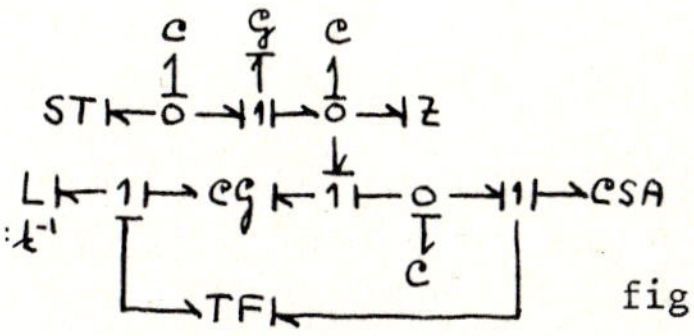

fig. 9

dualization

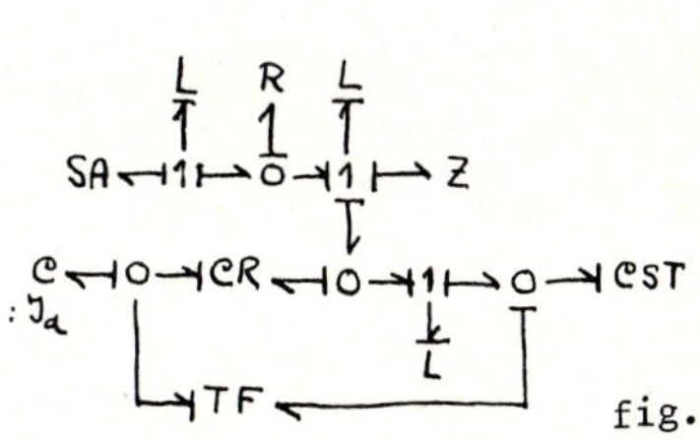

fig. 10

Where: 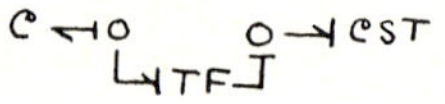

equals 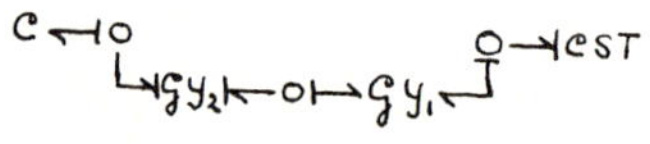

equals 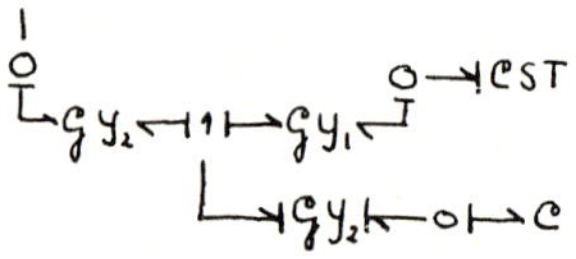

equals 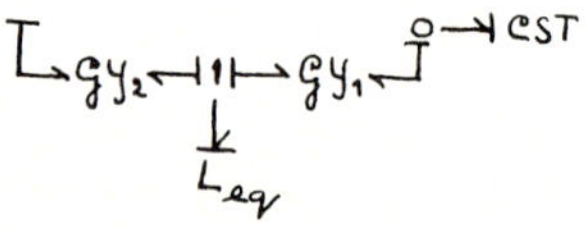

CR is dual of CG.

By finally combining 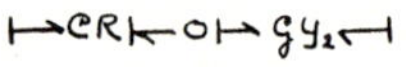

to 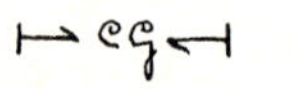 we get via fig. 11 the physical realization:

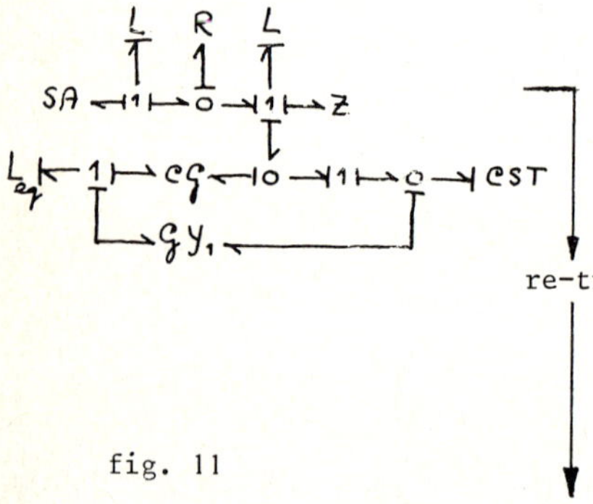

fig. 11

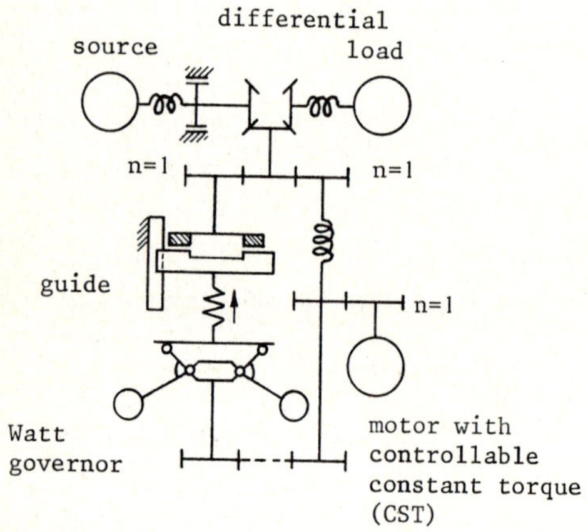

fig. 12 Centrifugal speed regulator

The purpose of this example is not to claim that every dual system component can be realized technically. It even seems that most duals are physically not realizable. From a technical point of view a dual from fig. 10 is more difficult to construct than from fig. 11.

7. CHOICE BETWEEN ANALOGY AND DUALOGY

A dualog is defined as an analog of a dual or a dual of an analog, as is seem in figure 13:

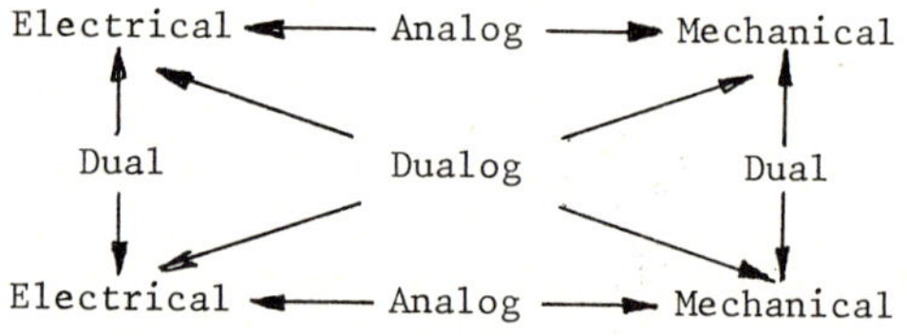

fig. 13 Relations between Duals, Analogs and Dualogs

In the theoretical framework as being presented here ατ analogies are (real) analogies and ef analogies are dualogies! The message of this contribution is to prefer analogies above dualogies. Firstly the process of "analog seeing" with analogies is essentially easier than with dualogies, as is confirmed by the structural similarities between the systems in fig. 14

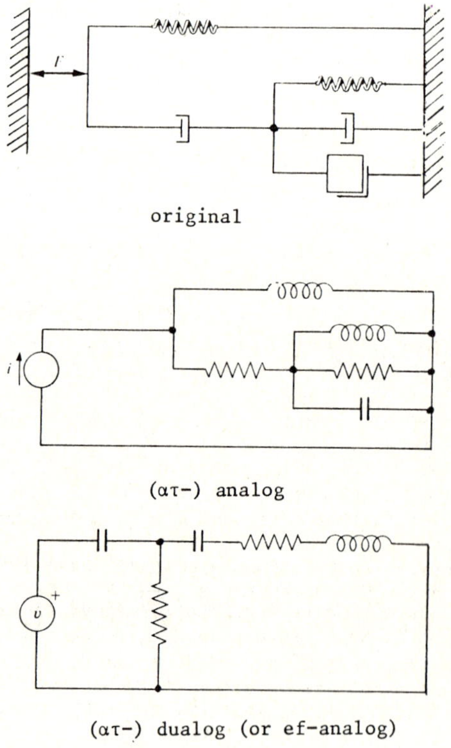

fig. 14 Process of "Analog seeing"

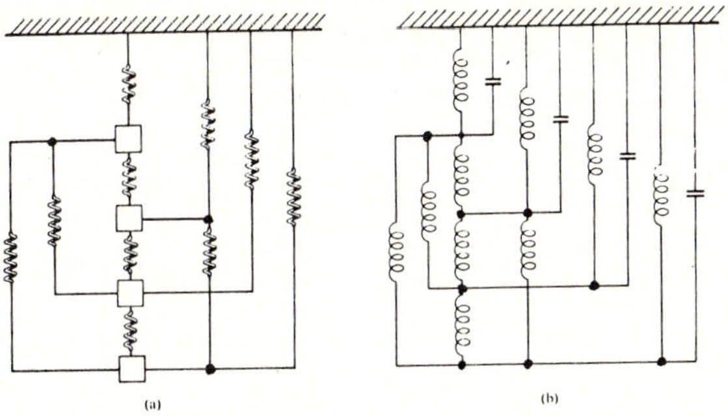

fig. 15 Mechanical system for which mass-inductance analog does not exist.

Secondly, for a "non-planar system" as is being portrayed in fig. 15 a dualog can not always be constructed while an analog always can.
In 1932 Whitney [10] proved that from a non-planar network one can never construct a dual. As long as a mechanical system can be represented by a planar graph, no practical difficulties arise from using analogs or dualogs because of the "planar graph theorem". When the graph is non-planar, however, the mass-inductance analogy is the one which fails. The mass-capacitance analogy always applies because of its topological consistence. "For pedagogical reasons, therefore, the mass-inductance analogy owing its existence solely to the planar graph theorem, should be discarded in favor of the mass-capacitance analogy which is fundamentally the correct one" [6]. According to Breedveld [11] this is too strong since "The planar graph theorem states that only planar graphs can be dualized. The fact however that some systems can be dualized and some not, does not provide a criterion to choose between the mass-inductance or the mass-capacitance analogy". Indeed no final verdict can be delivered, because the choice of analogs for mechanical systems depends on the requirements of the modelling techniques of the structure, which differ from application to application. While it is pointless to discuss about the "correct" analogy, as both are equally valid when they exist, the mass-capacitance (or ατ- or mobility) analog has considerable advantages, specially in the process of analogy seeing: (1) the mass-capacitance analogy is set up very easily; (2) for every possible mechanical combination (planar or non-planar) an electric analog exists, while an electric dualog does not necessarily exists; (3) the topological concept allows us to visually construct analogs; (4) in the modelling process it suffices to use only one procedure, while in mass-inductance analogy two procedures are necessary, one for the mechanical systems and one for the others [12]; (5) analogs of non-planar systems are always physically interpretable, while duals or dualogs are not.

REFERENCES

[1] Firestone, F.A., A New Analogy between Mechanical and Electrical Systems, Journal of the Acoustical Society, (Jan 1933), pag. 249-267.

[2] Trent, H.T., Isomorphisms between Oriented Linear Graphs Lumped Physical Systems, Journal of the Acoustical Society of America, 27 (May 1955), pag. 500-527.

[3] Kron, G., Diakoptics, MacDonald, London, 1963.

[4] Tonti, E., The reason for Analogies between physical theories, Appl. Math. Modelling, 1 (June 1976), pag. 37-50.

[5] Kouvaritakea and Macfarlane, Geometric approach to Analysis and Synthesis of System Zeros, Int. Journ. of Control, Vol 23, (Febr. 1976), no 2, parts I and III.

[6] Branin, F.H., The Algebraic-Topological Basis for Network Analogies and Vector Calculus, Symposium on Generalized Networks, Brooklyn Polytechnic Institute, (Apr 1966), pag. 453-491.

[7] Bowden, K.G., An Introduction to Homological Systems Theory: Topological Analysis of Invariant Systems Zeros, The matrix and Tensor Quality, (1980), pag. 36-52.

[8] Happ H.H., Piecewise Methods and Applications to Power Systems, Wiley, New York, 1980.

[9] Veblen, O., Analysis Situs, American Mathematical Society, Vol 31, (1931).

[10] Whitney, N., "Non-separable and planar graphs", Trans. of the American Mathemacal Society, Vol 34, (1932), pag. 339-362.

[11] Breedveld, P., Physical Systems Theory in terms of Bondgraphs, Thesis Twente University of Technology, Enschede, 1984.

[12] Karnopp, D.C., and R.C. Rosenberg, "System Dynamics: A Unified Approach", Wiley, New York, 1975.

SUGGESTED READING

[13] Roth, J.P., An Application of Algebraic Topology to Numerical Analysis: on the Existence of a Solution to the Network Problem, Proc. Nath. Acad. Sci, Vol 41 (1955), pag. 518-521.

[14] Kron, G., Tensors for Circuits, Dover, 1956.

[15] Kron, G., Camouflaging electrical 1-networks as graphs, Quarterly of Applied Maths, 1962, 20, 2, pag. 161-174.

[16] Nicholson, N., Structure of Interconnected Systems, Peter Peregrinus, Stevenage, 1978.

[17] Spillers, W.R., On diakoptics: tearing an arbitrary System, Quarterly of Appd Maths, XXIII, (1965), 2, pag. 188-190.

[18] Happ, H.H., Diakoptics and Networks, Academic Press, London, 1971.

[20] Hoffmann, B., The Nature of Primitive System in Kron's Theory, Am. j. Phys. Vol 23, (1955), page 341-355.

[21] System Structures in Engineering, edited by Bjørke, Ø., and O. Franksen, Tapir Publishers, Norway, 1978.

Complex and Distributed Systems: Analysis, Simulation and Control
S.G. Tzafestas and P. Borne (editors)
Elsevier Science Publishers B.V. (North-Holland)

A DEFINITION OF THE MULTIBOND GRAPH LANGUAGE

Peter C. Breedveld

Twente University, Electrical Engineering Department,
P.O.Box 217, 7500 AE Enschede, Netherlands,
phone: ..31-53-89-2792, telex: 44200

0) *Summary*

The multibond graph language is defined and its use is illustrated. The scope of this paper does not allow an exhaustive definition. For more details about multibond graph concepts, the reader is referred to [1,2,3], where most of the concepts were introduced. Readers unfamiliar with bond graphs, originated by Paynter [4], are referred to [5,6] and the literature listed in [7].

1) *Introduction to Multibond Graphs*

Languages evolve, fortunately, but it makes it hard to "define" them. After the "Proposition for an unambiguous vectorbond notation" [1], even a change in the name of the language itself took place: "vectorbonds" were renamed into "multibonds" [2], because many people not familiar with bond graph techniques appeared to have a wrong association with the former name. Herein, the reasons are outlined why possible changes with respect to the conventional notation, to the proposal mentioned and to other use of the notation in literature and in private communication, are deliberately made or not made.

In the sequel the reader is supposed to be familiar with the port-concept [8] and elementary bond graph symbolism [9]. Hence, there is no need to start "from scratch" with a formal mathematical definition, because the multibond graph language can be "defined in terms of" or rather "translated into" single bond graphs.

A multibond graph, as shown in basic form in Fig. 1a, consists by definition of multiports (MP) as "nodes" or "vertices" connected by "edges" called multibonds (Fig. 1b). The proper notation of these multibonds will be defined first. Secondly the notation principles of the fundamental MP elements will be defined. Next the important concept of "nesting" will be explained. Finally the elegance of the notation will be demonstrated by some examples.

Fig. 1a:

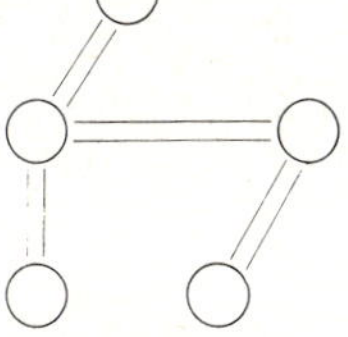

Fig. 1b:

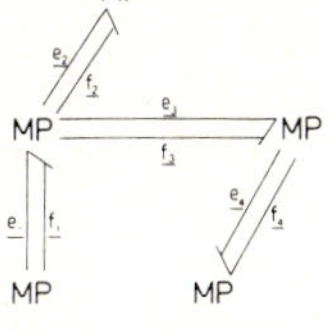

2) *Edges of the graph: Multibonds*

2a) *Multiports and multibonds*

The distinction between ports and bonds, as made in [9], has faded, mainly because the half-arrow symbol represents both concepts. The distinction cannot be observed from the symbol itself, but only from its context: if one end is connected to an element (node) it was called a port, if both ends are connected to elements, it was called a bond ("paired ports"). Nowadays most users call all half-arrows bonds. From this point of a view a multibond is defined.

An n-dimensional multibond connects an n-port of a sub-system or element with an n-port of another sub-system or element and represents the power exchanged through these ports (Fig. 1b). If one end of the multibond is not connected, the bond graph is not complete ("bond graph fragment"), but may be used to represent the ports of such a fragment, e.g. a multiport element. The elementary concept "multibond" and its explanation into single bonds is presented in Table Ia. It is trivial that a single bond is a degenerate multibond of dimension 1.

Table Ia:

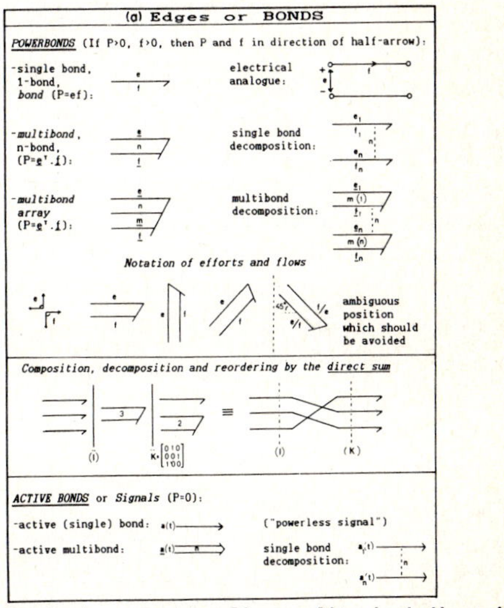

Computational (or operational) causality is indicated by the same causal stroke as in single bond graphs (Table Id). Multibonds with mixed causality are not possible: in case of a collection of single bonds with mixed causality the bonds should be reordered, e.g. with the use of the direct sum [2], and composed as two multibonds with opposite causality.

The "parallel line" notation has been chosen for several reasons:

1) it corresponds to the original notation of what has been called a vectorbond [10,1];
2) it corresponds to the n-dimensional notation of signals in blockdiagrams [11] and therefore is easy to interpret;
3) its multidimensional character can be observed along the whole length of the multibond, which is especially convenient if a multibond is rather long and "meanders" through the graph; by contrast, a multibond notation in which the multibond is distincted from the single bond by drawing a circle around the half-arrow (at one point!) does not have this feature and also has the disadvantage that the circles are easily mixed up with 0-junctions (this notation was proposed by prof. J.U. Thoma (Univ. of Waterloo, Canada) in private

communication);

4) the dimension of the multibond can be written between the parallel lines, thus not interfering with other symbols; this feature is optional: if not put in the graph, it has to be clear from the context; thus, in case of omission in graphical computer input, the program has to ask the user to specify the dimensions separately;
5) in literature this notation has been used successfully until now, which may indicate that most people active in this area consider it to have the best mnemonic value for its multidimensional character;
6) the (rather small) disadvantage that it is "difficult" and time consuming to draw parallel lines by hand, will be soon overruled by graphic I/O facilities of modelling and simulation software [12].

As Table Ia shows, the multibond represents the conjugate 1xn-dimensional column matrices (sometimes also called "vectors", which suggested the former name "vectorbond") of efforts and flows $\underline{e}$ and $\underline{f}$ respectively. The product of these matrices (the first has to be transposed; the superscript T indicates the transposed form) represents the total transmitted power $P = \underline{e}^T.\underline{f}$, along the multibond [4,3], with:

$$\underline{e}^T=[e_1,\ldots,e_i,\ldots,e_n]; \qquad \underline{f}^T=[f_1,\ldots,f_i,\ldots,f_n]$$

The positive orientation (not the "direction") of the power P and of the flow variables $\underline{f}$ (and consequently of the variables $\underline{e}$) with respect to the connected ports are defined by the half arrow. The flow matrix is written beneath or to the right of the multibond as much as possible and the effort matrix at the opposite side, but since this practical rule contains ambiguities (Fig. 2a: the "135 degrees position" and the "effort-flow switch" of a "right-bending" multibond [1]), it will be a formal rule that the half arrow always points to the "flow side" of the multibond (Fig. 2b: a rule recently proposed by prof. J.U. Thoma in private communication). Finally it should be stressed that multibonds are not called "vectorbonds" anymore, because the column matrices of efforts and flows are no "real" vectors. In other words: the multibond does not represent a direction in space.

Fig. 2a:

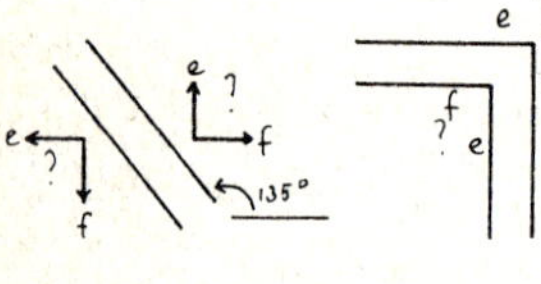

Fig. 2b:

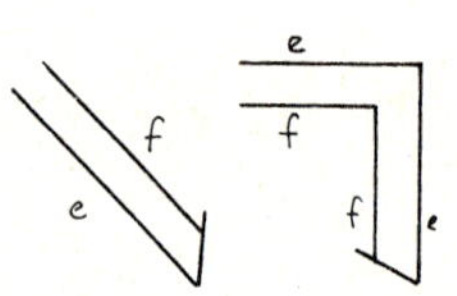

2b) *Multibond activation*

Like single bonds multibonds can be activated, i.e. the effort or the flow variables are suppressed, such that there is no power transmission and the multibond becomes an array of signals: the active multibond, represented by parallel lines with a full arrow, (Table Ia), in correspondence with the representation of signals in multidimensional block diagrams.

3) *Vertices of the graph: Multiport Elements*

3a) *General notation and word bond graphs*

Every multiport element (MP E) is a vertex of the multibond graph and is represented by a mnemonic code of one or more "shaded" alphanumeric characters (Table Ib), in contrast with the conventional characters of basic 1-, 2- and 3-port elements in single bond graphs. As discussed in sect. 2), the ports of an MP E are indicated by the multibond symbol. Apart from the fixed set of MP E's any MP subsystem may be given a shaded mnemonic code in case it is frequently used, although great care should be taken not to introduce all kinds of mnemonic codes at will. Usually, these subsystems are better represented by an ellipse containing a description of the vertex (subsystem) to be indicated, especially in an early phase of the modelling process. This results in a so-called word bond graph (Fig. 3). Of course, multibond graphs containing both word bond graph symbols and shaded mnemonic codes are also possible. In figures, i.e. in the multibond graphs themselves, shaded characters (or word bond graph elements) should always be used. If referred to in a text, the MP character of an element may be indicated, as in the present paper, by placing the characters MP (for MultiPort) in front of the mnemonic code of the element, all in conventional characters, mainly for typesetting convenience. It should be noted that multiport elements are sometimes called "fields" in literature (e.g. [9]).

Table Ib:

(b) Vertices or MULTIPORT ELEMENTS ($\mathbb{E}$)				
energic: $\underline{e}^T.\underline{f} = \frac{d}{dt}E$ (no modulation possible)	$\mathbb{C}$	$\Phi_C(\underline{e},\underline{q})=0$ *capacitor* $\underline{q} = \int\underline{f}dt$	$\mathbb{I}$	$\Phi_I(\underline{p},\underline{f})=0$ *inertance* $\underline{p} = \int\underline{e}dt$
entropic: $\underline{e}^T.\underline{f} = Tf_s^{irr}$ ($=P^{irr}$)	$\mathbb{RS}$	$\Phi_{RS}(\underline{e},\underline{f},T)=0$ *irreversible transducer* $f_s^{irr}=T^{-1}(\underline{e}^T.\underline{f})$	$\mathbb{R}$	$\Phi_R(\underline{e},\underline{f})=0$ *resistor*
boundary: $\underline{e}^T.\underline{f} = P^{ext}$ (flow-discontinuous)	$\underline{S_e}$	$\underline{e} = \underline{e}(t)$ *effort-source (array)*	$\underline{S_f}$	$\underline{f} = \underline{f}(t)$ *flow-source (array)*
junctionstructure: $\underline{e}^T.\underline{f} = 0$ (linear, possibly time-variant)	GJS (ES-type) (non-reciprocal):			
	$\mathbb{GY}$	$\underline{e} = G\underline{f}$, $G = -G^T$ *gyrator* + GJS (NES-type) (extended recipr.):		
	$\underline{SGY}$	$G = J$, J: symplectic; *symplectic gyrator* + WJS (non-mixing):		
	$\mathbb{TF}$	$\underline{e}_1 = T^T\underline{e}_2$, $-\underline{f}_2 = T\underline{f}_1$ *transformer* + SJS (PJS):		
	0	$e_i=e_j$; $\Sigma f_i=0$ *0-junction*		
	1	$f_i=f_j$; $\Sigma e_i=0$ *1-junction*		

3b) *Multiport element classification*

The MP E classification as extensively described in [3] is based on basic properties of a physical system, which are concentrated in the different types of MP E's.

In the next table this classification is represented in short form and also serves as a definition of the basic elements.

MNEMONIC CODES OF BASIC MULTIBOND GRAPH ELEMENTS

code	name	property
C	Capacitor	energic
RS	irreversible transducer (Resistor + Source)	entropic
R	Resistor	entropic
GY	GYrator	non-reciprocal
SGY	Symplectic GYrator	extended reciprocal
TF	TransFormer	non-mixing
WJS	Weighted Junction Structure	non-mixing
Se	effort Source	power-discontinuous

After introduction of the concept of (partial) dualization to eliminate a (S)GY [3,13], the following dual elements have to be defined.

MNEMONIC CODES OF DUAL MULTIBOND GRAPH ELEMENTS

code	name	consists of
I	Inertance	C + SGY
Sf	flow Source	Se + SGY

Fig. 3:

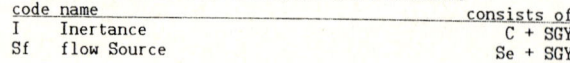

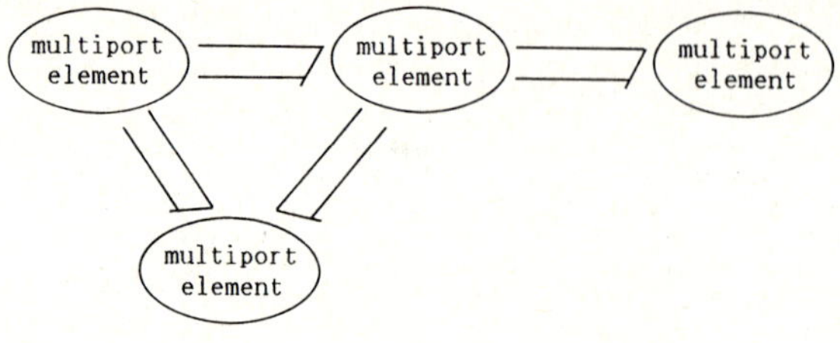

ADDITIONAL MNEMONIC CODES

code	name	special case of
AJS	Analytical Junction Structure	GJS
ES	unit ESsential junction 3-port	GJS
EJS	Eulerian Junction Structure	MP GY
GJS	Generalized Junction Structure (WJS + GY)	
GR	GyRistor (analytical element, due to transformation of a MP C or I over a MP MTF)	AJS
JS	Junction Structure	MP TF/WJS
M..	Modulated .. (e.g. MTF, MGY; cf. Table Ic)	
MP	MultiPort .. (e.g. MP MTF: Modulated MultiPort TransFormer)	
NES	unit Non-ESsential junction 3-port	MP TF/WJS
PJS	Physical Junction Structure	GJS
SJS	Simple Junction Structure (no TF's)	WJS
0	0-junction	SJS
1	1-junction	SJS

Like the junctions, there are some special types of MP TF's viz. "junction arrays" and "direct sums", which are given a special name and notation in the multibond graph language as proposed in [1]. Also the concept of "nesting", to be discussed in the next section, is important and useful. The direct sum is a special MP TF with an exceptional notation: it is not represented by an alphanumeric code, but by a line perpendicular to the connected multibonds (Table Ia). It has no physical meaning and disappears when the multibond graph is expanded into a single bond graph, but can be used everywhere to manipulate the representation of ("paired") ports by multibonds.

4) *Nesting of multibonds and multiport elements*

4a) *Multibond arrays*

Generally the hierarchical process of "nesting" is indicated by underlining of the multibond graph symbol. In case of a "multibond array" consisting of n multibonds which may have different dimensions m_i, the notation is shown in Table Ia for the first level of nesting: a line is added (for each level of nesting) to the "flow-" or "half-arrow"-side of the multibond. Of course, the multibond array is a multibond itself, but it is a handy tool in representing MP systems with clustered ports as shown in [2]. The column matrix $\underline{m}$ containing the dimensions of the multibonds forming the multibond array is written between the original multibond notation and the additional line. The dimension of the corresponding multibond (the total number of ports represented) equals the sum of the n entries of the matrix $\underline{m}$. If all entries of this matrix are equal, $\underline{m}$ can be replaced by the scalar entry m. In that case the total number of ports is nxm.

If the nesting process is proceeded, the result of the next step is an array of multibond arrays (Fig. 4). The matrix $\underline{k}$ contains the dimensions of the multibonds forming the n multibond arrays.

Fig. 4:

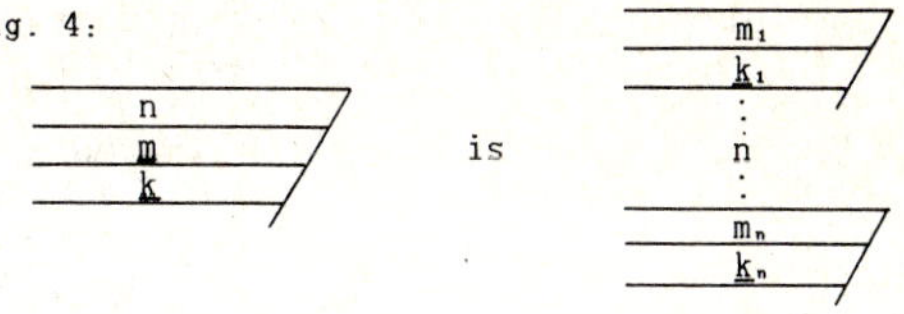

Active multibonds are nested into arrays in a similar way.

4b) *Multiport element arrays*

Multiport element arrays are best introduced by their linear constitutive form, because in that case the constitutive relation can be written in matrix form. In case this matrix is block-diagonal, the coupling of ports takes only place within the clusters represented by the blocks. Consequently, the MP E's may be considered to consist of an array of separate MP E's (Table Ic), indicated by underlining the MP E symbol and the dimension of the array written to the right of this line (optional). However, at the moment we have defined an MP E array this way, it also may be considered a generic form, of which a MP E is a degenerate case in which the number of diagonal blocks (=array dimension) is 1 as in [1]. However, if one does not want to constrain oneself to this first level of nesting, this is not a proper approach and the former generalization should be used, such that the underlining can be repeated.

Apart from this other viewpoint to enable the new concept of nesting and the new name multibond instead of vectorbond, all other definitions proposed in [1] still hold. As shown in that paper, the necessity of MP E arrays becomes especially clear in case a distinction between normal junctions and junction arrays has to be made.

5) *Examples of multibond graph notation*

5a) *Decomposition and Kron's method of tearing*

The decomposition of an MP R can be related to Kron's method of solving "the network problem" by tearing. The following problem may be considered as a simple form of the network problem: An MP R of which the constitutive relations are written in effort (impedance-) causality, is connected to an effort source array (Fig. 5a). The resulting causal conflict, the "network problem", may be solved by inverting the causality of the MP R, which means a matrix inversion in the linear case. Another way to attack this problem is to decompose the MP R first into its canonical form according to the algorithm presented in [2]. A multibond graph representation of the result for a nP R is presented in Fig. 5b. It shows that causal inversion is reduced to the inversion of the n-dimensional 1P array (Fig. 5c), i.e. the inversion of n scalars. The transformer array does not connect 1- and 0-junctions with the same index, such that no causal (algebraic) loops occur, although this representation may suggest that this is the case. The explicit addition of one port to form a (n+1)-port according to the decomposition algorithm of [2] illustrates this even better (Fig. 5d).

Table Ic,d,e:

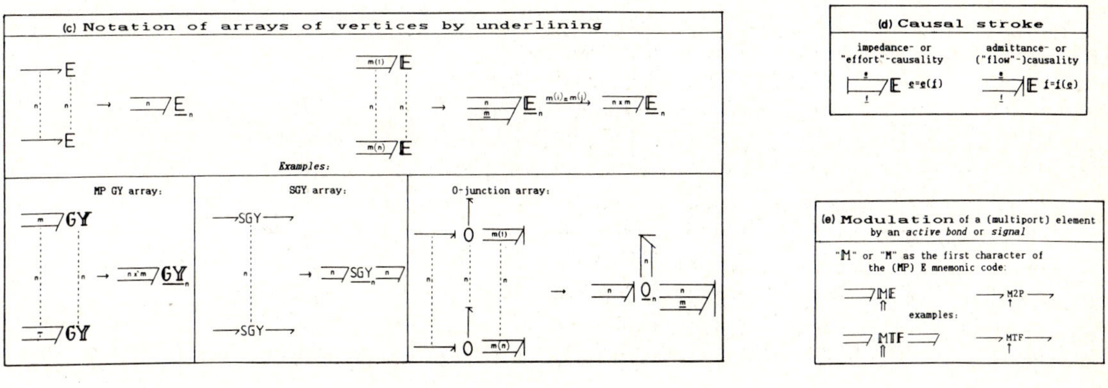

This example shows that (partial) causal inversion of a decomposed or "torn" multiport is much more simple than in the "untorn" case.

Fig. 5a:

Fig. 5b:

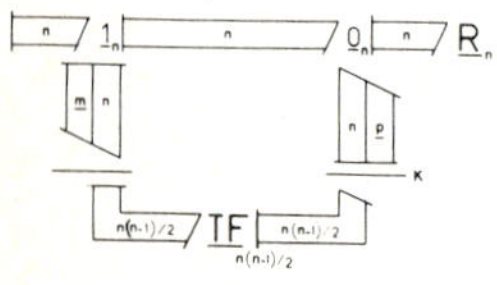

Fig. 5c:

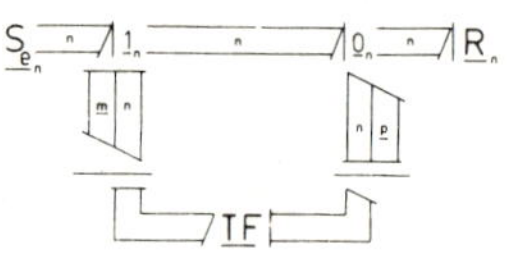

Fig. 5d:

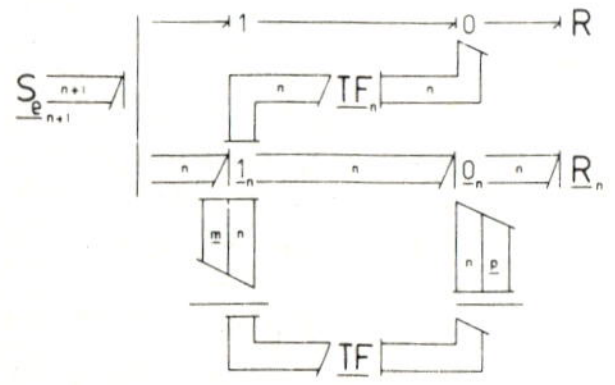

5b) *Multibond Graph representation of the Bernoulli-Euler beam*

The p.d.e. describing the Bernoulli-Euler beam can be solved by separation of variables and the solution has the form $w(x,t) = \Sigma Y_n(x)g_n(t)$, where g(t) satisfies a second order equation and is considered a "mode". The solution thus has the form of a weighted sum of these modes, with position dependent weighting factors Y(x). Margolis [14] presents a bond graph representation of a vehicle-guideway interaction problem, using this so-called "normal modes"- technique (Fig. 6a). This technique can be applied to all kinds of hyperbolic p.d.e.'s describing distributed systems. Lebrun [15], for instance, presents such a modal analysis of pipeline-networks in bond graph notation. Although he uses some of the advantages of the multibond notation, he does not use all features, with the consequence that too much information is hidden in the constitutive equations of the multiport elements. In order to illustrate this, Margolis' single bond modal representation (Fig. 6a), is worked out in the multibond graph language in Fig. 6b, with the addition of the active bonds and taking into account that there are two modes with zero frequency, related to mass and inertia of the beam. Note that the weighting factors of the vehicle interaction point are modulated by its changing horizontal position and that the arrays of 2-port TF's and MTF's are nested to the second level. Also note that the single bond representation is limited to a finite, relatively small, number of modes, whereas the multibond graph has the nice feature that it can even represent the exact solution, by replacing the dimension n by the infinity symbol. The direct sums characterized by K_1 and K_2 are needed to reorder the multibond arrays from the "spatial frame" to the "modal frame". In Fig. 6a m_1 and m_2 are 4 and 2 respectively.

Fig. 6a:

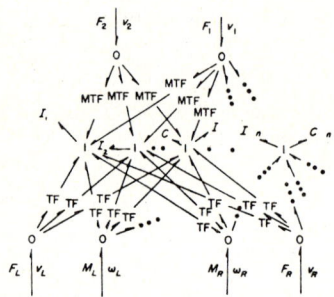

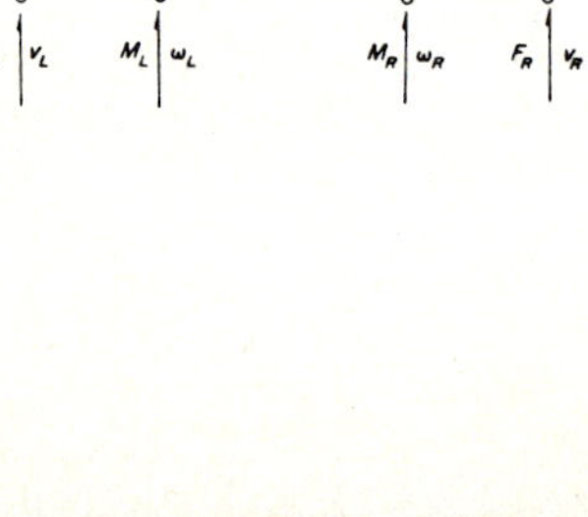

Fig. 6b:

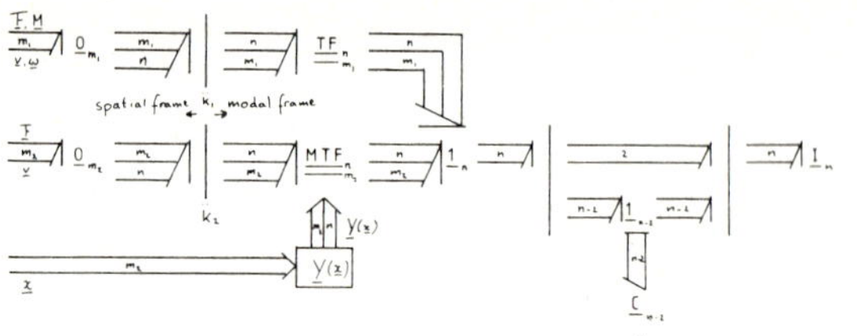

6) *Conclusion*

In principle Table Ia-e contains all information to define the multibond graph language and was, as such, already published in [3]. It appeared to be necessary, however, to pay a little more attention to the choice of the name multibond graphs, to the choice of the notation of multibonds and to the concept of nesting.

It is the authors hope that the reader has recognized advantages of this compact and unambiguous notation for his own applications, because its merits are only fully appreciated in direct use. If so, the language will stay "alive".

[1] Breedveld, P.C., "Proposition for an unambiguous Vector Bond Graph notation", Trans. ASME, J. of Dyn. Syst., Meas. & Control, Vol. 104, No. 3, pp. 267-270, 1982.
[2] Breedveld, P.C., "Decomposition of Multiport Elements in a Revised Multibond Graph Notation", J. Franklin Inst., Vol. 318, No. 4, pp. 253-273, 1984.
[3] Breedveld, P.C., "Multibond Graph Elements in Physical Systems Theory", J. Franklin Inst., Vol. 319, No. 1/2, pp. 1-36, 1985.
[4] Paynter, H.M., "Analysis and Design of Engineering Systems", MIT-press, Cambridge (Mass), 1961.
[5] Karnopp, D.C., and Rosenberg, R.C., "System Dynamics: A Unified Approach", Wiley, New York, 1975.
[6] Rosenberg, R.C., and Karnopp, D.C., "Introduction to Physical System Dynamics", McGraw-Hill, New York, 1983.
[7] Bos, A.M. and Breedveld, P.C., "1985 Update of the Bond Graph Bibliography", J. Franklin Inst., Vol. 319, No. 1/2, pp. 269-286, Jan./Feb. 1985.
[8] Wheeler, H.A., "Ten Years of Multiport Terminology for Networks", Proc. IRE, 47, p. 1786, 1959.
[9] Rosenberg, R.C., and Karnopp, D.C., "A definition of the bond graph language", Trans. ASME, J. Dynamic Syst. Measure. Control, Vol. 94, No. 3, pp. 179-182, 1972.
[10] Bonderson, L.S., "Vector Bond Graphs applied to one-dimensional Distributed Systems", Trans ASME J. Dyn. Syst., Meas., and Control, Vol.97, No. 1, pp. 75-82, 1975.
[11] Luenberger, D.G., "Introduction to Dynamic Systems, Theory, Models, and Applications", John Wiley & Sons, New York, 1979.
[12] Broenink, J.F., Nijen Twilhaar, G.D., "CAMAS, a computer aided modelling, analysis and simulation environment", proceedings fourth "Engsoft" conference on engineering software, june 18-20, London, 1985.
[13] Breedveld, P.C., "Thermodynamic Bond Graphs and the problem of thermal inertance", J. Franklin Inst., Vol. 314, No. 1, pp. 15-40, July 1982.
[14] Margolis, D.L., "Finite mode bond graph representation of vehicle-guideway interaction problems, J. Franklin Inst., Vol. 302, No. 1, pp. 1-17, 1976.
[15] Lebrun, M., "The use of modal analysis concepts in the simulation of pipeline transients", 1985 special issue of J. Franklin Inst. on "Physical structure in modelling", Vol. 319, No. 1/2, pp. 137-156, 1985.

Complex and Distributed Systems: Analysis, Simulation and Control
S.G. Tzafestas and P. Borne (editors)
Elsevier Science Publishers B.V. (North-Holland)

STRUCTURAL PROPERTIES OF SYSTEMS REPRESENTED BY BOND GRAPHS

Nobuhide SUDA and Toshio HATANAKA*

Department of Control Engineering,
Faculty of Engineering Science,
Osaka University,
1-1 Machikaneyama, Toyonaka, Osaka 560, Japan

Abstract In this paper we discuss the structural properties such as the reducibility and the structural controllability/observability of linear systems represented by bond graphs. In order to investigate such properties, the causal connection between the elements is the key concept. For a certain class of bond graphs including the multiport fields, the necessary and sufficient conditions for these structural properties are presented.

Keywords Bond graphs; causal connection; reducibility; structural controllability/observability.

1. INTRODUCTION

Bond graphs are often used as a tool to derive numerically the state equations or other representation of the system dynamics and to simulate the transient behavior of the systems [1]. They are also useful to investigate the structural properties of the systems. System properties are "structural" if they depend only upon the types of elements and the way they are interconnected and not on the numerical values of the parameters.

In this paper we discuss the reducibility and the structural controllability/observability of the systems represented by bond graphs. They are decided from the topology of the bond graphs without actually deriving the state equations:

$$\dot{x} = Ax + Bu \tag{1}$$

$$y = Cx + Du. \tag{2}$$

The reducibility is the existence of a permutation matrix P such that P^TAP, P^TB and CP have some specified forms. More detailed definitions will be given later. A system is structurally controllable if and only if there exists such a set of values of the parameters of the R-, C-, and I-elements involved that the pair (A, B) is controllable. The structural observability is similarly defined.

Using the concept of the causal connection between the elements, one of the authors has investigated such properties [2, 3]. Necessary conditions of the structural controllability/observability have been obtained for a certain class of bond graphs composed of multiport junctions and single-port elements. In the present paper we derive a necessary and sufficient condition. Multiport fields are also taken into consideration.

* Presently at Nippon Kokan K. K., Hiroshima

2. CLASS OF BOND GRAPHS INVESTIGATED

We only consider finite and connected bond graphs without self-loops. For simplicity's sake it is assumed that the bond graphs do not contain transformers (TF) nor gyrators (GY). Generalization to the cases containing these elements is possible. We consider multiport C-, I-, and R- fields as well as corresponding single-port elements.[1] Hereafter they are commonly designated as elements.

A bond is called <u>internal</u> if it connects one junction to another junction, otherwise it is called <u>external</u>. A <u>path</u> is an alternating sequence of junctions and bonds, in which all the junctions and the bonds are distinct and each bond is incident with the two junctions immediately preceding and following it. Closed paths are called <u>bond rings</u>. Note that the bond rings constitute of internal bonds only. A <u>latus</u> is a bond that lies on some bond ring.

Throughout the subsequent analysis, the following assumption is adopted.

<u>Assumption I</u> The number of lati incident with any 1-junction is either zero or two.

In modelling the physical systems by bond graphs, a 0-junction is first assigned to each point of distinct values of the across variable, and the elements bridging those points can in most cases be represented as shown in Fig.1. The resultant bond

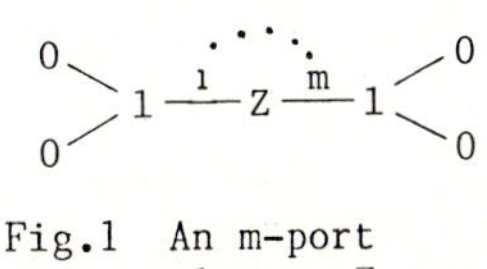

Fig.1 An m-port element Z

graphs, and those obtained by simplifying them, satisfies the assumption above. Hence the class of bond graphs satisfying the Assumption I is considered to be wide enough for the analysis of the physical systems.

It is further assumed that there is no single-port junction (i.e. 0—, 1—) and no pair of similar junctions connected together by a bond (i.e. —0—0—, —1—1—). Since they can be removed easily, there is no loss of generality due to these assumptions.

Here we define several conditions concerning the assignments of the strokes and the arrows.
(C1) At any n-port 0-(1-)junction, 1(n-1) bond(s) has the stroke directed towards the junction.
(C2) There exists no causal loop, that is, no bond ring exists, on which the strokes of all the constituting bonds are assigned in the same direction, in the clockwise-counterclockwise sense.
(C3) Each across variable source is attached to the end without stroke, while each through variable source is attached to the end with stroke.
(C4) Each through variable detector is attached to the end without stroke, while each across variable detector is attached to the end with stroke.
(C5) Each C-element is attached to the end without stroke, while each I-element is attached to the end with stroke.
(C6) Each I-element is attached to the end without stroke, while each C-element is attached to the end with stroke.
(C7) In any bond ring, an even number of bonds have arrows directed clockwise.
(C8) The arrows on the bonds incident with the C-, I- and R-elements are directed towards the elements.

It should be noted that, thanks to the Assumption I, the condition (C7) is readily satisfied by assigning arrows of the two lati incident with 1-junction in the same direction (i.e., ⇀1⇀). Throughout this paper, it is assumed that the conditions (C7) and (C8) are satisfied.

The constitutive laws of the elements are represented as follows

$$C\frac{dv_C}{dt} = i_C \quad \text{for C-elements} \tag{3}$$

$$I\frac{di_I}{dt} = v_I \quad \text{for I-elements} \tag{4}$$

and

$$\begin{bmatrix} G & H \\ K & R \end{bmatrix}\begin{bmatrix} v_R \\ i_G \end{bmatrix} = \begin{bmatrix} i_R \\ v_G \end{bmatrix} \quad \text{for R-elements} \tag{5}$$

where v and i stand for the across and through variables, respectively. For an m-port C-element, v_C and i_C are m-vectors and C is an m x m matrix, and similar remarks apply for I-elements. Suppose that the strokes are directed outwards at m_1 out of m ports of an R-element and inwards at the remaining $m_2 = m - m_1$ ports, v_R and i_R are m_1-vector of the variables at the first group of ports, v_G and i_G are m_2-vector at the second group of ports and G, H, K and R are matrices of appropriate size. The matrices have the following properties [4].

Lemma 1 The matrices C, I, G and R are symmetric, and

$$K = -H^T . \tag{6}$$

For the across (through) variable source element V (J), v (i) is a specified function of time. The across (through) variable detecting element is a through (across) variable source with $i \equiv 0$ ($v \equiv 0$).

If there are two or more elements of the same type, then they are numbered and distinguished with the subscript; C_k, v_{ck} and i_{ck} for the k-th C-element and so forth.

In the subsequent analysis, the numerical values of the coefficient matrices in Eqs.(3) through (5) are not fixed. We will investigate the structural properties which are common to all the coefficient matrices that observe the following conditions:
(C9) The matrices C, I, G and R are nonsingular.
(C10) There are no fixed zero elements in either C, I, G, R, H, K, or C^{-1}, I^{-1}, $\begin{bmatrix} G & H \\ K & R \end{bmatrix}^{-1}$.
(C11) The coefficient matrix of one element can vary independent of all the other coefficient matrices. For example, C_k can vary independent of C_j ($j \neq k$) and I_j, G_j, R_j, H_j and K_j for all j.

From Lemma 1 and the condition (C9), it follows that $\begin{bmatrix} G & H \\ K & R \end{bmatrix}$ is nonsingular and that C, I, G and R are positive definite.

3. CAUSAL CONNECTION

A pair of ports, P_1 and P_2, are causally connected, if the input variable at P_1 is influenced by the output variable of P_2, and vice versa. A path between P_1 and P_2 is a causal path, if the strokes on all the constituting bonds are in the same direction; namely either all of them on the ends near P_1 or all of them on the ends near P_2. An example is shown in Fig.2, the input i_1 at P_1 is influenced by the output i_2 at P_2,

$$i_1 = \cdots + m_{12}\, i_2 + \cdots$$

and also

$$v_2 = \cdots + m_{21}\, v_1 + \cdots$$

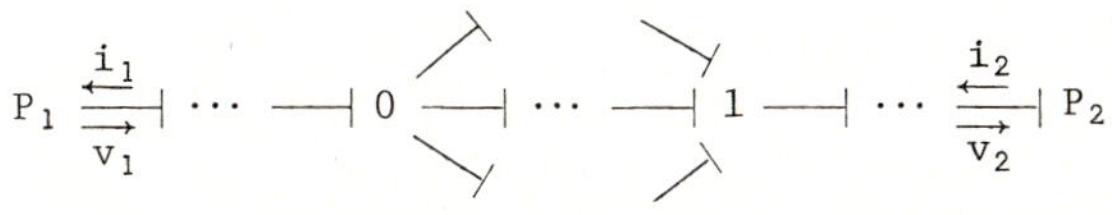

Fig.2 A causal path

The signs m_{12} and m_{21} are determined by

$$\left.\begin{aligned} m_{12} &= (-1)^{n_0} \\ m_{21} &= (-1)^{n_1} \end{aligned}\right\} \qquad (7)$$

where n_0 $[n_1]$ is the number of the arrow reversal at 0-[1-]junctions, that is either $\leftarrow 0 \rightarrow$ or $\rightarrow 0 \leftarrow$ $[\ \leftarrow 1 \rightarrow$ or $\rightarrow 1 \leftarrow\ ]$.

It can be shown that, under Assumption I, there are at most two distinct causal paths between any pair of ports. If there are three or more of them, then at least one of the 1-junctions has more than two lati incident with it. In the case of two distinct causal paths, it is easily noted that the sign, m_{12}, of one path is opposite to that of the other, m_{12}', due to the condition (C7), and hence, they cancel each other. Thus a lemma follows.

Lemma 2 A pair of ports are directly (or structurally) causally connected (i.e., causally connected without intervention of elements) if and only if there is a unique causal path between them.

Let Z be an element other than a source or a detector. If the ports P_1 and P_2 are directly causally connected with j-th and k-th ports of Z, respectively, then P_1 and P_2 are causally connected with intervention of the element Z. Here j and k may or may not be the same. Causal connection with intervention of two or more elements is similarly defined. Note that the outputs of sources and detectors are independent of the inputs and, hence, these elements cannot intervene the causal connection. If all the intervening elements are R-elements, then P_1 and P_2 are statically causally connected.

4. DERIVATION OF STATE EQUATIONS

It is assumed that the strokes on bonds are assigned so that the conditions (C1) through (C5) are satisfied. In view of the condition (C5), all the dynamic (C, I) elements are in integral causality.

Let us take, as an example, an input i_C to a port of a C-element. In order to determine i_C, we trace the causal path to this port. The condition (C1) guarantees that this trace never meets a deadlock at any junction. The condition (C2), on the other hand, guarantees that the trace never ends up in a perpetual circulation on a bond ring. Thus every branch of the trace arrives at one of the elements attached to the ends with stroke, namely, I-elements, R-elements with across variable input, through variable sources, and across variable detectors. Hence i_C is represented as an algebraic sum of the output of these elements. The detectors may be disregarded because they simply supply zero through variable. Repeating this procedure for all the ports of the elements except the sources, we obtain the following expression:

$$\begin{bmatrix} i_C \\ v_I \end{bmatrix} = \begin{bmatrix} 0 & M_{CI} \\ -M_{CI}^T & 0 \end{bmatrix} \begin{bmatrix} v_C \\ i_I \end{bmatrix} + \begin{bmatrix} 0 & M_{CG} \\ M_{IR} & 0 \end{bmatrix} \begin{bmatrix} v_R \\ i_G \end{bmatrix} + \begin{bmatrix} 0 & M_{CJ} \\ M_{IV} & 0 \end{bmatrix} \begin{bmatrix} v_V \\ i_J \end{bmatrix} \qquad (8)$$

$$\begin{bmatrix} i_R \\ v_G \end{bmatrix} = \begin{bmatrix} 0 & -M_{IR}^T \\ -M_{CG}^T & 0 \end{bmatrix} \begin{bmatrix} v_C \\ i_I \end{bmatrix} + \begin{bmatrix} 0 & -M_{GR}^T \\ M_{GR} & 0 \end{bmatrix} \begin{bmatrix} v_R \\ i_G \end{bmatrix} + \begin{bmatrix} 0 & M_{RJ} \\ M_{GV} & 0 \end{bmatrix} \begin{bmatrix} v_V \\ i_I \end{bmatrix} \qquad (9)$$

$$\begin{bmatrix} i_D \\ v_{\tilde{D}} \end{bmatrix} = \begin{bmatrix} 0 & M_{DI} \\ M_{\tilde{D}C} & 0 \end{bmatrix} \begin{bmatrix} v_C \\ i_I \end{bmatrix} + \begin{bmatrix} 0 & M_{DG} \\ M_{\tilde{D}R} & 0 \end{bmatrix} \begin{bmatrix} v_R \\ i_G \end{bmatrix} + \begin{bmatrix} 0 & M_{DJ} \\ M_{\tilde{D}V} & 0 \end{bmatrix} \begin{bmatrix} v_V \\ i_J \end{bmatrix} \qquad (10)$$

Here

$$i_*^T = [i_{*1}^T \cdots i_{*N_*}^T], \qquad v_*^T = [v_{*1}^T \cdots v_{*N_*}^T]$$

where * represents C, I, R or G and N_C, N_I and $N_R = N_G$ are the numbers of the C-, I- and R-elements, respectively. The vectors v_V and i_J are composed of the outputs of the across and the through variable sources, respectively, while i_D and $v_{\tilde{D}}$ are the vectors of inputs to the through and the across variable detectors, respectively. The matrices M_{XY} describe the causal connection among the ports. If there exists a causal path between X and Y the element of M_{XY} is ±1, the sign is determined as in Eq.(7). It is equal to 0 if there is no causal path, or there exist two of them, between the pair of ports.

The constitutive laws of all the elements are collected to form the following expression:

$$\begin{bmatrix} \hat{C} & 0 \\ 0 & \hat{I} \end{bmatrix} \begin{bmatrix} \dot{v}_C \\ \dot{i}_I \end{bmatrix} = \begin{bmatrix} i_C \\ v_I \end{bmatrix} \qquad (11)$$

$$\begin{bmatrix} G & H \\ K & R \end{bmatrix} \begin{bmatrix} v_R \\ i_G \end{bmatrix} = \begin{bmatrix} i_R \\ v_G \end{bmatrix} \qquad (12)$$

where

$$\hat{C} = \mathrm{diag}[C_1, \cdots, C_{N_C}], \quad \hat{I} = \mathrm{diag}[I_1, \cdots, I_{N_I}]$$
$$G = \mathrm{diag}[G_1, \cdots, G_{N_R}], \quad R = \mathrm{diag}[R_1, \cdots, R_{N_R}]$$
$$H = \mathrm{diag}[H_1, \cdots, H_{N_R}], \quad K = \mathrm{diag}[K_1, \cdots, K_{N_R}]$$

Combining Eqs.(8) through (12), we obtain the following:

$$F_1 \dot{x} = M_1 x + M_2 w + M_3 u \qquad (13)$$
$$F_2 w = -M_2^T x + M_4 w + M_5 u \qquad (14)$$
$$y = M_6 x + M_7 w + M_8 u \qquad (15)$$

where

$$x = \begin{bmatrix} v_C \\ i_I \end{bmatrix}, \quad w = \begin{bmatrix} v_R \\ i_G \end{bmatrix}, \quad u = \begin{bmatrix} v_V \\ i_J \end{bmatrix}, \quad y = \begin{bmatrix} i_D \\ v_{\tilde{D}} \end{bmatrix},$$

$$F_1 = \begin{bmatrix} \hat{C} & 0 \\ 0 & \hat{I} \end{bmatrix}, \quad F_2 = \begin{bmatrix} G & H \\ K & R \end{bmatrix}$$

and the matrices, M_k ($k=1,\cdots,8$), are those composed of M_{XY}'s. From Lemma 1 and the positive definiteness of R and G, it follows that $F_2 - M_4$ is nonsingular, and the following equations are readily derived:

$$\left. \begin{array}{l} F_1 \dot{x} = A_0 x + B_0 u \\ y = C_0 x + D_0 u \end{array} \right\} \qquad (16)$$

where

$$A_0 = M_1 - M_2(F_2 - M_4)^{-1} M_2^T, \quad B_0 = M_3 + M_2(F_2 - M_4)^{-1} M_5,$$
$$C_0 = M_6 - M_7(F_2 - M_4)^{-1} M_2^T, \quad D_0 = M_8 + M_7(F_2 - M_4)^{-1} M_5.$$

The state equations (1) and (2) follow immediately from Eq.(16) with the definitions:

$$A = F_1^{-1} A_0, \quad B = F_1^{-1} B_0, \quad C = C_0 \text{ and } D = D_0 \qquad (17)$$

5. REDUCIBILITY

A *fixed zero element* is defined as an element of the coefficient matrices, A, B, C, D, which is always zero for every choice of the numerical values of F_1 and F_2 satisfying (C9), (C10) and (C11) and the condition in Lemma 1.

A system (1) is defined to be **A-reducible** if and only if there exists a permutation matrix P such that $P^T A P$ is block triangular, namely, for some integer p ($1 \le p \le n-1$)

$$P^T A P = \begin{bmatrix} A_{11} & 0 \\ A_{21} & A_{22} \end{bmatrix} \begin{array}{l} \}p \\ \}n-p \end{array} \qquad (18)$$
$$\underbrace{\phantom{A_{11}}}_{p} \underbrace{\phantom{A_{22}}}_{n-p}$$

where 0 is a submatrix of fixed zero elements only.

A system (1) and (2) is defined to be **(A, B)-reducible** [**(A, C)-reducible**] if and only if there exists a permutation matrix P such that, in addition to Eq.(18), the following Eq.(19) [Eq.(20)] holds.

$$P^T B = \begin{bmatrix} 0 \\ \hdashline B_2 \end{bmatrix} \begin{array}{l} \}p \\ \}n-p \end{array} \qquad (19)$$

$$C P = [\underbrace{C_1}_{p} \vdots \underbrace{0}_{n-p}] \qquad (20)$$

As is obvious from Eq.(16), every state variable, x_k, is the output of one of the ports of a dynamic element, that is the across variable at a port of C-element or the through variable at a port of I-element. Let $P[x_k]$ be a port at which x_k is the output variable, and $S[x_k]$ be a set of all the ports of the element to which $P[x_k]$ belongs.

From Eqs.(11) and (16), it follows that

$$\dot{x}_i = \Sigma_k \tilde{f}_{ik} x (\text{input to } P[x_k])$$

where $\tilde{f}_{ik}$ is the (i, k) element of F_1^{-1}, and summation is made over all the k's such that $P[x_k] \in S[x_i]$. Hence the i-th equation in Eq.(1) reads

$$\Sigma_k \tilde{f}_{ik} x (\text{input to } P[x_k]) = \cdots + a_{ij} x_j + \cdots + b_{ip} u_p + \cdots$$

and the following propositions are derived.

Proposition 1 The (i, j) element of A, a_{ij}, is a fixed zero if and only if none of the ports in $S[x_i]$ is statically causally connected with $P[x_j]$.

Proposition 2 The (i, p) element of B, b_{ip}, is a fixed zero if and only if none of the ports in $S[x_i]$ is statically causally connected with the port whose output is u_p.

The following proposition is easier to derive.

Proposition 3 The (q, i) element of C, c_{qi}, is a fixed zero if and only if $P[x_i]$ is not statically causally connected with the port whose input is y_q.

Based upon these propositions, we obtain the results on reducibility. The proof is omitted.

Theorem 1 A system is A-reducible if and only

if there exists a pair x_1 and x_2 such that no port in $S[x_1]$ is causally connected with $P[x_2]$.

Theorem 2 A system is (A, B)-reducible if and only if there exists x_1 such that no port in $S[x_1]$ is causally connected with any source element.

Theorem 3 A system is (A, C)-reducible if and only if there exists x_2 such that $P[x_2]$ is not causally connected with any detector.

The above theorems imply that the necessary and sufficient condition for (A, B)-irreducibility is existence of causal connection from at least one port of each dynamic element to some source and that for (A, C)-irreducibility is existence of causal connection from each port of each dynamic element to some detector.

An example is shown in Fig.3. Two cases are examined.

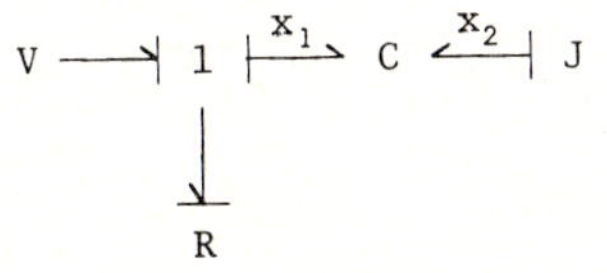

Fig.3 An example

Case A Let V be an across variable source, and J be an across variable detector (i.e. zero through variable source, $i \equiv 0$). It is observed that $P[x_1]$ is causally connected with V via R, while $P[x_2]$ is not causally connected with source V. The port $P[x_2]$ is directly causally connected with the detector J. The port $P[x_1]$ is also causally connected with J via R and C itself. Thus the system is (A, B)-irreducible and (A, C)-irreducible. In fact, the state equations are given by

$$\begin{bmatrix} \dot{x}_1 \\ \dot{x}_2 \end{bmatrix} = \begin{bmatrix} -\tilde{C}_{11}/R & 0 \\ -\tilde{C}_{21}/R & 0 \end{bmatrix} \begin{bmatrix} x_1 \\ x_2 \end{bmatrix} + \begin{bmatrix} \tilde{C}_{11}/R \\ \tilde{C}_{21}/R \end{bmatrix} u$$

$$y = \begin{bmatrix} 0 & 1 \end{bmatrix} \begin{bmatrix} x_1 \\ x_2 \end{bmatrix}$$

where $\tilde{C}_{ij}$ is (i, j) element of $\hat{C}^{-1}$.

Case B Let J be a through variable source, and V be a through variable detector (i.e. zero across variable source, $v \equiv 0$). Both $P[x_1]$ and $P[x_2]$ are causally connected with the source. The port $P[x_1]$ is causally connected with the detector via R, while $P[x_2]$ is not causally connected with the detector V. Hence the system is (A, B)-irreducible but (A, C)-reducible. In fact, the state equations now become

$$\begin{bmatrix} \dot{x}_1 \\ \dot{x}_2 \end{bmatrix} = \begin{bmatrix} -\tilde{C}_{11}/R & 0 \\ -\tilde{C}_{21}/R & 0 \end{bmatrix} \begin{bmatrix} x_1 \\ x_2 \end{bmatrix} + \begin{bmatrix} \tilde{C}_{12} \\ \tilde{C}_{22} \end{bmatrix} u$$

$$y = \begin{bmatrix} 1/R & 0 \end{bmatrix} \begin{bmatrix} x_1 \\ x_2 \end{bmatrix}$$

and clearly the system is (A, C)-reducible.

6. STRUCTURAL CONTROLLABILITY/OBSERVABILITY

The notions of controllability and observability of the system (1) and (2) are well known. The necessary and sufficient conditions are

$$\text{controllability: } \operatorname{rank} \begin{bmatrix} A-sE & B \end{bmatrix} = n \qquad (21)$$

$$\text{observability : } \operatorname{rank} \begin{bmatrix} A-sE \\ C \end{bmatrix} = n \qquad (22)$$

for every complex number s, where n is the dimension of the state vector, x, and E is an identity matrix.

A parameter set (F_1, F_2) is admissible if and only if it satisfies conditions (C9) through (C11) and the condition in Lemma 1. A system is structurally controllable [observable] if and only if there exist an admissible parameter set (F_1, F_2) such that Eq.(21) [Eq.(22)] holds. In other words, a system is structurally uncontrollable [unobservable] if and only if it is uncontrollable [unobservable] for every admissible parameter set (F_1, F_2).

In this section necessary and sufficient conditions of the structural controllability and observability are derived. Towards this end, several lemmata are needed, which are stated below without proof.

Lemma 3 There exists a port of a dynamic element which is not causally connected with any source if and only if either or both of the conditions (3-A) and (3-B) are satisfied:
(3-A) There exists a dynamic element, none of whose ports are causally connected with any source.
(3-B) There exists a port of a dynamic element, whose input is identically zero.

The following lemma is easily derived from well known results in the linear systems theory.

Lemma 4 A single-input system, $\dot{x} = Ax + bu$, is controllable if and only if, for an appropriate selection of output variable, $y = cx + du$, the transfer function,

$$G(s) = d + c\,(sE - A)^{-1}b$$

is irreducible, that is, the degree of its denominator is equal to the dimension of the state vector, x.

The transfer function, G(s), is directly obtained from the bond graphs by applying the Mason's rule. The basic idea is the same as Brown's [5]. However, we will not decompose multiport fields into networks consisting of simpler elements.

The input to a port of an element influences the outputs of all the ports of the elements. The influence of across [through] variable is determined by tracing a causal path in the direction from the end without stroke to the end with stroke [from the end with stroke to the end without stroke]. Such traces will be called causal traces. A Mason loop is defined as a causal trace from the output of a port back to the input of the same port, without tracing the same bond in the same direction more than once. In Fig.4, $i_X = i_1 \longrightarrow i_2 \longrightarrow i_3 = i_Y \longrightarrow v_Y = v_3 \longrightarrow v_2 \longrightarrow v_1 = v_X$ is a Mason loop. Here the bonds 1, 2 and 3 are traced twice, but never in the same direction. It is noted that $i_X = i_1 \longrightarrow i_2 \longrightarrow i_3 = i_Y \longrightarrow v_Y = v_3 \longrightarrow v_4 \longrightarrow v_5 \longrightarrow v_6 \longrightarrow v_1 = v_X$ is another Mason loop. In general, there are four distinct Mason loops, if there are two distinct causal paths involved.

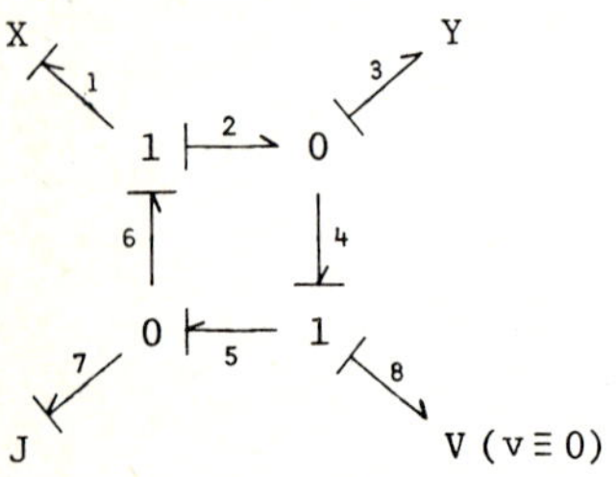

Fig.4 Examples of Mason loops

The transmittance from j-th port to i-th port of an element is the (i, j) element of

$$\frac{1}{s} C^{-1} \text{ for C-elements, } \quad \frac{1}{s} I^{-1} \text{ for I-elements}$$

and

$$\begin{bmatrix} G & H \\ K & R \end{bmatrix}^{-1} \quad \text{for R-elements,}$$

respectively. The loop gain, $L_{1k}(s)$, of the k-th Mason loop is defined by

$L_{1k}(s) = \sigma$ x (product of all the transmittance involved in the k-th Mason loop)

where the sign σ is determined as follows:

$$\sigma = (-1)^{n_0+n_1}$$

n_0 = total number of the arrow reversal at 0-junctions while tracing the through variable.

n_1 = total number of the arrow reversal at 1-junction while tracing the across variable.

For example, in the first Mason loop in Fig.4, $n_0 = 0$, $n_1 = 1$, (arrow reversal in $v_2 \longrightarrow v_1$) hence $\sigma = -1$. In case of the second Mason loop, $n_0 = n_1 = 0$ and hence $\sigma = 1$.

A forward path is defined as a causal trace from a source to a detector, and the path gain, $P_k(s)$, of the k-th forward path is

$P_k(s) = \sigma$ x (product of all the transmittance involved in the k-th forward path)

where σ is obtained in the same way as for $L_{1k}(s)$.

Mason loops, forward paths, or a Mason loop and a forward path touches each other if the same bond is traversed in the same direction in both traces. In the above example, where J is a through variable source and V ($v \equiv 0$) is a through variable detector, $i_J = i_7 \longrightarrow i_5 \longrightarrow i_8 = i_V$ is a forward path. The first Mason loop obviously does not touch this forward path. Although the bond 5 is common to the second Mason loop and the forward path, they do not touch because the bond 5 is traversed in different directions.

Let $L_{rk}(s)$ be the product of the loop gains of k-th set of r Mason loops which do not touch one another. The graph determinant D(s) is defined by

$$D(s) = 1 - \Sigma_j L_{1j}(s) + \Sigma_j L_{2j}(s) - \Sigma_j L_{3j}(s) + \cdots . \quad (23)$$

The definition of $D_k(s)$ is the same as that of D(s) except for an additional condition that the Mason loops should not touch the k-th forward path.

The Mason's rule asserts that G(s) is given by

$$G(s) = \frac{\Sigma_k P_k(s) D_k(s)}{D(s)} . \quad (24)$$

The following lemma is obvious from Eq.(24) and the definition of the relevant terms.

Lemma 5 The degree of the denominator of G(s) is structurally equal to n if and only if both of the following conditions are satisfied,
(5-A) Structurally there is no common factor, other than constants, between D(s) and $\Sigma_k P_k(s) D_k(s)$.
(5-B) Either or both of D(s) and $\Sigma_k P_k(s) D_k(s)$ structurally includes a term, K/s^n, when K is a constant.

The next three lemmata correlate the conditions (5-A) and (5-B) with the structure of bond graphs. It is assumed that the strokes are assigned observing the condition (C1) through (C5).

Lemma 6 The condition (5-A) is satisfied if and only if each dynamic element in Mason loops has at least one port causally connected with the source and also at least one port causally connected with the detector.

A set of nontouching causal paths is <u>maximal</u> if it is not a proper subset of another set of nontouching causal paths.

<u>Lemma 7</u> The $D(s)$ structurally includes a term K/s^n if and only if there exists a maximal set of nontouching causal paths which connects all the ports of dynamic elements and none of the sources or detectors and there is no parallel causal paths satisfying the nontouching condition between any pair of ports.

In the bond graph shown in Fig.5(a), a causal path $C \overset{1}{\longrightarrow\!|} 0 \overset{2}{\longrightarrow\!|} 1 \overset{3}{\longrightarrow\!|} I$ is a maximal set connecting all the dynamic elements (C and I) and none of the sources or detecters (V and J). There is a parallel causal path

$C \overset{1}{\longrightarrow\!|} 0 \overset{4}{\longrightarrow\!|} 1 \overset{5}{\longrightarrow\!|} 0 \overset{6}{\longrightarrow\!|} 1 \overset{3}{\longrightarrow\!|} I$ and thus the condition of Lemma 7 is not satisfied. In fact, $D(s) = 1 + R_1/sI$ and does not include the term $1/s^2$. In Fig.5(b), on the other hand, the above causal path is no longer a maximal set, since it is a proper subset of a set of nontouching causal paths constituting of

$C \overset{1}{\longrightarrow\!|} 0 \overset{2}{\longrightarrow\!|} 1 \overset{3}{\longrightarrow\!|} I$ and $R_1 \overset{7}{\longrightarrow\!|} 1 \overset{5}{\longrightarrow\!|} 0 \overset{8}{\longrightarrow\!|} R_2$, which

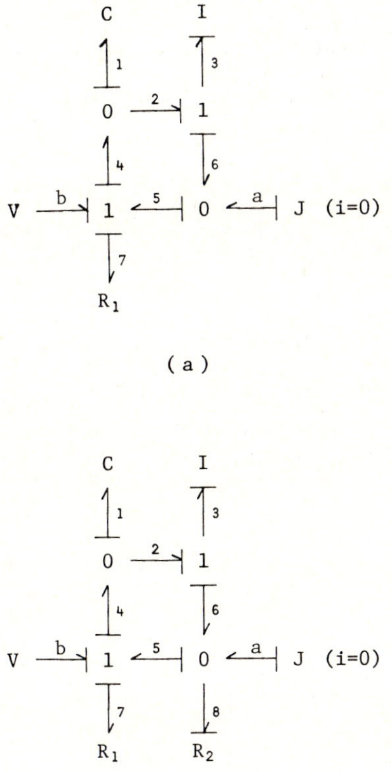

Fig.5 Examples of sets of nontouching causal paths

also connects all the dynamic elements and none of the sources or detecters. The latter set is obviously maximal and there is no parallel causal paths satisfying the nontouching condition. Hence the condition of Lemma 7 is satisfied in this case. In fact, $D(s) = 1 + 1/sCR_2 + R_1/sI + R_1/R_2 + R_1/s^2CIR_2$.

Now we disregard the detector and define the output, y, of the system as the input to the source element, that is the through variable at the across variable source, or the across variable at the through variable source.

<u>Lemma 8</u> For the particular selection of y mentioned above, $\Sigma_k P_k(s) D_k(s)$ includes a term K/s^n if and only if there exists a maximal set of nontouching causal paths which connects the port of source element and all the ports of dynamic elements and there is no parallel causal paths satisfying the nontouching condition between any pair of ports.

Now we state the necessary and sufficient conditions of the structural controllability and observability. They are proved first for the single input/output case with the aid of the Lemmata 5 through 8, and then extended to the multi-input/output case. The details of the proof are omitted. It is assumed that the strokes can be assigned observing the conditions (C1) through (C5).

<u>Theorem 4</u> A system represented by bond graph is structurally controllable if and only if both of the following conditions are satisfied.
(A_c) Under the causality observing the conditions (C1) through (C5), each port of every dynamic element is causally connected with some source.
(B_c) It is possible to assign strokes observing the conditions (C1), (C2), (C4) and (C6).

<u>Theorem 5</u> A system represented by bond graph is structurally observable if and only if both of the following conditions are satisfied.
(A_o) Under the causality observing the conditions (C1) through (C5), each port of every dynamic element is causally connected with some detector.
(B_o) It is possible to assign strokes observing the conditions (C1), (C2), (C3) and (C6).

The example in Fig.3 is examined. In Case A, it is readily noted that (C4) and (C6) cannot be satisfied simultaneonsly. Hence the system is structurally uncontrollable. In fact,

$$\tilde{C}_{21} x_1 + \tilde{C}_{11} x_2 = \text{const} .$$

whatever the input u may be. The conditions (A_o) and (B_o) are satisfied, and the system is structurally observable. In Case B, the system is structurally controllable. It is, however, structurally unobservable, because neither the condition (A_o) nor (B_o) are satisfied.

The examples in Fig.5 are now examined. In the case of Fig.5(a), the C-element is not causally connected with the source V, neither directly nor with the intervention of other elements. Hence the condition (A_c) is not satisfied, and the system is structurally uncontrollable. In fact, the state equation is obtained as follows,

$$\begin{bmatrix} \dot{v}_C \\ \dot{i}_I \end{bmatrix} = \begin{bmatrix} 0 & 0 \\ 0 & -R_1/I \end{bmatrix} \begin{bmatrix} v_C \\ i_I \end{bmatrix} + \begin{bmatrix} 0 \\ 1/I \end{bmatrix} v_V \quad (25)$$

$$v_J = [\ 1 \quad R_1\ ] \begin{bmatrix} v_C \\ i_I \end{bmatrix} - v_V \quad (26)$$

and obviously the state $x_1 = v_C$ is not controllable whatever the numerical values of the parameters may be. Both of the dynamic elements are causally connected with the detector J; the C-element directly and the I-element with the intervention of R_1. Hence the condition (A_o) is satisfied. The condition B_o can be satisfied by inverting the causality of the bonds 1, 2, 3, 5, 7 and a. The system is, therefore, structurally observable. This fact is confirmed by the state equations (25) and (26).

In the case of Fig.5(b), a similar reasoning reveals that all the four conditions are satisfied and, hence, the system is structurally controllable and observable. It is readity confirmed that the system is controllable and observable, for any $C > 0$, $I > 0$, $R_1 > 0$ and $R_2 > 0$.

7. CONCLUSIONS

The reducibility and the structural controllability/observability are investigated for the systems represented by bond graphs. The bond graphs may contain multiport as well as singleport C-, I- and R- elements. It is assumed that their structure satisfies the condition stated in Assumption I.

Using the concept of the causal connection between the elements, necessary and sufficient conditions of the structural properties are derived. It is noted that the conditions for (A, B)- and (A, C)-irreducibility are not dual to each other, because of nondiagonality of the matrix F_1. The condition for structural controllability is dual to that of structural observability. The condition (A_o) is equivalent to (A, C)-irreducibility. The condition (A_c), a dual of (A_o), is stronger than (A, B) -irreducibility. Thus a system may be structurally uncontrollable even if it is (A, B) -irreducible. The Case A of the example in Fig.3 is an example of this situation.

The structural controllability was first discussed by Lin [6]. He divided the entries of matrices A and B into two categories: fixed zeros and free elements. The system is structurally controllable if and only if there exists a set of numerical values of free elements for which Eq.(21) is satisfied. In the present study, the free parameters are F_1 and F_2, and some elements of A and B are constrained. For example, $a_{11} = -\tilde{c}_{11} / R = -b_1$ in Case A. The parameters F_1 and F_2 are assumed to vary freely except for the constraint by (C9), (C10) and (C11), and the condition in Lemma 1. In some physical systems, however, further constraints are imposed. In case of an elastic beam, for example, the coefficient matrix is given by [1, p.235],

$$F_1 = \frac{L^3}{243EI} \begin{bmatrix} 4 & \frac{7}{2} \\ \frac{7}{2} & 4 \end{bmatrix}$$

and hence $f_{11} = f_{22} = 8\ f_{12} / 7$ should additionally be satisfied. In such cases, the conditions given here are necessary but may not be sufficient for the structural controllability/observability.

The results derived here can be generalized to the bond graphs containing transformers and gyrators, with appropriate modifications in definitions of the causal connection and the entries of matrices M_{XY}.

FOOTNOTES

1 In the present paper, the force-current analogy is adopted. The variables are classified into across and through variables. In defining 1- and 0-junctions, C- and I- elements, causal strokes and so on, the effort and the flow in textbooks, e.g. [1], are interpreted as the across and the through variables, respectively.

REFERENCES

[1] Karnopp, D. and Rosenberg, R., System Dynamics: A Unified Approach (John wiley, 1975).

[2] Suda, N. and Kodama, S., J. of Soc. Instrument and Control Engrs, 16, 4 (1977) pp. 348-355.

[3] Suda, N., ibid, 20, 7 (1981) pp. 655-662.

[4] Desoer, C.A. and Kuh, E.S., Basic Circuit Theory (McGraw-Hill, 1969) Chap. 16-17.

[5] Brown, E.T., Trans. ASME Series G, 14, 3 (1972) pp. 253-261.

[6] Lin, C.T., IEEE trans. on Auto. Cont. AC-19, 3 (1974) pp. 201-208.

Complex and Distributed Systems: Analysis, Simulation and Control
S.G. Tzafestas and P. Borne (editors)
Elsevier Science Publishers B.V. (North-Holland)

SIDOPS, A BOND GRAPH BASED MODELLING LANGUAGE

JAN F. BROENINK

Twente University of Technology,
Department of Electrical Engineering,
P.O.Box 217, 7500 AE Enschede, Netherlands,
phone: ..31-53-89-2793, telex: 44200

Here, SIDOPS (Structured Interdisciplinary Description Of Physical Systems), a language for structured description of physical systems is introduced. SIDOPS is based on the bond graph systems theory, and thus has a clear link with physics. Modelling in SIDOPS enables the user to concentrate on the physical modelling process exclusively, leaving computational problems to be solved in a later phase. For a proper support, a SIDOPS interpreter is necessary, resulting in a high level, hands-on interactive tool for modelling of physcial systems. With a brief discussion of an example, the appearance of SIDOPS is shown.

1. INTRODUCTION

Now, the bond graph modelling theory, founded by Paynter [1], and brought to application by Karnopp and Rosenberg [2], is evolved to a worthwhile systems theory after the formulation of a thermodynamically based framework by Breedveld [3]. But, it can only grow to a powerful tool for modelling, analysis and simulation of physical systems, when this theory is supported by sophisticated computer aids.

There are a lot of simulation languages [4], each with its own applications. Among them there are some languages which accept bond graphs as input [5,6,7]. However, these bond graph simulation languages do neither support all possibilities of recent developments in the bond graph systems theory, nor have support for the man machine dialogue on a contemporarely adequate level.

So, there is a need for a software package that can process bond graphs which are fed in graphically (just as one draws it). In order to work on this problem meaningfully, the simulation language SIDOPS (Structured Interdisciplinary Description Of Physical Systems) is developed. With SIDOPS, bond graph models can be described robustly in a way that shows much resemblance with high level general purpose programming languages like Pascal or ALGOL60. In order to create a useful tool for modelling and simulation, a software environment, called CAMAS (Computer Aided Modelling, Analysis And Simulation) is being developed [8]. CAMAS provides for model editing, analyzing and simulation tools with a high degree of user friendliness and interactiveliness.

With the approach of defining a special language like SIDOPS, it is possible to develop the software necessary for processing models, while the problem of processing graphically input bond graphs is not yet solved, and can be studied mutually.

The topic of this paper is the language SIDOPS. In the next section, the design aims of SIDOPS are formulated. After that, a survey of SIDOPS will be given, and an example is presented.

2. DESIGN AIMS

While doing research for the requirements of SIDOPS, two important trends, worth mentioning here, were recognized:

1. The development of simulation languages: The general ones [4], i.e. non bond graph supporting, tend towards interactive, userfriendly software. The official standard, however, only defines a framework for batch oriented simulation software [9,10]. The special bond graph supporting programs, p.e. ENPORT [5], CAMP [6], TUTSIM [7], only accept bond graphs containing single bonds.

2. The evolution of general purpose programming languages: With the development of Pascal-like languages as successors of FORTRAN, there is a trend towards structured and robust programming. The compilers do perform more sophisticated checks, whereas the input must contain extra, redundant information. This enhances the robustness of the software.

These trends, together with the demands of the bond graphers at our institute were the main guide for the design aims of SIDOPS, resulting in two groups of design aims, as presented below:

1. Every part of the bond graph systems theory must be supported: This means power bonds and active bonds, both categories in single, multi or multibond arrays [11], and all the (standard) elements, including the so-called "word bond graph". Other ways of describing systems, which are not according to bond graphs, must not be possible in SIDOPS. This must lead to clear error messages.

2. It should be possible that the physical dimensions of bond variables of every macro physical domain can be entered. In this way SIDOPS can check the consistency of them, which will enhance the robustness of the language.

3. Structured model description must be encouraged. In bond graph terms, this means an emphasis on the use of the word bond graph. The problem to be analyzed, can be split up in parts, each of them represented by a word bond graph. Every part can be investigated separately resulting in a more structured approach. It is stated that such an approach will help to avoid errors. An implication of this is that SIDOPS has to contain possibilities to call (sub)models within model descriptions.

4. In bond graph modelling, generation and manipulation of models can be separated from activities necessary for preparing models for simulation (p.e. causal analysis). This phenomenon must be emphasized in SIDOPS, in order to let the modeller realize that two different stages in the modelling process are concerned.

These four demands can be satisfied because SIDOPS is based on bond graphs. Due to this direct link with physics, SIDOPS has the "semantics of physics" and offers the possibility to perform semantic tests automatically to some extend. It is expected, that this will lead to less error-prone models.

The second group of demands which is originated from trends towards more userfriendly and robust software can be formulated as follows:

1. The use of SIDOPS should be such, that errors will be prevented as much as possible. This implies that (a) every variable must be declared before it is used, and (b) the structure of a bond graph must be entered in an unambigious way: Errors really remain unrecognized when there is a difference between what is written and what is thought to be written.

2. Redundancy, like declarations, must exist in order to enhance robustness, but must not lead to the user typing the same information more than necessary. This means that the software must be as intelligent as to recognize this and to present the information to be confirmed by the user.

3. The error messages must be consistent and concise. Provisions should be made to have different levels of explanation for p.e. the experienced users and novices.

To satisfy the above three demands, an interactive model manipulation environment is necessary (cf. [12]). This means, that there must be a SIDOPS interpreter as a part of the environment CAMAS. In this way, it is possible to serve the user on an adequate level as early as possible (p.e. model checking on a per-line basis).

All of these demands, together with the idea to develop a powerful tool for bond graphers, resulted in the design of the language SIDOPS and the software package CAMAS. The first version of SIDOPS will be presented here, whereas the minimal version of CAMAS is expected to be operational after about a year and a half.

3. DEFINTION OF THE SIDOPS LANGUAGE

SIDOPS (Structured Interdisciplinary Description Of Physical Systems) is a language for the description of simulation models of physical systems. A single model consists of a class description and a parameterset. A class description is a description in SIDOPS of a model whereby all constants are parameters. A parameterset is a list of names of parameters, used in the associated class description whereby numerical values are assigned to each parameter.

An occurence of a model will be created by combining its class description and one of its parametersets. In these parametersets, the values of all parameters are stored. It is possible, of course, that there is more than one occurence of the same class. With the use of CAMAS, the declaration of models (class description), and the realisation of models are treated separately. For instance, a constitutive relation of a bond graph element is described by a class description. Using bond graph elements in a bond graph, means referencing the bond graph elements as submodels, and assigning values to their parameters in the parameterset of the bond graph model.

The description of a class consists of the following parts: heading, interface, subclasses and body.

Heading: The heading serves as an identification of the classdescription, it consists of:

1. A class name and a version number, to serve as the identification. The version number is added in order to distinguish p.e. different versions of a class.

2. An (optional) parameter list, to parmeterize the dimensions of the class description. P.e.,

when this mechanism is used, multiport elements can be described without knowing the dimension at the time of generating the class description.

Interface: The connections with other classes are described here. The class description can only have connections with other classes via the connections declared in this part. So every power bond or signal which connects to the class description must be described here. A power bond consists of three signals: effort, flow and power. The connection of power bonds are called power ports, the connection of signals are called inputs and outputs respectively according to the direction of the signal flow.

In this part, nothing can be said about the causality of the port. This is not necessary, because the modelling process performed here is the physical modelling process, in which power bonds are treated as the physical connections, and not as flow of signals. It is emphasized that only the name of the power bond must be given.

Subclasses: The models that are described by a class may be decomposed into several submodels each of them described by a class, which all must be enumerated in the subslasses part. The syntax is: Name of the class to which the used subclasses belong (together with its version number), followed by the names of all occurences of that used subclass. These names are formal names and are only valid within this description.

Body: The body of a class description can take two possible forms: a topological description or a description by formulas. The former describes a system as a network of interconnected subsystems, while the latter describes a system as relations between signals. These two shapes describe a system on a level of energy exchange by bonds (p.e. a bond graph), or on a signals level respectively (p.e. a constitutive relation of a bond graph element).

Topological description: Here, the connections of the subclasses with each other and the interface are listed; it is called "structure part". It describes the class as a network of subclasses There are a number of possible connections: bonds and signals, each of them having a direction and a dimension. The structure part consists of connection lines, which describe all connections to the subclass that is denoted by the name at the beginning of that line. Every subclass and interface connection (ports, inputs and outputs, referred to as nodes) must appear exactly once in that position, and is called the current node of that line.

In a connection line, all connections to the current node are enumerated, whereby the interface of the current node must completely be connected. Parts of the interfaces of the other connecting nodes must fit on the parts of the current subsystem, in the same order as they appear in the connection line.

In this way, a one-to-one relation exists between a connection line and that part of the bond graph which describes the connections to and from the current node. In a connection line, besides the orientation of a connection, also the causality can be mentioned, by a vertical bar, just as is the case with bond graphs. As the example in figure 1 shows, the syntax of a connection line is quasi graphical.

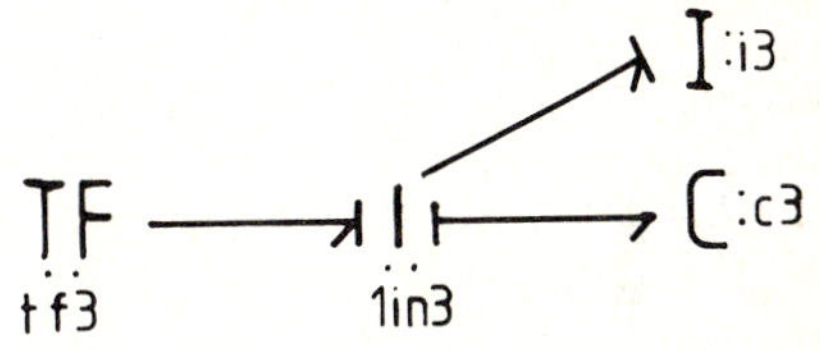

1in3 |<-tf3,|->c3,->|i3

Fig. 1: Example of a connection line and the corresponding part of the bond graph.

While the orientation and causality of the connections of the interface are not known, neither orientation nor causality may be mentioned when a subclass is connected to ports of the interface.

Especially in this part, an interpreter can be very helpful: The nodes (subclasses and interface connections) are already known by the interpreter because of the declarations in the interface and subclasses parts. Connections of the current node which were already typed in are shown for confirmation, so there is no need for the user to type redundant information. At the moment of completion of a connection line, the interpreter checks if the connections described here, are permitted according to the used class descriptions. After the structure part is completed, an overall check is performed of the model consistency on a semantic level.

Description by formulas: Here, classes are described on a signals level: Using assignment statements with some extensions, constitutive relations of a system can be described. A description by formulas consists of three parts: a declaration part, a restrictions part and a dynamic part.

As the description is Pascal-like, for reasons of reliability and robustness, every used variable must be declared. This declaration part is split up into four parts, to declare each of the four groups of variables separately: Parameters and initial conditions, to be constant during a simulation run. Numerical values will be assigned to them via the associated parametersets and initial condition sets.

The other two are constants and local variables. Both are local to their class description. Constants declarations enable the definition of short hands for (may be complex) constant expressions. (These expressions do not change value during a simulation run.) Local variables are variables which serve as intermediates during the simulation run.

In the restrictions part, causality and orientation restrictions of the power ports of the interface can be formulated. Although these restrictions could be extracted from the formulas in the dynamic part, they must be given explicitely. The reason for this is twofold:

1. To let the modeller realize that these restrictions exists, as a consequence of the formulas. This could be compared with the declaration of variables.

2. To get these restrictions easily for the processing software. Otherwise they have to be retrieved using complex interpretation algorithms.

The consistency between restrictions and formulas must be a task of the interpreter.

Dynamic part: This part contains the formulas which describe the relations between signals. It consists of a sequence of statements. In order to get a powerful set of expression capabilities, four types of statements are implemented. Besides the standard assignment statement, there are the directed relation statement, the alternative statement and the repetitive statement.

The assignment statement consists of a (single) acceptor and a (single) expression. The value of the expression is assigned to the acceptor. This assignment is carried out dynamically, i.e. for every time t. Every variable, output and effort or flow of a port must serve exactly once as an acceptor. Whether effort or flow must serve as an acceptor depends on the causality of the port.

The directed relation statement associates an input-output relation with zero or more expressions and one or more acceptors. Permitted directed relations are:

1. Standard functions like arithmetic functions, integration and differentiation with respect to time and time delay.

2. Tables, to describe a function with tuples of numbers.

3. Subsystems of which the interfaces contain no ports.

By use of the alternative statement, it is possible to write both causal forms of a relation as the two alternative clauses, and choose one as result of the causal analysis. This selection must be done before simulation runs are performed. This construction prevents the necessity of formula manipulation at a rather early stage. Of course, the alternative statement can be used for any case at which between two alternatives must be choosen.

Constructs containing repetitive parts can be described very compactly using the repetitive statement. The number of repetitions needs only to be known just at the beginning of simulation runs. P.e. the constitutive relations of junctions can be written in a compact way.

The definition of SIDOPS, as described here in a rather superficial way, is exhaustively described by Welleweerd [13]. It is emphasized that SIDOPS is designed to provide a formalism to perform modelling of physical systems in a structured and robust way on a physical level. The physical backbones are assured by the bond graph systems theory. So, SIDOPS enables the user to concentrate on the physical aspects of modelling exclusively.

In order to simulate a model, it is of course necessary to describe it mathematically. The steps which are necessary to translate a SIDOPS description to a simulatable one are almost automatically generated by CAMAS, and need only a little concern of user. (These steps are actions like submodel expansion, performing causal analysis and sorting of statements. After this, a simulation run can be performed by evaluating the generated set of equations using an adequate integrating routine.)

4. EXAMPLE

In the following, an example is presented to illustrate the rather formal explanation of SIDOPS in the former section. A bond graph representation of vehicle guideway interaction modelled with normal modes is worked out [14]. The number of modes and constraints are arguments of the class description, to have a description that is as general as possible.

The whole problem is divided into three parts: (1) the guideway modes, modelled using Bernouilli-Euler beams, (2) the fixed constraints representing the support locations and (3) the moving constraints, modelling the movement of a vehicle. The bond graph, drawn in multi bond notation [11] shows the grouping in the three submodels, called 'bebeam', 'FixedCon' and 'MovingCon' respectively (fig. 2).

The model descriptions use standard bond graph elements which are avaliable in the public part of the database of CAMAS. Two SIDOPS descriptions are presented (fig. 3 and 4), showing the

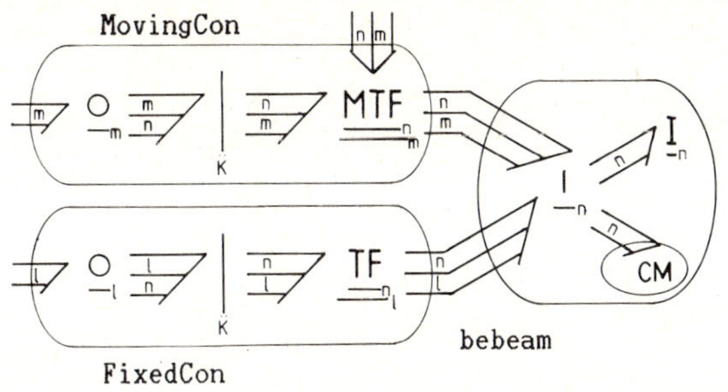

Fig. 2: Bond graph of the vehicle guideway, showing 3 subsystems.

two types of bodies: a topological description (bebeam) and a description by formulas (CM).

Line number 1 of the listing of figure 3 denotes that a class of systems, named 'bebeam' is defined. The arguments n, r, l and m denote the total number of modes, the number of rigid modes and the number of the fixed and the moving constraints respectively. The interface part describes the two physical connections (ports) called 'FixIn' and 'MovIn', which represent the connection to the fixed and moving constraints respectively. The expression between the brackets denote the dimension of the connection. In the subsystems part, the subsystems are declared, e.g. line 5 means that n subsystems named '1in' of the class '1J' (1-junction) will be used. Line 7 shows a declaration of a parameterized subsystem.

The structure part describes the connections between the subclasses, e.g. on line 9 and 10 the connections of the 1-junction-array '1in' are described: the two multi bonds FixIn and MovIn, and the modal elements. The '\' denotes extension of one line. The rest describes the

```
SYNTACTIC ANALYSIS OF A SIDOPS CLASS

1  class bebeam(n,r,l,m) version 2
2     interface
3       ports:FixIn[n,l],MovIn[n,m]
4     subsystems
5       1J             1in[n]
6       I              Mmass[n]
7       CM(n,r)        Mcom
8     structure
9       1in          =n,l=FixIn,=n,m=MovIn\
10                   =n=>Mmass,=n=>Mcom
11      Mmass        <=n=1in
12      Mcom         <=n=1in
13      FixIn        =n,l=1in
14      MovIn        =n,m=1in

0 error(s) detected.
```

Fig. 3: Class description of 'bebeam'

connection of the other subclasses and the interface parts, so every declared node has its connection line. As a consequence, every bond is mentioned twice. This is done deliberately for reasons of robustness. The '<' and '>' denote the orientation of the bond, the '=' denotes multi bond, with the expression between two '='-characters being the dimension of the bond.

```
SYNTACTIC ANALYSIS OF A SIDOPS CLASS

1  class cm(n,r) version 1
2    interface
3      ports:inp[n]
4    parameters
5      real cmp[n-r]
6    initial conditions
7      real e0[n-r]
8    causality restrictions
9      for i=1..n do
10       and causality inp[i]=ecy
11     od
12   dynamic
13     for i=1..r do
14       inp[i].e:=0
15     od
16     for i=r+1..n do
17       inp[i].e:=\
18         int(e0[i-r],inp[i].f)/cm[i-r]
19     od

0 error(s) detected.
```

Fig. 4: Class description of 'CM'

In figure 4, where the description of 'CM' is given, the efforts and flows of the connecting bonds are denoted as 'inp[i].e' and 'inp[i].f' meaning the i-th effort and flow respectively (see p.e. line 17 and 18). Also the use of the repetitive statement is shown (line 9 to 11 and lines 13 to 19). It is emphasized that systems with a description by formulas describe constitutive relations on a signals level, therefore it is necessary to distinguish the effort and flow of a bond.

The Bernouilli-Euler beam is modelled using a especially defined class for the modal compliance ('CM'), as shown in figure 2 where word bond graphs are used. This is necessary to model the rigid modes, when the solution presented here is chosen. It is possible to model Bernouilli-Euler beams with standard bond graph elements. Then, direct sums are needed to cope with the difference in dimension of the modal masses on one hand and the modal compliances on the other (cf. [11]).

With the use of the SIDOPS descriptions of the above classes, the description of the vehicle guideway interaction is completed. With the use of CAMAS, this system can now be analyzed or simulated.

5. CONCLUSION

At this moment, the language definition and the syntax checker are the only parts that are finished. But, despite this, the following remarks can be made:

As SIDOPS is especially defined to describe physical systems which are represented by bond graphs, it can be said that SIDOPS has a clear link with physics, it has the 'semantics of physics'.

To be able to work with SIDOPS convenientely, an interpreter is very important. This interpreter is being developed now, whereby special care is taken to build a software tool that supports the modeller in efficiently developing correct models.

REFERENCES

[1] Paynter, H.M., Analysis and Design of Engineering Systems (MIT-press, Cambridge (Mass), 1961).
[2] Karnopp, D.C., Rosenberg, R.C., Systems Dynamics: A Unified Apprach (Wiley, New York, 1975).
[3] Breedveld, P.C., Physical systems theory in terms of bond graps (Ph.D. Thesis, Twente Univ., Enschede, The Netherlands, 1984).
[4] Cellier, F.E., Simulation software: today and tomorrow, in: Mezencev, K., (ed.) Proc. IMACS int. symp. on simul. in engg. sci. (Nantes, France, 1983).
[5] Rosenberg, R.C., A user's guide to ENPORT-4 (Wiley, New York, 1974).
[6] Granda, J.J., Computer generation of physical system differential equations using bond graphs, J. Franklin Inst., 319, 1/2, pp 243-256, 1985.
[7] Dixhoorn, J.J. van, Simulation of bond graphs on minicomputers, Trans. ASME J. Dyn, Syst. Meas. Control, 99, pp 9-14, 1977.
[8] Broenink, J.F., Nijen Twilhaar, G.D., CAMAS, a computer aided modelling, analysis and simulation environment, in: Adey, R.A., (ed.), Engineering Software IV (Springer Verlag, Berlin, 1985).
[9] Strauss, J.C., et al., The SCi continuous system simulation language, Simulation, 9, 6, pp 281-303, 1967.
[10] Crosbie, R.E., Cellier, F.E., Progress in Simulation language standards, Proc. 10th IMACS world congress, 1:411-412, Montreal, Canada, 1982.
[11] Breedveld, P.C., A defintion of the multibond graph language, this volume.
[12] Astrom, K.J., Computer aided modelling, analysis and design of control systems - A perspective, IEEE Contr. Sys. Mag., 3, 5, pp 4-16, 1983.
[13] Welleweerd, A, Definition of the SIDOPS simulation language and realisation of the interpreter (M.Sc. thesis Twente Univ., Enschede, The Netherlands, 1985).
[14] Margolis, D., Finite mode bond graph representation of vehicle-guideway interaction problems, J. Franklin Inst., 302, 1, pp 1-17, 1976.

Complex and Distributed Systems: Analysis, Simulation and Control
S.G. Tzafestas and P. Borne (editors)
Elsevier Science Publishers B.V. (North-Holland)

MODEL SIMPLIFICATION USING BOND GRAPH TECHNIQUES

D. SINGER

Computer and Automation Institute, Hungarian Academy of Sciences,
P.O.Box 63, 1502-Budapest, Hungary

Bond graph technique can serve in itself or combined with other methods for simplification of large system models. The paper shows a systematic procedure for the reduction of bond graphs in simpler ones allowing to improve the efficiency of large systems analysis considerably. The application of this technique is demonstrated by the analysis of transients in a havaria situation of a water net.

1. INTRODUCTION

The demand for model simplification has its origin in the necessity of augmenting the efficiency of numerical calculations and in the finite capacity of available computers. The computation time rises polynomially with the number of model components, therefore, the need for simplification is of primary necessity for the analysis of large systems.

The basic condition for the model simplification is the insight in the physical nature of the forces and the topology of interactions between the system elements. This insight can be supported by graphical representations as block schemes, Mason-graphs, bond-graphs, etc. The aim of the paper is to show possibilities offered in this respect by bond graph technique.

2. ON BOND GRAPH MODELS AND ON THEIR STRUCTURAL SIMPLIFICATION

Systems fulfilling the conservation principles of physics can be described by conjugate power variables - by generalized potentials and flows - and can be represented pictorially in bond graph form, as pointed out by many authors beginning with the originator of the method, H.M. Paynter [1-4].

Bond graphs consist from a few energetic elements as resistors, capacitors, inductors, potential and flow sources, resp., denoted by the symbols R, C, I, S_E and S_F. The interactions between these elements are transmitted by the junction elements -0- representing a parallel coupling, -1- representing a serial coupling, TR the transducer, and GY the gyrator. All these elements are coupled together by the bonds representing the power flows through them. Each bond is characterized by a pair of potential and flow variables. The intensity of coupling between the elements is characterized by the value of the appropriate potential and flow variables.

In contrary to pure mathematical model simplification methods to be found in the literature, the bond graph method described in the followings is essentially a structural one, reducing the primary bond graph to one with less elements and/or less number of bonds. This is achieved by the following techniques:

- decomposing the bond graph into important and less important zones from the point of view of the goal of application;
- substituting some configuration of elements occurring periodically in the scheme with appropriately defined composite elements;
- omitting the irrelevant elements of the model;
- substituting parts of the scheme by constant source elements.

According to the near likeness relation of the bond graph representation to the system state space equations, from the simplified bond graph the simplified state space equations can be obtained. From the state space equations the eigenvalues of the simplified system needed e.g. for the investigations of system stability can be determined. Mainly by the stability investigation of very large systems, a model simplification by the bond graph technique can be very useful, because the calculation of eigenvalues represent here solving of algebraic equations of very high degree.

Naturally for numerical computation of large system transients, the simplified bond graph can be directly applied without the detour across the state space equation - using the near likeness of bond graphs to program schemes of simulation languages, as CSMP3 or others.

3. DECOMPOSITION OF THE BOND GRAPH INTO ZONES OF INFLUENCE

The decomposition of the bondgraph into zones is the main step of the simplification procedure determining in what extent can be used the remaining three techniques. E.g. in a zone of less significance, the elimination of an element may be allowed. The elimination is, however, not allowed, if the element belongs to the primary zone, having a predominant influence on the behaviour of the model.

The interactions between the input, state and output variables represent paths between the corresponding branches in the bond graph. The strength of the paths determined by the strength of their bonds can be regarded as the measure of influence among the appropriate variables. In general, elements coupled directly or by a few number of bonds, are in a much stronger relation as those coupled through long bond paths. If the goal of application in the simplified model is given and the input and output variables of interest are specified, one can mark out in the original bond graph a zone of primary interest with sufficient reliability. This holds mainly for large systems.

In a similar manner, a secondary zone of the bond graph can be defined containing the paths (and elements) having less influence on the input and output variables of interest. For very large systems, it is eventually appropriate to define a tertiary zone of influence. It is worth-while to note that the secondary and tertiary zones need not represent connected districts of the original bond graph. It must be accentuated once more that this subdivision is goal-oriented and is a function of the specified input and output variables.

The zone concept is useful because it signals in what an extent can be simplificated the several parts of the original bond graph without danger of augmenting the error of the resulting simplified model over the allowed limits.

The subdivision of the bond graph in interaction zones needs in general no considerable precision. The notion "zone" itself is in our case not sharply defined and can be considered as a fuzzy subset of the set of elements and functions creating the bond graph. The boundaries of the zones can be, therefore, drawn by inspection based on the understanding of the physical content of the original system.

The interaction zone boundaries can be marked out by computing the stationary state of the original system without and with the input (disturbance) variables, respectively. Variables affected with the change belongs relatively strongly to the primary zone of influence (of the input variable). The variables affected less belong to the secondary zone and those with weak or no influence, to the tertiary one.

The determination of the stationary state represents the solution of a system of nonlinear algebraic equations which can be accomplished effectively by modern software for many thousand variables [5]. The determination of the influence zones with this method is demonstrated on the example below.

4. COMPOSITE BOND GRAPH ELEMENTS

The structure of large system models, especially those of large energy and hydraulic distribution nets show symmetrically repeating groups of bond graph elements. Such groups can be considered as "composite elements" of the scheme. Such composite elements can be handled similarly as the primary elements R, C, I, TR and GY. For the purpose of the analysis of distribution nets, we introduce four composite bond graph elements:

- RI - element representing the serial coupling of a resistor and an inductor,
- GC - element representing the parallel coupling of a capacitor and a conductor,
- TME element for coupling two pipe flows,
- PITR - element representing the element of a real conductor (that of a pipe, an electric wire, etc.).

The constitutive relations of these elements are as follows:

RI-element: $e = (Ip+R)f$

GC-element: $f = (Cp+G)$

TME-element: $e_i-(Ip+R)f_a+E_g-e_b = 0$

$f_a-f_b-f_{by}-f_o = 0$

$f_b-(Cp+G)E_b = 0$

PITR-element: $e_{i1}-(I_1p+R_1)f_{a1}+E_{g1}-e_b = 0$

$e_{i2}-(I_2p+R_2)f_{a2}+E_{g2}-e_b = 0$

$f_{a1}+f_{a2}-f_{b1}-f_{by2}-f_o = 0$

$f_b-[(C_1+C_2)p+(G_1+G_2)]e_b = 0$

p is the derivation operator $p = d/dt$.

The bond graph representation of these elements are on Figs 1-4.

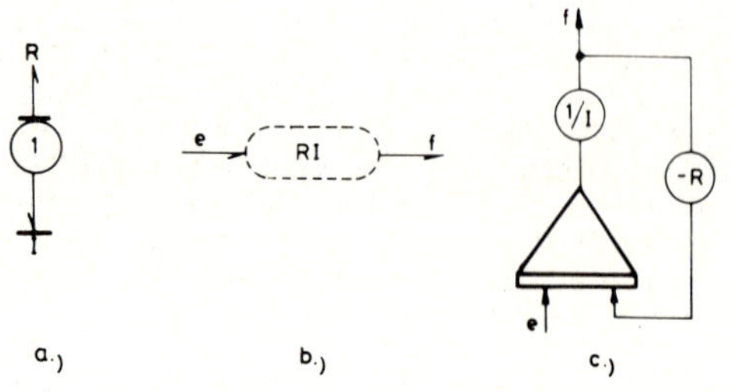

Figure 1

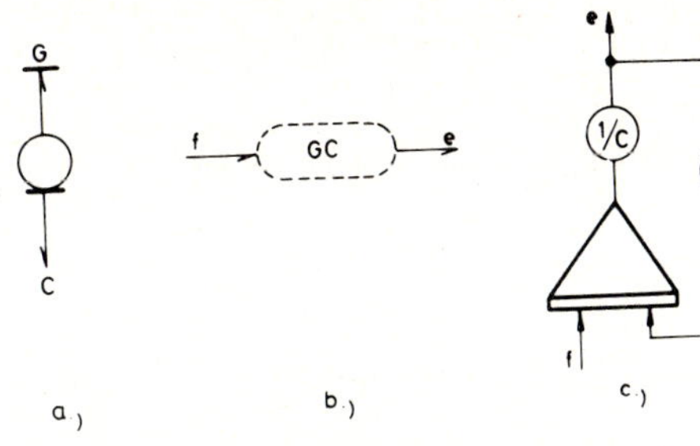

Figure 2

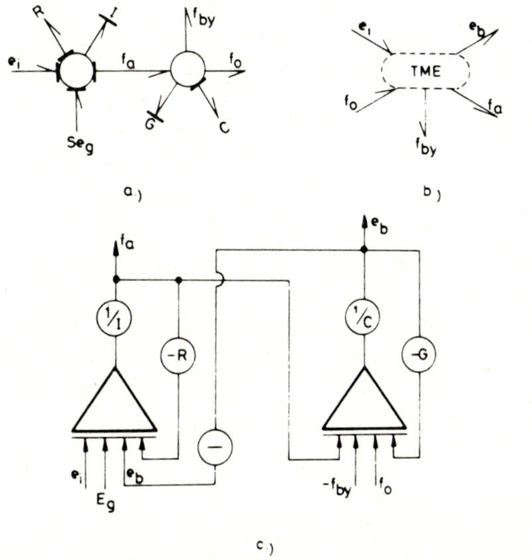

Figure 3

The denotations of the figures are:

R - resistance,
G - conductance,
C - capacitance,
E_g - distributed source potential, e.g., the gravitational force in hydraulic nets,
e_i - input potential,
e_b - potential of the capacitor,
f_a - input flow,
f_o - output flow,
f_{by} - flow of the medium toward the consumers.

To help the writing bond graph programs in a simulation language, the analogue scheme of the new composite elements are also to be seen on the figures.

In the case of distribution nets, the branches can be built up connecting an appropriate number of PITR-elements in series. The number of elements depends of the accuracy needed. The dependence of the model accuracy on the number of the PITR-elements needs deeper studies based

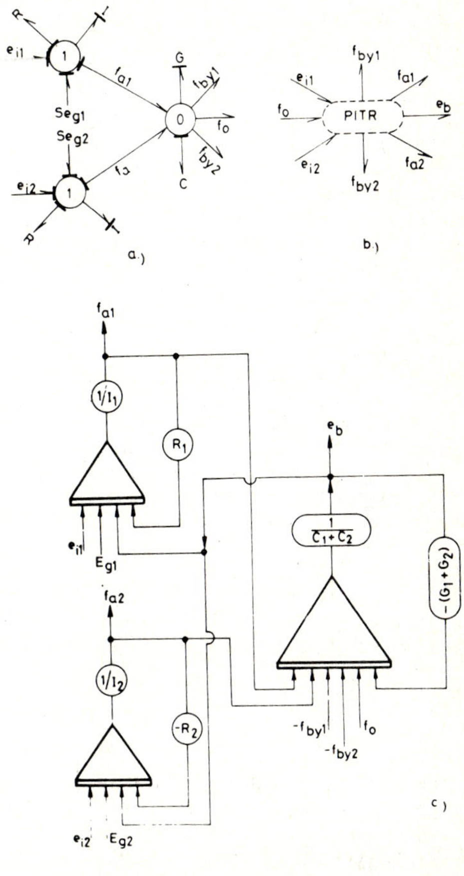

Figure 4

on the theory of 2-ports which cannot be given here. In practice, one can estimate the number needed of PITR-elements by trial and error. In all cases for the branches in the secondary zone a minor number of elements is needed.

By introducing composite elements, the transparency of the bond graph is considerably augmented and its programming for numerical calculation is made more easy.

5. OMITTING ELEMENTS AND SUBSTITUTING PARTS OF BOND GRAPHS BY CONSTANT SOURCES

As mentioned above, bond graphs can be simplified by omitting irrelevant elements, not concerning to the goals to which the simplified model will be used and by replacing parts of the scheme by pressure or flow sources.

From two energetic elements 1 and 2 adjacent to a 0-junction, the one figuring in the denominator can be omitted, if the following inequalities hold:

$C_2/C_1 \leq \varepsilon_z$ for capacitors

$I_1/I_2 \leq \varepsilon_z$ for inductors

$R_1/R_2 \leq \varepsilon_z$ for resistors

C, I and R are here the values of the appropriate constants. The values ε_z are chosen according to the accuracy needed in the simplified model. The index z=1,2,3 denotes the primary, secondary and tertiary zones. In the case of a 1-junction, the conditions for the elimination of the elements are duals of the foregoing inequality relations

$C_1/C_2 \leq \varepsilon_z$ for capacitors

$I_2/I_1 \leq \varepsilon_z$ for inductors

$R_2/R_1 \leq \varepsilon_z$ for resistors.

Parts of the bond graph differing only slightly in potential during the whole transient response time can be considered as a constant potential sources. The conditions for the applicability of this rule can be expressed by the inequality relation

$$(E_1-E_2)/E_1 \leq \varepsilon_z$$

where ε_z is, analogously to the foregoing cases, a constant differing for the primary, secondary and tertiary zones, E_1 and E_2 are the potentials of the most far points of the domain in consideration.

Using these rules for the simplification of bond graphs, one must take care for all directives concerning the causality directions at the energetic element and junction structures. The eventual contradictions with these directives - customary at large systems - can be resolved by using unit gyrators.

6. AN EXAMPLE: THE BONDGRAPH OF A PART OF A LARGE WATER NET

The considerations given in the foregoings will be used for preparing the simplified bond graph model of a part of a town water net (Fig. 5).

The net branches following the direction of the streets and forming a quadratic grid, are of equal length, 100 m. The net is feeded from the vertebral pipeline in the left, having significantly larger capacity as the other pipelines, assuring a constant input pressure of 6 bars for the pipelines. The diameter of these pipelines directed from left to right is 0.3 m, the pipes in the transversal direction have a diameter of 0.2 m. A bond graph model is to be built up to investigate what happens if a pipe breaks in the neighbouring of node No. 14.

The first step is to determine the primary zone of influence of this accident. This is done by determining the stationary pressure profile of the net with a nonlinear network analysis program for the state before and after the break [5].

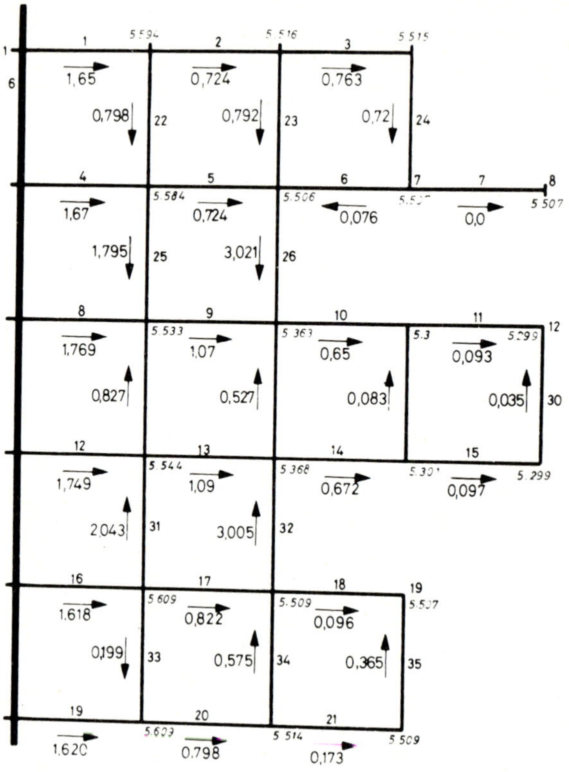

Figure 5

It is assumed that the water consumption in all points of the net is nearly equal. The results of the calculation, the pressure and flow values being given in bars and m^3/s, respectively can be seen on Fig. 5. A second calculation considering a break in the vicinity of node 14, shows that only the area bordered by the branches No. 8, 9, 10, 16, 17, 32 and the vertebral pipeline on the left is significantly affected.

One can consider this area as the primary zone of the bond graph to be constructed for studying the behaviour of the distribution net in the case of the mentioned havaria situation. The remaining part of the net - the secondary zone - can be considered with good approximation as a steady state net, not affected by the disturbance. The bond graph model for investigating the transient behaviour of the net can be, therefore, considerably simplified; see Fig. 6a.

The compressed bond graph model of the simplified system using the concept of composite bond graph elements is seen on Fig. 6b.

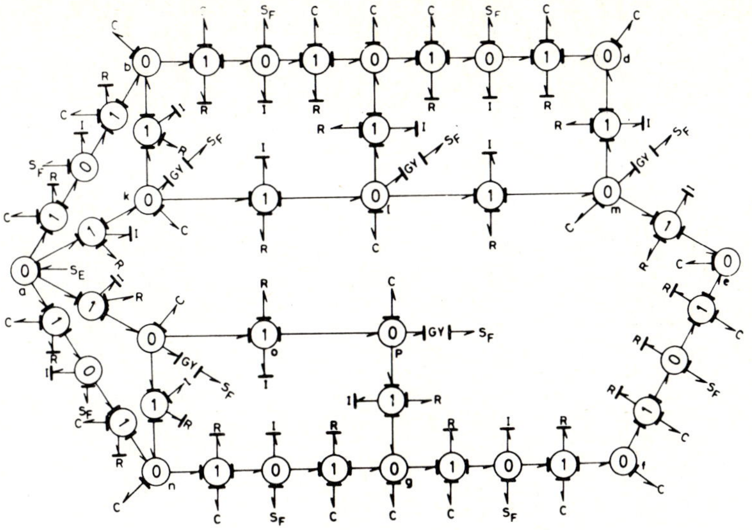

Figure 6a

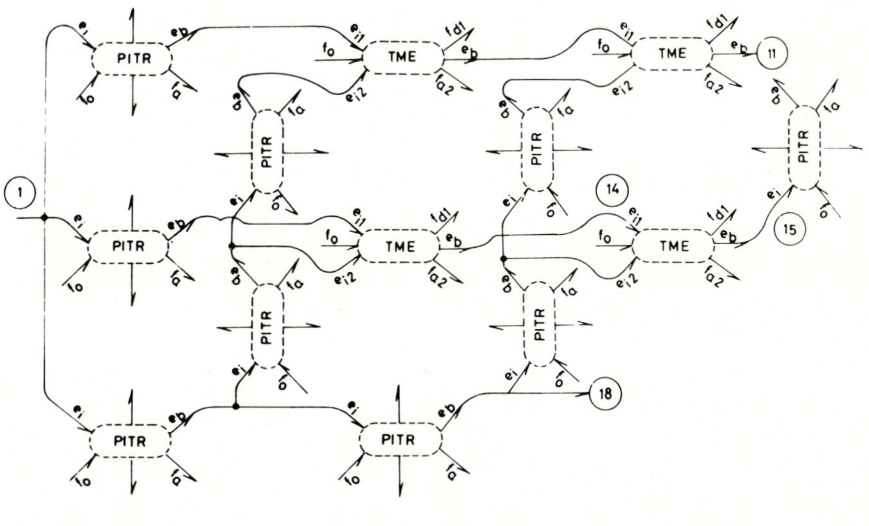

Figure 6b

The model consists of the composite bond graph elements of the type PITR, corresponding to the branches 8, 12, 16, 17, 27, 28, 29 and of elements of type TME corresponding to the branches 9, 10, 13, 14. The vertebral pipeline supplying the net, is assumed to be an ideal potential source (branch 1).

The R, C and I values for the PITR and TME composite bond graph elements have been calculated from material constants and the stationary pressure values given on Fig. 5 determined by the steady state analysis program.

7. CONCLUSIONS

The bond graph method outlined here can be a help for the simplification of very large systems on structural basis. The simplified model can be used directly for numerical calculation or eventually further can be simplified by a non-structural, e.g. eigenvalue method.

REFERENCES

[1] van Dixhorn, J.J., Evans, F.J. (Eds.), Physical Structure in System Theory (Academic Press, New York, 1974).

[2] Karnopp, D.C., Rosenberg, R.C., System Dynamics: A Unified Approach (J. Wiley and Sons, New York, 1974).

[3] Thoma J., Introduction to Bondgraphs and their Applications (Pergamon Press, Oxford, 1975).

[4] Karnopp, D., Pomerantz, M.A., Rosenberg, R. C., van Dixhorn, J.J. (Eds.), Spec. issue of the J. of the Franklin Institute (1979) 173.

[5] Singer, D., Bognår, G. Borossay, J., Elek, J., Int. J. System Sci. (1981) 1261.

Complex and Distributed Systems: Analysis, Simulation and Control
S.G. Tzafestas and P. Borne (editors)
Elsevier Science Publishers B.V. (North-Holland)

A GENERAL LAGRANGIAN BOND GRAPH AND ITS SIMULATION PROGRAM PACKAGE

Shang-Cai Zhang
Dept. of Mechanical Engineering
Zhejiang University
Hangzhou, China

Xing-Zhi Zhang
Institute of Computer Application
Ministry of Nuclear Industry
Beijing, China

SUMMARY

In this present study,it is shown how a GLBG (General Lagrangian Bond Graph) can be developed from a LBG (Lagrangian Bond Graph) to model nonlinear mechanical systems containing particles and rigid bodies interconnected by compliance and dissipation elements in translation and rotation. A Lagrangian bond graph orientated program, ZHLBS, is described in this paper. The function and stucture of ZHLBS are introduced in some detail. The applications of GLBG and ZHLBS are illustrated by several examples. Results of simulation are provided.

NOMENCLATURE

q_G, $\dot{q}_G$ generalized coordinates, the vector of kinematic velocity

F_I ; τ, F_R, F_C, F_S vectors of inertial force, inertial moment, dissipation force, compliance force, generalized force input on $\dot{q}_G$

V_{IL}, Ω linear velocity vector, angular velocity vector of I-field

V_R, $\dot{q}_C$, V_S velocity vector of R-field, C-field, S-field

T_{IL}, T_{IA} linear velocity transfer matrix, angular velocity transfer matrix of I-field

T_R, T_C, T_I, T_S velocity transfer matrices of R-field, C-field, I-field, S-field

C_C rotation matrix

p_I, H_o linear momentum, angular momentum of I-field

F_R', F_C', $F(t)$. vectors of dissipation force, compliance force, generalized force input

I, M, R, C, K inertial matrix, mass matrix, dissipation matrix, compliance matrix, stiffness matrix

1. INTRODUCTION

Lagrange's equation often provides a faster route to equation of motion for mechanical systems. But the numerical simulation procedure of its initial form is quite a bit complex. Bond graph invented by Paynter and further developed by Karnopp, Rosenberg, van Dixhoorn and a number of researches [1-4] is a powerfull tool in the modeling of systems.
A LBG for modeling nonlinear mechanical systems has been developed by combining these two powerfull tools [5]. However, the LBG is only suitable for modeling nonlinear systems composed of particles.
In this present study, a GLBG which appears with more qeneral form than that of LBG is developed on the basis of LBG to model nonlinear mechanical systems containing particles and rigid bodies interconnected by compliance and dissipation elements in translation and rotation, The GLBG can also be used to model electric, hydraulic systems etc. In the GLBG, a general form of expression which contains the generalized coordinates, parameter matrices, velocity transfer matrices, and generalized force input vector is implied.
It is needed to develop a program package which accepts a GLBG model directly for simulating dynamic systems. Such a program, ZHLBS, can be developed from the general expression implied in the GLBG model.

2. GENERAL LAGRANGIAN BOND GRAPH

Generally speaking, the modeling of a system containing rigid bodies is more complicated than that composed of particles, because the motion of a rigid body may involve translation as well as rotation. For this reason, both of linear momentum and angular momentum should be calculated in this case. Since the major difference between systems composed of particles and systems containing rigid bodies is the manner of motion of inertial elements, the GLBG can be constructed by modifying the part of inertial field of the LBG presented in reference [5].
The evaluation of linear momentum for a rigid body is the same as that for a particle. But the evaluation of angular momentum is quite complex. Although the angular momentum can be defined at any moment center, the expression of angular momentum will take the simplest form when the moment center is either a point fixed in inertial space or the center of mass of the body.In general case, the mass center is taken as such a moment center. It is shown in Fig. 1

that a body rotates in the inertial space at angular velocity ω. The body is attached to an intermediate frame $Cxyz$ with origin at its mass center C. Assume that the inertial matrix of

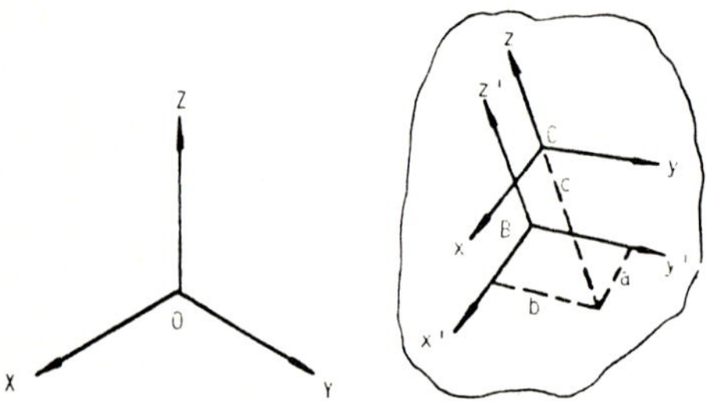

Fig. 1 Rigid body with angular velocity

the body with respect to the intermediate frame $Cxyz$ is I_C; the components of angular velocity and angular momentum about point C along the axices of the frame are ω_x, ω_y, ω_z, and H_x, H_y, H_z, respectively. Let

$$\Omega_c = \begin{pmatrix} \omega_x \\ \omega_y \\ \omega_z \end{pmatrix}, \qquad H_c = \begin{pmatrix} H_x \\ H_y \\ H_z \end{pmatrix} \tag{1}$$

then

$$H_c = I_C \Omega_c \tag{2}$$

and inertial moment about the point C of the rigid body is

$$\tau_c = \begin{pmatrix} \tau_x \\ \tau_y \\ \tau_z \end{pmatrix} = \dot{H}_c \tag{3}$$

If the rigid body is fixed to an intermediate frame $Bx'y'z'$ with an arbitrary point B in the body as the origin, the linear velovity of the point B will not be zero and will not be parallel to the linear velocity of mass center either. Assume that I_B, the inertial matrix of the body with respect to the intermediate frame $Bx'y'z'$; ω'_x, ω'_y, ω'_z, the components of angular velocity of the body along the axises of the frame $Bx'y'z'$; a, b, c, the coordinates of mass center of the body relative to the frame $Bx'y'z'$ are known, by setting up a frame $Cxyz$ with origin at mass center C and parallel the frame $Bx'y'z'$, then the inertial matrix of the body with respect to the frame $Cxyz$, I_C, can be obtained. Now let's set up an inertial frame XYZ shown in Fig. 1.
It is assumed that the components of the angular velocity of the body along the axises of the frame XYZ are ω_x, ω_y, ω_z, and the inertial matrix of the body with respect to the frame XYZ is I_0. Let

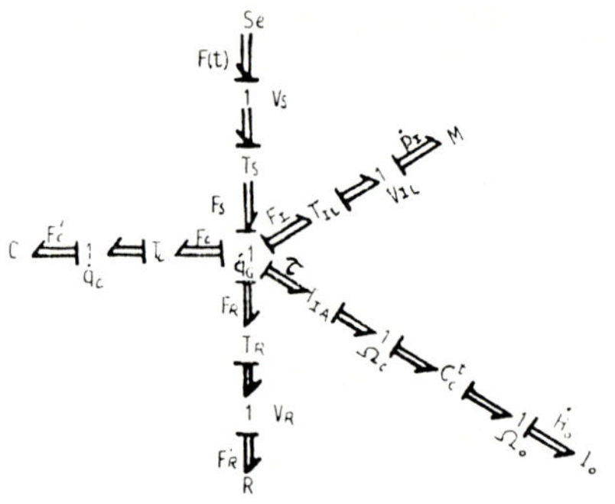

Fig. 2 General Lagrangian bond graph

$$\tau_0 = \begin{pmatrix} \tau_x \\ \tau_Y \\ \tau_z \end{pmatrix}, \qquad \Omega_0 = \begin{pmatrix} \omega_x \\ \omega_Y \\ \omega_z \end{pmatrix}, \qquad H_0 = \begin{pmatrix} H_x \\ H_Y \\ H_z \end{pmatrix} \tag{4}$$

then

$$I_0 = C_c^t I_c C_c \tag{5}$$

$$\Omega_0 = C_c^t \Omega_c \tag{6}$$

$$H_0 - I_0 \Omega_0 \tag{7}$$

$$\tau_0 = \dot{H}_0 \tag{8}$$

where C_C is the rotation matrix, which consists of direction consines of the frame $Cxyz$ relative to the frame XYZ. Let T_{IA} be the angular velocity transfer matrix defined by

$$\Omega_c = T_{IA} \dot{q}_G \tag{9}$$

then the GLBG can be constructed as shown in Fig. 2 by considering equations (4) to (9) and modifying the LBG.

3. VECTOR SYSTEM EQUATION

According to the principle of bond graph and considering equations (4) to (9), we know that the inertial moment on $\dot{q}_G$, τ, is

$$\tau = T_{IA}^t I_c (T_{IA} \ddot{q}_G + \dot{T}_{IA} \dot{q}_G) + T_{IA}^t C_c \dot{C}_c^t I_c T_{IA} \dot{q}_G \tag{10}$$

Finally, the vector system equation describing the dynamics of nonlinear mechanical systems and other systems can be founded on the basis of GLBG shown in Fig. 2 as follows

$$T_I^t I (T_I \ddot{q}_G + \dot{T}_I \dot{q}_G) + T_{IA}^t C_c \dot{C}_c^t I_c T_{IA} \dot{q}_G + T_c^t K f(q_G) = T_s^t \phi_s(t) - T_R^t R T_R \dot{q}_G \tag{11}$$

The form of matrices T_I, I, and C_C depends on

the system. For systems composed of particles, T_{IA} is equal to zero and T_I is the same as T_{IL}. If a system contains more than one (for instance, 2) rigid bodies which rotate about a fixed point, the velocity vector of inertial field of the system takes the form

$$V = [\ \Omega_{C_1} \ \vdots\ \Omega_{C_2}\ ]^t \tag{12}$$

then T_{IA} and T_I are the same and can be expressed as

$$T_I = T_{IA_1} = (T_{IA1} \ \vdots\ T_{IA2})^t \tag{13}$$

The inertial matrix I and I_C are given by

$$I = I_C = \begin{bmatrix} I_{C_1} & 0 \\ 0 & I_{C2} \end{bmatrix} \tag{14}$$

The rotation matrix C_C is determined by

$$C_C = \begin{pmatrix} C_{C_1} & 0 \\ 0 & C_{C_2} \end{pmatrix} \tag{15}$$

In above equations, subscript 1 denotes body 1 and 2 denotes body 2.
If a rigid body rotates and translates in the inertial space, the velocity vector of I-field of the body is

$$V = (V_{IL} \ \vdots\ \Omega_C\ )^t \tag{16}$$

then T is obtained in the following form:

$$T_I = (T_{IL} \ \vdots\ T_{IA}\ )^t \tag{17}$$

The inertial matrix I becomes

$$I = \begin{bmatrix} M & 0 \\ 0 & I_C \end{bmatrix} \tag{18}$$

It is easy to determine the rotation matrix C_C in this case.

4. ZHLBS PROGRAM PACKAGE

The ZHLBS described in this paper is a Lagrangian bond graph orientated simulation program package. ZHLBS accepts GLBG input directly and generates system equations automatically. ZHLBS also accepts a set of first-order differential equations as the input data. If the computer accepts a GLBG model by accepting parameter matrices, velocity transfer matrices and generalized force input, ZHLBS will change the GLBG model into a set of first order differential equations by following steps. Let

$$J = T_I^t I T_I$$
$$L = -(T_I^t IP - T_{IA}^t C_C Z I_C T_{IA} - T_R^t R T_R)$$
$$N = T_S^t F(t) - T_C^t Kf(q_G)$$
$$Y = q_G$$

where

$$P = \dot{T}_I \qquad Z = \dot{C}_C^t$$

then the GLBG model is equivalent to equation

$$J\frac{d^2Y}{dt^2} = L\frac{dY}{dt} + N$$

if

$$|J| \neq 0$$

then

$$\frac{d^2Y}{dt^2} = J^{-1}(L\frac{dY}{dt} + N)$$

Let

$$Y^t = [\ y_1 \quad y_2 \text{------} y_n\ ]$$

$$\frac{dY^t}{dt} = [\ y_{n+1} \quad y_{n+2} \text{-----} y_{n+n}\ ]$$

we get a set of first-order differential equations as follows:

$$\frac{dy_1}{dt} = y_{n+1}$$

$$\frac{dy_2}{dt} = y_{n+2}$$

......

$$\frac{dy_n}{dt} = y_{n+n}$$

$$\frac{dy_{n+1}}{dt} = \sum_{j=1}^{n} J_{1j}^{-1}(\sum_{i=1}^{n} L_{ji}\, y_{n+i} + N_j)$$

$$\frac{dy_{n+2}}{dt} = \sum_{j=1}^{n} J_{2j}^{-1}(\sum_{i=1}^{n} L_{ji}\, y_{n+i} + N_j)$$

......

$$\frac{dy_{n+n}}{dt} = \sum_{j=1}^{n} J_{nj}^{-1}(\sum_{i=1}^{n} L_{ji}\, y_{n+i} + N_j)$$

ZHLBS works in an interactive mode. The syntax error checking can be carried out for the input model. If the error exists, the relevant information will be displayed. Commands can be used to correct errors, modify or delete the input model. If a correct model has been confirmed, it will be stored automatically.
The numerical calculation of relevant expressions are accomplished by the interpreter. For the sake of saving memorial spaces and giving users convenience, the memorial spaces are arranged in an indefinite length access way for the terms of elements in the mathematical model, and zero element terms will not occupy the memory. The source file of ZHLBS written in FORTRAN-77 language possesses more than 2500 statements, 20 subroutine modules. The object file compiled by FORTRAN-77 compiler generates an executable image file, ZHLBS.TSK builder, which can be run on PDP-11/23 minicomputers with RSX-11M operating system. Four algorithms,

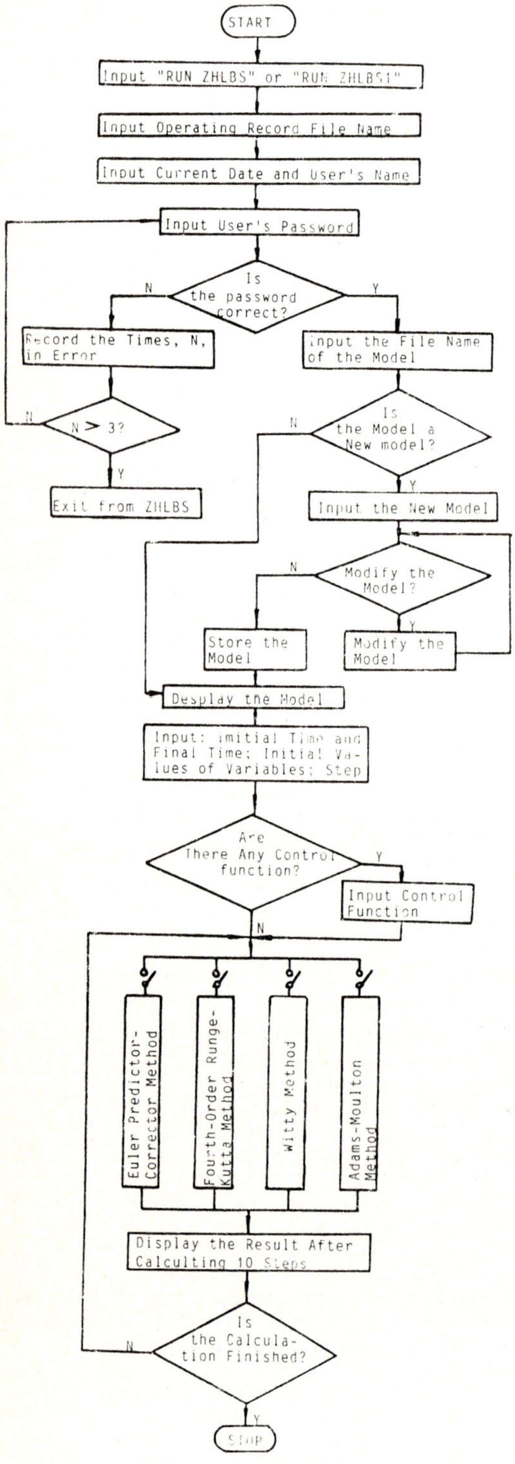

Fig. 3 Simulation flowchart of ZHLBS

such as Euler predictor-corrector method, Witty method, Fourth-order Runge-Kutta method, and Adams-Moulton method [6], can be chosen for a given system. Solutions are given in time domain. The response data can be displayed in both tabular and graphical form. Six curves can be drawn in a picture in a same or different scale. The simulation flowchart of ZHLBS is shown in Fig. 3. ZHLBS can easily be implemented on other computers with a FORTRAN-77 compiler, such as APPLE-II and IBM-pc computer.

5. MODELING AND SIMULATION EXAMPLES

In order to demonstrate the applications of GLBG and ZHLBS, several examples including mechanical [7], hydraulic [8], and electromechanical systems are illustrated.
As the first example, consider the electromechanical configuration of the capacitor microphone shown in Fig. 4. Sound waves pass through the mouthpiece exerting a force F(t) on the movable plate. The voltage E appears in the

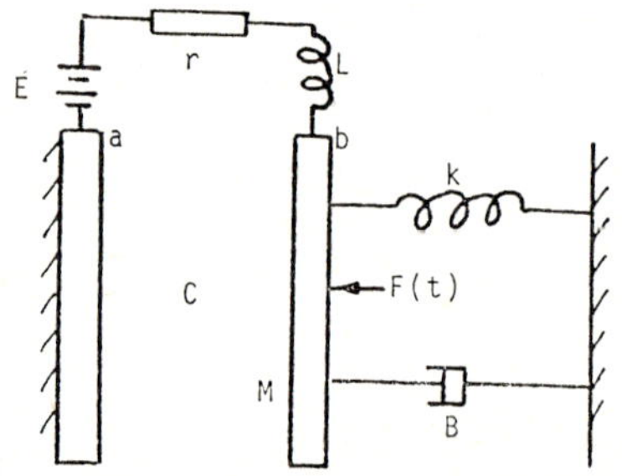

Fig. 4 Capacitor microphone

equilibrium state, causing a charge q_o. The charge q_o causes a force of attraction and stretches the spring an amount denoted by x_1. Let x_o denote the equilibrium distance between the plates; C denote the equilibrium value of capacitance C. Motion from the equilibrium point is denoted by X. The change in charge on the plates is denoted by q. Assume the plates are sufficiently close together, then

$$C = \varepsilon A/(x_o - x), \qquad C_o = \varepsilon A/x$$

where A is the plate area and is the dielectric constant for air. For this system, we have

$$q_G = [q \quad x]^t, \qquad f(q_G) = [(q_o + q)\ (x_1 + x)]^t$$

$$F(t) = [E \quad f(t)]^t$$

$$T_C = \begin{bmatrix} 1 & 0 \\ 0 & 1 \end{bmatrix} \quad T_I = \begin{bmatrix} 1 & 0 \\ 0 & 1 \end{bmatrix} \quad T_R = \begin{bmatrix} 1 & 0 \\ 0 & 1 \end{bmatrix}$$

$$T_S = \begin{bmatrix} 1 & 0 \\ 0 & 1 \end{bmatrix} \quad I = \begin{bmatrix} L & 0 \\ 0 & M \end{bmatrix} \quad R = \begin{bmatrix} r & 0 \\ 0 & B \end{bmatrix}$$

$$K = \begin{bmatrix} 1/C & 0 \\ -(q+q_0)/2\varepsilon A & k \end{bmatrix} \qquad C_C = 0$$

The system equations

$$L\ddot{q} + r\dot{q} + (x_0 - x)(q_0 + q)/\varepsilon A = E$$

$$M\ddot{x} + B\dot{x} - (q_0 + q)^2/2\varepsilon A - k(x_1 + x) = f(t)$$

can be generated automatically and system can be simulated by entering above eata into computer using ZHLBS.

The flyball governor shown in Fig. 5 is the second example. For simplicity sake, it is assumed that the only significant inertial elements are two flyballs with mass m, respectively. Due to the symmetry of the system, two flyballs can be regarded as one particle with

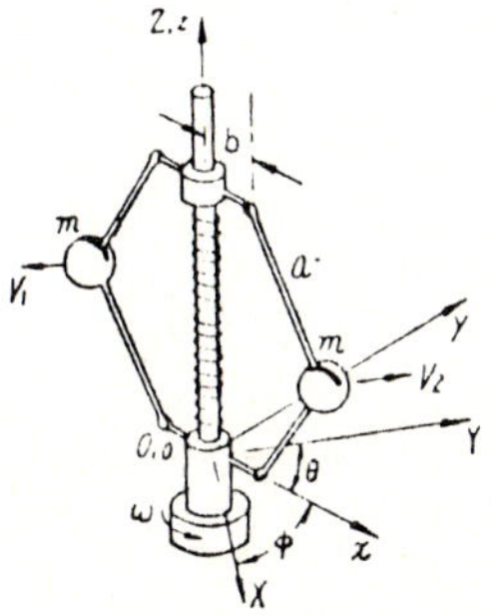

Fig. 5 Flyball governor

mass 2m. The free length of the spring is 2a. The spring constant is k. The viscous friction coefficient cooresponding to the angular displacement θ is B. We set an intermediate frame XYZ and an inertial frame XYZ as indicated and choose the angle θ of the linkage and angle ϕ, the angular displacement of the frame XYZ with respect to the inertial frame XYZ, as the generalized coordinates. Assume a=10, b=5, m=1, k=200, B=3000, we have the key vectors and matrices:

$$q_G = [\theta \quad \phi]^t = (y_1 \quad y_2)^t \qquad q_c = 20(1-\sin y_1)$$

$$q_I = (X \quad Y \quad Z)^t \qquad C_C = 0$$

$$T_C = [-20\cos y_1 \quad 0], \quad T_R = [1 \quad 0] \quad T_S = [-10\cos y_1 \quad 0]$$

$$K = 200 \qquad R = 3000 \qquad F(t) = 1960$$

$$M = \begin{bmatrix} 2 & 0 & 0 \\ 0 & 2 & 0 \\ 0 & 0 & 2 \end{bmatrix}$$

$$T_I = \begin{bmatrix} -10\sin y_1 \cos y_2 & -(5+10\cos y_1)\sin y_2 \\ -10\sin y_1 \sin y_2 & (5+10\cos y_1)\cos y_2 \\ 10\cos y_1 & 0 \end{bmatrix}$$

Let

$$\dot{y}_1 = y_3 \qquad \dot{y}_2 = y_4$$

$$y_4 = \begin{cases} 2t & 0 \leq t < 0.5 \\ 1.0 & 0.5 < t \leq 1.0 \end{cases}$$

and initial condition

$$y_{10} = 0.8557, \qquad y_{20} = y_{30} = y_{40} = 0$$

then the system can be simulated by ZHLBS after inputing above data into computer. Result of simulation is given in tabular form shown in Fig. 6.

TIME	Y1	Y2	Y3
0.00000E+00	0.85570E+00	0.00000E+00	0.00000E+00
0.50000E-01	0.85569E+00	0.25000E-02	-0.46427E-03
0.10000E+00	0.85566E+00	0.10000E-01	-0.12675E-02
0.15000E+00	0.85557E+00	0.22500E-01	-0.25975E-02
0.20000E+00	0.85541E+00	0.40000E-01	-0.43668E-02
0.25000E+00	0.85515E+00	0.62500E-01	-0.63996E-02
0.30000E+00	0.85479E+00	0.90000E-01	-0.85375E-02
0.35000E+00	0.85431E+00	0.12250E+00	-0.10679E-01
0.40000E+00	0.85374E+00	0.16000E+00	-0.12778E-01
0.45000E+00	0.85305E+00	0.20250E+00	-0.14822E-01
0.50000E+00	0.85227E+00	0.25000E+00	-0.16822E-01
0.55000E+00	0.85138E+00	0.30000E+00	-0.14210E-01
0.60000E+00	0.85078E+00	0.35000E+00	-0.85681E-02
0.65000E+00	0.85048E+00	0.40000E+00	-0.34719E-02
0.70000E+00	0.85040E+00	0.45000E+00	-0.24008E-03
0.75000E+00	0.85044E+00	0.50000E+00	0.11770E-02
0.80000E+00	0.85052E+00	0.55000E+00	0.13869E-02
0.85000E+00	0.85059E+00	0.60000E+00	0.10275E-02
0.90000E+00	0.85063E+00	0.65000E+00	0.54769E-03
0.95000E+00	0.85065E+00	0.70000E+00	0.17241E-03
0.10000E+01	0.85065E+00	0.75000E+00	-0.38935E-04

Fig. 6 Simulation result of flyball system

The last example is the hydraulic system shown in Fig. 7. The system equations can easily be derived from the bond graph presented in reference [9] as follows:

$$dy_1/dt = 5.223(45-4y_1)^{0.5} - 0.5y_2$$

$$dy_2/dt = 200y_1 - 5y_2 - 1300$$

where y is the volume increment of oil in the upstream ram chamber due to compressibility. y is the momentum of moving unit. The initial condition is

$$y_1 = 3 \qquad y_2 = 60$$

when t=1.

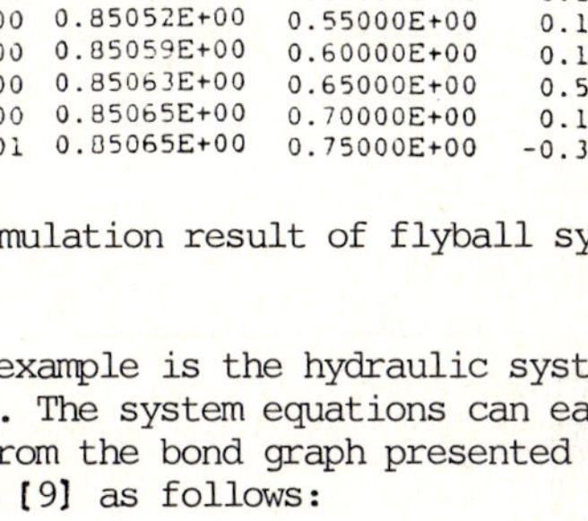

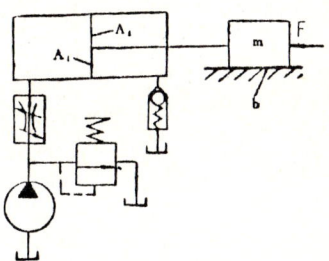

Fig. 7 Hydraulic system

The simulation result we got using ZHLBS is given in both tabular and graphical form (Fig. 8)

TIME	Y1	Y2
0.10000E+01	0.30000E+01	0.60000E+02
0.10500E+01	0.36251E+01	0.16251E+02
0.11000E+01	0.50204E+01	-0.72797E+01
0.11500E+01	0.65852E+01	-0.11206E+02
0.12000E+01	0.78549E+01	-0.96843E+00
0.12500E+01	0.85932E+01	0.16032E+02
0.13000E+01	0.87944E+01	0.33093E+02
0.13500E+01	0.86029E+01	0.45885E+02
0.14000E+01	0.82092E+01	0.52761E+02
0.14500E+01	0.77818E+01	0.54134E+02
0.15000E+01	0.74360E+01	0.51604E+02
0.15500E+01	0.72279E+01	0.47156E+02
0.16000E+01	0.71613E+01	0.42552E+02
0.16500E+01	0.72044E+01	0.38992E+02
0.17000E+01	0.73090E+01	0.37004E+02
0.17500E+01	0.74278E+01	0.36547E+02
0.18000E+01	0.75261E+01	0.37208E+02
0.18500E+01	0.75866E+01	0.38436E+02
0.19000E+01	0.76070E+01	0.39730E+02
0.19500E+01	0.75961E+01	0.40743E+02
0.20000E+01	0.75676E+01	0.41320E+02

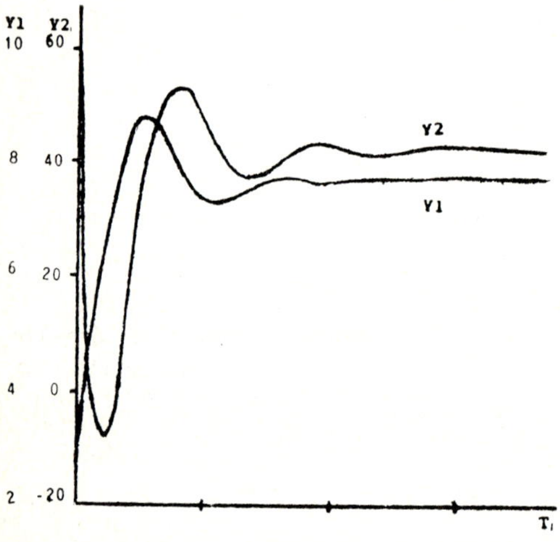

Fig. 8 Simulation result of hydraulic system

6. CONCLUSION

A GLBG model which can be used to model nonlinear mechanical systems containing particles and rigid bodies, electrical, electromechanical, hydraulic systems etc. is developed by modifying the part of the I-field of LBG. In GLBG, a general form of vector system equation is implied. Once the generalized coordinates, parameter matrices, velocity transfer matrices, rotation matrix, and generalized force input are determined, the GLBG for the system studied is constructed.

The simulation can be carried out using the ZHLBS which accepts GLBG input directly and generates system equations automatically. The simulation results can be displayed in both tabular and graphical form.

ACKNOWLEDGEMENT

The financial support of the Science Fund of the Chinese Education Ministry is gratefully acknowledged.

REFERENCES

1. Paynter, H. M., Analysis and Design of Engineering Systems, MIT Press, Cambridge, Mass., 1961.
2. Gebben, V., Bond Graph Bibliography, Journal of Franklin Inst., Vol. 303, No. 3, 1979.
3. Karnopp, D. C. and Rosenberg, R. C., System Dynamics:A Unified Approach, Wiley, New York, 1975.
4. Rosenberg, R. C., and karnopp, D. C., Introduction to Physical System Dynamics, McGraw-Hill Book Co., 1983.
5. Rosenberg, R. C., and Zhang, S. C., Modeling of Nonlinear Mechanics Problems Using Lagrangian Bond Graph, Michigan State University, 1982.
6. Gerald, Curtis F., Applied Numerical Analysis, Addison-Wesley Publishing Company, 1978.
7. Crandall, Stephen H., Dynamics of mechanical and Electromechanical Systems, McGraw-Hill Book Co.. 1972.
8. Eveleigh, Virgil W., Introduction to Control Systems Design McGraw-Hill Book C ., 1972.
9. Rosenberg, R. C., and Zhang, S. C., Simulation of Hydrostatic Speed-Control System using Bond Graph, Computer in Engineering, ASME, Voluime One, 1983.

Complex and Distributed Systems: Analysis, Simulation and Control
S.G. Tzafestas and P. Borne (editors)
Elsevier Science Publishers B.V. (North-Holland)

IMPLICIT SOLUTIONS OF EQUATIONS DERIVED FROM MECHANICAL BOND GRAPHS

A.M. BOS

Twente University of Technology,
Department of Electrical Engineering,
P.O. Box 217, 7500 AE Enschede, The Netherlands.

In this paper it is shown that equations of motion of systems of rigid bodies can be derived from a bond graph. Due to constraints in rigid body systems, mechanical bond graphs will show several causal conflicts, which lead to implicit differential equations. It is shown that the equations consist of an algebraic part and a differential part. It is proposed to solve the equations by embedding these in an integration algorithm and solve both system equations and integration equations simultaneously. Moreover it is shown that an explicit differential equation can be derived from the implicit equations as well.

0. INTRODUCTION

Generating explicit differential equations of a system of rigid bodies is difficult, independent of the method used. Some programs for deriving equations exists, e.g. MEZAVERDE by Wolz [1], and NEWEUL by Schiehlen [2]. However, these programs are only capable of generating the differential equations of mechanical systems, due to the d'Alembert-like solution techniques on which these are based. Mechanical systems often include other physical domains as well (eg. control loops and actuators). In the bond graph representation other physical domains can be included and the solution technique can be chosen independent of the bond graph model. However no programs exist to derive equations of motion from bond graphs of systems of rigid bodies, because these bond graphs contain several causal conflicts and many state dependent transformers.

In this paper it will be shown that equations of motion can easily be derived from a bond graph. The equations will appear to be partly explicit, partly implicit. It will be shown that these equations can be solved numerically if the equations are embedded in an implicit integration algorithm.

The bond graph concepts will not be discussed in this paper. Karnopp and Rosenberg [3] introduced bond graphs extensively. Breedveld [4,5] gave a definition of multibond graphs. Tiernego and Bos [6] described the modelling of rigid bodies by multibond graphs.

In this paper it will be assumed that the systems have a tree structure. Closed kinematic loops are byond the scope of this paper.

Although in this paper only the mechanical domain will be described the proposed method is not restricted to mechanical systems only.

No distinction is made in the notation of scalars and matrices. Generally matrices are ment. In some cases, which will be clear from the context, matrices will be reduced to scalars. In the appendix is a list of symbols.

1. SOLUTION TECHNIQUES FOR MULTIBOND GRAPHS OF SYSTEMS OF RIGID BODIES

Bond graphs of mechanical systems with n bodies have 2n 3-port Inertances, with in total 6n ports. The inertances are related to the inertias and to the masses of the rigid bodies. As the number of degrees of freedom reduce to N then an amount of (6n-N) I-ports with derivative causality will occur.

A causal bond graph with storage elements in integral form only, yields an explicit set of first order differential equations. Deriving equations from a bond graph containing elements with derivative causality, results in a set of implicit differential equations. For a system to be solved by a Continuous Systems Simulation Language, storage elements should have integral causality. Derivative elements lead to numerical instability if the corresponding equations are solved by a solution technique for explicit differential equations.

Two methods exist to manipulate the bond graph in such a way that only elements with integral causality occur:

- Relaxation of the system by adding compliances, so as to expand the number of degrees of freedom from N to 6n. However, realistically small compliances will lead to a (numerically) stiff system, cf. Margolis and Karnopp [7].
- Transformation of the I-elements with differential causality to the other I-elements. For small systems this can be done by hand [3]. For not very large systems this is quite well possible using programs for formula manipulation (see Allen [8] and Bos and Tiernego [9]). For large systems code optimizers (van Hulzen [10] have to be used. The transformation requires an excessive amount of core and computation time, because sets of linear equations need to be solved [9].

Instead of the above mentioned methods, in section 2 of this paper it will be proposed to derive an implicit set of differential equations and solve the implicit system equations numerically. The time integration of these equations will be explained in section 3.

2. DERIVATION OF IMPLICIT SYSTEM EQUATIONS

Bond graphs with dependent elements correspond with a set of implicit differential equations. In this section these equations will be derived from the bond graph and a symbolic representation of the equations will be given.

A simple example will be used to illustrate the procedure of deriving the equations and to illustrate the form of the equations. The example system is a planar single pendulum. Friction is assumed in the hinge, the pendulum mass (m) is assumed not to be point mass but to have rotational inertia (J) as well with respect to its centre of mass.

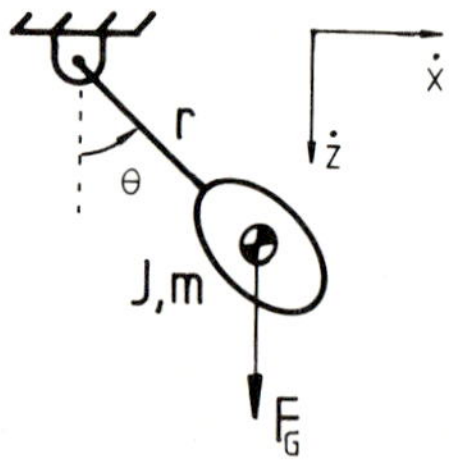

Symbols (scalars)

J inertia
m mass
R damper
r distance from the axis of rotation to the centre of mass
g gravity
Ω angular velocity
θ angle, integral of Ω

Figure 1. Pendulum

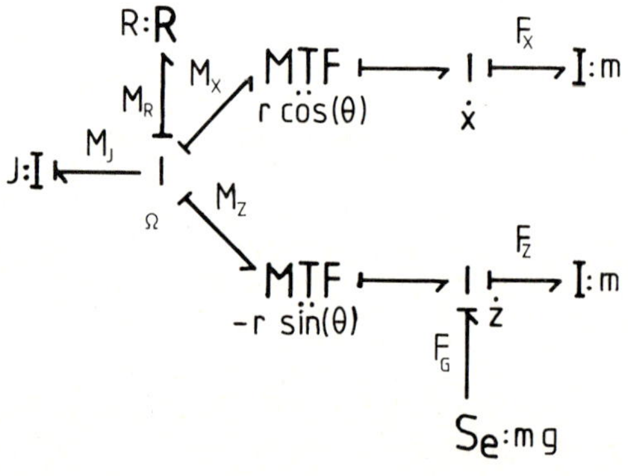

Figure 2. Bond graph of the pendulum

In this case it is easy to find the explicit differential equations. The causal strokes are assigned by putting the inertia in integral causality. The translational inertias, which represent the mass, then obtain derivative causality. Both translational inertias with derivative causality can be transformed to the angular velocity, taken together and combined with J to form an inertia $J+r^2m$, as would be found with the Huygens-Steiner rule, and the equations (1) can be derived. As stated, such transformations become difficult for large bond graphs and the explicit equations are not easily derived.

$$\dot{\Omega} = -(\sin(\theta)\ m\ g + R\ \Omega) / (J + r^2\ m)$$
$$\dot{\theta} = \Omega \qquad (1)$$

The implicit system equations can be derived as follows:
- Assign causality by the standard procedure as is described e.g. by Karnopp and Rosenberg [3]. If possible, integral causality is assigned to storage elements, else derivative causality is assigned.
- Write the mixed equations directioly from the bond graph. It is important that no variables are eliminated by substitution, so many simple equations appear. The equations are written in accordance with their causality, except the equations of storage elements with derivative causality. The relations of all storage elements are devided into two seperate equations. One equation contains only the integration, the other equation contains the rest of the constitutive relation. In this way both the state variable and its derivative occur in the equations. The integral equations of all storage elements are written in integral form.

The example system will be used to illustrate this procedure. The causality has already been assigned. The translational inertias obtained derivative causality.

First the equations of the junction structure will be written, secondly the relations of the elements. See the bond graph in fig. 2 for the meaning of the identifiers.

Junction structure, consisting of the two transformers:

$$\dot{x} = r\ \cos(\theta)\ \Omega$$
$$\dot{z} = -r\ \sin(\theta)\ \Omega$$
$$M_x = r\ \cos(\theta)\ F_x$$
$$M_z = -r\ \sin(\theta)\ (F_z - F_g)$$
$$M_J = -(M_R + M_x + M_z)$$

Constitutive relations of the elements (source, dissipator and inertias):

F_g	$= mg$	source
M_R	$= R\ \Omega$	dissipator

Constitutive relations of the storage elements excluding the integral relation:

p_x	$= m\ \dot{x}$	linear momentum of mass
p_z	$= m\ \dot{z}$	linear momentum of mass
Ω	$= p_J / J$	angular velocity

Integral part of the constitive relations of the storage elements:

p_x	$= \int F_x\ dt$	the three momenta	
p_z	$= \int F_z\ dt$	related to their	
p_J	$= \int M_J\ dt$	derivative	(2)

If the variables, which appear in this example are identified by a symbolic name, is becomes clear that the equations can be represented in a symbolic way by two functions (3). The first function states that the time derivative of the state variable of the elements with integral causality $\dot{x}_i$ are a function of the state variables x_i and of the derivative of the state variable of the dependent elements $\dot{x}_d$. The second function states the the state variable of the dependent elements x_d is a function of the independent state variable x_i.

$$\dot{x}_i = f_i(x_i, \dot{x}_d)$$
$$x_d = f_d(x_i) \qquad (3)$$

These system equations are partly explicit, partly implicit. The first function is implicit, the second is explicit.

For the dimensions of the functions and variables follows:

$\dim(x_i) = \dim(f_i) =$ the number of independent state variables

$\dim(x_d) = \dim(f_d) =$ the number of dependent state variables

(4)

In (5) the functions f_i and f_d are shown for the example system.

The function f_i:

$$\Omega = p_J / J$$

$$F_x = \dot{p}_x$$

$$F_z = \dot{p}_z$$

$$M_R = R\,\Omega$$

$$M_x = r\cos(\theta)\,F_x$$

$$M_z = -r\sin(\theta)\,F_z$$

$$\dot{p}_J = -(M_R + M_x + M_z)$$

$$\dot{\theta} = p_J / J \qquad (5a)$$

The function f_d:

$$\Omega = p_J / J$$

$$\dot{x} = r\cos(\theta)\,\Omega$$

$$\dot{z} = -r\sin(\theta)\,\Omega$$

$$p_x = m\,\dot{x}$$

$$p_z = m\,\dot{z} \qquad (5b)$$

In the following section solution methods for these equations will be proposed

3. SOLUTION OF THE SYSTEM EQUATIONS

The system equations have to be solved in time by numerical integration. It is advantagous to use an implicit integration method because it has a larger stability area then an explicit integration method. In this section it will be proposed to join the iteration for system solving and the iteration for integration. Therefore the system equations will be embedded in the integration algorithm.

First an implicit integration alogrithm will be combined with an explicit state function. Secondly the implicit system equations and the integration algorithm will be combined.

A Backward Euler algorithm will be used to demonstrate the solution method because it is the simplest implicit integration method. The choice of a more sophisticated algorithm would distract from the aim of this paper.

3.1. Implicit integration method

Equation 6 shows a symbolic notation of an implicit integration method of order m:

$$\text{int}(\,x_k,\ \dot{x}_k,\ x_{k-1},\ \ldots\ldots,\ x_{k-m}\,) = 0 \qquad (6)$$

This equation states that the state variable at timestep k is calculated from previous values and its derivative at timestep k. It is an implicit method because both the state variable and its derivative are calculated at the same instant k.

In this paper an abbreviate notation, in which the previous values of x are not mentioned, will be used:

$$\text{int}(\,x_k,\ \dot{x}_k\,) = 0 \qquad (7a)$$

With the Backward Euler algorithm, (7a) becomes:

$$F = x_{k-1} + h\,\dot{x}_k - x_k = 0 \qquad (7b)$$

The function F can be considered as an error fuction. Embedding explicit system equations into the integration algorithm (7a) results in (8): the implicit integration of explicit differential equations which is in general used in Continuous Systems Simulations Languages.

$$\text{int}(\,x_k,\ g(x_k)\,) = 0 \qquad (8)$$

in which g is of the form: $\dot{x}_k = g(x_k)$

This equation can be solved for x by iteration. A general Newton-like iteration routine may be used: the state variables are independent, so the Jacobian is regular.

3.2. Implicit integration method and implicit system equations

It has been stated that the equations of motion of mechanical systems have the form of (3). In this equations the time derivative of both independent and dependent state variables have to be integrated. In (9) the state variables are substituted in the integration algorithm.

$$\mathrm{int}\left(\begin{bmatrix} x_i \\ x_d \end{bmatrix}, \begin{bmatrix} \dot{x}_i \\ \dot{x}_d \end{bmatrix} \right) = 0 \qquad (9)$$

In (10) the full equations are embedded in the integration algorithm. This equation can be solved for each timestep to perform the integration in time.

$$\mathrm{int}\left(\begin{bmatrix} x_i \\ f_d(x_i) \end{bmatrix}, \begin{bmatrix} f_i(x_i, \dot{x}_d) \\ \dot{x}_d \end{bmatrix} \right) = 0 \qquad (10)$$

Using the function F of the Backward Euler integration algorithm, (10) results in (11).

$$F = \begin{bmatrix} F_1 \\ F_2 \end{bmatrix} =$$

$$\begin{bmatrix} x_{i,k-1} \\ x_{d,k-1} \end{bmatrix} + h \begin{bmatrix} f_i(x_{i,k}, \dot{x}_{d,k}) \\ \dot{x}_{d,k} \end{bmatrix} - \begin{bmatrix} x_{i,k} \\ f_d(x_{i,k}) \end{bmatrix} = 0 \qquad (11)$$

Four possible solution techniques of this equation will be proposed:

1. Solve F by iteration. The variables that should be solved in F are $x_{i,k}$ and $x_{d,k}$, because the functions f_i and f_d can be calculated from these two variables.

2. Calculate $\dot{x}_d$ from F_2, by writing the integration algorithm F_2 in an explicit form i.e. calculate $\dot{x}_d$ by numerical differentiation, (12a). Now only the function F_1 is solved for the variable $x_{i,k}$, (12b)

$$\dot{x}_{d,k} = (f_d(x_{i,k}) - x_{d,k-1}) / h \qquad (12a)$$

$$F_1 = x_{i,k-1} + h\, f_i(x_{i,k}, \dot{x}_{d,k}) - x_{i,k} = 0 \qquad (12b)$$

3. Differentiate f_2 analytically (13a) and substitute this into f_1 (13b). Now only F_1 needs to be solved for x_i.

$$\dot{x}_d = \frac{\partial f_d}{\partial x_i} \dot{x}_i = f_d'(x_i)\, \dot{x}_i \qquad (13a)$$

$$\dot{x}_{i,k} = f_i(x_{i,k}\, ,\, f_d'(x_{i,k})\, \dot{x}_{i,k}) \qquad (13b)$$

In this equation only the independent state variable and its time derivative occur. The time derivative of the independent state variable becomes implicit. The dependent state variable disappeared by substitution. Gear [11] solved this particular type of equation by substituting an integration algorithm.

Substitution of an integration algorithm into (15b), e.g. the Backward Euler algorithm (14), yields an equation in which only the independent state variable needs to be solved.

$$\dot{x}_{i,k} = (x_{i,k} - x_{i,k-1}) / h \qquad (14)$$

4. At last an explicit form will be proposed. In (13b) the state derivative appears on both sides of the equal sign, but in a special form. In the bond graph and therefore in f_i, the variable $\dot{x}_d$ appears only in linear form. Due to the properties of a matrix product, the time derivative of the independent state variable appears in linear form, both in (13a), and in (13b). Now two new functions can be defined such that the state derivative only appears in a matrix product, (15). The coefficients in the terms with $\dot{x}_i$ are combined in the square matrix function g_2. All other terms are combined in the column matrix function g_1. The state derivative may be calculated explicitly by inverting a matrix. This should be done numerically.

$$\dot{x}_i = g_1(x_i) + g_2(x_i)\, \dot{x}_i \qquad (15)$$

$$\dot{x}_i = (E - g_2(x_i))^{-1}\, g_1(x_i) \qquad (16)$$

4. DISCUSSION

It has been shown how implicit equations can be derived from the bond graph by standard techniques, including causal analysis. It has been shown that these equations can be solved implicitly.

Four solution methods have been proposed to solve the system equations. The first three are base on implicit solution methods, the fourth on an explicit solution method. The third and fourth method require anlytical differentiation. Solution of equations with the specific form as appears in the third method, have been studied before by Gear. In the fourth solution method it is demonstrated that an explicit differential equations can be derived as well. The explicit form is equivalent to the form which should be obtained by transformation of dependent inertias to the independent inertias. This transformation is complicated and is not needed for solving the system equations in time.

Analytical differentiation increases the number of equations. This differentiation can very well be performed by computer, because the equations are kept simple.

Implicit solving is not necessarily efficient, but deriving implicit system equations is straightforward and the equations can be kept simple. So the formula manipulation effort is relatively small and no small compliances, and thus high eigenfrequencies, need to be introduced.

The proposed derivation of equations and the proposed solution methods have succesfully been used for small systems, as e.g. the pendulum. Further promising reseach on complex mechanical systems is in progress.

APPENDIX List of symbols

Variables

e effort
f flow
p generalized momentum
q generalized displacement
x column matrices with state variables
F general iteration function
f,g general state function
m order of integration method
h integration step length
n number of bodies
N number of degrees of freedom
E unit matrix of proper dimension

Subscripts

i independent
d dependent
k k-th integration step

REFERENCES

[1] Wolz, U., "MEZA VERDE preliminary manual", Dept. of mech. engng., Univ. of Karlsruhe, Germany, Aug. 1984.

[2] Schiehlen, W.O., "Computer generation of equations of motion", Proceedings of the NATO advanced study institute on computer aided analysis and optimization of mechanical system dynamics, ed. E.J. Haug, IOWA City, 1-12 Aug. 1983, Springer-verlag, 1984.

[3] Karnopp, D.C., Rosenberg, R., "System dynamics: a unified approach", John Wiley, New York, 1975.

[4] Breedveld, P.C., "Proposition for an unambiguous vector bond graph notation", Trans. ASME, J. of Dyn. Syst. Meas. & Control, Vol. 104, No. 3, pp. 267-270, Sept. 1982.

[5] Breedveld, P.C., "A definition of the multibond graph language", Proceedings IMACS conf., Oslo Norway, Aug. 1985.

[6] Tiernego, M.J.L., Bos, A.M., "Modelling the dynamics and kinematcis of mechanical systems with multibond graphs", J. Franklin Inst., Vol. 319, No. 1/2, pp. 37-50, 1985.

[7] Margolis, D.L., Karnopp, D.L., "Bond graphs for flexible multibody systems", Trans. ASME, J. Dyn. Syst. Meas. Control, Vol. 101, pp.550-77, 1979.

[8] Allen, R.R., "Multiport representation of inertia properties of kinematic mechanisms", J. Franklin Inst., Vol. 308, pp 235-253, Sept. 1979.

[9] Bos, A.M., Tiernego, M.J.L., "Formula manipulation in the multibond graph modelling and simulation of large mechanical systems", J. Franklin Inst., Vol. 319, No. 1/2, pp. 51-66, 1985.

[10] Hulzen, J.A. van, "Code optimization of multivariable polynomial schemes: a pragmatic approach", Lecture notes in computer science, ed. J.A. van Hulzen, no. 162, pp. 283-300, 1983.

[11] Gear, C.W., "Simultaneous numerical solution of differential-algebraic equations", IEEE Transactrions on circuit theory, Vol. CT-18, No. 1 january 1971.

Complex and Distributed Systems: Analysis, Simulation and Control
S.G. Tzafestas and P. Borne (editors)
Elsevier Science Publishers B.V. (North-Holland)

BONDGRAPH VALIDATION OF EXPERIMENTAL SYSTEM IDENTIFICATION

Chen, Meng Luo, Visiting Scholar; D. Richter, Graduate Student;
J. Thoma, Professor; H.R. Martin, Professor;
University of Waterloo, Waterloo, Ontario, Canada

ABSTRACT

There are a variety of experimental methods for identifying the dynamics of a system. However, there is really only one method available, that allows system identification to be carried out during the normal operation of the system. The method requires the use of a random input signal of constant power spectrum. This white noise signal, unfortunately, requires long sample times in order to obtain good quality data. An alternative approach is to use an artificial signal which is created electronically from a sequence of pulses which have identical statistical properties to white noise, but is periodic. This is called a Pseudo Random Binary Sequence, and because of its periodicity, requires much shorter sampling times.

The objective of this presentation is, firstly, to demonstrate how this type of signal can be injected into a fluid filled pipe and successfully used to identify the pipe's dynamic characteristics.

In order to validate the results from this experimental approach, a bondgraph model was developed. This simulation was examined using TUTSIM on a personal computer. In addition a more accurate simulation of the pipe dynamics was carried out by solving the wave equations on a VAX 11/750 computer.

INTRODUCTION

There are a variety of methods available for identifying, experimentally, the dynamics of a system. However, there is really only one method available that allows system identification to take place during normal operation of the system. This is achieved with the use of a random signal of constant power spectrum. A signal of this type is commonly known as white noise and its random characteristics require long sample times to ensure good quality data. It has been well established by other researchers that an artificial signal can be created from a sequence of pulses, which have identical statistical properties to that of white noise, called a Pseudo Random Binary Sequence (P.R.B.S.). This excitation signal perform the identical function to that of white noise, but is periodic. Here the required sampling times are considerably reduced (Davis 1970, Martin 1983). A more traditional method of transfer function identification, is the use of sinusoidal excitation. This method has the danger of exciting discrete resonances in the system, and it requires the system to be shut down during testing.

When a signal is constrained in such a way that the only possible levels are ±V volts; that a change between these levels can only occur at regularly spaced intervals in time; that changes between these levels can only occur at regularly spaced intervals, and that there is an equal probability of either level occurring; then this signal is called a Random Binary Sequence. If now this sequence is made to repeat after a specific number of event points, it becomes a P.R.B.S.

Referring to Figure 1, the sequence repeats after N events (bits), so that the period of the sequence is

$$T = N \cdot \Delta t$$

where N is constrained to be always an odd number.

The power spectral density envelop is defined by,

$$G_x(f) = \left[\frac{\sin \pi f/fc}{\pi f/fc}\right]^2 \tag{1}$$

and consists of a series of spectral lines, separated by (fc/N), with zero power at fc. Since N is always chosen as odd, there is a component at zero (Hz) equal to V^2/N; fc is the clock frequency and is equal to $1/\Delta t$, where Δt is the time length of one bit (Figure 1). The usable portion of the spectrum is the bandwidth up to the -3dB point, as indicated. It can be shown that the maximum frequency used in system identification should be less than 0.45 fc.

The object of this paper is to demonstrate how this type of signal can be generated in hydraulic systems, in particular a fluid carrying pipe, and successfully used to identify that pipe's dynamic characteristics.

Validation of the results is then carried out using a classical model, and a bondgraph model of the system.

BACKGROUND

The input-output relationship for any type of system is often expressed in terms of the transfer function H(f). This is expressed in terms of the amplitude ratio and phase angle over a range of frequencies. The power spectral density approach, using P.R.B.S for the input spectrum, will produce the amplitude spectrum from the relationship

$$G_y(f) = |H(f)|^2 \, G_x(f) \tag{2}$$

where $G_x(f)$ and $G_y(f)$ are respectively the input and output power spectrums. Hence for any physically realizable system, the characteristics of that system's amplitude response can be determined. However, by using the cross spectral density, it is possible to obtain the phase spectrum also, since

$$G_{xy}(f) = H(f) \, G_x(f) \tag{3}$$

The term $G_{xy}(f)$ is a complex quantity, containing the phase angle of the cross spectral density, $\theta_{xy}(f)$. If $\phi(f)$ is the phase angle of the required transfer function, then,

$$|H(f)| = \frac{|G_{xy}(f)|}{G_x(f)} \tag{4a}$$

$$\phi(f) = \theta_{xy}(f) \tag{4b}$$

Since it is arranged to have $G_x(f)$ represented by a P.R.B.S. signal, which has a constant power spectral density function over the frequency range of interest, then the amplitude and phase characteristics can be easily extracted.

All these functions are extracted from the collected data by an FFT analyser, and since they are available it is good practice to evaluate the quality of the data using ordinary coherence functions.

The coherence between input and output spectrum is defined as,

$$\gamma^2_{xy}(f) = \frac{|G_{xy}(f)|^2}{G_x(f) \, G_y(f)} \tag{5}$$

where

$$0 < \gamma^2_{xy}(f) < 1.0$$

Perfect data would therefore exhibit a coherence of unity over the frequency range of interest. This is a measure of how much the output signal is dependent on the actual input, and how much is due to extraneous noise contamination. There will normally be poor coherence at resonant and anti-resonant frequencies, but this is due to the digital processing in the FFT analyser and not poor data.

EXPERIMENTAL TEST RIG

The experimental test rig used in these tests is shown in Figure 2. The pipeline used for evaluation was a steel pipe 3.0 m in length, and 16 mm (exactly 15.8 mm) in diameter. The end of this standard hydraulic tubing was terminated with an orifice with coefficient R set at 2.25×10^{10} N.s/m. The test fluid was standard industrial hydraulic oil of density 830 kg/m^3 and dynamic viscosity of 54.10^{-3} Pa s at room temperature.

The P.R.B.S. signal was obtained from a Wavetek function generator, which in turn drove a Moog series 74 servo valve. This allowed the P.R.B.S. to produce identical pressure pulses in the pipeline, superimposed on the normal flow. The servo amplifier also served to introduce a bias signal so that the valve would move in one direction only and about its linear region of operation. The PCB pressure transducers, maximum range of 70 MPa and 300 kHz bandwidth were 70 MPa used to collect the upstream and downstream data for processing. The analysis of the data was carried out using a HP 3582A spectrum analyser.

It was decided to limit the measurements to the first three modes for this pipe, which the computer study showed to be below 500 Hz, hence fc = 1360 gives a usable bandwidth of 612 Hz which covers the pipe modes of interest.

COMPUTER MODEL

The classical computer model was based on assuming laminar flow in a pipe with non-elastic walls (Viersma 1980) and is derived from the Continuity and Navier-Stokes equations.

The pressure ratio between two points along the line is determined from,

$$\frac{P_2}{P_1}(jw) = \frac{1}{A_{L1} - C_{L1}\left(\dfrac{D^*_{L2} - RB^*_{L2}}{RA^*_{L2} - C^*_{L2}}\right)} \tag{6}$$

The terms are defined in the appendix. These relationships were then computed using a VAX 11/750.

COMPARISON OF RESULTS

The quality of the experimental results can be assessed from Figures 3 and 4. In Figure 3 the amplitude response is shown as a result of the P.R.B.S. excitation via the Moog servo valve. The coherence is reasonably good, noting that it will normally be poor at the resonances anyway. The data shows resonances at 88 Hz, 460 Hz. These results are consistent with the theoretically determined odd harmonics for this type of end loading. Two other significant reductions in coherence are seen to occur at 60 Hz and 180 Hz, these are related to contamination from the main power supply.

Since there will, inevitably, be noise present in the signal, especially in a field environment, it is important to establish the minimum input current to drive the servo valve and still maintain minimum interference with the process. Figure 4 shows the effects of this on the identified frequency response. It is interesting to note that for no P.R.B.S. input, the process attempts to identify the frequency response function. The explanation of this phenomena is simply that the input excitation is broadband random noise coming from other sources through the test rig structure.

Averaging of the collected data was also used to improve the signal-to-noise ratio. However, it was found that averages in excess of 32, did not significantly improve the data. At this frequency range the average process took 16 seconds.

A comparison between the experimental results and those calculated using equation 6, are shown in Figure 5. It can be seen that there is quite good correlation for the first two modes, with some discrepancy on the third mode. As the frequency is increased, the differences are probably due to the effect of the many fittings, their shape and dimensions, necessary in the construction of the test rig. This is not such a problem at low frequencies because the wavelength is large compared to the dimensions of the fittings. There are also some necessary simplifications used in the equation formulation.

BONDGRAPH MODEL FROM PARTIAL DIFFERENTIAL EQUATIONS

In order to build a Bondgraph model for further system analysis and validation, the pipe is not broken, as usual, into separate segments with lumped resistances, capacitances and inductances (mass effects). Following instead a method originally published by Karnopp (1968, see also Thoma 1976), the PDE (Partial Differential Equation) for the pressure waves is solved under open boundary conditions (zero pressure at the ends and flow determined by the pipe). As known from classical mathematical physics, we obtain an infinity of particular solutions, each representing a normal mode of eigenfunction. The total solution is then the sum of certain eigenfunctions.

Each eigenfunction or mode is representable by a combination of R,C, and I elements (resistance, capacitance and inertia elements). They do not correspond to a definite segment of the pipe, but represents certain modes. Certain values of the C elements decreases for higher modes proportionally to $1/i^2$ where i = mode number, whilst values of the r and i element are the same for all modes. Consequently the resonance frequency of the modes increase with higher orders and the fundamental idea is to use a small number of modes for representing FRF (Frequency Response Function) up to a certain frequency.

As illustrated on Figure 6 on top the procedure is to inject pressures after the end of the pipe, where this injection represents a disturbance term in the original PDE. On the bottom of Figure 6 the pipe is reticulated (represented by a network of bondgraph) with five modes of the order C order 0 to 4, where the resonance frequency of the top mode is 1000 Hz.

The bondgraph of Figure 6 was programmed on an IBM PC using the TUTSIM program (Meerman 1982) for a time domain simulation. Subsequently a FFT (Fast Fourier Transform) was applied using the new TUTFIT program (Thoma/Fife 1986) in order to obtain certain frequency domain representation of Figure 7. Here we find peaks of the FRF at frequencies of 101, 297, 508 and 719 Hz, which correlate well with the experimental data.

CONCLUSIONS

The results have to be taken with some caution however, since it was found that additional capacity corresponding to the rubber hose 6 on Figure 2 would detune the resonance frequencies. Furthermore, since the resonance frequencies of the lowest neglected mode (i=5) is 1082 Hz and such relatively near to our measured frequencies some inaccuracies may be expected. We investigate at present the usefulness of replacing certain neglected modes by sum of inertias, as proposed already by Karnopp (1968). Finally the close interfacing of experimental results, Bondgraph simulation and FFT is a signpost for the future.

ACKNOWLEDGEMENTS

The authors wish to acknowledge the financial support provided by NSERC, and the computing equipment donated by DEC, through the WATDEC program.

REFERENCES

Davies, W.D.T. (1970) System Identification for Self Adaptive Control, John Wiley.

Karnopp, D., (1968), Analysis of Multipoint Systems, the MIT Press, Cambridge, U.S.A.
Martin, H.R.,(1983) Experimantal Aspects of

Using P.R.B.S. for System Identification. 9th Can/Cam, Saskatchewan.

Meerman, J.W. et al (1982) TUTSIM, Interactive Simulation on 8 Bit PC. Proceedings 10th IMACS Congress, Montreal, Canada.

Thoma, J.U. (1976) Introduction to Bondgraphs and Their Application, Pergamon Press, Oxford,U.K.

Thoma, J.U. and Fyfe, K. (1986), to be published at the 4th IMAC Conference (Los Angeles, U.S.A.).

Viersma, T.J. (1980) Analysis, Synthesis and Design of Hydraulic Servosystems and Pipelines. Elsevier Scientific Publishing Company, 1980.

APPENDIX

Details of the pipeline are given in Figure 2 where the matrix form of four pole equations representing the dynamics of the line are,

$$\begin{bmatrix} A_{L1} & B_{L1} \\ C_{L1} & D_{L1} \end{bmatrix} \begin{bmatrix} Q_1 \\ P_1 \end{bmatrix} = \begin{bmatrix} Q_2 \\ P_2 \end{bmatrix}$$

$$\begin{bmatrix} A_{L3} & B_{L3} \\ C_{L3} & D_{L3} \end{bmatrix} \begin{bmatrix} A_{L2} & B_{L2} \\ C_{L2} & D_{L2} \end{bmatrix} \begin{bmatrix} Q_2 \\ P_2 \end{bmatrix} = \begin{bmatrix} Q_o \\ P_o \end{bmatrix}$$

and for the end load on the line

$$P_o = RQ_o$$

where

$$A_{Li} = D_{Li} = \cosh T_i s\sqrt{N_i}$$

$$B_{Li} = -[\mathrm{Sinh}\ T_i s\sqrt{N_i}\ ]/Z_{ci}\sqrt{N_i}$$

$$C_{L1} = z_{ci}\ \sqrt{N_i}\ \mathrm{Sinh}\ T_i s\ \sqrt{N_i}$$

$$A_{Li}D_{Li} - B_{Li}C_{Li} = 1$$

$$T_i = L_i/V_{ai}\ :\ V_{ai} = \sqrt{E_i/P_i}\ :\ \alpha_i$$

$$= 32\ v_i/D_i^2$$

$$z_{ci} = \rho_i v_{ai}/A_i$$

$$N(s) = -\frac{J_o\ (\alpha R)}{J_2\ (\alpha R)} \doteq 1 + \frac{\alpha}{s} + \sum_{i=1}^{3} \frac{K_i}{\tau_{is} + 1}$$

$$= 1 - j\,\frac{\alpha}{\omega} + \sum_{i=1}^{3} \frac{K_i(1-j\omega\tau_i)}{\omega^2 T_i^2 + 1} \qquad \omega < 300\alpha$$

$$\sqrt{N} = \sqrt{a + jb} = c + jd$$

$$\mathrm{Sinh}\ T_i s\ \sqrt{N} = -\ \mathrm{Cosc}_i\omega T_i \mathrm{Sinh}\ d_i\omega T_i + j\mathrm{Sinc}_i\omega T_i \cosh\ d_i\omega T_i$$

$$\mathrm{Cosh}\ T_i s\sqrt{N} = \mathrm{Cosc}_i\omega T_i \cosh\ d_i\omega T_i - j\mathrm{Sinc}_i\omega T_i \sinh\ d_i\omega T_i$$

$$c^2 = 0.5(a + \sqrt{a^2 + b^2})\ :\ d^2 = 0.5(-a + \sqrt{a^2 + b^2})$$

$$a = \mathrm{Re}[N] = 1 + \sum_{i=1}^{3} \frac{K_i}{w^2 T_i^2 + 1}$$

$$b = \mathrm{Im}[N] = 1 - \frac{\alpha}{\omega} - \sum_{i=1}^{3} \frac{K_1\ T_i}{\omega^2 T_i^2 + 1}$$

$$K_1 = 0.5/3.3,\ K_2 = 4.05/25,\ K_3 = 2/100$$

$$\tau_1 = 1/3.3\alpha,\ \tau_2 = 1/25\alpha,\quad \tau_3 = 1/1000\alpha$$

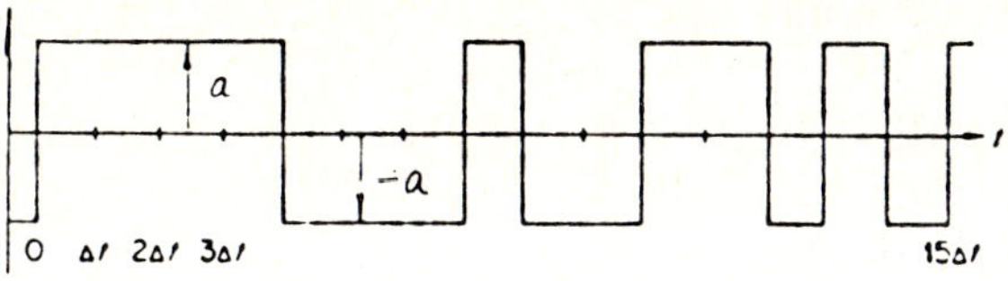

Fig. 1 P.R.B.S. Signal

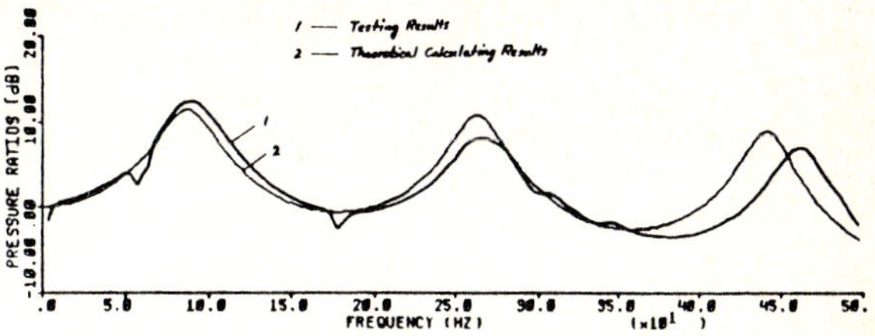

Fig. 5 Comparison of Test Data with Model

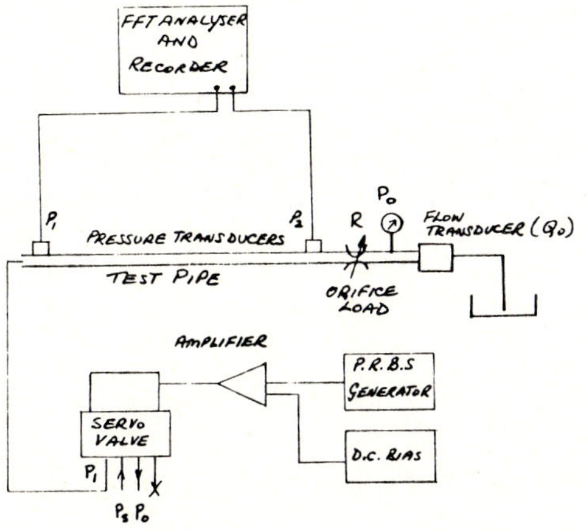

Fig. 2 Schematic Diagram of Test Rig

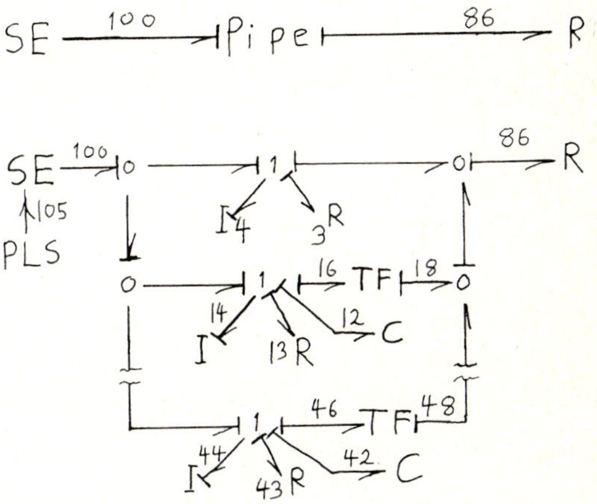

Fig. 6 Bondgraph Model. Top: simple word BG driven by effort source. Bottom: model with modes 0 through 4 obtained from solving the acoustic PDE including resistance effects. Modes 2 and 3 have been left out between the cut lines.

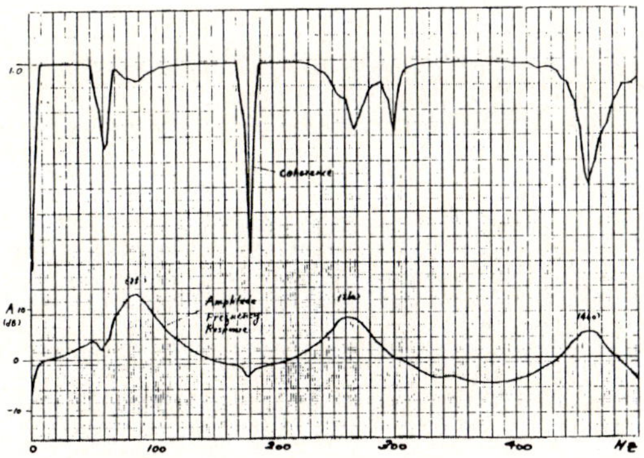

Fig. 3 Amplitude Response and Coherence

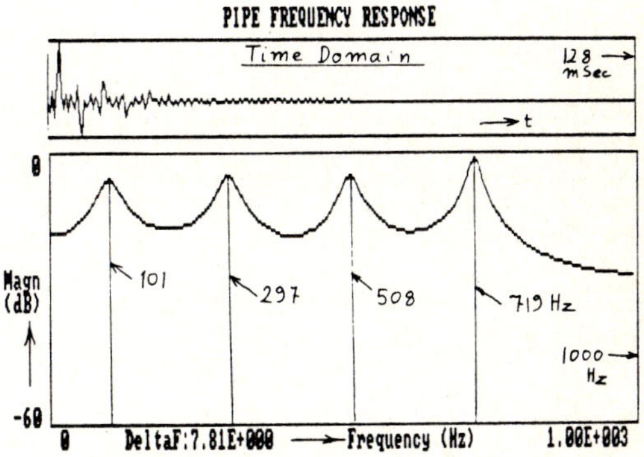

Fig. 7 Pipe frequency response obtained from FFT of the time domain simulation of the BG on Fig. 6.

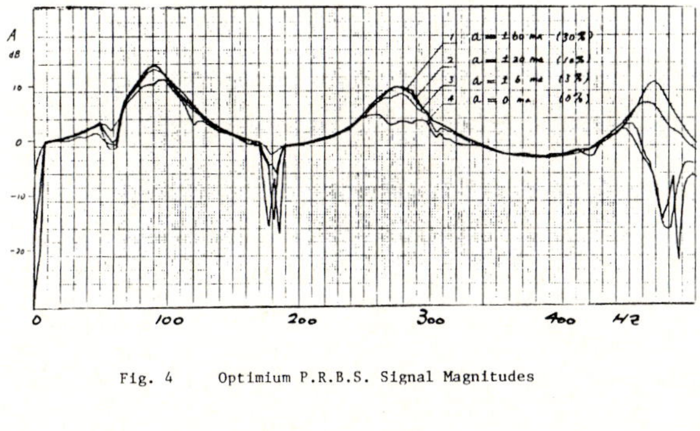

Fig. 4 Optimium P.R.B.S. Signal Magnitudes

Complex and Distributed Systems: Analysis, Simulation and Control
S.G. Tzafestas and P. Borne (editors)
Elsevier Science Publishers B.V. (North-Holland)

PSEUDO BOND GRAPH REPRESENTATION OF UNSTEADY STATE HEAT CONDUCTION

Hallvard Engja

Department of Marine Technology
The Norwegian Institute of Technology
7034 TRONDHEIM-NTH Norway

Heat transfer problems confront investigators in nearly every branch of engineering but have not been considered much using bond graphs as a modelling aid. Constructing dynamic models of such phenomena is often not a trivial task since nonlinear phenomena and partial differential equations are involved.

In this paper normal modes and pseudo bond graphs are used to represent the distributed nature occurring in heat conduction problems. An indication is given how this type of analysis allows coupling of subsystem models to construct system models incorporating either temperature or heat flux sources or interactions.

1. INTRODUCTION

Problem formulation or mathematical modelling of dynamic systems is often more difficult than the solution of the resulting differential equations. One reason for this difficulty is that modelling projects may cut through many disciplines and it is difficult to be an expert in all fields. The interdisciplinary nature of mathematical modelling has urged system analysts to look at the modelling process itself in a more systematic and rigorous way.

The bond graph modelling technique is such a systematic rigorous method for building mathematical models for a broad spectrum of physical systems. Researchers in many fields [1] have added to the number of systems that can be modelled in the style of bond graphs.

One branch of engineering which is important but has not been considered much by bond graphs is engineering heat transfer. The litterature of heat transfer generally recognizes three distinct modes of heat transmission: conduction, radiation and convection. Convection is the name given to the process by which thermal energy is transferred between a solid and a fluid flowing past it. Radiation heat transfer is the net exchange of thermal energy between two surfaces obeying the laws of electromagnetics. Conduction is the transfer of heat through materials without net mass motion of the material but resulting from molecular motion. The equations describing the temperature distribution in a solid due to that conduction are partial differential equations, because the temperature is a function of time as well as space.

The majority of engineering heat transfer problems are concerned with steady-state systems. However, transient and periodic variations are increasingly important as systems become more complex and interactive. Periodic heat flow is of particular importance in internal-combustion engines, airconditioning, instrumentation and process control [2].

In the field of partial differential equations, even more than in ordinary differential equations, the limitations of analytical and numerical methods are quickly apparent. The degree of difficulty increases when sub-systems described by partial differential equations are coupled to larger nonlinear systems. Approximate methods of solution, mainly based on finite differences or variational principles, have been used in practical work since the days of hand calculation.

Bond graphs have been shown to be useful as a unified modelling tool for a wide variety of physical dynamic systems and this is a strong motivation for modelling distributed systems using bond graphs [3].

In this paper modeling of distributed heat conduction is presented and it is shown how such a model can be coupled with heat convection and radiation. The modelling method discussed is based on normal mode analysis in which the distributed nature is represented by a finite number of basic bond graph elements. This type of analysis allows a coupling of sub-system models to construct system models incorporating either temperature or heat flux sources or interactions.

2. DERIVATION OF THE HEAT CONDUCTION EQUATIONS

Many practical heat transfer problems involve cases where one-dimensional analysis will suffice. They occur also very often in conjunction with thermo fluid systems.

Many practical heat transfer problems involve cases where one-dimensional analysis will suffice. They occur also very often in conjunction with thermo-fluid systems. Traditionally in engineering heat transfer work temperature and heat flow are used as basic variables. Since these variables are common in practice it is shown in [4,5,6] that these variables are useful when using bond graphs as a modelling aid. The resulting bond graph is called a pseudo bond graph since the effort, e, and flow, f, are not power variables. Heat flow itself has the dimensions of power. It is therefore, convenient to derive the required equations with temperature, T, as effort variable and heat flux, Q, as flow variable.

Consider a small element of material in a solid body, as shown in Figure 1.

It is assumed that heat will flow only in the direction of x. The direction of increasing distance x is assumed to be the direction of positive heat flow.

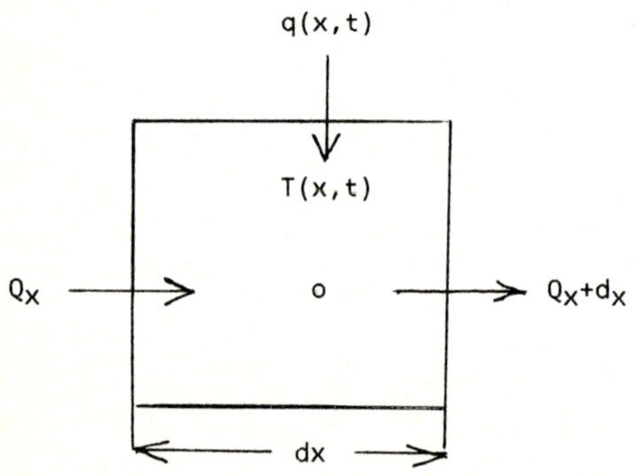

Figure 1. Infinitesimal element for derivation of equations.

To obtain an equation for the temperature distribution we write an energy balance for the element

$$Q_X - [Q_X + \frac{\delta Q}{\delta x} dx] + q(x,t)dx = A\rho \cdot c_V \cdot \frac{\delta T}{\delta t}$$

or

$$-\frac{\delta Q}{\delta x} + q(x,t) = A \cdot \rho \cdot c_V \cdot \frac{\delta T}{\delta t} \qquad (1)$$

where

$q(x,t)$ = externally applied heat flow rate per unit length.

A = area
ρ = mass density of material
c_V = specific heat of material

According to the second law of thermodynamics heat will automatically flow from points of higher temperature to points of lower temperature, and this means heat flow will be positive when the temperature gradient is negative. Accordingly, the elementary equation for one-dimensional conduction is written

$$Q = -k \cdot A \cdot \frac{\delta T}{\delta x}$$

or

$$\frac{\delta Q}{\delta x} = -k \cdot A \cdot \frac{\delta^2 T}{\delta x^2} \qquad (2)$$

where k is the thermal conductivity of the material.

Substituting equation (2) into (1) yields

$$\alpha^2 \frac{\delta^2 T}{\delta x^2} + \frac{q(x,t)}{A\rho c_V} = \frac{\delta T}{\delta t} \qquad (3)$$

where

$$\alpha^2 = \frac{k}{\rho c} \text{ (thermal diffusivity)}$$

Assuming that the forcing is the heat flow rate at the ends, $Q_o(t)$ and $Q_e(t)$, then equation (3) takes the form

$$\frac{\delta T}{\delta t} - \alpha^2 \frac{\delta^2 T}{\delta x^2} = \frac{1}{A\rho c_V} [Q_o(t)\delta(x) - Q_e(t)\delta(x-L)] \qquad (4)$$

where $\delta(\cdot)$ stands for the Dirac delta function.

The eigenfunctions of equation (4) are found by temporarily setting $Q_o(t)=Q_L(t)=0$. Using the classical approach of separation - of - variables by assuming a product solution of the form

$$T(x,t) = \sum_{i=0} X_i(x) \cdot U_{i^\bullet}(t) \qquad (5)$$

Substituting equation (5) into equation (4) and using the notation

$$(\cdot) = \frac{\delta}{\delta t} \text{ and } (\prime) = \frac{\delta}{\delta x}$$

yields the result that each eigenfunction must satisfy the following equation

$$\frac{\dot{U}_i}{U_i} = \alpha^2 \frac{X_i''}{X_i} = -\lambda_i^2 \qquad (6)$$

The solutions to the component differential equations are then

$$U_i = e^{-\lambda_i t}$$

and

$$X_i = A \sin \frac{\lambda_i}{\alpha} x + B \cos \frac{\lambda_i}{\alpha} x \qquad (7)$$

From the conditions that $Q_0 = Q_e = 0$ and equation (2) the following end conditions are given

$$\frac{\delta T}{\delta x}(0,t) = \frac{\delta T}{\delta x}(L,t) = 0$$

Applying the end conditions to equation (7) gives the following eigenvalues and eigenfunctions or normal modes

$$\lambda_i = \frac{i\pi\alpha}{L} \quad \text{and} \quad X_i(x) = \cos \frac{\lambda_i}{\alpha} x \qquad (8)$$

for i = 0, 1, 2

The series solution of the particular solutions takes the following form

$$T(x,t) = \sum_{i=0} Be^{-\lambda_i^2 \cdot t} \cos \frac{\lambda_i}{\alpha} x \qquad (9)$$

Now when equation (5) is substituted into equation (4) and using the condition of orthogonality

$$\int_0^L X_i \cdot X_j \cdot dx = 0 \qquad i \neq j$$

the following decoupled equations for the U_i are obtained

for i = o

$$\dot{U}_0 = \frac{1}{A\rho \cdot C \cdot L} [Q_0 - Q_L] \qquad (10)$$

for i ≠ o

$$\dot{U}_i + \lambda_i^2 U_i = \frac{2}{A\rho \cdot C_v \cdot L} [Q_0 - Q_L(-1)^i] \qquad (11)$$

Redefining variables such that

$$q_0 = A \cdot \rho \cdot C_v \cdot LU_0 \quad \text{and} \quad q_i = \frac{1}{2} A\rho C_v \cdot L \cdot U_i$$

and equation (10) and (11) then takes the form

$$\dot{q}_0 = [Q_0 - Q_L] \qquad (12)$$

and

$$\dot{q}_i + \lambda_i^2 \cdot q_i = [Q_0 - Q_L(-1)^i] \qquad (13)$$

Equation (12) is the equation one would obtain by representing the bar as a single lumped capacitance, and equation (13) can be represented by lumped capacitance and resistance elements.

The individual capacitance and resistance values are

$$C_0 = A\rho c_v \cdot L \quad \text{and} \quad R_0 = \infty$$

$$C_i = \frac{1}{2} A\rho \cdot c_v \cdot L \quad \text{and} \quad R_i = \frac{2}{\lambda_i^2 A\rho \cdot c_{v_L}}$$

Figure 2 shows a bond graph representation of the normal mode model.

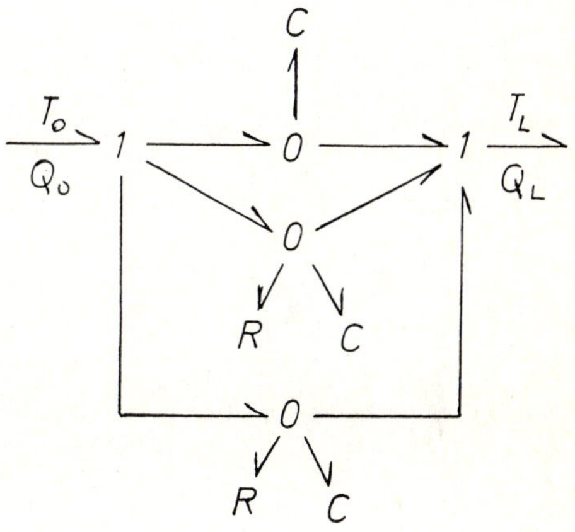

Figure 2. Normal mode model of heat conduction.

In practice, only a finite number of modes will be retained in a model as shown in Figure 2. The temperatures at the two ends can be computed using equation (5) as follows

$$T_0 = \sum_{i=0}^{n} \frac{q_i}{C_i} \quad \text{and} \quad T_L = \sum_{i=0}^{n} \frac{q_i}{C_i} (-1)^i \qquad (14)$$

The temperature at any interior point can accordingly be computed from

$$T_x = \sum_{i=0}^{n} \frac{q_i}{C_i} \cos \frac{\lambda_i}{\alpha} x \qquad (15)$$

How the computation of the temperature of a interior location is performed on a bond graph is shown in Figure 3. The heat flow source is set to zero.

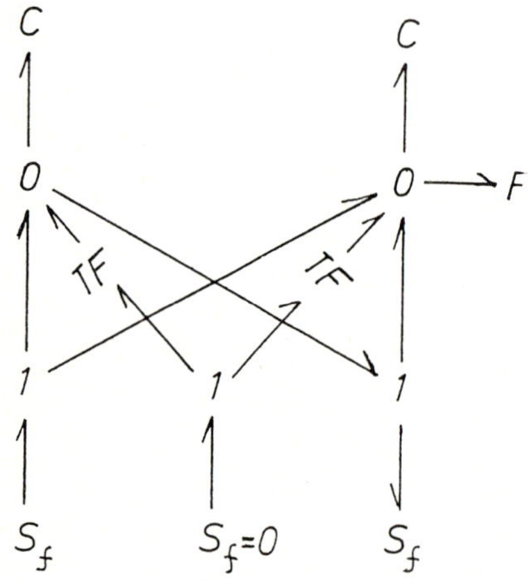

Figure 3. Computation of temperature at interior point

From the above discussion and the bond graph shown in Figure 2 the modal analysis is performed in a causal manner with heat flow considered as natural inputs. On the other hand the system itself should be well represented independent of the nature of the inputs. That is, whether the system is regarded as being supplied with heat flow or temperature, the fundamental characterization of the system should remain valid. The causality of the bond graph will be decided when the total system model is assembled.

If the bond graph shown in Figure 2 is considered with temperatures at the two ends as inputs, two C-elements must have differential causality, which means that two algebraic loops occur.

In most practical problems the system shown in Figure 2 exchanges energy with its environment by convection and radiation.

The equation for heat energy transfer by convection which is in general use is given as

$$Q_C = h \cdot A \cdot (T_W - T_a) \qquad (16)$$

where

h = coefficient of heat transfer
A = surface area
T_W= Wall temperature
T_a= ambient temperature

The convective heat transfer coefficient is usually a nonlinear function of ambient fluid properties and wall temperature.

The heat energy transfer by radiation is generally given as

$$Q_R = \tau \cdot A \cdot (T_W{}^4 - T_a{}^4) \qquad (17)$$

where τ is a constant.

Equation (16) and (17) can be represented as a two port R field indicated in Figure 4.

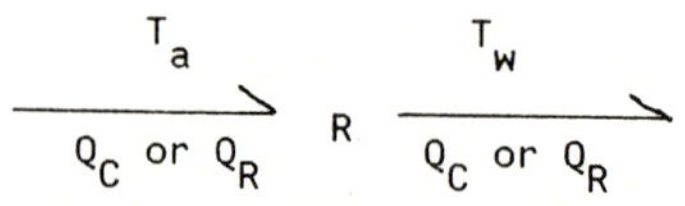

Figure 4. Bond graph model of convection or radiation heat transfer.

Thus a bond graph representation of heat transfer consist of merging together the two bond graphs in Figure 2 and Figure 4.

3. EXAMPLE PROBLEM

In this section the application of the developed theory is illustrated on a problem of transient heat conduction in a 0.3 m thick concrete wall. The wall originally at 30°C, perfectly insulated at one side and suddenly exposed to a hot gas at 850°C at the other face. With a heat-transfer coefficient on the hot side of 28 W/m²K the problem is to determine time required to raise the temperature at the insulated face to 590°C.

The bond graph model of the given problem is shown in Figure 5.

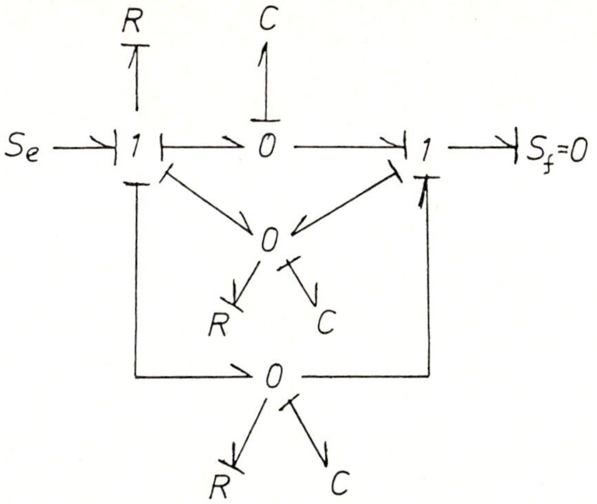

Figure 5. Bond graph of example problem

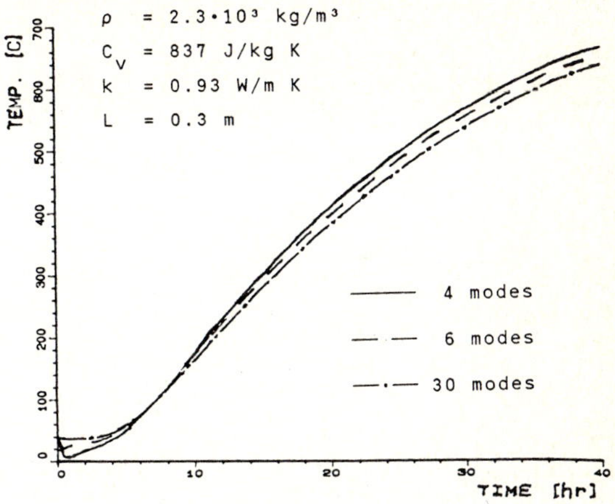

Figure 6. Simulation results

The simulation results is shown in Figure 6. In practice only a finite number of modes are retained, and the response with various number of modes is illustrated. As can be seen, fairly accurate results are obtained with only a few modes. The error is largest at the first time increments.

4. CONCLUSION

In this paper the use of normal modes and pseudo bond graphs for modelling transient heat conduction is presented. An indication is given of how the bond graph representation of the distributed system can be coupled with convective and radiation heat transfer. The modelprocedure is demonstrated with an example. It has been shown how one can easily produce the desired results by using existing well known programs.

REFERENCES

[1] A.M. Bos and P.C. Breedveld, "Bond Graph Bibliography Update", J. Franklin Inst., Vol. 319, No. 1/2, 1985.

[2] H. Valland and G.K. Wyspianski, "A Theoretical Analysis of Thermal Barriers in Diesel Engine Cylinders", Norwegian Maritime Research, Vol. 10, No. 2, 1982.

[3] D. Margolis, "A Survey of Bond Graph Modelling for Interacting Lumped and Distributed Systems", J. Franklin, Inst., Vol. 319, No. 1/2, 1985.

[4] D.C. Karnopp and R.C. Rosenberg, "System Dynamics: A Unified Approach", Wiley, New York, 1975.

[5] D.C. Karnopp, "State Variables and Pseudo Bond Graphs for Compressible Thermofluid Systems", J. Dyn. Syst., Measure. Control, Trans. ASME, Vol. 101 Series G. No. 3, Sept. 1979.

[6] H. Engja, "Bond Graph Model of a Reciprocating Compressor", J. Franklin Inst., Vol. 319, No. 1/2, 1985.

Complex and Distributed Systems: Analysis, Simulation and Control
S.G. Tzafestas and P. Borne (editors)
Elsevier Science Publishers B.V. (North-Holland)

MODELING AND SIMULATION OF ADAPTIVE VEHICLE AIR SUSPENSIONS WITH PSEUDO BOND GRAPHS, CAMP AND ACSL

Dean Karnopp
Professor of Mechanical Engineering

University of California
Davis, California 95616 USA

ABSTRACT

Air suspensions involve interactions among mechanical, pneumatic, electrical and hydraulic elements. Pseudo bond graphs, true bond graphs and block diagrams can be used to model electronically controlled suspensions in a convenient and concise way. The CAMP program produces the equations of motion in the form of an input file for a general purpose simulation program such as ACSL. After editing of the file to insert nonlinear constitutive laws and control system equations, simulation of the system can be accomplished in the usual way. Using macros or sub-routines, commonly occurring elements such as pneumatic accumulations and restrictors can be conveniently represented.

1. INTRODUCTION

Air suspensions have long been in use for automobiles, trucks and busses but interest in such suspensions has increased recently with the production of electronically controlled systems. The ride height of the vehicle can be adjusted for load leveling or aerodynamic drag reduction and the effective volume of air can be varied to change the stiffness of the suspension.

Figure 1 shows a highly schematic sketch in which a 2 degree-of-freedom model represents one-quarter of an automobile. The body is represented by the mass M the wheel by m and the tire by the spring with constant k. The air suspension itself is shown as a rolling diaphragm with effective area A and variable volume V_1. When the valve with area A_{12} is open, the volume V_2 supplements V_1. The volumes V_2 and V_3 are supposed to be continuously variable as the piston with area A' is moved back and forth by an electric drive system. (The drive must be self-locking so that pressure differences across the piston do not cause movement.) The area A_{23} represents a leakage path between V_2 and V_3 which provides a long-term pressure equilization. Present day switched stiffness suspensions can be modeled by opening and shutting the area A_{12} and continuously variable systems are modeled by considering movement of the piston separating the volumes V_2 and V_3. Ride height control is achieved by activating the valve on the left side. Area A_{1a} allows air to escape to the atmosphere and A_{1s} allows air from a supply vessel to enter the volume V_1. The valve actuator varies A_{1a} and A_{1s} simultaneously. The damper between the wheel and the body would typically be a hydraulic shock absorber with electro-mechanical valves to vary the damping properties of the device.

2. BASIC PNEUMATIC SYSTEMS ELEMENTS

The construction of a bond graph representation for the mechanical, electromechanical or hydraulic parts of a system such as that in Fig. 1 is routine, [1]. Compressible fluid systems and thermal systems modeled using a control volume approach can be conveniently incorporated in a pseudo bond graph, [2-7]. In such a bond graph, effort and flow variables are used which are not complementary power variables. Except for this, however, the sign convention and causality operations as well as the basic classifications of R and C elements are the same as in the case of true bond graphs.

We consider now specific models for the two pneumatic elements which appear in systems such as that in Fig. 1. The first called an accumulator as shown in Fig. 2. A mass M of gas is contained in a variable volume V with energy E, pressure P and temperature T. We consider P and T to be efforts and $\dot{E}$, $\dot{M}$, and $\dot{V}$, to be flows. The bond with P and $\dot{V}$ is a true bond since $P\dot{V}$ is power expended during a volume change. However, the paired bonds with T and $\dot{E}$ and P and $\dot{M}$ are pseudo bonds. The "thermal bonds" with T and $\dot{E}$ variables are shown dotted merely for convenience to distinguish them from their P, $\dot{M}$. partners. The use of C for the 3-port implies that the efforts P, P, and T should be functions of the displacements E, M, and V which are the time integrals of the flows.

If we assume, as is common [8,9] that the ideal gas laws are sufficiently accurate

$$PV = MRT \quad , \qquad (1)$$

$$E = Mc_vT \quad , \qquad (2)$$

where R is the gas constant and c_v is the specific heat at constant volume, then the accumulator in integral causality has the constitutive laws

$$P = (R/c_v)E/V \quad , \qquad (1')$$

$$T = (1/c_v)E/M \quad , \qquad (2')$$

(Equation (1') yields the effort for two of the C's bonds.) The initial values of the state variables V_0, M_0, E_0 are related to initial pressure and temperature by

$$M_0 = P_0V_0/R\,T_0 \quad , \qquad (3)$$

$$E_0 = P_0V_0c_v/R \quad , \qquad (4)$$

The two zero junctions in Fig. 2 are used to enforce conservation of mass and energy as air flows into and out of the accumulator. The extra flow source is necessary to assure that the mechanical power $P\dot{V}$ is properly considered in the rate of change of energy stored in the gas. Other energy flow rates come from heat transfer or enthalpy flows of the form $\dot{m}c_pT$ where $\dot{m}$ is a mass flow rate, c_p is the specific heat at constant pressure and T is the gas temperature.

Figure 3 shows a psuedo bond graph for another common element in pneumatic systems. It is called a restrictor and it represents valves if the area A is variable or a fixed orifice if A is constant. Here we will use a model based on isentropic nozzle equations, [8,9], which will consider forward and reverse flow and choking.

We assume that the pressures P_a, P_b and temperatures T_a, T_b, on the two sides of the restrictor are inputs and the mass flow rate $\dot{m}_a$, $\dot{m}_b$ and energy flow rates $\dot{E}_a$, $\dot{E}_b$ are to be calculated. The procedure is as follows:

1. Decide on upstream and downstream variables.

If $P_a > P_b$, $P_u = P_a$, $T_u = T_a$, $P_d = P_b$,

If $P_b \leq P_a$, $P_u = P_b$, $T_u = T_b$, $P_d = P_a$ (5)

2. Compare the pressure ratio to the critical pressure ratio, P_{cr}

$$P_{cr} = [2/(\gamma-1)]^{[\gamma/(\gamma-1)]} \quad , \qquad (6)$$

where $\gamma = c_p/c_v$. (7)

If $P_d/P_u > P_{cr}$, $P_r = P_d/P_u$,

If $P_d/P_u \leq P_{er}$, $P_r = P_{cr}$. (8)

3. Compute a mass flow rate.

$$\dot{m} = \frac{A\,C_dP_u}{\sqrt{T_u}}\sqrt{\frac{2\gamma}{R(\gamma-1)}}\;\sqrt{P_r^{2/\gamma} - P_r^{(\gamma-1)/\gamma}} \qquad (9)$$

where C_d is a discharge coefficient.

4. Correct for direction of flow.

If $P_a > P_b$, $\dot{m}_a = \dot{m}_b = \dot{m}$

If $P_a \leq P_b$, $\dot{m}_a = \dot{m}_b = -\dot{m}$ (10)

5. Compute energy flow rate.

$$\dot{E}_a = \dot{E}_b = \dot{m}_a\,c_p\,T_u \qquad (11)$$

Note that it is possible to define a different discharge coefficient for forward and reverse flows in step 1.

With specific constitutive laws for accumulators and restrictors, it is possible to assemble a bond graph model for the suspension of Fig. 1 as well as a variety of other pneumatic systems.

3. AUTOMATIC EQUATION FORMULATION

An advantage of bond graphs lies in their compact nature. Figure 4, for example, represents the system of Fig. 1 in a much more precise way than the schematic diagram and yet it shows clearly the types of elements and their connection laws. The left side of the figure represents the mechanical dynamics with true bond graph elements. The MTF at the lower left converts forward velocity into vertical input velocity to the tire through the roadway unevenness slope function which depends on the position variable X. On the right side, there are three accumulators for the three volumns and four restrictors for the valves and leakage passages. Physical parameters and variables are indicated on this bond graph to aid in relating it to the schematic diagram. Also, a number of signal interactions are shown with full arrows as signals or activated bonds. The valve areas and the variable damper are to be varied by the electronic control system. Other signals arise because the pseudo bond graph interacts with a true bond graph.

Another version of the system bond graph showing and numbering all bonds but suppressing control system signals is shown in Fig. 5. This bond graph shows the basic energetic structure of the physical system and the equations of motion for this system. For nonlinear systems, it is convenient to write out the constitutive laws in causal form for each multiport element without elimination of internal variables, [10].

The analysis of causal interactions among multiport elements follows a strict procedure, [1]. In the CAMP program, [11] this process has been automated. CAMP functions as a preprocessor for a variety of continuous system simulators such as ACSL [12].

The input to CAMP is a line code as shown in Fig.6. It is a listing of the multiport type followed by the numbers of all attached bonds. The sign convention half arrow is assumed to point from the first listed multiport to the later listed multiport for any bond number. (When there are loops in bond graph junction structures, one may occasionally have to insert 2-port 0- or 1-junctions in order to achieve an ordering equivalent to a given set of sign convention half- arrows.) There are several projects to develop means to draw bond graphs on a graphics terminal and to generate the line code for CAMP automatically [13].

CAMP analyses the line code to make sure that it is complete and then checks for derivative causality and algebraic loops. If no problems occur, the causal equations for each multiport are written as if the multiport were linear and incorporated in an input file for the simulator program. For the multiports, which are actually linear, coefficient values must be supplied while for nonlinear multiports the linear relations are replaced by nonlinear ones in an editing process.

4. COMPLETION OF THE SYSTEM MODEL

The result of a CAMP run on the line code of Fig. 6 is an input file for ACSL consisting of a large number of simple, linear, causal equations. In the example, CAMP writes about 125 equations, many of which are trivial identities and sums from the junction structure while the physical system is only 13th order.

Figure 7 shows a small portion of the CAMP generated ACSL input file. Note the 4-port R relations for a restrictor, part of the 3-port C relations for an accumulator, some derivative definitions, some junction structure laws and some state variable integration definitions in the ACSL format.

In Fig. 8, an ACSL macro is shown for the accumulator, Eqs. (1') and (2'), and for the restrictor or valve, Eqs. (5) to (11). Figure 9 shows how the two macros are used as functions to replace the 4-port R and 3-port C linear relations of Fig. 7 with nonlinear versions.

The remainder of the system model is written using the ACSL language in a standard fashion. For the suspension example, one needs to develop a control algorithm for stroking the values and varying the volumes and one needs to decide on test inputs from the roadway unevenness or from external forces in order to exercise the model. It is often useful to construct a block diagram for candidate control systems particularly if nonlinear functions are to be incorporated.

5. CONCLUSIONS

The use of bond graph models, an equation formulator such as CAMP and a general purpose simulator such as ACSL in combination has resulted in increased productivity in modeling, analysis, and simulation of dynamic systems containing physical elements. Work of a routine nature such as writing equations for a rigorously defined model is aided to the maximum extent possible by the computer. The bond graph model and the power flow sign conventions implied by the line code order are documented in the CAMP output automatically so the results of previous simulations can be easily checked. The use of bond graph methods for the study of systems such as electronically controlled suspensions in which a variety of energy domains are involved and in which nonlinearities are of crucial importance has proved useful.

REFERENCES

1. Rosenberg, R. and Karnopp, D., Introduction to Physical System Dynamics, McGraw-Hill, NY, 1983.

2. Karnopp, D., "State Variables and Pseudo Bond Graphs for Compressible Thermofluid Systems", Trans. ASME, J. of Dynamic Systems, Measurement and Control, v. 101, n. 3, Sept. 1979, pp. 201-204.

3. Karnopp, D., "Pseudo Bond Graphs for Thermal Energy Transport", Trans. ASME J. of Dynamic Systems, Measurement and Control, v. 100, n. 3, Sept. 1978, pp. 165-169.

4. Tylee, J. L., "A Bond Graph Description of U-Tube Steam Generator Dynamics", J. of the Franklin Inst, v. 315, n. 3, March 1983, pp. 165-178.

5. Tylee, J., "Computationally Convenient State Variables for Bond Graphs of Two-Phase Accumulators", Trans. ASME J. Dynamic Systems, Measurement and Control, v. 105, n. 3, Sept. 1983, pp. 202-204.

6. Engja, H. and Xinle, J., "A Nonlinear Mathematical Model of a Marine Boiler Using Bond Graph Techniques", I.S.M.E., Tokyo, 1983.

7. Engja, H. and Strand, K., "Modelling for Transient Performance of Diesel Engines Using Bond Graphs", I.S.M.E., Tokyo, 1983.

8. Doebelin, E. O., System Modeling and Response, John Wiley, N.Y., 1980, pp. 48-53, 179-187.

9. Anderson, B. W., The Analysis and Design of Pneumatic Systems, John Wiley, N.Y., 1967, p. 19.

10. Karnopp, D., "Direct Programming of Continuous System Simulation Languages using Bond Graph Causality", Trans. Soc. for Computer Simulation, v. 1, n. 1, May 1984, pp. 49-60.

11. Granda, J., "CAMP User Manual", Dept. of Mechanical Engineering, California State University, Sacramento, CA 95819.

12. ACSL User Guide/Reference Manual, Mitchell and Gauthier, Assoc., Inc., 290 Baker Ave., Concord, MA 01842, 1981.

13. Granda, J. J. and Pourrahimi, F., "Computer Graphics Techniques for the Generation and Analysis of Physical System Models", Preprints, Computer Graphics and Simulation Conference, Society for Computer Simulation, San Diego, CA, Jan 23-26, 1984.

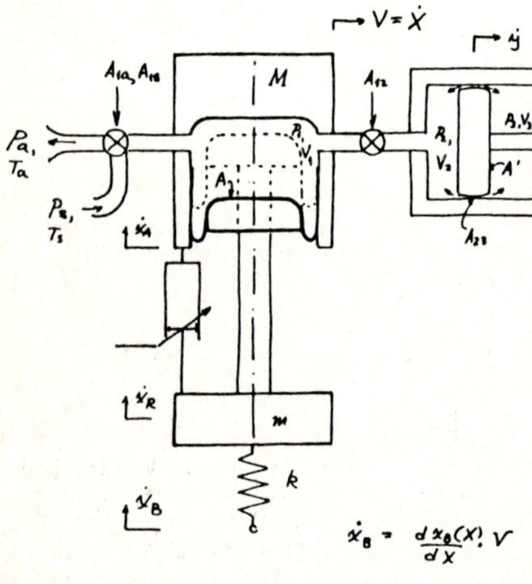

Figure 1 Schematic Diagram of Air Suspension Unit

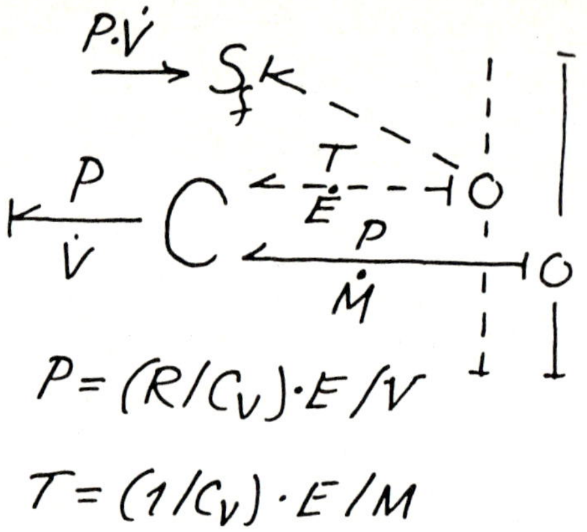

Figure 2 Ideal Gas Accumulator

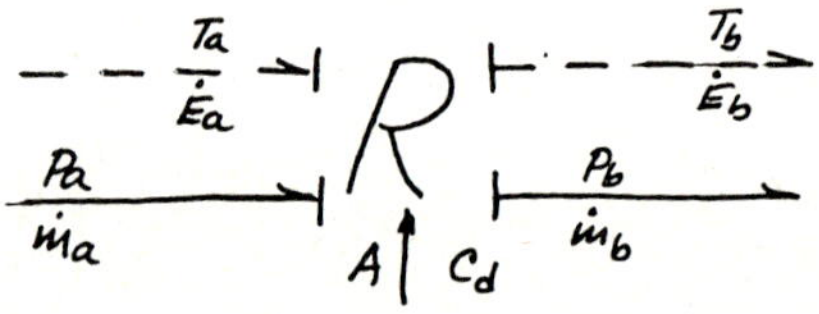

Figure 3 Pseudo Bond Graph for Restrictor

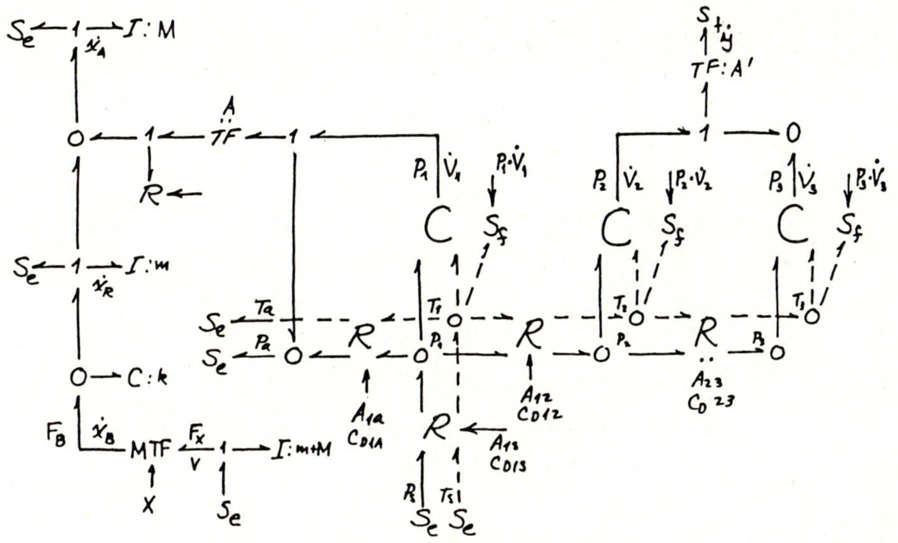

Figure 4 Bond Graph with Physical Variables Corresponding to Fig. 1

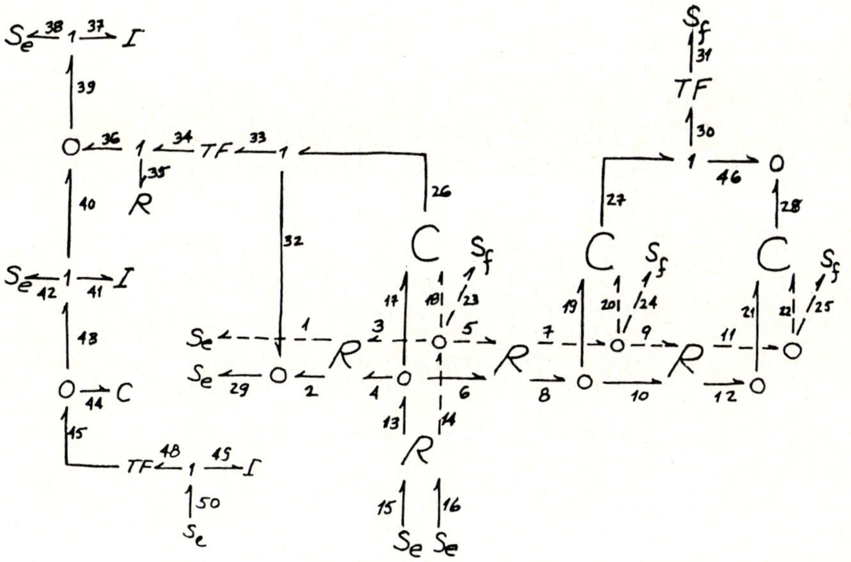

Figure 5 Bond Graph with Numbered Bonds

```
SE50,1 48 49 50,I49,TF48 45,
SE15,SE16,R13 14 15 16,
0 13 6 4 17,0 14 18 23 3 5,R1 2 3 4,
R5 6 7 8,C17 18 26,0 8 10 19,
0 7 20 24 9,R9 10 11 12,0 12 21,
0 11 22 25,C19 20 27,C21 22 28,
1 32 26 33,0 29 32 2,TF34 33,1 34 35 36,
0 43 44 45,1 40 41 42 43,0 39 36 40,
1 37 38 39,1 27 30 46,0 46 28,TF30 31,
SE38,I37,SE42,I41,C44,R35,SE1,
SE29,SF23,SF24,SF25,SF31
```

Figure 6 Line Code Input for CAMP Corresponding to the Bond Graph of Fig. 6

```
  F5=E5*R55+E6*R56+E7*R57+E8*R58
  F6=E5*R65+E6*R66+E7*R67+E8*R68
  F7=E5*R75+E6*R76+E7*R77+E8*R78
  F8=E5*R85+E6*R86+E7*R87+E8*R88
DQ17=F17                         $
DQ26=F26                         $
  E18=Q17/C1817+Q18/C1818+Q26/C1826
  E26=Q17/C2617+Q18/C2618+Q26/C2626
  F19=F8-F10                     $
  E10=E19                        $
  E7=E20                         $
  E9=E20                         $
'...... STATE VARIABLES ......'
    F49= INTEG (DF49,F49IN)
    Q17= INTEG (DQ17,Q17IN)
    Q18= INTEG (DQ18,Q18IN)
    Q26= INTEG (DQ26,Q26IN)
```

Figure 7 Portion of CAMP Output

```
MACRO ACCUM (PV,PM,TEMP,VOL,MASS,ENGY)
PROCEDURAL (PV,PM,TEMP=VOL,MASS,ENGY)
PV= (R/CV)*(ENGY/VOL)
PM= PV
TEMP= (1./CV)*(ENGY/MASS)
END $'OF PROCEDURAL'
MACRO END


MACRO VALVE(MA,MB,EA,EB,PA,PB,TA,TB,A,CD)
PROCEDURAL (MA,MB,EA,EB=PA,PB,TA,TB,A,CD)
PU=PB
PD=PA
TU=TB
IF(PA .GT. PB) PU=PA
IF(PA .GT. PB) PD=PB
IF(PA .GT. PB) TU=TA
PR=PD/PU
IF(PR .LE. PCR) PR=PCR
MDOT=A*CD*(PU/SQRT(TU))*SQRT(2.0*GA/(R*(GA-1.0))) ...
*SQRT(PR**(2.0/GA)-PR**((GA+1.0)/GA))
MA=MDOT
IF(PB .GE. PA) MA=-MDOT
MB=MA
EA=MA*CP*TU
EB=EA
END $'OF PROCEDURAL'
MACRO END
```

Figure 8 Accumulator and Restrictor Macros

```
  VALVE(F4,F2,F3,F1=E4,E2,E3,E1,A1A,CD1A)
  VALVE(F6,F8,F5,F7=E6,E8,E5,E7,A12,CD12)

DQ17=F17                          $ DQ18=F18
DQ26=F26

  ACCUM(E26,E17,E18=Q26,Q17,Q18)
```

Figure 9 Portion of Edited ACSL Input File

Complex and Distributed Systems: Analysis, Simulation and Control
S.G. Tzafestas and P. Borne (editors)
Elsevier Science Publishers B.V. (North-Holland)

BOND GRAPH AND ELECTROHYDRAULIC SYSTEMS MODELISATION OF THE FEEL FORCE SYSTEM USED ON FLIGHT SIMULATOR

Michel LEBRUN

Laboratoire d'Automatique, Université Claude Bernard Lyon I,
Bât 721, 43 Boulevard du 11 Novembre 1918,
69622 Villeurbanne Cedex, France.

Hydraulic and electrohydraulic systems are made up with sub-system with lumped-parameters exchanging energy by means of hydraulic lines. The "distributed parameter" characteristic of the lines yields modelisation and simulation difficulties of such systems. Many works dealt with this problem in which the researchers suggested to combine the concepts and the techniques that are specific to the representation and treatment methods of these two system categories. From an example, in this article, we show the applicability of the bond-graph technique so as to give a unified representation which could be numerically used by means of simulation program of the type CSMP, ACSL, TUTSIM etc. A feel force system, used in flight simulation, was chosen as an example. It is mainly constituted of an electrohydraulic servovalve, a hydraulic jack, a mechanical structure, a force captor and an electronic control. The developped model takes into account the principal physical phenomena presenting a significant influence.

1. INTRODUCTION

Historically, the study of hydraulic systems presents two aspects : one of them deals more particularly with the hydraulic lines analysis, whose distributed character is intrinsic, and the other one is orientated towards the study of the subsystems connected to the lines. They are generally described as lumped parameters. The numerous works done on the subject show a priority for one of the afore-mentioned aspects, introducing very restrictive hypotheses on the other aspect. In order to make a more accurate analysis, helping complex system optimization, mixing hydraulic lines with hydraulic components, the lastest works suggest to interface the treatment techniques of each category of the subsystems [1] and [2] suggest to use the characteristic methods as well as the integration procédures of the non-linear differential equations.

For some years, we have been confronted to that problem, so it seemed interesting to suggest a solution based on a bond-graph technique [3]. In previous works, it has been shown how this approach enabled to modelize hydraulic lines precisely [4] and [5]. Many articles show an interest of the bond graph technique in electrohydraulic systems [6] to [8]. In this paper, we suggest to illustrate this point of view through a specific example. The system studied is a "feel force system" used on flight simulator. The following effects of radial clearance, rounded corners of the spool-valve, the evolution of the flow coefficient in fonction to the flow number and the pressure differences relative to each spool-valve orifice. For the hydraulic jack and the mechanical structure, the coulomb friction, the clearance of the joints of the mechanical structure. For the supply circuit, the input and output hydraulic lines. The theoretical results obtained by treatment of the model with CSMP III are compared to experimental measurements in order to verify the model validity.

2. DESCRIPTION OF THE SYSTEM

The feel force system now examined is shown in mechanical schematic form in Fig. 1.

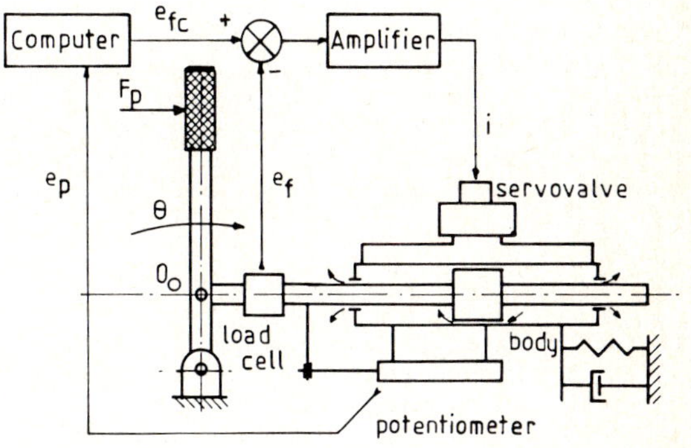

Figure 1. Electrohydraulic control loading system.

The control loading system senses the pilot input force by means of a load cell in the actuator-column linkage and transmits an electrical signal e_f to the amplifier which calculates the fluid-flow control signal by difference between

e_f and e_{fc}. In the system, the position feedback signal e_p in turn commands the effort system e_{fc} by computer. The computer can be programmed to provide a wide range of dynamic characteristics that are typical of control loads which occur in real aircraft.

As Fig. 1 indicates, the control loading comprises a double closed-loop system whose input force and output position are transmitted through the same element, that is, the pilot's control column. Ideally the inner loop, which contains the servovalve and actuator, should have a fairly flat frequency response over a wide bandwidth.

3. MODEL OF THE SYSTEM

The system represented on Fig. 1 is described by the bond graph of Fig. 2.

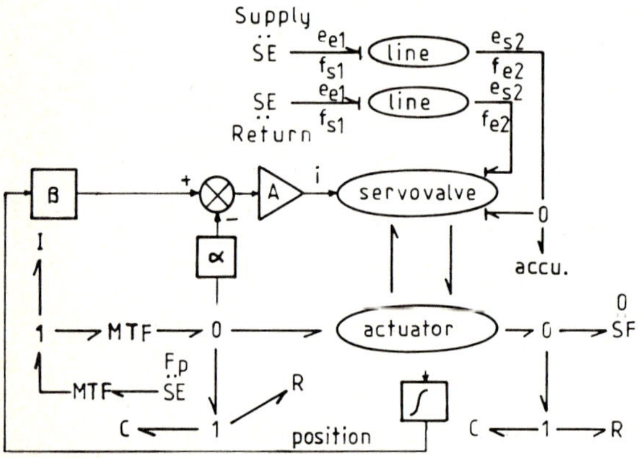

Figure 2. Bond graph model of the system.

On the figure, the various sub-system exchanging energy can be seen, supply pressure supposed constant, hydraulic lines input and output, accumulator, servovalve, hydraulic jack, mechanical linking elements and column linkage. On the bond graph, the control system is superposed by means of signals. This view point is adopted as that part is not energetic interaction with the system.

3.1. Modelisation of the different parts

Hydraulic lines. The sub-system lines are connected to a pressure supply on the hand, and to the servovalve on the other hand. The two sub-systems respectively present a "pressure" causality as well as a "flow" causality. Among the suggested formulations in the paper [4], the mixed causality is to be adopted. The state equations of the lines are thus written :

$$\dot{\underline{p}} = H1^t |e_{e1} e_{s2}|^t - K_1 \, q_1 - R \, \dot{\underline{q}}_1 \quad (1)$$

$$\dot{\underline{q}}_1 = M^{-1} \, \underline{p} \quad (2)$$

$$e_{s2} = C_{EQ}^{-1} \left(\int f_{e2} dt + L2H1 \, q_1 - C_R \, e_{e1} \right) \quad (3)$$

where

$$C_{EQ} = \frac{2AL}{r^2 B} \sum_{j=1}^{\ell} \frac{1}{(n+j)^2} \quad (4)$$

$$C_R = \frac{2AL}{r^2 B} \sum_{j=1}^{\ell} \frac{(-1)^{n+j}}{(n+j)^2} \quad (5)$$

L2H1 = second line of H1 matrix (6)

$$H1 = \begin{vmatrix} Y_0^1 & Y_1^1 & \dots & Y_n^1 \\ Y_0^2 & Y_1^2 & \dots & Y_n^2 \end{vmatrix} = \begin{vmatrix} 1 & 1 & \dots & 1 \\ 1 & -1 & \dots & (-1)^{n+1} \end{vmatrix} \quad (7)$$

where Y_j^k are the mode shape functions with j the jth mode and k the kth input and with :
the index ℓ must be as high as possible and n is the number of modes retained,
r = radius of the line,
ρ = specific mass, L = length of the line,
A = cross-sectional area of the line,
B = bulk modulus,
ν = kinematic viscosity,

$$\underline{e}_s = |e_{s1} \; e_{s2}|^t \text{ output pressure vector} \quad (8)$$

$$\underline{f}_e = |f_{e1} \; f_{e2}|^t \text{ input flow vector} \quad (9)$$

$\underline{p} = |p_0 \; p_1 \; \dots \; p_{n-1}|^t$ modal momenta for the n inertia elements inclued in the model where $\dot{p}_i = e_i$ = pressure acting in the ith element bond (10)

$\underline{q} = |q_0 \; q_1 \; \dots \; q_{n-1}|^t$ modal displacements for the n modal compliances associated with the n included dynamic modes where $\dot{q}_i = f_i$ = flow acting on the ith compliance element bond (11)

$$M = \begin{vmatrix} I & & 0 \\ & I_i & \\ 0 & & I_{n-1} \end{vmatrix}$$ where I_i is the modal mass from Table 1 (12)

$$K_1 = \begin{vmatrix} 0 & & 0 \\ & K_i & \\ 0 & & K_{n-1} \end{vmatrix}$$ where K_i is the modal stiffness from Table 1 (13)

$$R = \begin{vmatrix} R. & & 0 \\ & R_i & \\ 0 & & R_{n-1} \end{vmatrix}$$ where R_i is the dissipative term from Table 1 (14)

Table 1.

	I	C	R
i = 0	$I_0 = \frac{\rho L}{A}$	∞	$R_0 = F \frac{\rho L}{A}$
i = 1,2,∞	$I_i = \frac{\rho L}{2A}$	$C_i = \frac{2AL}{\pi^2 \rho c^2 i^2}$	$R_i = F \frac{\rho L}{2A}$

$$c^2 = B/\rho \qquad K_i = 1/C_i \quad (15)$$

3.2. Accumulateur

In Fig. 3 a typical example of a gas-accumulator is depicted. A compressed gas (mostly N_2) is separated from the oil by an elastic bladder inside a steel housing.

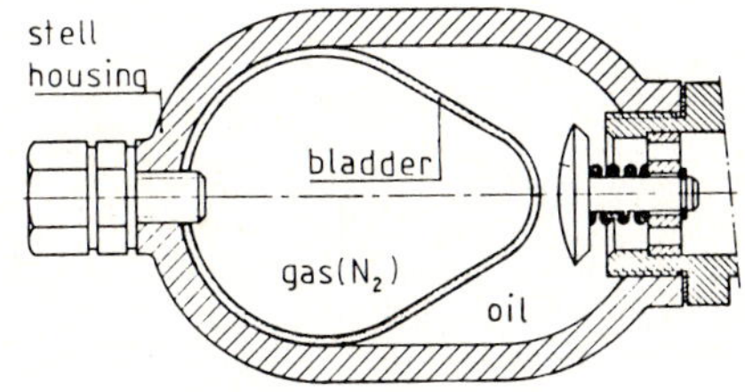

Figure 3. Gas-accumulator

Assuming an adiabatic process in a perfect gas with pressure P_g and volume V_g, we have $P_g V_g^{\gamma}$ = constant. This algebraic relation between pressure and volume variables express a capacitive element [3]. If we taking into account the resistive effect and inertial effect at the short line linked the gas-accumulator and the hydraulic line, the bond graph is describe in Fig. 4.

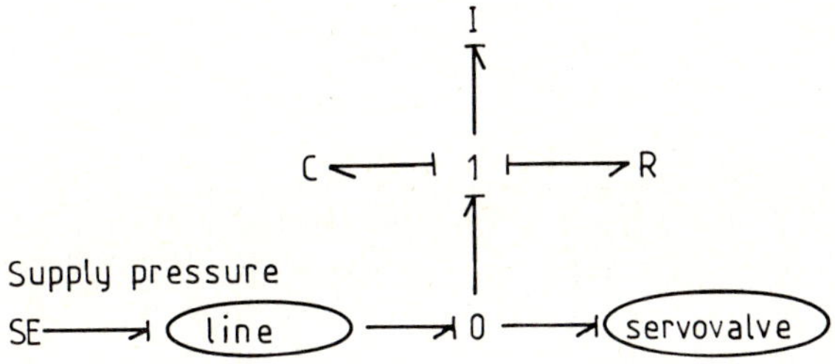

Figure 4. Bond graph model of the gas accumulator

3.3. Servovalve

The servovalve studied, represented on Fig. 5 is a two-stage model fabricated by SOPELEM. (French Society). (See description in the Annexe A).

Bond graph model of the servovalve. Adopting the hypothesis of a rotation and translation deformation of the flexure tube, we showed in [9] that the model of the electromechanic part of the first stage can be modelized by bond graph. After we applied the singular perturbation technique, the model is reduced to that of Fig. 6 in which the armature translation is algebraically linked to its rotation. On the bond graph os the Fig. 6 we have the source of the flux, SF(29), modulated by the input control current. This current is applied with an hysteresis which enable to take into account the influence of the armature magnetic hysteresis of the torque-motor [7]. On the bond graph of the Fig. 6, the bond graph of the electromechanic part is extended by the bond graphs of the flapper valve and the spool-valve. On the bond graph, the input and return pressures P_o and P are represented by the sources SE(105) and SE(96). The hydraulic resistance of restrictions N_1, N_2 and N_3 are modelled by $R_1(42)$, $R_2(59)$ and $R_3(62)$.

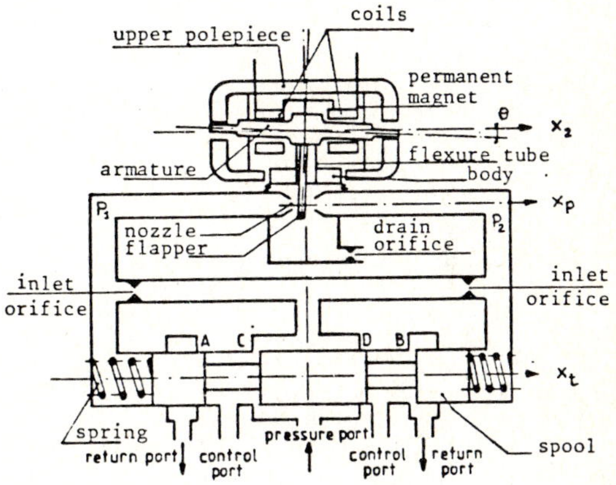

Figure 5. Two stage servovalve

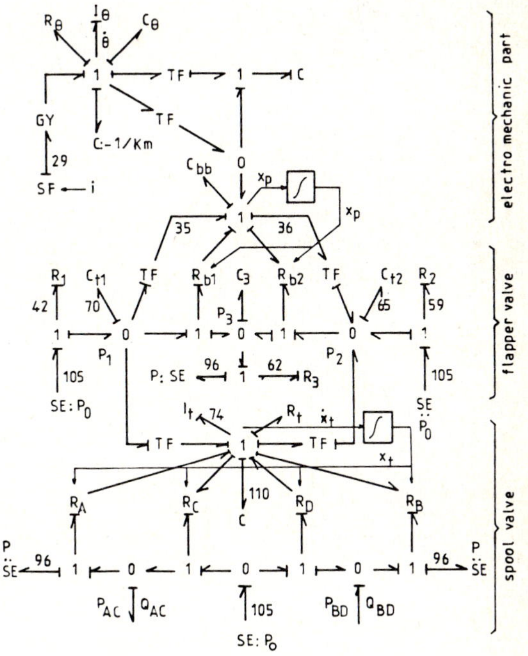

Figure 6. Bond graph model of the servovalve

The capacitances $C_{t1}(70)$, $C_{t2}(65)$ modulated by the spool displacement x_t models the influence of the compressibility of the fluid in the two cavities at the extremes of the spool. The static pressures P_{70} and P_{65} creates, through the two transformers TF, the efforts e_{35} and e_{36} acting on the flapper. The influence of the annular restrictions created by the flapper and the two nozzles is represented by the double port resistances R_{b1}, R_{b2} modulated by the flapper displacement x_p. Each of them has a hydraulic and mechanical ports, modelling the hydraulic losses across the nozzle as well as its mechanical effect on the flapper. The capacitance C_3 models the effect of the fluid compressibility in the return chamber. The spool whose inertia is $I_t(74)$ is subjected to the following actions ; the effort due to the pressure difference created by the flapper valve, the effort due to springs of centering C(110), the effort due to the viscous friction modelled by R_t, the flow forces in the double-ports of the four restrictions of the spool valve, R_A, R_B, R_C and R_D, modulated by the spool displacement x_t.

3.4. Flow laws in the spool-valve orifices

The four way spool-valve model has been developped where radial clearance was taken into account as well as rounded corners and the evolution of the flow coefficient in function of the flow number and the pressure difference applied to each orifice. A detailed work is presented in previous paper [10]. (See the principal equations in the Annexe B).

3.5. Hydraulic jack

The bond graph model of the jack is described in Fig. 7 where a mechanical resistive element expressing coulomb frictions can be seen, as well as resistive elements taking the internal and external leakage of the jack into account. The capacitive effects of the fluid in the volumes of the jack chambers are also considered.

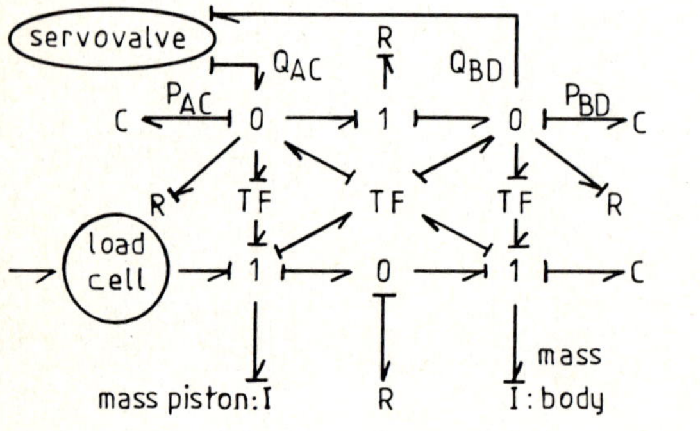

Figure 7. Bond graph model of the hydraulic jack

3.6. Mechanical links

The mechanical links are characterized by their stiffness and dissipative effects. In order to take clearance into account, the effort-displacement characteristic of the links includes a dead zone.

In associating the bond graph of the different subsystems, according to the structure of the Fig. 2, we obtained the model of the system.

4. SIMULATION AND EXPERIMENTAL RESULTS

Using the manufacturer's data as well as specific measures, the various parameters of the model have been quantified. It has been numerically calculated from a simulation program CSMP III. Among the various calculations ot the model, the result, concerning the step effort applied to the column linkage is presented in this paper. The dynamic behaviour of a feel force system fairly well characterized. It consists in applying abruptly or with drawing an effort F_{po} on the column linkage. The evolution of jack piston displacement is then observed as well as the effort in function of time. Ideally, the system response should be an undamping oscillation whose pulsation is written :

$$\omega_n = \sqrt{K/M} \tag{16}$$

where :
K is the simulated stiffness determined by the coefficients α and β of Fig. 2 such as :

$$K = \alpha/\beta \tag{17}$$

M is the column linkage equivalent mass linked to the point 0_o of the captor.

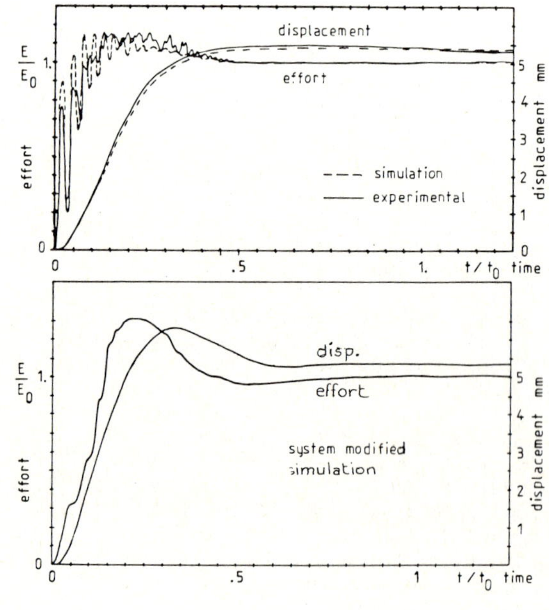

Figure 8(a,b). Responses of the system

The Fig. 8a shows a very different experimental result from the ideal case. In fact, the limited pass band of the closed loop of the effort servo mechanism explains this phenomenom qualitatively. As shown on Fig. 8 the theoretical result is compared to an experimental measure. The satisfactory correlation appearing between the curves show the validity of the model.

The model has been used to make a sensitiveness analysis to the various conception parameters. From this analysis, two important points appear:
- A model simplification cannot be conceived without the loss of good agreement between simulation and experimentation.
- The adaptation of certain functional parameters show that a very good improvement of the system behaviour can be obtained.

The simulated responses of the system thus reconsidered are shown of Fig. 8b. For reasons of industrial properties, the modified parameters will not be considered in this paper.

5. CONCLUSION

In this paper, from the modelization and the simulation of a feel-force system, it has been possible to show the applicability of bond graphs in order to take into account the hydralic lines as well as the complex subsystems connected at their ends.

This unified approach is justified by the validity of the model of the feel force system where a good correlation between theoretical and experimental results appears.

ACKNOWLEDGMENT

The author is indebted to SIGNABIEUX and CERNEAU for developing experimental apparati and the data used in this paper. This work was supported by the THOMSON-CSF (division simulateur) Société (TRAPPES-FRANCE).

REFERENCES

[1] Sharbek-Wazinski, C.M., "Wave propagation by the method of characteristics", Phd Thesis, University of Bath, (1981).
[2] Bowns, D.E., Bonson, L.A., Richards, C.W., and Caney, K. "The simulation of hydraulics systems", IFAC Pneumatic & Hydraulic Components, Warsaw, Poland (1980).
[3] Karnopp, D.C. and Rosenberg, R.C., "System dynamics : A unified Approach", (Wiley, New York, 1975).
[4] Lebrun, M., "Normal modes in hydraulic lines," Proc. A.C.C., San Diego, California, (6-8 June 1984).
[5] Lebrun, M., "The use of modal analysis concepts in the simulation of pipeline transients", J. Franklin Inst. Vol. 319, n°1/2, pp. 137-156, (1985).
[6] Dransfield, P., "Using Bond Graphs in simulating an electro-hydraulic system", J. Franklin Inst., Vol. 308, n°3, September (1979).
[7] Rabie, G., Lebrun, M., "Modélisation par les graphes à liens et simulation d'une servovalve électrohydraulique à deux étages", R.A.I.R.O., Automatique - Systems Analysis and Control, Vol. 15, n° 2, pp. 97 à 129 (1981).
[8] Grabowiecka, A. and Grabowiecki K.A., "Application of bond graphs to the digital simulation of a two-stage relief valve dynamic behaviour", I.F.A.C., Pneumatic & Hydraulic Components, Warsaw, Poland (1980).
[9] Lebrun, M. and Dauphin-Tanguy, G., "Study of the first stage servovalve using bond graphs and singular perturbation method". Submit to the ASME Revue.
[10] Lebrun, M., "A model for a four-way spool valve applied to a pressure control system". Submit to the ASME Revue.

ANNEXE A

Description of the servovalve

The servovalve studied, represented on Fig. 5 is a two-stage model, fabricated by SOPELEM, commounly used in industry. The servovalve consists of a polarized electrical torque-motor and two stages of hydraulic power amplification. The motor armature extends into the air gaps of the magnetic flux circuit and is supported in this position by a flexure tube member. The flexure tube acts as a seal between the eletromagnetic and hydraulic sections of the valve. The two motor coils surround the armature, one on each side of the flexure tube.
The flapper of the first-stage hydraulic amplifier is rigidly attached to the mid point of the armature. The flapper extends through the flexure tube and passes between two nozzles, creating two variable orifices between the nozzle tips and the flapper. The pressures P_1 end P_2 controlled by the flapper-and-nozzle variable orifices are fed to the end areas of the second stage-spool.
The second stage is a conventional four-way spool valve.

ANNEXE B

Principal equations of the spool valve

Flow laws in the spool-valve orifices. The flows Q_A, Q_B, Q_C and Q_D, relative to the hydraulic resistors R_A, R_B, R_C and R_D of the spool valve, Fig. 5, are deduced from the following expressions :
in underlap position Fig. 1B(a)

$$Q = C_q\, S(x_t) \sqrt{(2/\rho)(P_u - P_d)} \qquad (1.B)$$

in overlap position Fig. 1B(b)

$$Q = \frac{n\, L_o\, c_o^3}{12\mu(x_t + K_g)} (P_u - P_d) \qquad (2.B)$$

where

$$C_q = f(\lambda,\ P_u - P_d) \qquad (3.B)$$

$$S(x_t) = n\, L_o(\sqrt{x_t^2 + (c_o + x)^2} - x_o \qquad (4.B)$$

$$\lambda = \frac{4\, S(x_t)}{2\,\pi\, D_t\, \nu}\sqrt{2(P_u - P_d)/\rho} \qquad (5.B)$$

with : P_u = upstream static pressure,
P_d = downstream static pressure,
ρ = specific mass of the fluid,
x_t = spool displacement,
L_o = length of an orifice,
n = number of orifices,
c_o = radial clearance,
D_t = the spool diameter,
ν = kinematic viscosity,
K_g is determined so as to ensure the flow continuity in the underlap position.

The curve network $C_q(\lambda,\ P_u - P_d)$ is represented by the arbitrary function generator of two variables of the simulation program CSMP III. A detailed determination method of $C_q(\lambda,\ P_u - P_d)$ is presented in previous paper [10].

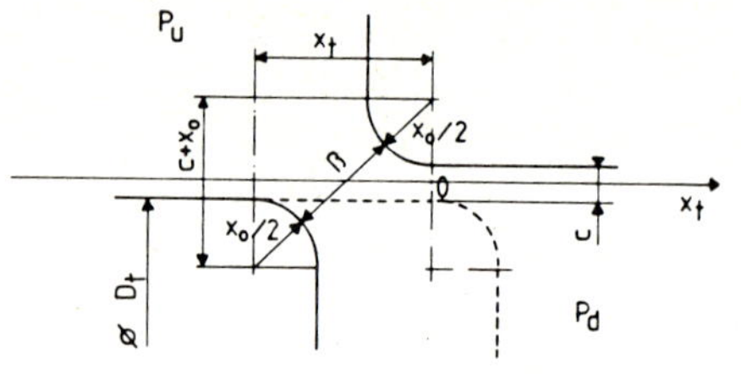

(b)

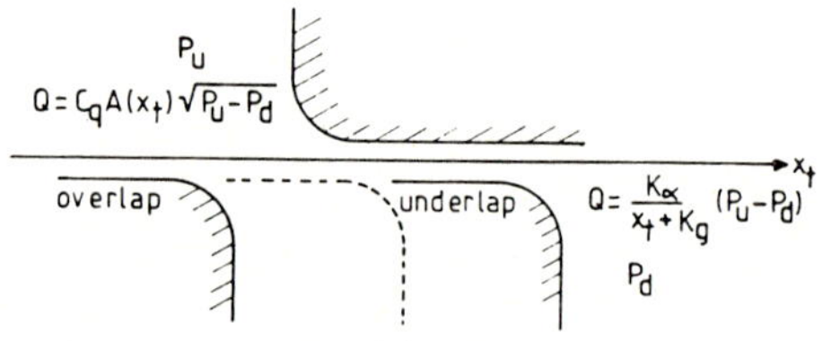

(c)

Figure 1B(a,b). Detail of on orifice of the spool-valve.

Complex and Distributed Systems: Analysis, Simulation and Control
S.G. Tzafestas and P. Borne (editors)
Elsevier Science Publishers B.V. (North-Holland)

MODELLING OF RIGID BODY DYNAMICS AND ITS COMPUTER SIMULATION

Masato Outa, Hideo Nakano and Takehiko Kawase

Department of Mechanical Engineering, Waseda University
3-4-1 Okubo, Shinjuku, Tokyo 160, Japan

A symbolic generation scheme of the system equations of multibody systems is proposed. To support this, the required kinematical and dynamical information are categorized into the physical, structural and causal. And they are all incorporated into the bond graph. It is shown that the above symbolic generation scheme is quite easily formulated by using the above bond graph. The scheme utilizes MACSYMA, and some numerical simulation results are illustrated to show the effectiveness of the proposed scheme.

1. INTRODUCTION

Major stumbling blocks we encounter in model making of rigid body systems like artificial manipulators lie in the processes of 1) deriving great number of governing equations of the system in any correct form and 2) the numerical integration. The first comes from the increase of the degrees of freedom as the number of the involved bodies increases, and the second from the strong stiffness of the resultant implicit nonlinear algebraic differential equations.

One possible way to overcome the first is to structurize the required kinematical and dynamical information. To do this, in the present paper, a categorization of such information into the structural, causal and physical is proposed using the bond graph concepts. The bond graph enables us to incorporate these into a graph and above all to envisage the system structure. In doing so, the non-energic multiport which is the mathematical expression of the system structure plays a dominant role.

The system equations are first derived for the fundamental pair or the pair of a mechanical joint and one of the adjacent bodies, and then tabulated for the primitive system. Here by the term primitive, we simply mean the totally torn apart system. The final form of the system equations is obtained by eliminating the inner variables from the above system equations for the primitive system. The process is much laborious and tedious. So a method to do this symbolically on the computer is proposed. The method consists of two steps; 1) symbolic reduction of the equations of the primitive system into those of the interconnected and 2) symbolic substitution of the functions representing the coordinate transformations, kinematical constraints, mechanical joints and other system parameters all of which are stored a priori in data bases.

On the other hand, for the second problem stated previously, the BDF method utilizing the sparse matrix algorithm is proposed.

2. FUNDAMENTAL BOND GRAPH TOOLS

Fundamental tools of the bond graph to be used in the paper are listed in Table 1. Any modelling method of dynamical systems, implicitly or explicitly, requires three categories of information, that is, the *structural, causal* and *physical*. The first describes how the system components are interconnected, the second the consistent allocation of input and output variables among the system variables and the last represents dynamical properties of the system components.

Any element in the system is evaluated as a multiport whose dynamical interaction with the other multiport is expressed by the *power bond* ⇀ . The power bond represents the bond through which dynamical energy is transmitted toward the direction indicated by the half arrow and whose dynamical states are represented by the dual variables, that is, velocity and force. The double lined bond ⇒ indicates the multibond which is the juxtaposition of power bonds and whose dynamical states are described by a pair of vector variables of velocity and force.

Table 1 Symbols of the bond graph

Symbols	Meanings
SE⇒	External force source
SF⇒	External velocity source
I ⇒	Moment of inertia
M ⇒	Mass
⇓ ⇒1⇐	1-junction
⊢MTF⊣	MTF
⊢EJS	EJS

The first four elements in Table 1 represent one ports two of which are the independent power sources and the remaining two the inertia and mass elements. The *physical* properties of these one ports are determined by their dynamics.

Now we easily see that any numerical simulation scheme requires, implicitly or explicitly, the consistent allocation of input and output variables throughout system variables. This is called the *causality* in the BG theory and symbolically designated by the stroke attached to one of the two sides of each power bond($\vdash\!\!\rightharpoonup$ or $\rightharpoonup\!\!\dashv$). Thus A $\vdash\!\!\rightharpoonup$ B indicates that, for the element A, velocity is the output and the input is then force, and for the element B vice versa.

The consistency of such allocation of causal strokes is checked by the following rules;

1) Any independent power source uniquely determines the stroke depending upon whether it is a velocity source or a force source.
2) Each bond has to be stroked at just one side of it.
3) Every 1-junction has to have the causality indicating the velocity input at just one bond.
4) Every MTF and EJS has the causality determined by their causal relations.
5) The remaining bonds attached to the inertia or mass are free from the causal stroke assignment.

The principal tool of the BG is the *non-energic multiport* which describes the way of interconnection of the system components. It is a so-called workless constraint in mechanics. Let us denote the generalized coordinate and its conjugate force by q and R respectively. The non-energicness[1] is then described as

$$\langle \dot{\bar{q}}, \bar{R} \rangle = 0 \quad \text{for every time t} \qquad (1)$$

where < , > indicates the scalar product of vectors, $(\bar{\ })$ the column vector and $(\dot{\ })$ the time derivative. In the paper, we will use four categories of such non-energic multiports.

1. *1-junction*

The 1-junction represents the equilibrium condition of forces or torques, which is the D'Alembert's principle.

2. *Modulated transformer(MTF)*

2.1 MTF for the coordinate transformation
Any coordinate transformation, integrable or non-integrable, is modelled as an MTF. Let us describe the coordinate transformation of our concern as

$$d\bar{\pi} = \underline{\bar{A}}(\bar{q};\bar{\sigma})\,d\bar{q} \quad \text{or} \quad \dot{\bar{\pi}} = \underline{\bar{A}}(\bar{q};\bar{\sigma})\dot{\bar{q}} \qquad (2)$$

where $\bar{\pi}$ stands for the new coordinate and $(\underline{\ })$ the matrix. In the above, $\bar{q}$ is an (n+1)-vector where q_{n+1} = t(time), and σ indicates the parameter like Euler angle. Now the non-energicness means the power invariant $\langle \dot{\bar{q}}, \bar{R} \rangle = -\langle \dot{\bar{\pi}}\ \bar{\Pi} \rangle$ where Π denotes the force observed along the new coordinate. By the above power identity, we easily find the relation between the forces;

$$\bar{R} = -\underline{\bar{A}}(\bar{q};\bar{\sigma})^{T}\bar{\Pi} \qquad (3)$$

In the above, $(\ )^{T}$ indicates the matrix transpose.

2.2 MTF for the kinematical constraint
Any smooth kinematical constraint is described by the following form which is not necessarily integrable;

$$\underline{\bar{\alpha}}(\bar{q})d\bar{q} = 0 \quad \text{or} \quad \underline{\bar{\alpha}}(\bar{q})\dot{\bar{q}} = 0 \qquad (4)$$

where $\underline{\bar{\alpha}}(\bar{q})$ denotes a k x (n+1) coefficient matrix(k<n). The kinematical constraint described above can be modelled as an MTF by permitting the appropriate quasi-coordinate $\bar{\pi}$ such that

$$d\bar{\pi} = \begin{pmatrix} \underline{\bar{\alpha}}(\bar{q}) \\ \text{---} \\ \underline{\bar{\beta}}(\bar{q}) \end{pmatrix} d\bar{q} \qquad (5)$$

where the matrix $\underline{\bar{\beta}}$ is arbitrary except that it is so chosen that the resultant coefficient matrix of eq.5 is non-singular.

2.3 MTF for the mechanical joint
Let $\dot{\bar{\pi}}^{*}$ and $\dot{\bar{\pi}}^{\circ}$ be the velocities of the (i-1)-th and i-th bodies respectively, and $\bar{\Pi}^{*}$ and $\bar{\Pi}^{\circ}$ the corresponding conjugate interactive forces.

Table 2 Excerpt from the catalog of ideal joints

Joint		Coefficient matrices
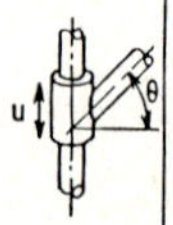	symbol 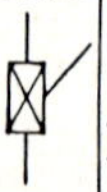 deg. of freedom 1	$\underline{\bar{C}} = \begin{pmatrix} 1 & 0 & 0 & 0 & -r_1 c\theta & -(u+r_0+r_1 s\theta) \\ 0 & s\theta & -c\theta & -(u+r_0)c\theta & 0 & 0 \\ 0 & c\theta & s\theta & (u+r_0)s\theta+r_1 & 0 & 0 \\ 0 & 0 & 0 & 1 & 0 & 0 \\ 0 & 0 & 0 & 0 & s\theta & -c\theta \\ 0 & 0 & 0 & 0 & c\theta & s\theta \end{pmatrix}$ $\underline{\bar{c}} = (0\ \ s\theta\ \ c\theta\ \ 0\ \ 0\ \ 0)^{T}$
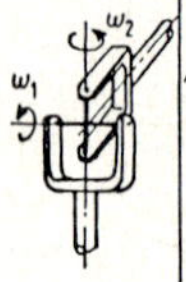	symbol  deg. of freedom 2	$\underline{\bar{C}} = \begin{pmatrix} -s\varphi c\theta & -s\varphi s\theta & -c\varphi & r_1 s\theta - r_0 c\varphi & -r_1 c\theta & r_0 s\varphi c\theta \\ -c\varphi c\theta & -c\varphi s\theta & s\varphi & r_0 s\varphi & 0 & r_0 c\varphi c\theta \\ -s\theta & c\theta & 0 & -r_1 s\varphi c\theta & -r_1 s\varphi s\theta & r_0 s\theta - r_1 c\varphi \\ 0 & 0 & 0 & -s\varphi c\theta & -s\varphi s\theta & -c\varphi \\ 0 & 0 & 0 & -c\varphi c\theta & -c\varphi s\theta & s\varphi \\ 0 & 0 & 0 & -s\theta & c\theta & 0 \end{pmatrix}$ $\underline{\bar{c}} = \begin{pmatrix} 0 & 0 & -r_1 c\varphi & -c\varphi & s\varphi & 0 \\ -r_1 & 0 & 0 & 0 & 0 & 1 \end{pmatrix}^{T}$ $\theta = \theta_0 + \int_0^t \omega_1 dt\ ,\quad \varphi = \varphi_0 + \int_0^t \omega_2 dt$

r_0, r_1 stand for the constants.

The relations which describe the kinematical relations of the mechanical joints are expressed below;

$$\dot{\bar{\pi}}^{\circ}_{i(rot)} = \dot{\bar{\pi}}^{*}_{i-1(rot)} + \bar{\Omega}$$

$$\dot{\bar{\pi}}^{\circ}_{i(tr)} = \dot{\bar{\pi}}^{*}_{i-1(tr)} - \dot{\bar{\pi}}_{i-1(rot)} \times \bar{r}_{i-1,i} + \dot{\bar{\pi}}^{\circ}_{i(rot)} \times \bar{r}_{i,i} + \bar{V} \qquad (6)$$

where $\bar{\Omega}$ and $\bar{V}$ indicate the driving velocities and $\bar{r}_{i,i}$ the distance between the i-th joint and the i-th body. In the above, the suffix rot and tr stand for the rotational and linear motions respectively. Then we have the following relations[3].

$$\begin{pmatrix} \dot{\bar{\pi}}^{\circ} \\ \bar{\Pi}^{*} \\ \bar{R} \end{pmatrix} = \left(\begin{array}{c|cc} \bar{0} & \bar{C}(\bar{u}) & \bar{c} \\ \hline -\bar{C}(\bar{u})^{T} & & \\ -\bar{c}^{T} & \multicolumn{2}{c}{\bar{0}} \end{array}\right) \begin{pmatrix} \bar{\Pi}^{\circ} \\ \dot{\bar{\pi}}^{*} \\ \dot{\bar{u}} \end{pmatrix} \qquad (7)$$

In the above, $\dot{\bar{u}}$ and $\bar{R}$ indicate the driving velocity and the required torque at the joint. Some illustrative examples from the catalog of mechanical joints are shown in Table 2.

3. *Eulerian Junction Structure(EJS)* [2]

Any free body, when its motion is observed from the local frame whose origin is fixed to the mass center and whose axes coinside with the principal axes, is subjected to the interacting inertia forces which satisfy the non-energicness. The set of such forces is then modelled as a 3-port called Eulerian junction structure whose dynamical relation is described as $\bar{f} = \bar{S}\,\bar{p}$ where $\bar{f}$ and $\bar{p}$ stand for the interacting force and momentum respectively, and the coefficient matrix $\bar{S}$ is given below.

$$\bar{S} = \left(\begin{array}{c|ccc} \bar{0} & & \bar{0} & \\ \hline & 0 & 0 & \alpha p_5 \\ \bar{0} & 0 & 0 & \beta p_6 \\ & -\alpha p_5 & -\beta p_6 & 0 \end{array}\right) \qquad (8)$$

The parameters used in the matrix $\bar{S}$ are given below; $\alpha = 1/I_3 - 1/I_2$, $\beta = 1/I_1 - 1/I_3$. In the above expression, the canonical variables q and p are used instead of the dual variables for we will use the Hamilton form of equations of motion.

3. MULTIBODY DYNAMICS

3.1 System equations of the primitive system

Let us consider the manipulator illustrated in Fig.1. The manipulator, when it does not have any kinematical loop, consists of *fundamental pairs* each of which is the pairing of a mechanical joint and one of the adjacent bodies. Now such a fundamental pair is symbolically represented as in Fig.2. The upper half indicates the linear motion of the mass center of the body and the lower half the rotational motion. The former is usually observed from the local frame fixed to the i-th body and hence to get the required expressions observed from the inertial frame, we need the successive coordinate transformations which are modelled as MTFs. On the other hand, the observation of the rotational motion from the body fixed frame induces the quasi-coordinate and this brings us the EJS mentioned previously.

To get the final form of the system equations

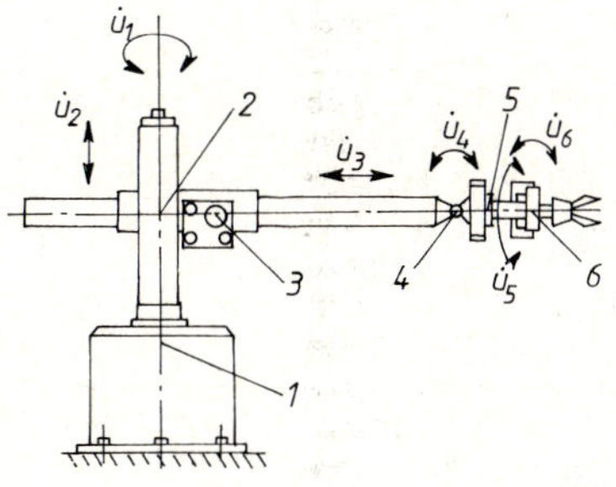

	joint type	mass kg	θ rad	r_0 m	r_1 m	I_1 kgm	I_2 kgm	I_3 kgm
1	RJ1	15	π/2	0.16	0.25	0.81	0.06	0.81
2	TJ1	4.2	0	0.15	0.1	0.3	0.04	0.31
3	TJ2	7.3	π/2	0.45	0.1	0.4	0.05	0.4
4	RJ2	1.3		0.06	0.1	0.007	0.003	0.098
5	RJ1	2.1	π/2	0.06	0.06	0.009	0.006	0.009
6	RJ2	1.6		0.06	0.06	0.007	0.003	0.007

Fig. 1 Manipulator and its dimensions

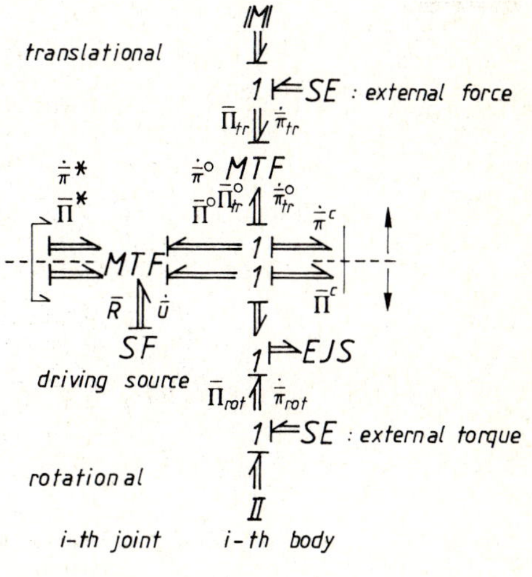

Fig. 2 Bond graph model of the fundamental pair

of the interconnected system of our concern, the system is first understood as a *primitive system* which means the juxtaposition of such fundamental pairs, and they are interconnected using the 1-junction representing the continuity relations of velocity and force between the adjacent fundamental pairs.

As to the equations of motion, we can not say, definitely, as Kane pointed out [6], that the particular form is most advantageous over the others in solving various problems which come up with the multibody dynamics. So we here use the Hamilton form.

The equations of motion of the i-th body are easily written down by inspecting the bond graph (Fig.2) together with the relations obtained so far.

$$\frac{d}{dt}\bar{x} = \bar{J}\,\nabla H + \bar{g} \qquad (9)$$

where the state vector $\bar{x}^T = (\bar{\pi}^T, \bar{p}^T)$ is made up with the translational coordinate observed from the inertial frame and the rotational coordinate observed from the body fixed frame together with the correspondent momentum. In the above, H indicates the Hamiltonian, $\bar{g}$ the forcing term and $\bar{J}$ the so-called symplectic matrix and ∇ the gradient operator. Furthermore the forcing term $\bar{g}^T = (\bar{0}^T, \bar{g}_2^T)$ is given below.

$$\bar{g}_2 = -\bar{f} + \bar{V} + \bar{\Pi}^c \qquad (10)$$

where $\bar{f}$ stands for the coupling force, $\bar{V}$ the external force and $\bar{\Pi}^c$ the interacting force with the (i+1)-th body. Thus the set of the system equations for the i-th fundamental pair is described as follows;

$$\begin{aligned}
&\dot{\bar{\pi}} - \bar{C}(\bar{u})\dot{\bar{\pi}}^* - \bar{cm}(t) = 0 && \text{joint constraint}\\
&\dot{\bar{\pi}} - \bar{B}\,\dot{\bar{\pi}}^\circ = 0 && \text{coordinate transformation}\\
&\dot{\bar{\pi}} - \dot{\bar{\pi}}^\circ = 0 && \text{1-junction} \qquad (11)\\
&\begin{pmatrix}\bar{\Pi}^*\\ \bar{R}\end{pmatrix} + \begin{pmatrix}\bar{C}^T\\ \bar{c}^T\end{pmatrix}[\,\bar{B}^T(\dot{\bar{p}} + \bar{H}_\pi^T + \bar{f} - \bar{V}) - \bar{\Pi}^c] = 0 &&\\
&\dot{\bar{\pi}} - \bar{H}_p^T = 0 && \text{equation of motion}\\
&\bar{m}(t) - \dot{\bar{u}} = 0 && \text{kinematic relation}
\end{aligned}$$

In the above, $\bar{H}_\pi$ and $\bar{H}_p$ indicate the partial derivative of the Hamiltonian with respect to $\bar{\pi}$ and $\bar{p}$ respectively and $\bar{m}(t)$ the function representing the driving velocity at the joint.

Now to get the system equations of the interconnected system, we need the continuity conditions of velocities and forces between i-th and (i+1)-th bodies, and they are described as

$$\begin{aligned}
&\dot{\bar{\pi}}^*_{i+1} - \dot{\bar{\pi}}^c_i = \bar{0}\\
&\bar{\Pi}^c_i - \bar{\Pi}^*_{i+1} = \bar{0}
\end{aligned} \qquad (12)$$

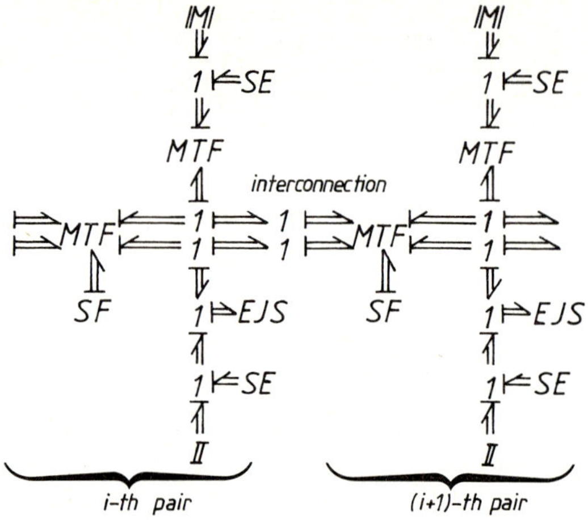

Fig. 3 Interconnection of two fundamental pairs

These equations, equ.9~12, have to be augmented by the following boundary conditions.

For the 0-th body; velocity constraints

For the end effector; force conditions

3.2 System equations of the interconnected system

The set of the total system equations of the fundamental pairs give us the original set of the system equations. These equations are deemed to describe the dynamics of the *primitive system* which consists of the separated fundamental pairs. Now to get those of the *interconnected system* of our present concern, we have to eliminate large number of velocities and forces. To do this by hand requires us laborious and tedious work. So it is strongly required to find any symbolic scheme to generate the required system equations on the computer. This will be the subject of the next chapter.

Thus we have more than approximately 370 equations in total number for the present system which has six degrees of freedom together with the relations representing the continuity conditions of force and velocity. Theoretically speaking, we ultimately have six equations whatever state variables we may employ. However from the stand point of symbolic generation of the system equations, it is not necessarily advantageous to have directly such ultimate form of the system equations. For it is more desirable to retain the original expressions of functions and equations as much as possible in order to meet the requirement of the change of system parameters and also the functions representing elements of the coefficient matrices of mechanical joints. So we set the three levels of *system reticulation*;

1) The original primitive system
2) The system corresponding to the step obtained by eliminating the valiables associated with the inner bonds of each fundamental pair and the valiables assigned with the interacting bonds between the adjacent fundamental pairs
3) The interconnected system

Depending upon such system reticulation we have 370, 54 and 6 equations in total number. Thus we have, for example, the final form of the system equations listed below.

$$
\begin{bmatrix}
\underline{B}_1^T \underline{\Gamma}_1 \bar{P}_1 - \underline{C}_1 \dot{\bar{U}}_1 \\
\begin{bmatrix} \bar{\pi}_1^* \\ \bar{R}_1 \end{bmatrix} + \begin{bmatrix} \underline{C}_1^T \\ \underline{C}_1^T \end{bmatrix} \Big[\underline{B}_1^T \bar{S}_1 + \underline{C}_2^T \big[\underline{B}_2^T \bar{S}_2 + \underline{C}_3^T [\underline{B}_3^T \bar{S}_3 + \underline{C}_4^T [\underline{B}_4^T \bar{S}_4 + \underline{C}_5^T [\underline{B}_5^T \bar{S}_5 + \underline{C}_6^T \underline{B}_6^T \bar{S}_6]]]\big]\Big] \\
\underline{B}_2^T \underline{\Gamma}_2 \bar{P}_2 - \underline{C}_2 \underline{B}_1^T \underline{\Gamma}_1 \bar{P}_1 - \underline{C}_2 \dot{\bar{U}}_2 \\
\bar{R}_2 + \underline{C}_2^T \big[\underline{B}_2^T \bar{S}_2 + \underline{C}_3^T [\underline{B}_3^T \bar{S}_3 + \underline{C}_4^T [\underline{B}_4^T \bar{S}_4 + \underline{C}_5^T [\underline{B}_5^T \bar{S}_5 + \underline{C}_6^T \underline{B}_6^T \bar{S}_6]]]\big] \\
\underline{B}_3^T \underline{\Gamma}_3 \bar{P}_3 - \underline{C}_3 \underline{B}_2^T \underline{\Gamma}_2 \bar{P}_2 - \underline{C}_3 \dot{\bar{U}}_3 \\
\bar{R}_3 + \underline{C}_3^T \big[\underline{B}_3^T \bar{S}_3 + \underline{C}_4^T [\underline{B}_4^T \bar{S}_4 + \underline{C}_5^T [\underline{B}_5^T \bar{S}_5 + \underline{C}_6^T \underline{B}_6^T \bar{S}_6]]\big] \\
\underline{B}_4^T \underline{\Gamma}_4 \bar{P}_4 - \underline{C}_4 \underline{B}_3^T \underline{\Gamma}_3 \bar{P}_3 - \underline{C}_4 \dot{\bar{U}}_4 \\
\bar{R}_4 + \underline{C}_4^T \big[\underline{B}_4^T \bar{S}_4 + \underline{C}_5^T [\underline{B}_5^T \bar{S}_5 + \underline{C}_6^T \underline{B}_6^T \bar{S}_6]\big] \\
\underline{B}_5^T \underline{\Gamma}_5 \bar{P}_5 - \underline{C}_5 \underline{B}_4^T \underline{\Gamma}_4 \bar{P}_4 - \underline{C}_5 \dot{\bar{U}}_5 \\
\bar{R}_5 + \underline{C}_5^T \big[\underline{B}_5^T \bar{S}_5 + \underline{C}_6^T \underline{B}_6^T \bar{S}_6 \big] \\
\dot{\bar{\pi}}_6^c - \underline{C}_6 \underline{B}_5^T \underline{\Gamma}_5 \bar{P}_5 - \underline{C}_6 \bar{U}_6 \\
\bar{R}_6 + \underline{C}_6^T \underline{B}_6^T \bar{S}_6 \\
\underline{B}_6 \dot{\bar{\pi}}_6^c - \underline{\Gamma}_6 \bar{P}_6
\end{bmatrix} = \bar{0}
$$

$$\bar{S}_i \equiv \dot{\bar{P}}_i - \bar{H}_\pi + \bar{f}_i - \bar{V}_i$$

4. SYMBOLIC GENERATION OF THE SYSTEM EQUATIONS

As started previously, the reduction process of the unnecessary system variables included in the original set of the system equations requires quite a laborious work. In the present paper, an algorithm which enables us to generate the set of the required system equations *symbolically* on the computer is presented. It utilized MACSYMA.

Main part of the algorithm is illustrated in Fig. 4. The algorithm has four data bases;

1) The form of the equations of the fundamental pair(DATA BASE 1)
2) The required form of the system equations (DATA BASE 2)
3) Functions representing the elements of the coefficient matrices of MTFs and EJSs (DATA BASE 3)
4) The form of the discretized system equations, tableau error vector and the coefficient Jacobi's matrix of the discretized system equations(DATA BASE 4)

The first two store the forms of the equations, the third the parameters and functions characterizing any paticular mutibody system and the last the form of the equations and matrices which will be needed in numerical integration.

The algorithm consists of the two steps. In the first step, the reduction of the original system equations of the primitive system into those of the interconnected is acomplished by eliminating the unnecessary variables determined depending upon the reticulation level mentioned previously. The most important operation is the elimination of the unnecessary variables. To make it easy, the following two matrices shown in Fig. 5 is used. The first matrix determines the allocation of system variables into the input and output variables. The matrix is easily computed by inspecting the causal strokes associated with each power bond. The second determines the

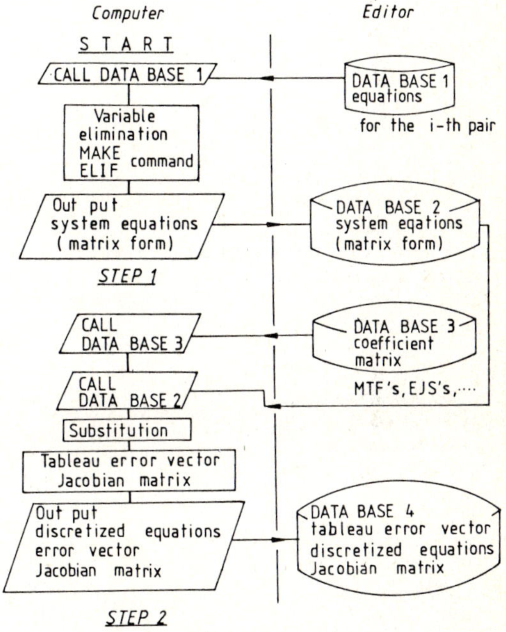

Fig. 4 Main part of the symbolic generation scheme

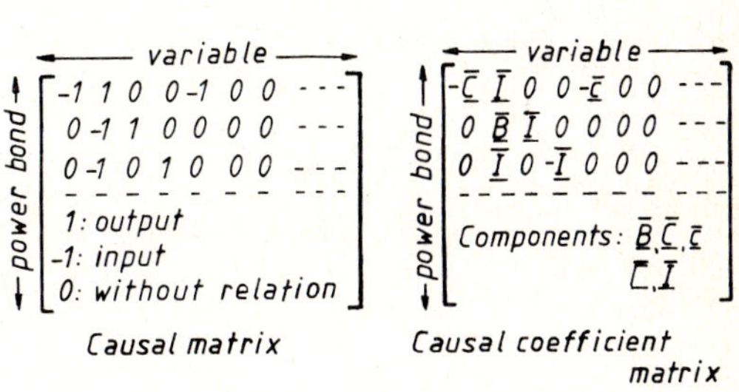

Fig. 5 Causal matrix and causal coefficient matrix

relations between the input and output variables so determined by the first matrix. These are all given in causal forms as stated previously. Thus the above elimination process of the system variables is easily systematized.

In the second, the substitution of the functions of the coefficient matrices of the MTFs and EJSs for the manipulator. Finally the contents of the DATA BASE 4 are symbolically generated using the results obtained so far. The last step is performed based on the relations described below.

The system equations obtained in the first step (DATA BASE 2) are formulated in the following implicit form.

$$\bar{G}(\bar{z},\dot{\bar{z}};t) = 0 \qquad (13)$$

where $\bar{G}$ stands for a vector function of the vectors $\bar{z}$ and its time derivative $\dot{\bar{z}}$ together with time t. In the above, the vector $\bar{z}$ represents

$$\bar{z}^T = (\bar{\pi}_1^{*T}, \bar{R}_1^T, \bar{P}_1^T, \bar{R}_2^T, \bar{P}_2^T, \bar{R}_3^T, \bar{P}_3^T, \bar{R}_4^T, \bar{P}_4^T, \bar{R}_5^T, \bar{P}_5^T, \bar{R}_6^T, \bar{P}_6^T, \bar{\pi}_6^{cT})^T. \qquad (14)$$

As will be stated later, we use the BDF method developed by Gear to integrate the above stiff nonlinear argebraic differential equation. To do this, the above equation is first discretized in the time domain. Then given the dynamical states $\bar{z}_k$ (k=0,...,n), we have the predictors

$$\dot{\bar{z}}_{n+1} = -(1/h)\sum_{i=0}^{k} \alpha_i \bar{z}_{n+1-i} \qquad (15)$$

$$\bar{z}^{p}_{n+1} = \sum_{i=1}^{k+1} \gamma_i \bar{z}_{n+1-i} \qquad (16)$$

where α_i and γ_i indicate the coefficients. Using the above expressions, we finally have the (p+1)-th corrector for $\bar{z}^{p}_{n+1}$, that is, Δz^{p}_{n+1}. This is shown in Fig. 6.

5. NUMERICAL SIMULATION

Before going into details of numerical simulation, we briefly discuss the results of the proposed symbolic scheme of the system equations. The algorithm has two obvious advantages;

1) The whole process is systematized by using the phsical, causal and structural information all of which are incorporated in the bond graph.
2) By the systematization of the reduction process, we can easily check the resultant equations. This is particularly important when the number of the involved bodies increases as in systems like space structures.

Preceding to do the work using MACSYMA, we tried the same work by REDUCE 3. According to our experience, MACSYMA is much more refined for symbolic operations of mathematical relations than REDUCE 3. However, it is operated under UNIX and hence the available machines are limited.

Now let us illustlate some simulation results. As is easily understood from the causalities

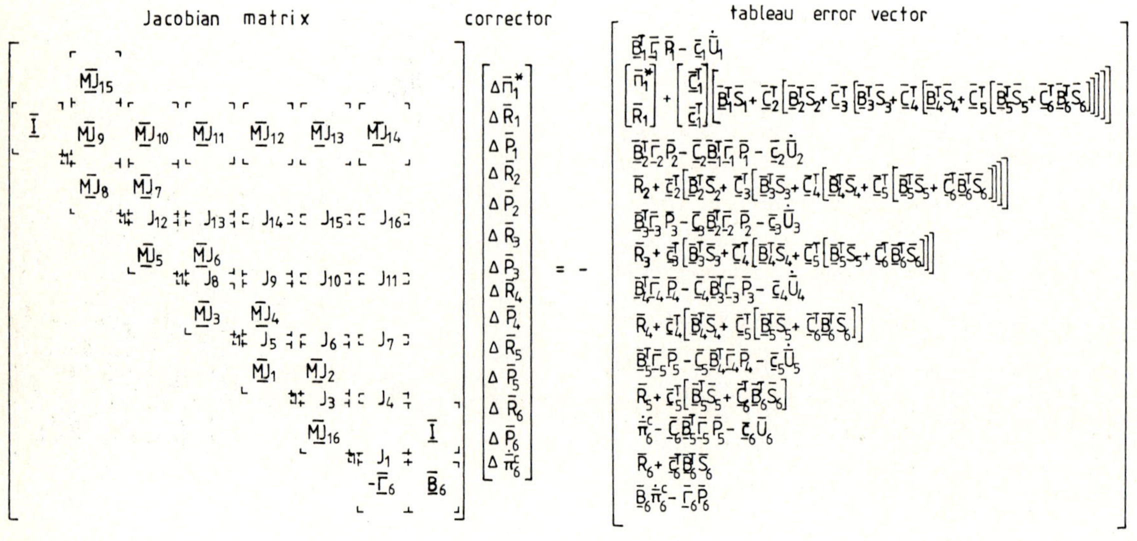

Fig. 6 Discretized system equations

shown in Fig. 2, the system inputs are the driving velocities at the mechanical joints, the velocity of the 0-th body and the loaded force and torque of the end effector. In the illustrative examples, we assume the followings;

1) Time functions giving the driving velocities of the joints are given as shown in Fig. 7.
2) The velocities of the 0-th body are assumed to be zero.
3) The end effector is not exerted any load force and torque.

Simulated results of the system behaviour under the above assumption are shown in Fig. 8 and Fig. 9 depicts the locus of the manipulator during the above operation. It is offen required to predict the system behaviour under some disturbance forces exerted upon bodies.
To illustrate this, we show the simulated results in Fig. 10. The disturbances are exerted upon the 5-th body.

$$d_{tr}^{T} = [\ 2.72\sin(104.7t),\ 0,\ 0\ ]$$
for linear motion

$$d_{rot}^{T} = [\ 0,\ 0,\ 0.16\sin(104.7t)\ ]$$
for rotational motion

Figure 10 illustratively shows the response torques and forces at 0-th body.

In the numerical simulation, the usual Gausian elimination method and the sparse matrix method are comparatively used. Figure 11 illustrates the sparsity of the Jacobi's matrix and the CPU time related to the order of the system equations which means the number of the system equations.

We easily see the significant reduction of the execution time by using the latter method as long as the system order is high, and when the reduction goes ultimate, the advantage of the sparse matrix method can not be so much expected. From the above results, it is concluded that we can not expect the effects of the sparsity when the sparsity is higher than approximately 15%. For comparison, we tried the same numerical integration on different

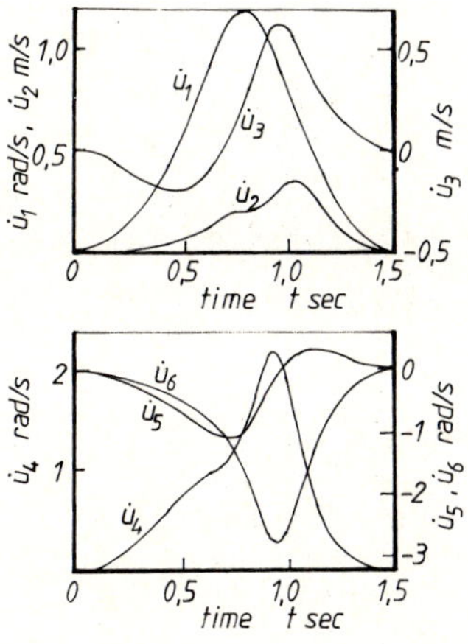

Fig. 7 Driving velocities

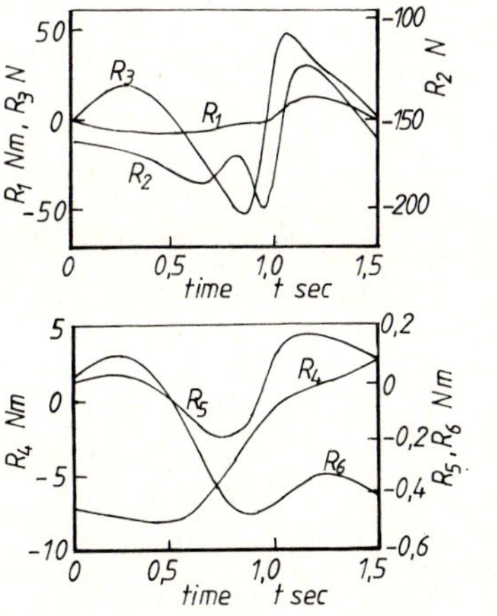

Fig. 8 Simulated responses

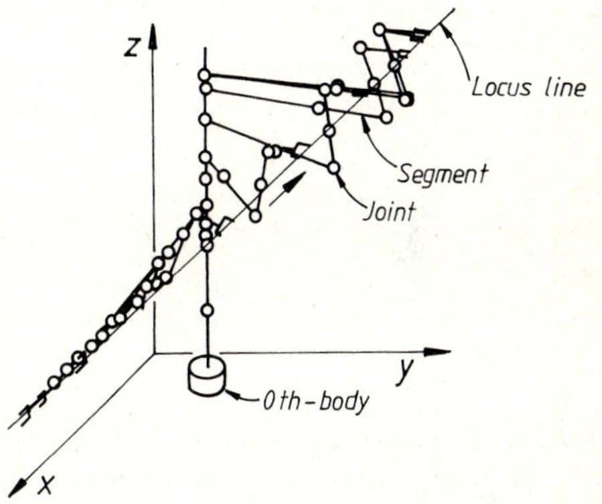

Fig. 9 Locus of the manipulator

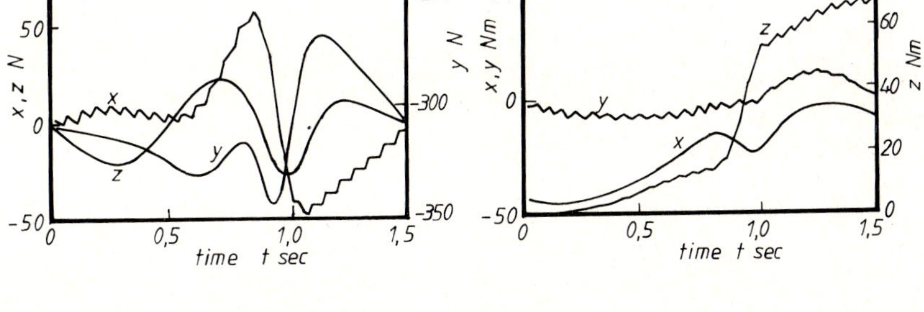

Fig. 10 Responses under the disturbances

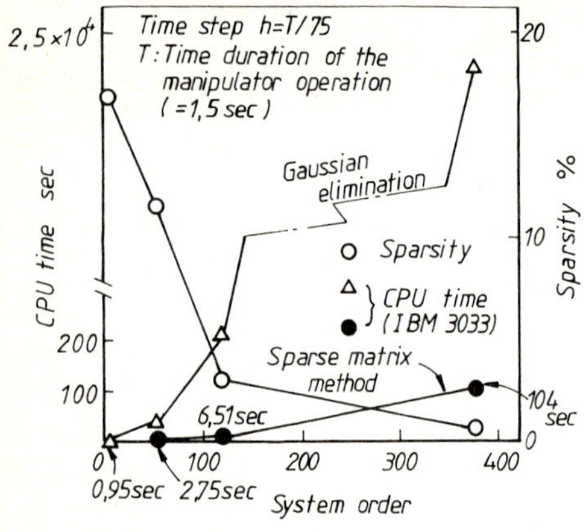

Fig. 11 Sparsity and CPU time vs systam order

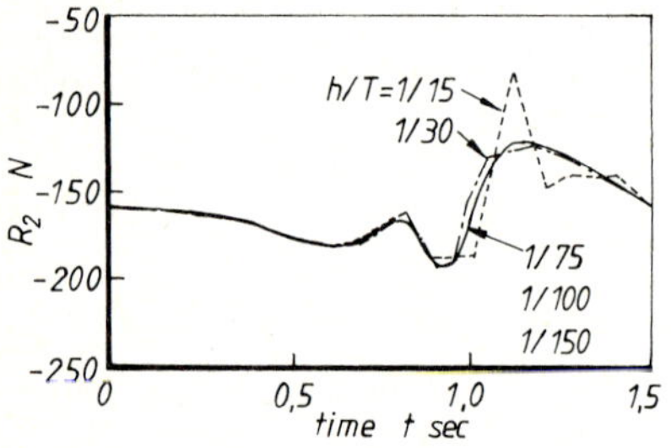

Fig. 12 Effects of the time step

computers, say, VAX11/780 and the super computer S-810(HITACHI). The execution time by the former is approximately 4 times longer and by the latter reduces to one tenth of the execution time by IBM3033 class machine. In addition, we illustrate the effects of the time step in the numerical integration. Figure 12 shows the driving torque at the joint R2. We easily see that the response converges to the most probable one by the selecting the time step h = T/75, where T(=1.5sec), indicate the time duration during which whole cycle is completed. Thus we can expect the execution time per step of roughly 15 msec for IBM3033 class machine.

7. CONCLUSIONS

A modelling method of multibody dynamics together with the symbolic generation scheme of the system equations is proposed and some simulation results are shown. In the above, we have the following results;

1) The categorization of the required kinematical and dynamical information into the physical, structural and causal greatly serves in order to construct the symbolic scheme to generate the system equations on the computer.
2) The above symbolic generation scheme is quite easily formulated by using the bond graph which installs the above mentioned information.
3) As to the numerical integration, the reduction of the system order is most effective to reduce the execution time per step.

The present work is partly supported by CARO Project organized by Waseda University and Japan IBM Corp..

REFFRENCES

[1] Birkhoff, G.D., Dynamical Systems, (AMS, New York, 1929).
[2] Karnopp, D.C., J. Franklin Inst., 288, 3(1969), 175.
[3] Kawase, T. et al, Modelling of Artificial Manipulators and Computer Simulation of their Dynamics, in: Morecki, A. et al (eds), Theory and Practice of Robots and Manipulators (Kogan Page, London, 1985) pp. 87-95.
[4] Kawase, T., Private note; Towards the modelling of multibody systems, (1983).
[5] Lur'e, L., Meçhanique Analytique, (Masson et Cie, 1968).
[6] Kane, T.R. and Levinson, D.A., Trans. ASME, J. Appl. Mech., Vol.50, No.11, (1983), 1071.

Complex and Distributed Systems: Analysis, Simulation and Control
S.G. Tzafestas and P. Borne (editors)
Elsevier Science Publishers B.V. (North-Holland)

A DYNAMIC ROBOTIC MODEL WITH MULTIBOND GRAPHS AND DECOUPLED CONTROL.

M.J.L. TIERNEGO

Twente University of Technology,
Department of Electrical Engineering,
P.O. Box 217, 7500 AE Enschede, The Netherlands.

Modelling robotic dynamics is, despite of the many developed methods, still a laborious job. The one and only reason is because a robot manipulator is a complex mechanism of which the dynamic equations are highly nonlinear. In this paper a method of modelling is used based on the multibond graphs. In order to get the set of equations necessary for control purposes formula manipulation is applied. So the control engineer can pay his attention to modelling and control while the tedious job is done by the computer. As an example decoupled control of a three axes driven camera will be used. Although decoupling seems to be a straight forward method to get independent control of each axis there are some problems in the realisation as will be shown by simulation.

1. INTRODUCTION TO MULTIBOND GRAPHS.

The bond graph modelling method, introduced by H.M.Paynter, has found in the recent years a growing field of applications. In some recent papers [1,2,3] special attention has been given to the modelling andsimulation of multibody systems with the use of multibond graphs. In [1,4] a systematic method is described for modelling such systems. The method will be described in this paper as far as necessary to understand to understand the symbolism of the multibond graphs used in this paper. In the sequel the reader is supposed to be familiar with the elementary bond graph symbolism [5]. Hence there is no need to start "from scratch" with a formal mathematical definition, because the multibond graph language can be "defined in terms of" or rather "translated into" single bond graphs.

The method is based on the fact that the absolute velocity of some point P can be calculated if the relative velocity and distance to the origin of a reference frame are known. From this reference frame the rotational and translational velocity with respect to the inertial frame have to be known. Consider figure 1, here is point P for instance the origin of a next frame, now yields in vector notation;

$$\overrightarrow{\dot{P}} = \overrightarrow{\dot{0}_i} + \overrightarrow{0_i\dot{P}} + \overrightarrow{\Omega_i} \times \overrightarrow{0_iP} \quad (1)$$

Or rewritten in matrix vector notation;

$$\dot{P}^0 = A_i^0 \,.\, \dot{P}^i = A_i^0(\dot{0}_i^i + 0_i\dot{P}^i + A(0_iP^i).\Omega_i^i) \quad (2)$$

in which A_i^0 a coordinate transformation from frame i to frame 0 and $A(0_iP^i)$ an anti-

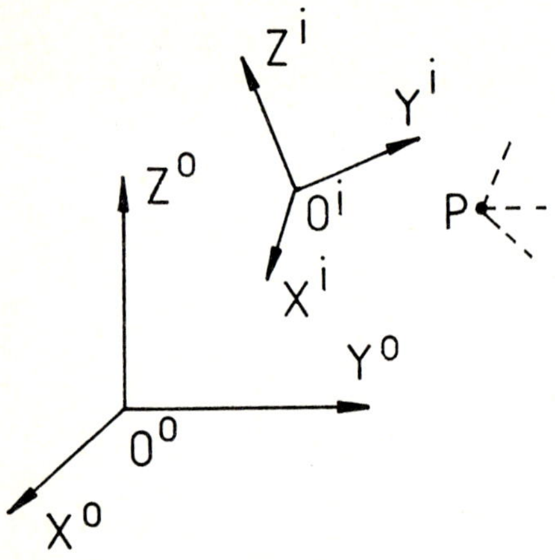

Frames

Figure 1

symmetric matrix containing the position elements of P with respect to 0_i in coordinates of frame i. In the multibond graph notation these velocity relations have the structure of figure 2. This structure can be repeated for any point within the same frame of which the velocity has to be calculated, Or to put it in another way, for any point with the same rotational velocity. These structures are then connected to the two 1-junctions in figure 2 as they all have these velocities in common.

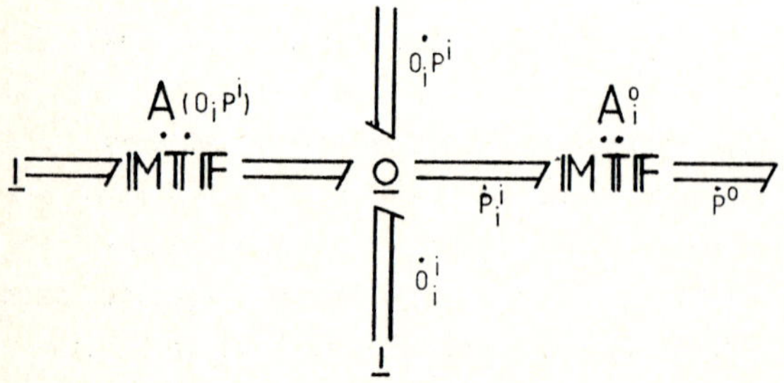

velocities in a moving frame

Figure 2

Upto now the body aspects are not considered. In this paper only rigid bodies are taken into account although the method can straightforward be used for systems with compliances. For rigid bodies the inertia matrix J and the center of mass have to be known. In this paper the inertia used is the inertia calculated along the axes of the chosen body fixed frame with respect to the center of mass. Another choice which has been made is that the origin of a frame is at an axis of rotation. The inertia could be calculated on beforehand with respect to this origin, according to the Huygens-Steiner rule, but this is only valid if the origin does not move. In that case the tranformations are more complex, and in such cases a formula manipulation program (e.g. REDUCE) is succesfully used.

By describing momenta and forces in body fixed frames new terms terms arise as shown by the Newton-Euler equations;

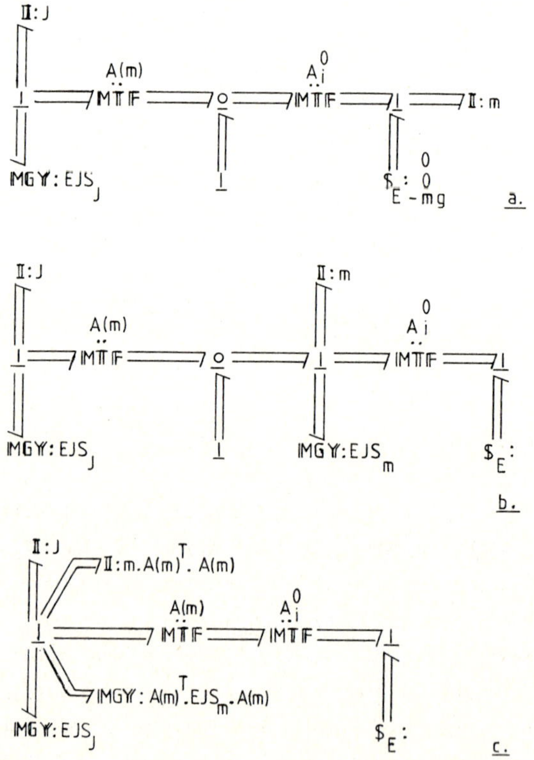

a. center of mass in the inertial frame

b. " " " in the body fixed frame

c. mass tranferred to the origin of the body fixed frame

Figure 3

$$M = \dot{H} + H \times \Omega$$
$$F = m\dot{v} + m v \times \Omega \qquad (3)$$
$$H = J \cdot \Omega$$

In the multibond graph the term $H \times \Omega$ has to be described by a MGY, as the force appears as a velocity multiplied by a modulating term. Sometimes this MGY is also called the Eulerian Junction Structure, EJS. As a result a body can be described in multibond graphs in three ways, shown in figure 3. The last step is only possible if there is no 0-junction present between the two 1-junctions. If there is a 0-junction then the transformation must go to both sides of the 0-junction.

In our description a body can be defined as a system of which all parts have the same orientation at any time. Sometimes in robotics a part of the system can be only translational with respect to the foregoing part. By the definition given here this will not be a separate body, but a part of a body. Connecting bodies together can be done by just a coordinate transformation and a 0-junction (summing rotational velocities), in the rotational velocities, as shown in figure 4.

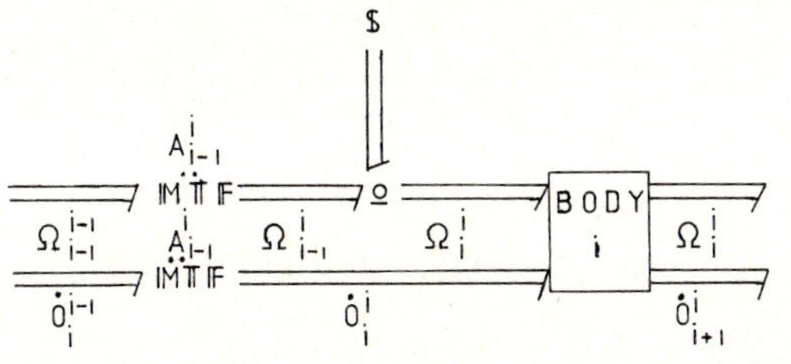

word bondgraph of a body in a moving frame

Figure 4

Upto this point only the velocities have been considered. But it will be clear from the power conserving properties of bondgraphs that the forces and momenta are explicit available in the presentation of the system by the multibond. If e.g. in figure 5 is calculated

$$v_2 = A \cdot v_1$$
$$\text{then} \qquad F_1 = A^T \cdot F_2 \qquad (4)$$

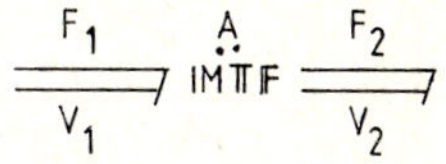

definition of a MTF

Figure 5

2. MODEL OF A TRHEE AXES DRIVEN CAMERA.

It was decided in our group to build a camera system which can be oriented over three perpendicular axes, although in many applications a two axes control of a camera will be sufficient. The purpose of the camera is twofold. The first application is the use as an external sensor in a robot control system (recogniton and positioning). The second application is to study the control of multiple input multiple output systems as they appear in robots. Although the spatial links of a robot ar missing the dynamic equations are, apart from some constants the same, as will be shown.

Some aspects of the system, shown in figure 6, will be briefly described and then

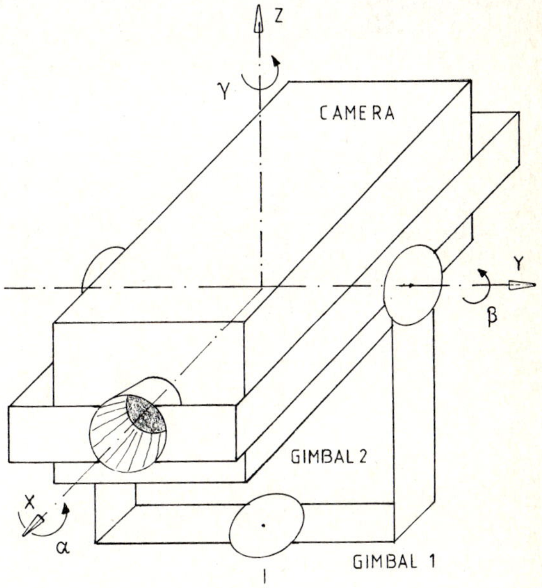

camera system

Figure 6

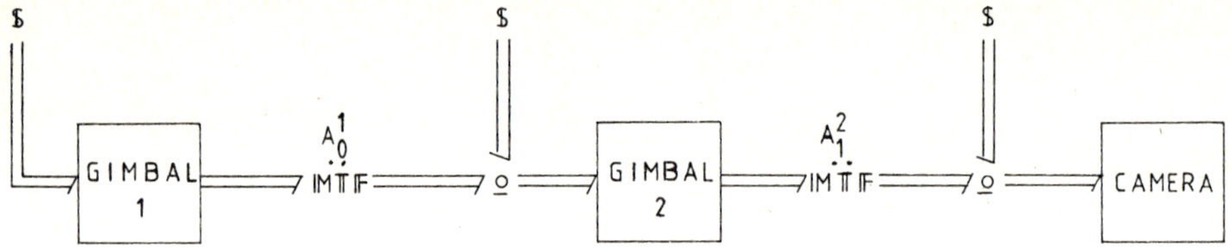

word bondgraph of the camera system

Figure 7

the model will be treated. The drives consist of three permanent magnet torque motors without gearing. Use is made of the hollow axis of such a motor as the camera looks through the axis of one of the motors. Opposit to the motors are dual speed resolvers (1x, 8x) placed to measure the angles. The resolvers are connected to resolver to digital converters (RD). The RD delivers additional to the digital angular value an analog speed signal, so no tachogenerator had to be build in. Where the axes could go through angles greater then 360 degrees use is made of sliding contacts. In order to stress the coupling effects between the axes no attempt is made to balance the camera and gimbal system. So the center of mass

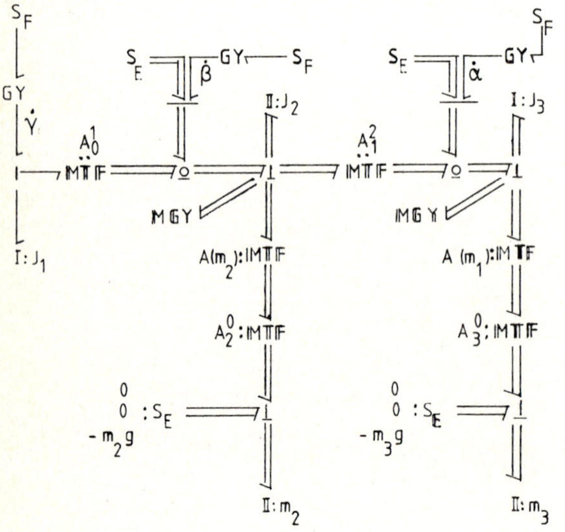

multibond graph of the camera system

Figure 8

of the camera and one of the gimbals do not coincide with the origings of their rotational frame. With this knowledge a dynamic model of the camera could be build. The model of such a system was described before [6] but in that case the centers of mass and the origins of the frames coincided.

In figure 7 the word bondgraph of the system is shown. By expanding each body (gimbals and camera) in the way shown in figure 3a the multibond graph of figure 8 arises. Each of the two vertical chains shows the effect of the center of mass and the corresponding gravity. The effect of earth rotation is neglected which means that the chosen inertial frame is connected to earth. From figure 8 the necessary equations can easily be derived.

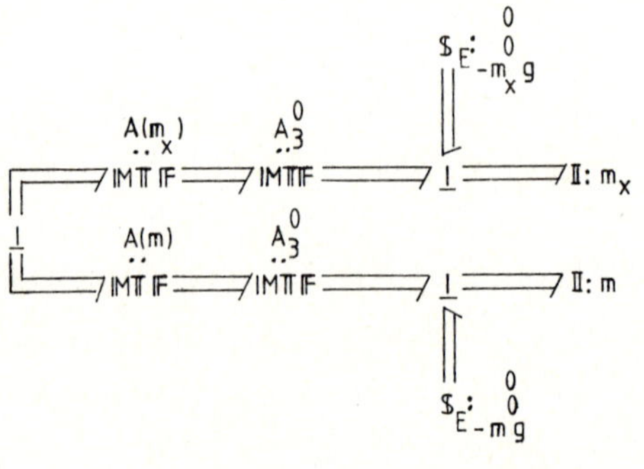

balancing a body

Figure 9

Some remarks have to be made as to the mechanical balancing of the camera. This can be done by adding a mass m_x such that the effect of gravity is a zero momentum. In the multibond graph this will mean an additional chain in frame 3, as illustrated in figure 9. The gravity effects to M_1 and M_2 can be calculated and the sum should be equal to zero. So

$$A(m_x) \;.\; A_3^0 \;.\; \begin{vmatrix} 0 \\ 0 \\ -m_x g \end{vmatrix} \;=\; -\;A(m)\;.\;A_3^0\;\begin{vmatrix} 0 \\ 0 \\ -mg \end{vmatrix} \qquad (5)$$

from which follows with $m_x = a \;.\; m$

$$A(m_x) = -\,a^{-1}\;.\;A(m) \qquad (6)$$

The result is that the static position is not ifluenced by gravity, so the force needed to compensate this effect has not to be delivered by the motors. But the dynamic part of the equations changes too. The force due to acceleration, coriolis and centripetal effects increase. These forces can only be minimized by more complex compensation, or beforehand by proper design.

3. DECOUPLED CONTROL.

One of our modelling aims was to obtain a set of differential equations from which the control scheme can be derived. The inputs to the system are the control signals for the electromotors and the outputs are the angular velocities. The three desired angles can be obtained by simply integrating the corresponding angular velocities. By inspection from the multibond graph of the system (figure 8) more velocities then the three independent ones occur. This indicates that a number of implicit equations exists. This means that the result from writing down the equations straight forward from the multibond graph will give rise to a mixture of integral and differential equations.

One way to obtain a set of explicit differential equations is transformation of all the elements to the three independent velocities. To do this by hand is only velocities. To do this by hand is only possible for relative simple systems. The way we have chosen is to use a formula manipulation program as described in [4]. This results in the general multibond graph of the system as shown in figure 10.

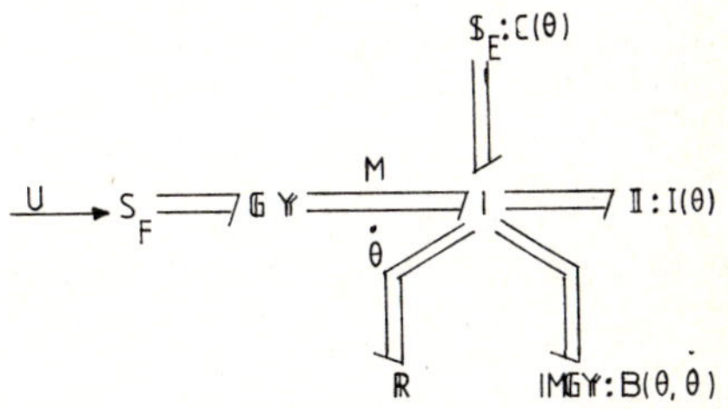

multibond graph with independent velocities

Figure 10

In formula form:

$$M = I(\theta).\ddot\theta + B(\theta,\dot\theta).\dot\theta + R.\dot\theta + C(\theta)$$

and (7)

$$M = C_m \;.\; I_a = C_m' \;.\; u$$

with $C(\theta)$ the momentum due to gravity, and $\theta^T = (\alpha\,,\beta\,,\gamma\,)$

The classical control of such a system consists of three positional servoloops with velocity an position feedback. This means that the torque applied by one of the motors in order to change the angle to which it is connected, will also move the other axes. These unwanted momenta can be seen as disturbances and can only be removed after their effects have been sensed by sensors for velocity and angle.

The general idea of decoupled control is that one control signal has effect on one output signal only. This means that the coupling momenta should be compensated by calculation of control signals for all the motors. Several control schemes have been proposed e.g. [7]. We have chosen a simple method not requiring additional sensors.

In order to achieve the desired goal, decoupled control, the following relation is desired:

$$v = \ddot{\theta} \tag{8}$$

where v is a new control signal. Now a relation between u, the actual control, and v has to be found. The most simple way to do this is inserting (8) into (7). This gives

$$u = C_m^{-1}(I(\theta).v + (B + R).\dot{\theta} + C(\theta)) \tag{9}$$

Formula (9) is now the relation between u and v which has to be performed in a computer. As can be seen in order to compute (9), the actual measurement of angles and angular velocities as well as the system parameters have to be known.

4. PRACTICAL ASPECTS AND SIMULATION.

The practical results which can be obtained by decoupled control highly depends on the knowledge about the system parameters. A method to avoid this problem is described in [7]. Another practical aspect is due to the realisation of (9) by the computer, because no general method is available to transform a set of nonlinear differential equations into a set of nonlinear difference equations. In this part a solution to this problem will be described and examined by simulation.

Suppose the velocities and angles are measured at time nT, where T is the fixed sampling time. The control signal u is available after the computations at (n+1)T. In the following equations the sampling period T is omitted. So from (9) follows with this delay for the computation

$$u(n+1)=C_m^{',-1}(I(\theta(n)).v(n)+(B(n)+R)\dot{\theta}(n)+C(\theta(n))) \tag{10}$$

A usual solution in practical applications, to neglect computation time, and to make the control signals available to the DA-converter as soon as they are calculated, is not allowable in this case. The calculation time due to (10) will in general even determine the minimum obtainable sampling time T. The chosen way to overcome this problem is to make an estimate of θ and $\dot{\theta}$ at (n+1) from the measurement at n. The chosen estimation is:

$$\begin{aligned}\theta(n+1) &= \theta(n) + T.\dot{\theta}(n)\\ \dot{\theta}(n+1) &= \dot{\theta}(n) + T.v(n)\end{aligned} \tag{11}$$

As the time required for the calculations should be as short as possible a simple prdeictor is used (11), in this prediction is made use of equation (8). Some of the results obtained in this way are shown in figure 11. The figure is obtained by simulation. As an inputsignal is chosen a sinus at the

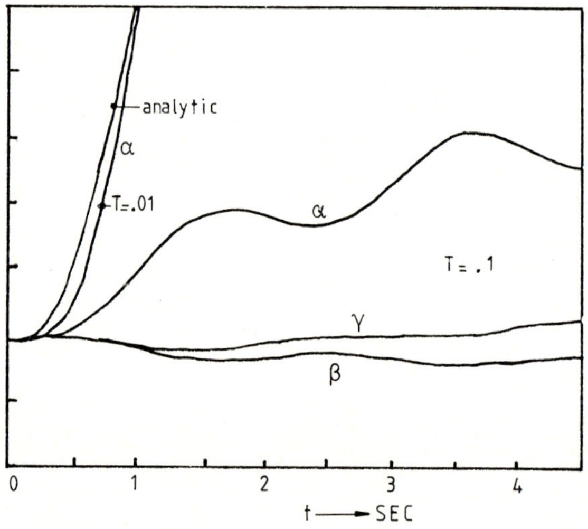

a simulation result

Figure 11

controlinput v_α, while the other two (v_β and v_γ) are equal to zero. As parameter in the figure is chosen the sampling time T. As can be seen for the large sampling time the result of the decoupling is rather poor, although there is a scale factor of five between α and β,γ. To compare the results also the analytcal calculated value of is plotted.

5. CONCLUDING REMARKS.

The formulas (7) and (9) are expanded with the formula manipulation program REDUCE in combination with a code optimizer. The use of formula manipulation in combination with multibond graph modelling proves to be a handsome way for the simulation and obtaining the control equations of large systems. The effect of parameter changes and of additions to the model can be obtained without matrix manipulations. The main disadvantage is the need of a large computer for the formula manipulations. Moreover a code optimizer should be able to recognise possible goniometric simplifications. As far as the decoupled control concerns the main problem is the knowledge of the systemparameters an the reduction of computing time.

REFERENCES.

[1] Tiernego, M.J.L., Bos, A.M., Modelling the dynamics and kinematics of mechanical systems with multibond graphs. J. Franklin Inst. vol. 319 no 1/2, pp. 37-51, 1985.

[2] Allen, R.R., Dynamics of mechanisms and machine systems in accelerating reference frames. Trans. ASME J. Dynamics Systems Measrurement and Control, vol. 103, pp. 395-403, dec 1981.

[3] Pacejka, H.B., Modelling complex vehicle systems using bondgraphs. J. Franklin Inst., vol 319 no 1/2, pp 83-92, jan 1985.

[4] Bos, A.M., Tiernego, M.J.L., Formula manipulation in the bond graph modelling and simulation of large mechanical systems. J. Franklin Inst., vol. 319 no 1/2, pp. 51-67, jan. 1985.

[6] Rosenberg, R.C., Karnopp, D.C., Introduction to physical system dynamics, McGraw-Hill, New York 1983.

[6] Tiernego, M.J.L., Dixhoorn, J.J. van, Three axes platform simulation: bondgraph and lagrangian approach. J. Franklin Inst., vol. 308 no 3, pp. 185-204, sept. 1979.

[7] Hewitt, J.R., Fast dynamic decuopled control of robots using active force control. In Robot Technology ed. A.Pugh, Perigrinus, London, 1983.

Complex and Distributed Systems: Analysis, Simulation and Control
S.G. Tzafestas and P. Borne (editors)
Elsevier Science Publishers B.V. (North-Holland)

BOND GRAPH SYNTHESIS AND ANALYSIS OF COVARIANT BILATERAL SERVO SYSTEMS IN MANUPULATOR CONTROL SYSTEMS

Dr. J.E.E. Sharpe and Mr. K.V. Siva

Queen Mary College, University of London, Mile End Road, LONDON. E1 4NS
Fairey Engineering Ltd., Crossley Road, Heaton Chapel, Stockport, SK4 5BD

Many different types of bilateral servo systems have been used with more or less success in computer controlled robots and operator controlled remote manipulators.

This paper describes how Bond Graphs have been used to synthesise a unique design of covariant bilateral servo system based on the use of the energy covariables.

Details are given of the Bond Graph simulations of these systems which are compared with their physical realisation in different energy domains. Bond Graph analysis has also been extended to include the human operator and highlight the behaviour of the covariant bilateral servos in tactile sensing.

The Bond Graph simulations were carried out using ENPORT-5 on a Vax 11/750 under UNIX operating system.

The development of the covariant bilateral system has been to meet the need to provide force reflection in operator controlled master-slave manipulators. The Bond Graph Synthesis and Analysis led to the construction of two experimental systems; one to study the compliant control of grasping; the other to study the effects of changing manipulator geometry stiffness and inertia.

In remote manipulator systems, the operator performance depends on his kinesthetic and visual interaction with the system and his task. Bond Graph techniques offer a unique insight into these relations.

1. INTRODUCTION

A number of different types of remote robotic manipulators have been developed and used in hostile nuclear, space and under-sea environments. Despite some considerable measure of success there remain problems related to the stability, dexterity and operational performance of these force reflecting or bilateral systems. Limitations of the currently used systems have been principally due to a lack of understanding of the influence of the operator on the overall dynamic and, the correct causal behaviour of the system.

This paper outlines the main features of the force reflecting bilateral control servo system. The need for a new approach to the design of manipulators based on a correct causal understanding of the manipulative task is derived from study of 'Ideal Manipulators'.

The understanding of the importance of causal relationships that may be derived from the concept of the Bond Graph is used to define the 'ideal' conceptual manipulator structure expressing the correct causal relationships between the operator and the manipulator and the manipulator and the environment. From this 'ideal' the simplest physical realisation in any appropriate energy domains may be synthesised.

In the case of the remote manipulator, the energy covariables may be replaced by a pair of directed signals which maintain the correct causal relationship and knowledge of the energy interactions of the slave system without the need to transmit that energy.

The importance of casualty and the representation of energy dissipation in control systems is discussed comparing the modulated resistor and the commonly used, linear, controlled source.

2.1 The Bilateral Servos

Forces encountered by a remote manipulator are fed back to the operator for smooth and efficient operation of remote tasks. Two kinds of bilateral servo systems are presently used to provide this feature of force reflection. Studies of operator performance [1] in remote manipulation have shown 40 to 50% improvements in the task completion time when low performance force reflection is provided. The majority of these remote manipulator systems are implemented on a master-slave basis.

2.2 Problems in Bilateral Servo Systems

The control problem of a bilateral servo system is more complex than that of a similar unilateral control system. The dynamic problems in bilateral master-slave systems can be observed from existing systems in use today [2], [3], [4], [5], [6]. The ideal bilateral system shall have no non-linearities (such as backlash, friction etc) between position and force in the master and slave units, and behave like two units connected by a very stiff mechanical linkage. Early mechanically connected 'through wall' master-slave manipulators used in the nuclear industry possessed some of these features. However, the performance was degraded by the high friction levels in the cable/tape transmissions and limited remoteness. A pair of pliers may be thought of as a good example of an ideal bilateral system having high stiffness and negligible non-linearities [7].

The quality of bilateralness in bilateral servo systems depends on the frequency bandwidth and force resolution achievable in such systems. The noise inputs, from sensors, truncation errors if any digital control used, and friction threshold degrades the forces perceived by the operator.

2.3 Need for a new system approach

Conventional control system analysis using block diagrams and signal flow graphs do not explicitly describe how the signals are attenuated in a bilateral servo system. The use of Bond Graph simulation of bilateral servo has opened a new approach to create new systems and analyse their complex dynamics.

A powered manipulator in general is a physical system composed of energy storages, dissipators and sources of different media. The manipulator constantly interacts with the environment, gravitationally, inertially, frictionally and contactually. A bilaterally servoed manipulator in addition to the normal internal dynamics and external disturbance is also subjected to the operator's kinesthetic neuro-muscular interactions.

Bond graph techniques permit these physiological systems to be modelled in a unique way irrespective of different kinds of energy (mechanical, hydraulic, muscular, electric, pneumatic etc.) transactions. Bond graphs of these bilateral servo systems can be constructed by decomposing the overall bilateral system into separate components that exchange energy or power through identifiable connections or ports.

Bond graphs are based on power or energy flows, with generalised variables effort (e) and flow (f) associated with each bond. This pair of oppositely directed variables in each bond suggests that a suitable performance index for bilateral servo systems can be based on the impedance between them. The lowest order of impedance, the compliance is important in remote grasping operations, while the highest order impedance or inertance is important in capturing freely moving objects, such as capturing a satellite from an orbit.

An important consequence of dynamic interactions between two physical systems such as a manipulator and its environment is that one must physically complement the other. All energy storing elements of a physical system always possess complementary energy functions termed the co-energy stores. The relationship between the effort and flow of the slave manipulator during energy interactions with its environment must be the same as that between the operator and the master unit.

In a master-slave manipulator control system, the "flow" of the slave can be controlled by the "effort" of an operator on a force fed-back master handle or arm, thus providing bilateralness. Therefore the associated co-energy variables (flow of slave and effort of master) form the basis of this kind of kinesthetic coupling between an operator and a remote manipulator. Hence, the more he perceives this impedance of manipulation, the better he performs the task remotely.

3. SYSTEMS OF COVARIANT BILATERAL SERVO

Existing systems, particularly those using a common error as a basis for the reflection of tactile information are unsatisfactory on several counts. It is not practically possible to produce two identical dynamic systems. It is frequently not desirable to have the master and slave parts of the system physically the same size and of the same power. More importantly, it is essential for true tactile reflection that the causal relationship of the operator and the manipulator is the same as that of the manipulator and its environment or work piece. The actual causalty in itself does not appear to matter.

This question of causalty formed the basis for the synthesis of the unique covariant bilateral servo controlled manipulator.

The most ideal manipulator must appear as a direct extension of the operator and interfere as little as possible with his feel and perception. Most common hand tools such as a screw driver, wrench, pliers etc. achieve this most satisfactorily. In the ideal manipulator we are trying to reproduce as nearly as possible the system characteristics of the hand tool but entirely remotely and physically scaled down in the case of the micro-manipulator or up in the case of an arm for space operations.

The simplest tool manipulator having these characteristics is a pair of pliers. These act as a simple transformer providing a forward velocity or position path and reflecting the forces at the interface with the workpiece or environment back to the operator, Fig. 1.

The pliers (manipulator) may be represented in bond graph terms as a transformer with inertia and stiffness, see Fig. 1. The success of pliers is principally due to their low inertia and high stiffness which give rise to very high natural frequencies. So much so that during tests undertaken to investigate the dynamic response of operators using simple tools [7] it was found that the frequency range 500-1500 Hz contained significant tactile information about the task being performed. Almost all the damping in the system is provided by the operator system.

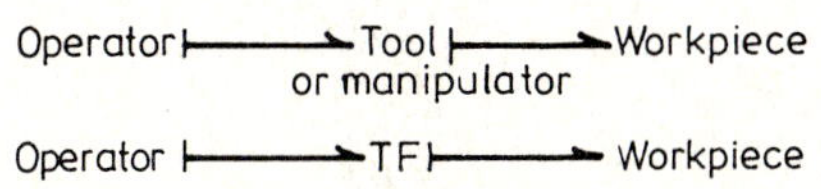

Fig. 1

The first extension of the system is to replace the transformer by a pair of gyrators in series as shown in Fig. 2. Such a system using the permanent magnet DC motors connected back to back was used for the remote alignment of optical components. The initial casualty is retained although the gyrators may be replaced by a pair of transformers to achieve the same effect in the hydrostatic domain. The automobile hydraulic braking system operates in this way.

Fig. 2

If the manipulator is to be entirely remote, the power supplies for the master and slave must be remote. To achieve this it is possible to replace the energy transfer between the master and slave represented by the bond in Fig. 2 by two signals representing the energy covariables and directed so as to maintain the correct causality. The energy bonds are reconstituted at the master and slave forming the inputs to a velocity (positional) control system at the slave and a force control system at the master. The signals representing the energy covariables may be directly measured or inferred. Fig. 3 shows a diagram of the system.

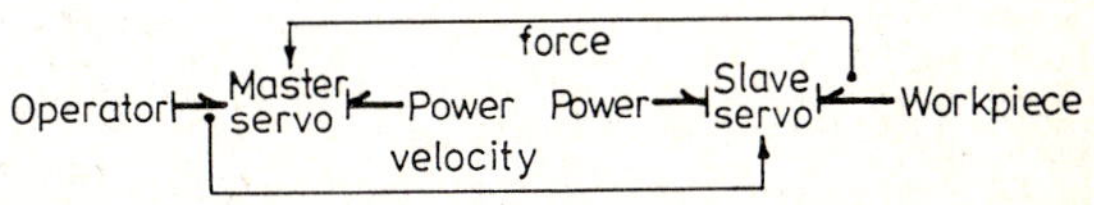

Fig. 3

The correct causalty is maintained and, the system is inherently stable as shown in physical studies and extensive simulations using Bond Graph techniques. It is also possible for different scaling factors to be used in the master and slave for each of the covariables allowing a high powered slave to be controlled by a low powered master.

The use of bond graphs to synthesise a practical system from an ideal system has a number of unique advantages. One is the fact that any system that has been created can automatically be analysed with very little further work. Another is that the causality has been correctly maintained and therefore the overall system behaviour unaltered. The ability of bond graphs to represent a wide range of different energy systems allows the easy development of alternative systems.

The design synthesis of the bilateral control system for remote manipulators relies on the use of a pair of control systems, one for position, the other for force, each requiring a different causality. In the case of the electro-mechanical systems this implies the use of a current controlled amplifier for the remote slave and a voltage controlled amplifier for the master.

4. BOND GRAPH SIMULATION OF PHYSICAL SYSTEMS

The bond graph of a physical sytem is constructed according to fundamental physical considerations of energy storage, energy supply or source, energy dissipation and energy exchanges of the system. Generation of the bond graph relies on the use of four generalised variables: effort (e), flow (f), momentum (p) and displacement (q).

4.1 Bond Graph Model of a Unilateral Servo System

Consider an electric permanent magnet d.c. motor used in a simple position control system. The schematic diagram Fig. 4, below illustrates such a system. The word bond graph of the system can be described as follows:

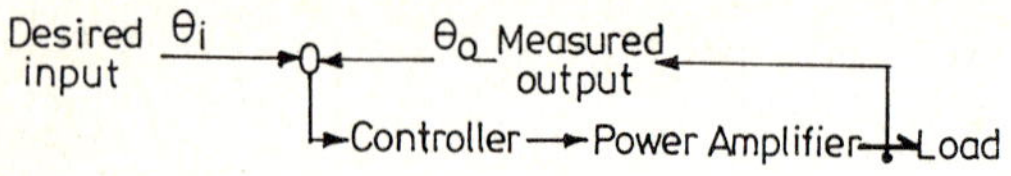

Fig. 4

The simplest bond graph of the position control system can now be derived as follows:

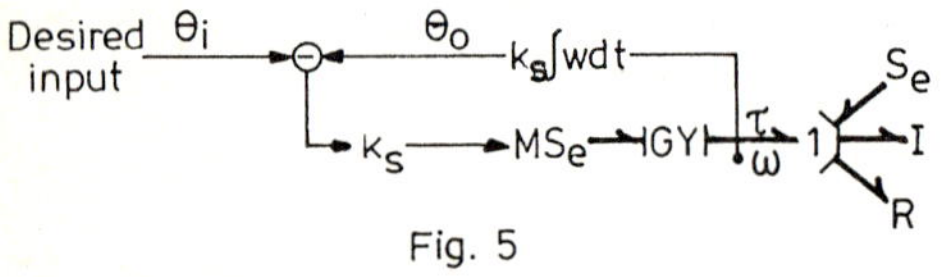

Fig. 5

Bond activation in which one of the two signals (effort and flow) associated with a bond is suppressed, is indicated in the usual way with an arrow in the centre of the bond [8].

The feedback scheme is represented by the use of activated bonds (instead of using modulated resistor element) to enable simulation using ENPORT-5. For instance, the velocity of the output shaft(ω) is integrated and scaled (k_f), and the torque imposed on the output shaft by this bond is suppressed.

Special bond graph techniques are required to model servo control systems. The use of modulated one ports or activated bonds are the two widely used methods in bond graph modelling of servo systems.

The bond graph model of the position control system using a modulated resistor element to represent a proportional control action can be constructed as follows:

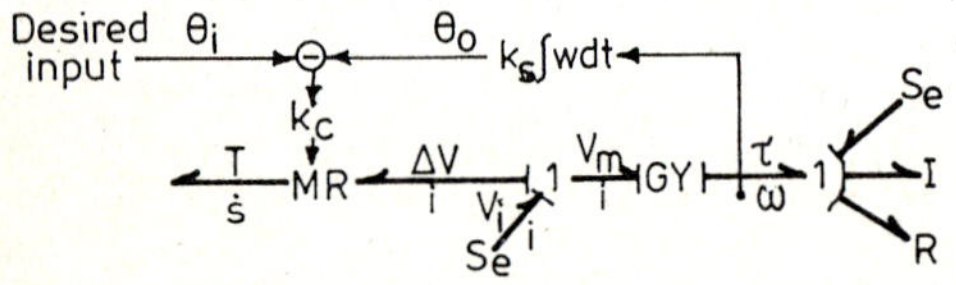

Fig. 6

It can therefore be seen from this model that some traditional function blocks and modulated ports (resistor) are required in the simulation program to complete bond graph models of servo control systems (These facilities are available in THTSIM, TUTSIM AND CAMP programs). Activated bonds on the other hand can also be used to simulate servo control systems (This facility is available in ENPORT modelling program).

4.2 Bond Graph Model of a 'Common-Error' Bilateral Servo System

Two positional servo systems are connected in this class of bilateral servo so that the positional error is common to both servo loops. A typical master-slave arrangement is shown in Fig. 7.

This system provides force reflection to the operator without the use of any force tranducers but requires the master and slave servo drives to be identical and capable of being 'back driven'. A simplified bond graph model is shown in Fig. 8.

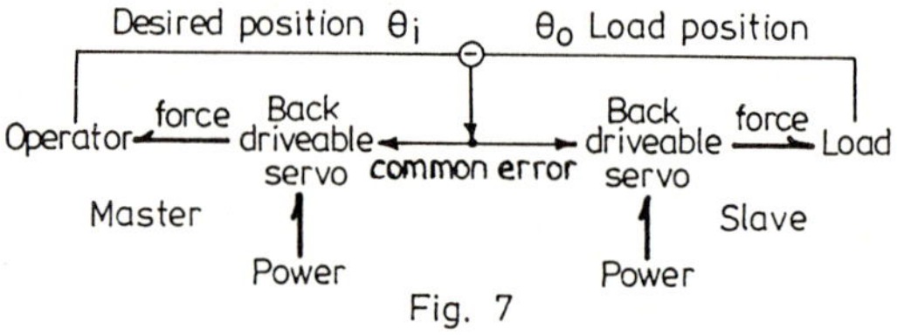

Fig. 7

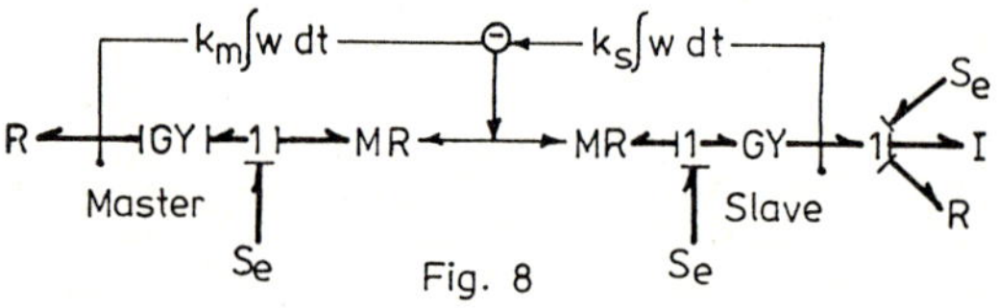

Fig. 8

The human operator (modelled as an effort source S_e) exerts a force necessary to maintain a command position input via the master unit to the slave position servo system. The force reflected back to the operator in this scheme is therefore dependent on the positional error between the master and slave units and the identical gains and time constants of the master and slave servos.

4.3 Bond Graph Model of a Covariant Bilateral Servo System

Unlike the Common Error System this scheme derived by Bond Graph Synthesis uses the covariables of the master and slave to implement the force reflecting bilateral servo loop. That is, the "flow" of the slave can be controlled by the "effort" of an operator on a force fed back master handle or arm. Though this scheme does not require back drivability in the servo drives the efforts need to be measured. An advantage of this scheme is that different energy media can be mixed in the master-slave configuration. The bond graph of this class of bilateral servo system can be described as follows:

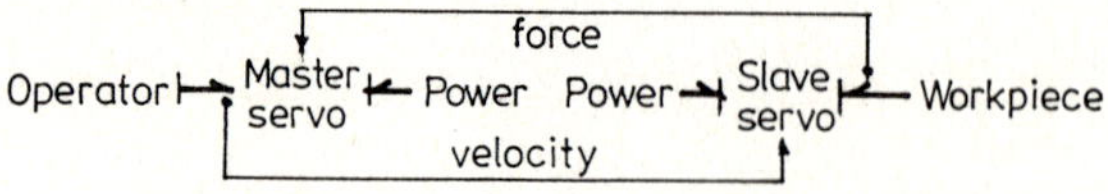

Fig. 9

The bond graph model using D.C. electric servo drives can be constructed as follows:

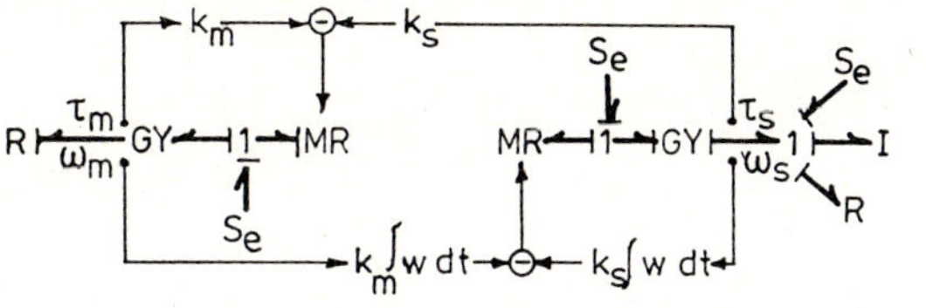

Fig. 10

In this sytem it can be noted that the slave drive is controlled with reference to the velocity covariable while the master drive is controlled with reference to the Effort Covariable. The force reflected back to the operator in this scheme is proportional to the forces acting on the slave rather than positional error between the master and slave units while the position of the slave is proportional to the position of the master. The signals between the master and slave representing the two energy covariables of the output of master and slave amd maintaining the all important causal relationships.

5 Model of Human Operator in System

Human operator controlled machine in general can be broadly classified into those with visual, audial or kinesthetic sensory-stimuli.

Position tracking can be performed with vision and kinesthetic information. However, reflex control in remote manipulation requires a good kinesthetic interface between man and machine as found in the high performance bilateral servos. Experiments have shown that oscillations imposed on bilateral servo systems can be dissipated by the controlled grip of the operator. This suggests that the human operator can be modelled as a resistor element. However, the process of kinesthesis is more complex involving neuro muscular interactions with the above stimulii.

The manipulations performed by human operators can be broadly classified into two groups: large motions performed by arms, and small motions performed by hands and wrists. The arm motions tend to be dominated by inertial effects whilst hand motions, such as grasping, would be more compliant. The operator perceives the impedance of manipulator to make corrections and the relationship between the effort to flow covariables form the basis of kinesthesis in an operator controlled master-slave manipulator system.

A simplified bond graph model of the operator - master arm (man - machine) kinesthetic interface can be constructed as follows :

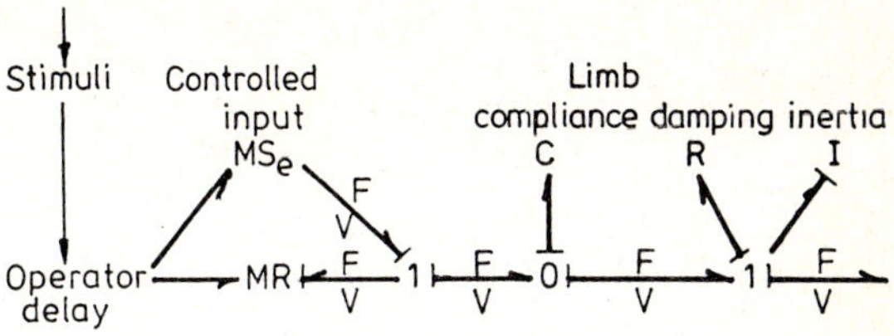

Fig. 11

CONCLUSIONS

Bond graphs have been used with considerable succes in analysing and simulating the dynamic behaviour of complex multi-disciplinary systems. It is the authors' opinion that in addition the Bond Graph concept has an important part to play in the creation or synthesis of optimal system designs beginning with the notion of the simplest 'ideal' system. 'Ideal' in the sense that it expresses the desired causal structure of the system. From this 'ideal' a bond graph representing the best physical realisation in the chosen energy domains is established. During this development phase the bond graph is used qualitatively to identify areas of particular concern such as high energy dissipation, the modal strucuture and problems of causal conflict.

The developed bond graph is then used in the normal qualitative manner to analyse the proposed system and, as indicated in the examples given, may be used as input to one or other of the established computer based simulation packages such as TUTSIM or ENPORT.

The example of the design of the covariant bilateral servo for remote manipulation control has demonstrated the power of Bond Graph Synthesis by establishing a novel system with many advantages compared with those developed by normal design methods. Further recent work in the field of the thermodynamic and electro-mechanical fields is confirming these advances.

ACKNOWLEDGEMENTS

We would like to thank the Science and Engineering Research Council for their support of the work described in this paper.

REFERENCES

[1] "Study to design and develop remote manipulator systems" by J. Hill, Quarterly Report, (1977).

[2] "Manipulator systems development at ANL" by R. Geortz, Proc. of 11th Conf. on Remote Systems Technology, (Nov. 1964).

[3] "A compact and flexible servosystem for master-slave electronic manipulators" by L. Galbiati et. al., Proc. of 12th Conf. on Remote Systems Technology.

[4] "Allocation of control between man and computer in Remote Manipulation" by A.K. Bejczy, 2nd International CISM-IFTOMM Symposium on Theory and Practice of Robots and Manipulators (1976).

[5] "Generalisation of Bilateral Force-Reflecting Control of Manipulators" by A.K. Bejczy and M. Handlykken, 4th International CISM-IFToMM Symposium, (1980)

[6] "A Bilateral Servo Manipulator for Remote Maintenance in Nuclear Facilities" by M. Suzuki et. al., Proc. of the 30th Conf. on Remote Systems Technology (1983).

[7] "Application of Bond Graphs to the Synthesis and Analysis of Telechir and Robots" by J.E.E. Sharpe, 3rd Int. CISM-IFToMM Symposium, (1978).

[8] "Direct Application of the Loop Rule to Bond Graphs" by F.T. Brown, Journal of Dynamic Systems, Measurement, and Control, ASME (Sept. 1972).

[9] "Multiport Models in Mechanics" by R.C. Rosenberg, Journal of Dynamic Systems Measurements and Control, ASME (Sept. 1972).

[10] "Modelling and Adaptive Control of a Mechanical Manipulator" by R.P. Anex, Jr. et. al., Journal of Dynamic Systems, Measurements and Control, ASME, (Sept. 1984).

Section III

NONLINEAR OSCILLATORS
AND CHAOTIC SYSTEMS

Complex and Distributed Systems: Analysis, Simulation and Control
S.G. Tzafestas and P. Borne (editors)
Elsevier Science Publishers B.V. (North-Holland)

SINGULAR NONLINEAR OSCILLATIONS: METHOD OF HARMONIC BALANCE

Ronald E. Mickens

Department of Physics
Atlanta University
Atlanta, Georgia 30314 U.S.A.

We investigate a new class of nonlinear oscillators for which the nonlinear terms may become unbounded for finite values of x and/or $\dot{x}$. These oscillators correspond to large amplitude generalizations of the van der Pol oscillator. The method of harmonic balance is used to obtain approximate analytic solutions. A number of general results are presented.

1. INTRODUCTION

For our purposes, nonlinear, singular oscillators are those whose defining equation of motion can be written

$$\ddot{x} + f(x) = \frac{\varepsilon g(x,\dot{x})}{1 - x^2}, \qquad (1)$$

where the "dot" represents differentiation with respect to the independent variable t; f(x) is an odd function of x with the property that

$$xf(x) > 0, \quad \text{for } x > 0; \qquad (2)$$

ε is a positive parameter satisfying the condition

$$0 < \varepsilon << 1; \qquad (3)$$

and g(z,z) is an odd function of z.

Note that the right-side of Eq. (1) is unbounded for $x = \pm 1$. Physically, this means that the dynamic system described by Eq. (1) exists for amplitudes $|x| < 1$; however, as the amplitude approaches the value one, the system is destroyed. Another possibility is that for "large" amplitudes, near one in value, the equation of motion changes from that given by Eq. (1).

In any case, the purpose of this article is to investigate the possible solution behaviors of dynamic systems whose equation of motion have the form given by Eq. (1). A particular example of such a system is the new negative resistance oscillator model of Walker and Connelly [1]. A second example is a non-linear oscillator equation derived by Mathews and Lakshamanan [2] to model certain phenomena in elementary particle physics.

Earlier work in this area is summarized in the paper of Bota and Mickens [3].

We wish to indicate that the analytical method used in this paper, harmonic balance [4] can be applied to a more general class of singular oscillator equations. This larger class can be represented in the form

$$\phi(x,\dot{x},\varepsilon)\ddot{x} + \psi(x,\dot{x},\varepsilon) = 0, \qquad (4)$$

where ϕ and ψ are polynomial functions of x and $\dot{x}$; $\phi(z,z,\varepsilon)$ and $\psi(z,z,\varepsilon)$ are even functions of z; and ε is a small positive parameter. We also assume that in the limit that $\varepsilon \to 0$, Eq. (4) takes the form

$$\ddot{x} + f(x) = 0, \qquad (5)$$

where f(x) is an odd function and $xf(x) > 0$ for $x > 0$. It should be noted that under these conditions, all the solutions to Eq. (5) are bounded and periodic.

Finally, since the right-side of Eq. (1) depends on both x and $\dot{x}$, we expect this equation to have limit cycle, limit point behavior. Similar expectations hold for dynamic systems whose equation of motion is given by Eq. (4).

2. GENESIS OF SINGULAR EQUATIONS

We now show how nonlinear, singular oscillator equations can arise. To begin, consider the function $f(x)$ of Eq. (1). The condition, of Eq. (2), implies that the potential energy corresponding to $f(x)$ is an even function having a single (absolute) minimum. Consequently, all the motions in this potential are bounded and periodic.

The simplest possibility is $V(x) = kx^2$, where k is a constant. This leads to a linear form for $f(x)$, i.e., $f(x) = 2kx$. Next, we have $V(x) = kx^4$ and $f(x) = 3kx^3$. This latter form does actually occur in a number of applications [5,6].

To obtain limit cycle behavior, we write the function $g(x,\dot{x})$ as follows

$$g(x,\dot{x}) = h(x)\dot{x}, \tag{6}$$

where $h(x)$ is an even function of x. The simplest, nontrivial case is obtained by expanding $h(x)$ and retaining the first two terms; doing this gives

$$h(x) = h_1 - h_2 x^2 , \tag{7}$$

where h_1 and h_2 are constants. Substitution of Eqs. (6) and (7) into Eq. (1) gives

$$\ddot{x} + f(x) = \frac{\varepsilon(h_1 - h_2 x^2)\dot{x}}{1 - x^2} . \tag{8}$$

Note that for small x, i.e., $|x| << 1$, we have on expansion of the denominator of Eq. (8), the result

$$\ddot{x} + f(x) = \varepsilon[h_1 - (h_2 - h_1)x^2]\dot{x}. \tag{9}$$

The right-side is the van der Pol nonlinear term provided

$$h_1 > 0, \quad h_2 > h_1 . \tag{10}$$

Thus, we see that the nonlinear, singular oscillators Eqs. (1) and (8) are large amplitude generalizations of the van der Pol equation [1,5].

If we redefine ε to be

$$\varepsilon \to \varepsilon h_2 , \tag{11}$$

and let

$$1-\beta = \frac{h_1}{h_2} , \tag{12}$$

then Eq. (8) becomes

$$\ddot{x} + f(x) = \frac{\varepsilon[(1-\beta) - x^2]\dot{x}}{1 - x^2} , \tag{13}$$

or

$$\ddot{x} + f(x) = \varepsilon\left[1 - \frac{\beta}{1-x^2}\right]\dot{x}. \tag{14}$$

This last equation is just the dimensionless form of the nonlinear, negative resistance circuit model derived by Walker and Connelly [1] if $f(x) = x$. Therefore, their circuit is the simplest possible example of the class of singular, nonlinear oscillators.

3. HARMONIC BALANCE

A number of analytic techniques [5] exist for constructing approximations to the solutions of nonlinear oscillatory systems whose equations of motion is

$$\ddot{x} + x = \varepsilon f(x,\dot{x}), \tag{15}$$

where ε is a small positive parameter and $f(x,\dot{x})$ is a polynomial function of its arguments. These procedures include perturbation theory [5], slowly varying amplitude and phase [5], multi-time methods [5], and harmonic balance [4]. However, consideration of special cases of Eq. (1) show that the only method that provides good results for singular, nonlinear equations, as that of Eq. (1), is harmonic balance [3]. We now give a brief review of the method of harmonic balance [4] and use it in the next section to analyze several particular examples of singular, nonlinear equations. More details are to be found in the paper by Mickens [4].

The method of harmonic balance can be used to determine approximations to the solutions of nonlinear differential equations that have periodic solutions that are close to being harmonic, i.e., the Fourier series expansion is dominated by the lowest harmonic.

Consider the differential equation

$$F(\ddot{x},\dot{x},x,\alpha) = 0, \tag{16}$$

where $F(z,z,z,\alpha)$ is an odd polynomial function in z and α represents the collection of parameters which occur in the differential equation. The method of harmonic balance consists in assuming that a good approximation to the solution of Eq. (16) is

$$x(t) = A \cos \omega t, \tag{17}$$

where the amplitude A and angular frequency ω are a priori unknown. Substitution of this result into Eq. (16) and expanding in a trigonometric series, gives, after some algebraic manipulations, the following

$$M(A,\omega,\alpha)\cos \omega t + N(A,\omega,\alpha)\sin \omega t + (\text{higher-order harmonics}) = 0. \tag{18}$$

Neglecting the higher-order harmonics terms gives the following two equations that can be solved for A and ω in terms of the parameters α:

$$M(A,\omega,\alpha) = 0, \quad N(A,\omega,\alpha) = 0. \tag{19}$$

Thus,

$$A = f_1(\alpha), \quad \omega = f_2(\alpha). \tag{20}$$

Note that the harmonic balance method gives only the (approximate) values of the steady-state solutions. Further, these steady-states correspond to limit cycles. For the situation where Eq. (19) has more than one set of solutions, then the sets of (A_i,ω_i), where k is the number of solutions and i = 1,2,...,k, correspond to the parameters of the k limit cycles.

Thus, an approximation to the steady-state solution of Eq. (16) is given by

$$x(t) \simeq f_1(\alpha)\cos[f_2(\alpha)t]. \tag{21}$$

Finally, it should be understood that the harmonic balance method is a general technique that can be applied to any nonlinear differential equation as long as the behavior of the motion is close to harmonic.

4. TWO EXAMPLES

In this section, we will use the method of harmonic balance to determine approximations to steady-state (limit cycle) solutions of the following two equations

$$\ddot{x} + x = \varepsilon\left[1 - \frac{\beta}{1-x^2}\right]\dot{x}, \tag{22}$$

$$\ddot{x} + x^3 = \varepsilon\left[1 - \frac{\beta}{1-x^2}\right]\dot{x}. \tag{23}$$

Putting these equations in "rational" form gives, respectively, the results

$$(\ddot{x}+x)(1-x^2) = \varepsilon[(1-\beta)-x^2]\dot{x}, \tag{24}$$

$$(\ddot{x}+x^3)(1-x^2) = \varepsilon[(1-\beta)-x^2]\dot{x}. \tag{25}$$

(Note that these equations are of the form given by Eq. (16).) The substitution of the assumed solution, Eq. (17), into Eqs. (24) and (25), and applying the method of harmonic balance gives the following for the respective amplitudes and angular frequencies:

$$A = 2\sqrt{1-\beta}, \quad \omega^2 = 1, \tag{26}$$

$$A = 2\sqrt{1-\beta}, \quad \omega^2 = \frac{(1-\beta)(7-10\beta)}{(2-3\beta)}. \tag{27}$$

Thus, an approximate solution to Eq. (22) is

$$x(t) \simeq 2\sqrt{1-\beta} \cos t, \tag{28}$$

while for Eq. (23), we have

$$x(t) \simeq 2\sqrt{1-\beta}\cos\left[\frac{(1-\beta)(7-10\beta)}{(2-3\beta)}\right]^{1/2} t. \quad (29)$$

Note, first, that both amplitudes have the same value, namely, that given by $A = 2\sqrt{1-\beta}$; however, the frequencies differ greatly. Second, the amplitude and frequency do not depend on the (small) parameter ε.

In the next section, we will discuss the implications of these results.

5. DISCUSSION

From Eqs. (22) and (23), we conclude that the equations of motion are not defined for $x = \pm 1$. Physically, these singular points correspond to the systems having an infinite amount of energy. For this reason, we place the following restriction on the amplitudes

$$|A| < 1. \quad (30)$$

Since the range of A values, $0 \le A < 1$ and $-1 < A \le 0$, are equivalent, up to a phase, we take our restriction to be

$$0 \le A < 1. \quad (31)$$

Substitution of the result for the amplitude, from either Eq. (26) or (27), gives the following conditions on the parameter β

$$\frac{3}{4} < \beta < 1. \quad (32)$$

Since, the frequency squares, ω^2, has to be positive, we have, from Eq. (27)

$$\omega^2 = \frac{(1-\beta)(7-10\beta)}{(2-3\beta)} > 0. \quad (33)$$

There are several cases to consider. First, if $2-3\beta > 0$, then

$$(1-\beta)(7-10\beta) > 0. \quad (34)$$

Since $\beta < 1$, we conclude that

$$7-10\beta > 0, \quad (35a)$$

or

$$\beta < \frac{7}{10}. \quad (35b)$$

However, this result contradicts the lower bound of Eq. (32). Therefore, we must have

$$7-10\beta < 0, \quad (36a)$$

which leads to

$$\beta > \frac{7}{10}, \quad (36b)$$

which agrees with the lower bound of Eq. (32).

Combining the inequalities of Eqs. (32) and (36b), we conclude that β must obey the condition

$$\frac{3}{4} < \beta < 1, \quad (32a)$$

if the method of harmonic balance is to apply to the singular, nonlinear Eqs. (22) and (23). We have numerically integrated Eqs. (22) and (23) for a wide range of initial values $x(0)$ and $\dot{x}(0) = 0$, and parameter values ε and β. Our general conclusion is, that for (almost) arbitrary values of $\varepsilon > 0$ and values of β in the interval given by Eq. (32a), the long-time behavior of the solutions is a steady-state (limit cycle) whose amplitude and frequency is given to a good approximation by the expressions of Eqs. (26) and (27). In more detail, we expect from theoretical considerations, based on the method of harmonic balance, that

$$A = 2\sqrt{1-\beta} + O(\varepsilon), \quad (37)$$

$$\omega^2 = \omega_0^2 + O(\varepsilon), \quad (38)$$

where ω_0^2 is given by Eqs. (26) or (27). The numerical integrations investigations are completely consistent with these expectations [3, 7].

To summarize, to terms of order ε, the amplitude and phase do not depend on ε.

This last result can be generalized to the following class of singular, nonlinear oscillators

$$\ddot{x} + f(x) = \varepsilon\left[1 - \frac{\beta}{1-x^2}\right]\dot{x}, \qquad (39)$$

where f(x) is an odd function of x and $xf(x) > 0$ for $x > 0$. To see this, rewrite Eq. (39) to the form

$$[\ddot{x} + f(x)](1-x^2) = \varepsilon[(1-\beta)-x^2]\dot{x}. \qquad (40)$$

Since the assumed solution and its derivatives, according to harmonic balance, is

$$x = A\cos\omega t, \quad \dot{x} = -\omega A\sin\omega t, \quad \ddot{x} = -\omega^2 A\cos\omega t, \qquad (41)$$

the substitution of Eq. (41) into Eq. (40) gives rise to trigonometric series such that the left-side is a cosine series, while the right-side is a sine series. Further, note that for all f(x), the right-sides are the same. A moments reflection shows that the left-side determines the frequency and the right-side the amplitude. We conclude that to terms of order ε, all the members of the class of Eq. (39) have the same amplitude which is given by the result of Eq. (37); however, the angular frequencies will differ. Further, we can conclude that β must be restricted to the range given by the inequality of Eq. (32a). (The lower limit might increase, but the upper limit will not.)

6. FURTHER INVESTIGATIONS

Recently, a number of physical systems in electrical and mechanical engineering have arisen whose dynamic equations of motion lead to singular, nonlinear oscillatory differential equations of the form given by Eq. (16). Often these equations are polynomials in the second derivative, x. Investigation of these equations using the harmonic balance method and comparison with the results of numerical integration is desirable.

As noted above, the harmonic balance method gives only the steady-state (limit cycle) behavior. No direct information is obtainable from this technique on the transient behavior of the motion as a given limit cycle or limit point is approached. It would be extremely useful if a Krylov-Bogoliubov [5] type of technique could be constructed to calculate this behavior.

ACKNOWLEDGMENTS

The author thanks K. B. Bota for many useful conversations on the topic of this paper and for his help with the various numerical procedures. This research was supported in part by grants from DOE and NASA.

REFERENCES

[1] S. S. Walker and J. A. Connelly: Circuits, Systems, Signal Processes 2, 213-238 (1983).

[2] P. M. Mathews and M. Lakshamanan: Quarterly of Applied Mathematics 32, 215-218 (1974).

[3] K. B. Bota and R. E. Mickens: Journal of Sound and Vibration 96, 277-279 (1984).

[4] R. E. Mickens: Journal of Sound and Vibration 94, 456-460 (1984).

[5] R. E. Mickens: Nonlinear Oscillations (Cambridge, New York, 1981).

[6] L. Hill and R. E. Mickens, Bulletin of the American Physical Society 29, 63 (1984).

[7] J. Frantz Huggins, MS Thesis, Atlanta University, Department of Physics (December 1985).

Complex and Distributed Systems: Analysis, Simulation and Control
S.G. Tzafestas and P. Borne (editors)
Elsevier Science Publishers B.V. (North-Holland)

ANALYSIS OF STOCHASTIC VAN DER POL OSCILLATORS USING THE DECOMPOSITION METHOD

Ida BONZANI

Dipartimento di Matematica - Politecnico di Torino - Corso Duca degli Abruzzi, 24
10129 Torino (Italia)

An analytical approximated solution of a generalized stochastic Van der Pol equation is here obtained, using the Adomian's decomposition method. Statistics of the solution process are compared with the statistical results obtained by extending the Krylov-Bogoliubov method to the stochastic case.

1. INTRODUCTION

A new mathematical method in the field of nonlinear analysis has been recently proposed by Adomian in his book [1] and developed and applied in several papers (see ref. [2, 3] and related bibliography). According to such a method, the solution of a general operator equation is decomposed into a parametrized form and each term of this decomposition is determined as a function of the preceding ones by executing a sequence of quadratures.

The method shows advantages with respect to the classical ones. In fact, since the decomposition parameter is not, in general, a perturbation parameter, it follows that strong nonlinearities in the operator equation can be handled and accurate approximated solutions may be obtained even in this cases. Moreover, since the considered operator equations have a very general form, the method may be applied to study various classes of problems including, for instance, the one described by partial differential equation [4, 5] and stochastic operator equations. In particular, as it will be shown also in this paper, the decomposition method can be very helpfull in the computation of the statistical properties of the solution process in stochastic problems, since it do not require the suitable limitations on the nature of stochastic processes, which are required by others methods.

The aim of the paper consists in an analysis of the transient behaviour of those stochastic mechanical systems which are described by generalized Van der Pol equations. The mathematical definition of the physical systems is presented in section 2, where approximated solutions of the stochastic equation are deduced using the said decomposition method. These solutions are then utilized to derive the first and second-order moments of the solution process. The results are compared both with numerical ones and with the statistical results obtained by extending to the stochastic case the well-known averaging method of Krylov Bogoliubov, which is widely used in the analysis of deterministic differential equations with small nonlinearities.

2. ANALYSIS

The following preliminary definition are supplied:

$T = [0,t) \subset R$ is the domain of the independent variable t;

$\underline{x} = \underline{x}(\omega,t) : (\Omega \times T) \to \mathbf{D} \subset R^2$ is a random process with initial state $\underline{x}_o(\omega)$, defined in a complete probability space (Ω,β,p); D is an open bounded subset of R^2;

$\underline{r} = \underline{r}(\omega) = \{\alpha(\omega), \beta_1(\omega), \beta_2(\omega)\}$ is a random vector whose components are known random variables;

$P(\underline{x}_o, \underline{r})$ is the probability density induced by p and is given as an assigned joint density function of the random variable $\underline{x}_o(\omega)$ and $\underline{r}(\omega)$.

The class of stochastic mechanical systems described by a generalized Van der Pol equation of the form

$$\ddot{x}(\omega,t) - \left(\beta_1(\omega) - \beta_2(\omega)^2 x(\omega,t)\right)\dot{x}(\omega,t) + \alpha(\omega)x(\omega,t) = 0 \quad (1)$$

is here considered.

Introducing the vector random process

$\underline{x}(\omega,t) = \{x_1(\omega,t) = x(\omega,t); x_2(\omega,t) = \dot{x}(\omega,t)\}$, with

$\underline{x}(\omega,0)=\{x_o(\omega),\dot{x}_o(\omega)\}$, the vectorial form of eq. (1) is

$$\dot{\underline{x}}(\omega,t)=\underline{f}\left(\underline{x}(\omega,t);\alpha(\omega),\beta_1(\omega),\beta_2(\omega)\right),\ \underline{x}(\omega,0)=\underline{x}_o(\omega),$$

or explicitely

$$\begin{cases}\dot{x}_1(\omega,t)=x_2(\omega,t) & (2)\\ \dot{x}_2(\omega,t)=\left(\beta_1(\omega)-\beta_2(\omega)x_1^2(\omega,t)\right)x_2(\omega,t)-\alpha(\omega)x_1(\omega,t) & \end{cases}$$

and the random process $\underline{x}(\omega,t)$ has the meaning of state variable of the system.

The objective of our analysis consists in the study of the transient approximated solution of the above equation which will be determined by means of the decomposition method proposed by Adomian. Accordingly introduce a decomposition parameter λ, $[\lambda\in(a,b),(a,b)\subseteq R]$ and write $\underline{x}(\omega,t)$ in the form

$$\underline{x}=\underline{x}(\omega,t;\lambda);$$

$\underline{x}$ is supposed to be infinitely differentiable with respect to λ and such that $\underline{x}(\omega,t;\lambda=1)=\underline{x}(\omega,t)$.

The integral formulation of eq. (2) can now be written in the form

$$\underline{x}(\omega,t;\lambda)=\underline{x}_o(\omega)+\lambda\int_0^t\underline{f}\left(\underline{x}(\omega,s;\lambda);\underline{r}(\omega)\right)ds,\quad \lambda=1 \qquad (3).$$

In order to search for the solution of eq. (3), the solution process $\underline{x}(\omega,t;\lambda)$ and the nonlinear function $\underline{f}$ can be decomposed in the form

$$\underline{x}=\sum_k\underline{x}^{(k)}\lambda^k,\quad \underline{x}^{(k)}=\frac{1}{k!}\left(\frac{d^{(k)}\underline{x}}{d\lambda^k}\right)_{\lambda=o} \qquad (4a)$$

$$\underline{f}=\sum_k\underline{f}^{(k)}\lambda^k,\quad \underline{f}^{(k)}=\frac{1}{k!}\left(\frac{d^{(k)}\underline{f}}{d\lambda^k}\right)_{\lambda=o}\ . \qquad (4b)$$

Casting the above expression into eq. (3) and equaling the terms with the same power of λ, the following sequence of equations is obtained:

$$\begin{aligned}\underline{x}^{(o)}&=\underline{x}_o(\omega)\\ \underline{x}^{(1)}&=\int_o^t\underline{f}^{(o)}ds\\ &\text{-----------------} \qquad\qquad (5)\\ \underline{x}^{(k+1)}&=\int_o^t\underline{f}^{(k)}ds,\end{aligned}$$

where each term $\underline{x}^{(k)}$, can be computed by quadrature, if all the preceeding ones are known. For the class of the stochastic equation (1) a useful recurrent formula can be found. In fact, since the k-th term of the sequence (4b) is given by

$$\begin{cases}f_1^{(k)}=x_2^{(k)}\\ f_2^{(k)}=\dfrac{1}{k!}\left(\dfrac{d^{(k)}f_2}{d\lambda^k}\right)_{\lambda=o}=\\ =-\alpha x_1^{(k)}+\beta_1x_2^{(k)}-\beta_2\sum_{j=o}^{k}x_2^{(k-j)}\sum_{i=o}^{j}x_1^{(i)}x_1^{(j-1)},\end{cases}$$

it follows from eq. (5) that the k-th term of the decomposition (4a) is given by

$$\begin{cases}x_1^{(k)}(\omega;t)=\int_o^t x_2^{(k-1)}ds=\gamma^{(k+1)}(\omega)\dfrac{t^k}{k!}\\ x_2^{(k)}(\omega;t)=\int_o^t f_2^{(k-1)}ds=\gamma^{(k+2)}(\omega)\dfrac{t^k}{k!}\end{cases},\qquad (6)$$

where the constants $\gamma^k(\omega)$ are polynomials of order $(3k+1)$ of the random variables $\alpha(\omega)$, $\beta_1(\omega)$, $\beta_2(\omega)$ and $\underline{x}_o(\omega)$ and are defined by the following sequence:

$$\begin{aligned}&\gamma^{(1)}(\omega)=x_{1,0}(\omega),\\ &\gamma^{(2)}(\omega)=x_{2,0}(\omega),\\ &\gamma^{(3)}(\omega)=-\alpha(\omega)x_{10}(\omega)+\beta_1(\omega)x_{20}(\omega)-\beta_2(\omega)x_{20}(\omega)x_{10}^2(\omega) \qquad (7)\\ &\text{-----------------------------}\\ &\gamma^{(k+1)}(\omega)=-\alpha(\omega)\gamma^{(k-1)}(\omega)+\beta_1(\omega)\gamma^{(k)}(\omega)-\\ &\quad-\beta_2(\omega)\sum_{j=o}^{k-2}\gamma^{(k-j)}(\omega)\sum_{i=o}^{j}\gamma^{(i+1)}(\omega)\gamma^{(i-j+1)}(\omega)\\ &\qquad\qquad (k\geq 3)\ .\end{aligned}$$

Consequently substitution into eq. (4a) for $\lambda=1$ yields:

$$x_1(\omega,t;\underline{x}_o(\omega))=\sum_{k=o}^{\infty}\gamma^{(k+1)}(\omega)\frac{t^k}{k!} \qquad (8)$$

where the actual expression of γ^{k+1} is supplied by eq. (7). This expression is effectively the solution of eq. (1) in a sufficiently bounded time interval, $t\varepsilon T$, if the following conditions hold:

i) $t\varepsilon T$, $\lim_{k\to\infty}\left(c^{k+1}\dfrac{t^k}{k!}\right)=0$, where $c^{k+1}=\sup|\gamma^{k+1}(\omega)|$;

ii) $t\varepsilon T$, $\lim_{n\to\infty}\left\|\underline{f}-\sum_{k=1}^{n}\underline{f}^{(k)}\right\|=0$, where $||\underline{g}|| = \sup_{\substack{i=1,2 \\ \omega \in \Omega}} |g_i|$.

In fact condition (i) assures that the solution (8) is "the solution" of eq. (3) where $\underline{f}$ has been replaced by the expression (4b), namely

$$\underline{x}(\omega,t;\lambda) = \underline{x}_o(\omega) + \lambda\int_o^t \Sigma_k \underline{f}^{(k)}(\underline{x}(\omega,s;\lambda);\underline{r}(\omega))ds,$$

$$(\lambda = 1) \tag{9}$$

whereas condition (ii) assures that eq. (3) can be replaced by eq. (9). It can be shown that condition (i) implies condition (ii). In fact, if condition (i) holds, the series (8) is uniformly convergent; therefore it is termwise differentiable and also the series $\Sigma_k \dot{\underline{x}}^{(k)}$ is convergent. Since from eq. (5) we have $\underline{f}^{(k)} = \dot{\underline{x}}^{(k+1)}$, it follows that

$$\lim_{k\to\infty} f_1^{(k)} = \lim_{n\to\infty} \dot{x}_1^{(k+1)} = 0$$

and analogously for the component $f_2^{(k)}$. Therefore it follows from eq. (4b), with $\lambda=1$, that

$\lim_{k\to\infty} \underline{f}^{(k)} = \underline{f}$ and condition (ii) holds.

The above result proves that, if the initial conditions $\underline{x}_o(\omega)$ and the known vector $\underline{r}(\omega)$ are suitably assigned in order to satisfy the condition (i), then the solution of eq. (1) is defined by eq. (8). As shown in the numerical example, condition (i) is satisfied also in cases of strong nonlinearities. In fact by assuming (fig. 1)

$$\alpha = \beta_1 = \beta_2 = 1 \ ; \quad x_{10} = x_{20} = 1.$$

the convergence of the solution (8) to the exact numerical one, obtained by numerical integration of the Van der Pol equation, is reached, in a conveniently bounded interval of time, by using only few terms of the decomposition (4a).

The above analytical results can be used to calculate the first and second-order moments of the solution process $x(\omega,t)$, in terms of the known statistics of $\underline{x}_o(\omega)$ and $\underline{r}(\omega)$.

The q-order moment is given by the expression

$$E\{x^q\}=\int(\Sigma_k x^{(k)}(t;\underline{x}_o(\omega),\underline{r}(\omega)))^q P(\underline{x}_o(\omega),\underline{r}(\omega)d\underline{x}_o d\underline{r},$$

where the $x^{(k)}(t;\underline{x}_o(\omega),\underline{r}(\omega))$ are expressed in eq. (8) and it is $P(\underline{x}_o,\underline{r}) = P_o(\underline{x}_o)P_r(\underline{r})$ if statistical indepencence of $\underline{x}_o$ and $\underline{r}$ is assumed.

Let us consider now only the first n terms of the sequence. The solution process is then approximated by the expression

$$x^*(\omega,t;\underline{x}_o(\omega)) = \sum_{k=o}^{n} x^{(k)}(\omega,t;\underline{x}_o(\omega))$$

and the q-moment is given by

$$E\{x^{*q}\}= \int_{D_o}\int_{D_r} [\Sigma_k x^k(t;\underline{x}_o(\omega),\underline{r}(\omega))]^q P(\underline{x}_o)P_r(\underline{r})d\underline{x}_o d\underline{r},$$

where $D_o \subset (\Omega \times R^2)$ and $D_r \subset (\Omega \times R^3)$ are the domains of $\underline{x}_o$ and $\underline{r}$ respectively.

In particular the mean value and the variance are expressed in the following form:

$$<x^*> = \Sigma_{k=o}^n \int_{D_o}\int_{D_r} x^k P_o(\underline{x}_o)P_r(\underline{r})d\underline{x}_o d\underline{r} =$$

$$= \Sigma_{k=o}^n <x^k>, \tag{10}$$

$$\sigma^2(x^*)= \int_{D_o}\int_{D_r} [x^*- <x^*>]^2 P_o(\underline{x}_o)P_r(\underline{r})d\underline{x}_o d\underline{r} =$$

$$= E\{x^{*2}\} - <x^*>^2 \tag{11}$$

To obtain these expressions in terms of the statistics of $\underline{x}_o$ and $\underline{r}$, we observe that

$$E\{x^{*2}\} = \int_{D_o}\int_{D_r} [\Sigma_{k=o}^n x^k]^2 P_o(\underline{x}_o)P_r(\underline{r})d\underline{x}_o d\underline{r} =$$

$$= \int_{D_o}\int_{D_r} [\Sigma_{i,j=o}^n x^i x^j] P_o(\underline{x}_o)P_r(\underline{r})d\underline{x}_o d\underline{r} =$$

$$= \Sigma_{i,j=o}^n \int_{D_o}\int_{D_r} x^i x^j P_o(\underline{x}_o)P_r(\underline{r})d\underline{x}_o d\underline{r} =$$

$$= \Sigma_{i,j=o}^n <x^i,x^j> .$$

On the other hand

$$<x^k> = \frac{t^k}{k!} <\gamma^{(k+1)}>, \quad k = 0, \ldots, n ,$$

where

$$<\gamma^{(1)}> = <x_{10}>$$

$$<\gamma^{(2)}> = <x_{20}>$$

$$<\gamma^{(3)}> = -<\alpha><x_{10}>+<\beta_1><x_{20}>-<\beta_2><\beta_2><x_{20}x_{10}^2>$$

....

$$<\gamma^{(k+1)}> = -<\alpha\gamma^{(k-1)}>+<\beta_1\gamma^{(k)}>-<\beta_2\Sigma_{j=o}^{k-2}\gamma^{(k-j)}.$$

$$\Sigma_{i=o}^j \gamma^{(i+1)}\gamma^{(j-i+1)}> \qquad (k \geq 3) \quad \text{and}$$

$$<\gamma^{(1)2}> = <x_{10}^2> ,$$

$$<\gamma^{(1)}, \gamma^{(2)}> = < x_{10}, x_{20}>,$$

....

Therefore the mean value and variance can be given in terms of the statistics on $\underline{x}_o(\omega)$ and $\underline{r}(\omega)$ as follows:

$$\begin{cases} \langle x^* \rangle = \sum_{k=0}^{n} \frac{t^k}{k!} \langle \gamma^{(k+1)} \rangle & (12) \\ \sigma^2(x^*) = \sum_{k,j=0}^{n} \frac{t^{k+j}}{k!j!} \langle \gamma^{(k+1)}, \gamma^{(j+1)} \rangle - \langle x^* \rangle^2 \end{cases}$$

where the moments of $\gamma^{(k+1)}(\omega)$ are deduced from eq. (7) in terms of the known joint moments of $\underline{x}_o(\omega)$ and $\underline{r}(\omega)$.

3. THE STOCHASTIC AVERAGING METHOD: COMPARISON AND DISCUSSION

To test the validity of the above results we consider the Krylov-Bogoliubov method; well known in non linear deterministic analysis and generalize it to the random case, in order to derive statistics on the first moment of the solution process, to be compared with the ones obtained in the last section.

The considered nonlinear second-order equation is now

$$\ddot{x} - \varepsilon\,(\beta_1(\omega) - \beta_2(\omega)\,x^2)\dot{x} + \alpha x = 0, \qquad (13)$$

where ε is assumed to be a small deterministic parameter, α and $\underline{x}_o$ (with $\alpha>0$) are deterministic constants and β_1, β_2 are the components of a known random vector

$$\underline{\beta}(\omega) = \{\beta_1(\omega), \beta_2(\omega)\} : \Omega \in B \subset \mathbb{R}^2$$

According to the deterministic Krylov-Bogoliubov solution, the solution process of eq. (13) may be searched in the form:

$$x(t;\omega,x_o) = a(t;\omega,x_o)\cos(\sqrt{\alpha}t+\theta(t;\omega,x_o)) = a\cos\phi,$$

$$(\phi = \sqrt{\alpha}t + \theta) \qquad (14)$$

with the condition

$$\frac{dx}{dt} = -a(t;\omega,x_o)\sqrt{\alpha}\,\sin[\phi(t;\omega,x_o)] \qquad (14')$$

Differentiating eq. (14) with respect to the time t gives

$$\frac{da}{dt}\cos\phi - a\sin\phi\frac{d\theta}{dt} = 0 \qquad (15)$$

whereas time differentiation of (14') gives

$$\frac{d^2x}{dt^2} = -a\alpha\cos\phi - \sqrt{\alpha}\frac{da}{dt}\sin\phi - a\sqrt{\alpha}\frac{d\theta}{dt}\cos\phi, \qquad (15')$$

Substituting this expression into eq. (13) and using eq. (15') we obtain

$$a\sqrt{\alpha}\frac{d\theta}{dt}\cos\phi + \sqrt{\alpha}\frac{da}{dt}\sin\phi = \varepsilon(\beta_1 - \beta_2 a^2\cos\phi)a\sqrt{\alpha}\sin\phi \qquad (16)$$

Solving (15') and (16) for $\frac{da}{dt}$ and $\frac{d\theta}{dt}$ yields

$$\begin{cases} \frac{da}{dt} = \varepsilon a\sin^2\phi(\beta_1 - \beta_2 a^2\cos\phi) \\ \frac{d\theta}{dt} = \varepsilon\sin\phi\cos\phi\,(\beta_1 - \beta_2 a^2\cos\phi). \end{cases} \qquad (17)$$

These two first-order equations replace the original second order differential equation (13). To solve them, we note that the right-hand sides of eqs. (17) are time-periodic with period $T = 2\pi/\sqrt{\alpha}$; moreover, since ε is a small parameter, a and θ are slowly varying functions both of time and of the random vector $\underline{\beta}(\omega)$. Consequently, averaging both over the interval (t,t+T) and over the probability space, a (ω) and θ (ω) can be taken to be constants on the right-hand side of these equations. Remembering also that $\underline{\beta}(\omega)$ is constant with respect to time, the above said double averaging yields

$$\langle\frac{da}{dt}\rangle = \frac{d}{dt}\langle a\rangle = \frac{\varepsilon}{T}\int_B\int_0^T a\sin^2\phi(\beta_1-\beta_2a^2\cos\phi)\cdot$$

$$P(\underline{\beta})dtd\underline{\beta} = \varepsilon\langle a\rangle\;\frac{\langle\beta_1\rangle}{2} - \frac{\langle a\rangle^2\langle\beta_2\rangle}{8}, \qquad (18)$$

$$\langle\frac{d\theta}{dt}\rangle = \frac{d}{dt}\langle\theta\rangle = 0 \qquad (19)$$

Eq. (18) is an ordinary differential equation for $\langle a\rangle$, whose solution can be expressed in terms of the known mean values of $\langle\underline{\beta}\rangle$, as follows:

$$\langle a\rangle = m_a(t;\langle\underline{\beta}\rangle,\underline{x}_o) = \frac{a_o\,ex\,[\varepsilon\langle\beta_1\rangle t|2]}{\{1+\frac{a_o^2\langle\beta_2\rangle}{4\langle\beta_1\rangle}[\exp(\varepsilon\langle\beta_1\rangle t)-1]\}^{\frac{1}{2}}} \qquad (20)$$

whereas from eq. (19) it is obtained

$$\langle\theta\rangle = \theta_o = \theta(t=0;\langle\underline{\beta}\rangle,\underline{x}_o) \qquad (21)$$

The constants a and θ_o are known functions of the initial conditions x_o and $\dot{x}_o$.

The above results, which extend the ones reported in [7,8] to the analysis of the first-order statistics of the stochastic Van der Pol equa -

tion, can be used in order to derive, in the considered limit of a small perturbation technique, an analytical expression of the "averaged" solution

$$\bar{x}(t;\omega,x_o) = \langle a\rangle \cos(\sqrt{\alpha}t + \langle\theta\rangle) \tag{22}$$

of eq. (13), which may be obtained by considering its solution process as a deterministic function of the mean values of the functions $a(\omega)$ and $\theta(\omega)$. The solution (22) may be compared with the results on the first-order moments which have been obtained by means of the decomposition method.

The above said comparisons are shown by fig. 2. The Runge-Kutta numerical solution "in the mean" $x_{num}(t;\langle\underline{\beta}\rangle)$, of the differential equation (13), in which $\underline{\beta}(\omega)$ is replaced by its mean, with $\langle\beta_1\rangle = \langle\beta_2\rangle = 1$, is here considered as the reference "exact" solution of the same equation. The values of $\langle x^*\rangle$, as given by eq. (12), for n=10, have been compared whith the numerical ones $\bar{x}_{num}$ in the bounded interval of time [0,1] and their difference

$$\eta_A = |\bar{x}_{num} - \langle x^*\rangle| \tag{23}$$

is reported as functions of time in fig. 2, for $\varepsilon = 0.1$ and $\varepsilon = 0.2$.

The "averaged" Krylov-Bogoliubov solution is also compared with $\bar{x}_{num}$ in the same conditions and the fig. 2 shows the behaviour of the difference

$$\eta_B - |\bar{x}_{num} - \bar{x}(t;\omega,\underline{x}_o)|, \tag{24}$$

where $\bar{x}(t;\omega,\underline{x}_o)$ is given by eq. (22). It results $\eta_A < \eta_B$ for any $t\varepsilon(0,1)$.

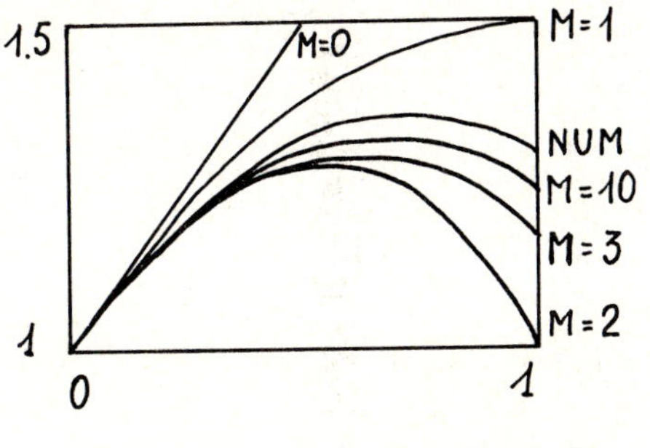

Approximations to the numerical solution, obtained by using the decomposition method.
$(x^* = x^{(o)} + \text{-------} + x^{(M)}$.

Figure 1

In other words, though the above stochastic generalization of the Krylov-Bogolibov method is shown to be valid in the framework of the small perturbation theory, yet it may be observed that the mean value $\langle x^*\rangle$, supplied by the decomposition method, gives a better approximation of the numerical solution "in the mean", not only for larger nonlinearities, but also in those cases for which the small perturbation techniques supply a valid approximated solution of the given problem.

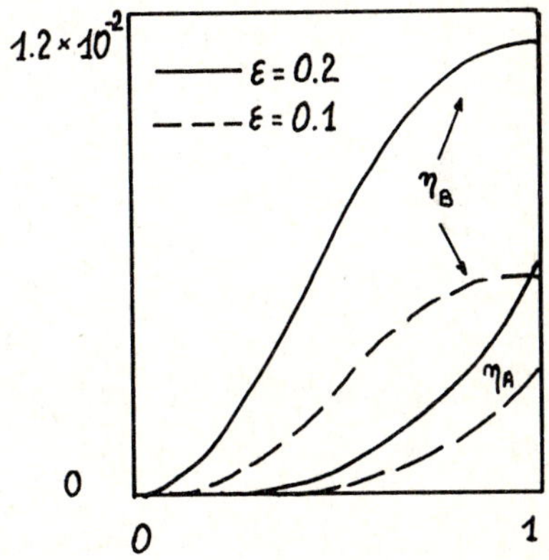

Behaviour of the functions η_A and η_B.

Figure 2

Concluding let's summarize the following results, obtained studying the transient stochastic solution of a generalized Van der Pol equation:

a) the Adomian decomposition method supplies a solution process $x(\omega,t)$ whose validity range is defined by an analytical condition on $x_o(\omega)$, $\dot{x}_o(\omega)$ and the random coefficients. It is shown by an example that such a condition is satisfied also in cases of strong nonlinearities;

b) the Krylov-Bogoliubov method, well known in non linear deterministic analysis is generalized to the random case and the "averaged" solution $\bar{x}(t,\langle\beta\rangle)$ is determined for small nonlinearities;

c) the Adomian decomposition method supplies in simple form the nth- order moments of the solution process, which, on the other hand, cannot be easily deduced from the extension of the Krylov-Bogoliubov method and from numerical procedures;

d) a comparison between $\langle x^*\rangle$ and $\bar{x}(t;\langle\underline{\beta}\rangle)$ is studied in two particular cases of small nonlinearities, showing that the decomposition method yields more satisfactory results in the approximation of the solution process of equation (1).

REFERENCES

[1] Adomian,G., Stochastic System, (Academic Press, New York, 1983).

[2] Adomian,G. and Roch, R., Inversion of Nonlinear Stochastic Operators, J. Math. Analysis and Appl., 91 (1983), 39-46.

[3] Bellomo, N. and Riganti, R., Time evolution of the Probability Density and of the Entropy Function for a Class of Nonlinear Stochastic Systems in Mathematical Physics, Computers and Mathematics with Applications, in print.

[4] Adomian, G. and Malakian, K. Inversion of Stochastic Partial Differential Equations - The Linear Case, J. Math. Analysis and Appl. 77 (1980), pp. 505-512.

[5] Adomian, G. and Bellman, R. Partial Differential Equations: New Methods for their Treatment and Solution, Reidel (1984).

[6] Bogoliubov, N. and Mitropolsky, I. Les Methodes Asymtotiques en Theorie des Oscillations Non Lineares, Gauthier Villars, Paris (1962).

[7] Riganti, R., Analytical Study of a Class of Nonlinear Stochastic Autonomous Oscillators with One Degree of Freedom, Meccanica, 14, No. 4 (1979), pp. 180-86.

[8] Nayfeh, A.H., Perturbation Method, (Wiley New York, 1973).

Complex and Distributed Systems: Analysis, Simulation and Control
S.G. Tzafestas and P. Borne (editors)
Elsevier Science Publishers B.V. (North-Holland)

A COMPUTATIONAL METHOD FOR THE CANARD-EXPLOSION

Christian Kaas-Petersen and Morten Brøns

Lab. of Applied Mathematical Physics and Mathematical Institute
The Technical University of Denmark
Building 303
DK-2800 Lyngby, Denmark

A periodic solution of small amplitude may explode into a periodic solution of large amplitude in a narrow parameterinterval. This instability stems from a change of position of two invariant manifolds followed closely by the periodic solutions. We determine the two manifolds numerically. Their change of position is formulated as a zero point problem.

1. INTRODUCTION

Periodic solutions to van der Pol's equation with an unexpected duck-like shape, see Fig. 2c, were by the french mathematician Diener [1] called canards. Later [4], the definition of a canard was generalized to be a trajectory that follows a slow manifold along its stable part, and then continued to follow the slow manifold when it has become unstable.

We shall invoke center manifold theory, and examine the canards in terms of invariant manifolds. This leads to a computational method which we present. The method is applied to the canards in the van der Pol equation.

Earlier, canards have been examined with non-standard analysis [1] and with asymptotic methods [5].

2. VAN DER POL'S EQUATION

We shall study the second order non-linear equations

$$\varepsilon\ddot{x} + (x^2-1)\dot{x} + x - c = 0 , \qquad (*)$$

$$(\dot{\ }) \equiv d(\ )/dt , \quad 0 < \varepsilon << 1 .$$

Since the system is invariant to the transformation $(x,\dot{x},c,\varepsilon) \to (-x,-\dot{x},-c,\varepsilon)$, it is sufficient to consider $0 \le c$. Using the Liénard transformation [12], we can formulate (*) as a system of two first order equations

$$\dot{x} = \frac{1}{\varepsilon}(y - (\frac{1}{3}x^3 - x)) \qquad (**)$$

$$\dot{y} = c - x .$$

Since ε is small, then $|\dot{x}| >> |\dot{y}|$ for most points (x,y) . For that reason x is called the fast variable, the $\dot{x}$-equation is called the fast equation, y is called the slow variable, the $\dot{y}$-equation is called the slow equation, and ε is the slow/fast-ratio.

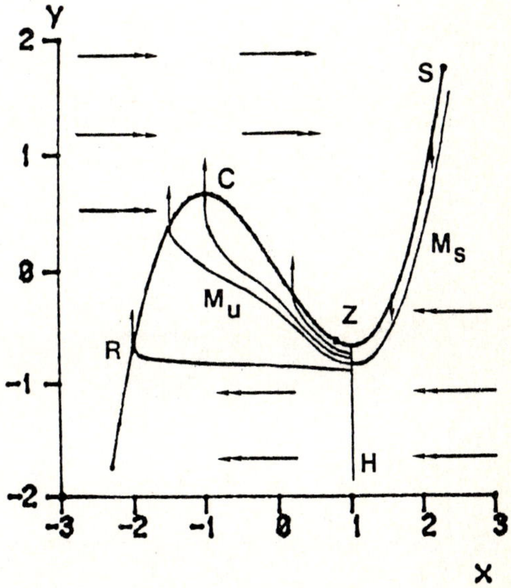

Fig. 1. Sketch of flow for van der Pol's equation when $c < 1$.

→ arrows means slow vector

↠ arrows means fast vector

C is the canard point

R is the relaxation point

S is the start point

Z is the unstable stationary point, and

H is a halfline at $x = 1$.

The manifolds M_S, M_C and M_R hit the half-line H . Let < denote 'beteween'. Then we define

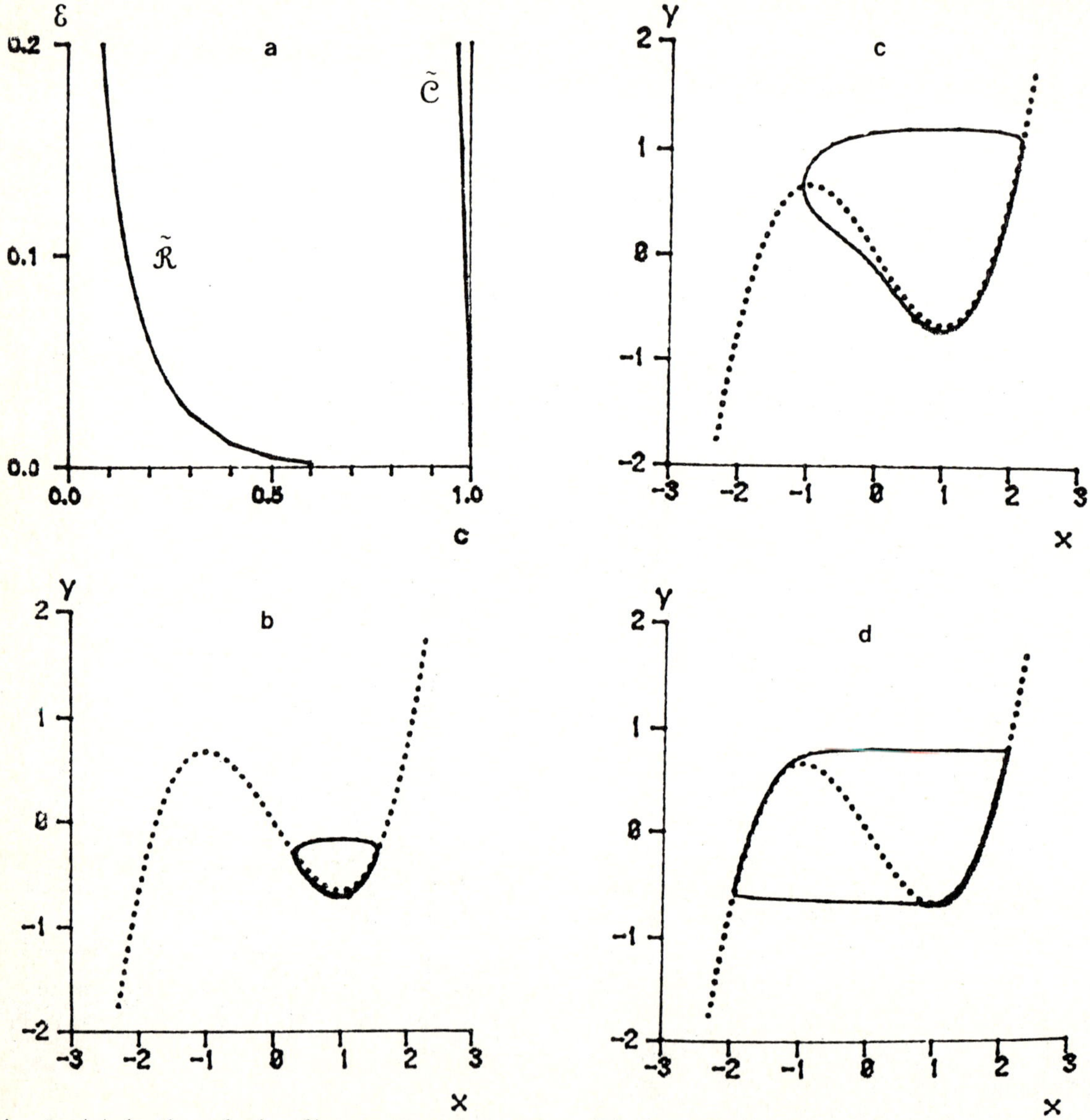

Fig. 2. (a) is the solution diagram. For $1 < c$ the stationary point $Z = (c, \frac{1}{3}c^3 - c)$ is the attractor. On the line $c = 1$ a Hopf bifurcation takes place generating small amplitude periodic solutions. The canard solutions live between the $\tilde{R}$-curve and the $\tilde{C}$-curve. The relaxation oscillations live to the left of the $\tilde{R}$-curve.

(b) The periodic solution for $c = 0.9868$ & $\varepsilon = 0.1$.
(c) The periodic solution for $c = 0.9863$ & $\varepsilon = 0.1$.
(d) The periodic solution for $c = 0.9863$ & $\varepsilon = 0.01$.

a) $M_C < M_s <$ slow manifold:
the periodic orbit has small amplitude, or is "a canard without head" [1].

b) $M_R < M_s < M_C$:
the periodic orbit is a canard (with head).

c) $\infty < M_s < M_R$:
the periodic orbit is a relaxation oscillation.

In van der Pol's equation the points $C = (-1, \frac{2}{3})$ and $R = (-2, -\frac{2}{3})$ are both independent of the parameters c, ε and H is the halfline $x = 1$, $y < -\frac{2}{3}$. On the other hand, the manifolds M_s, M_C and M_R depends on c and ε.

To determine the onset of canard solutions, we therefore fix c and ε. We integrate the differential equations forward in time from a point S on the slow manifold until the trajectory hits H. We integrate backwards in time from the point C until the trajectory hits H. The distance d between the two points in H is computed. The distance d is a function of the two variables c and ε. The distance is zero at the onset of the canard solutions with head.

A similar procedure can be formulated for the relaxation oscillations, this time involving the trajectory starting in S and the trajectory starting in R.

The solution diagram is shown in Fig. 2.

Here we shall give some details on the computations done. To calculate the stable manifold M_s, we choose $S = (3,6)$ on the slow manifold as an initial value, and start a numerical integration. The trajectory is attracted to the stable manifold very fast. The choice of S turned out to be unimportant.

The mathematical slow/fast property of a relaxation oscillation entails a stiff/nonstiff property in the numerical solution. We used the ODE-solver described in [11] which automatically switches between stiff and nonstiff methods as needed. We solved the ODEs with absolute error bound of 10^{-8}. The trajectory was said to be in H when $|x-1| \le \varepsilon_R$ with $\varepsilon_R = 10^{-11}$. The distance d between the two trajectories was said to be zero when $|d| \le \varepsilon_Z$ with $\varepsilon_Z = 10^{-6}$. Note, that we compute manifolds, not periodic solutions.

3. ADDITIONAL COMMENTS

The equation (**) is but a special example of the more general system

$$\dot{x} = \varepsilon^{-1} F(x,y;c)$$
$$\dot{y} = G(x,y;c) \; .$$

In this system the slow manifold and the points C and R will in general depend on c. An even more general system will be

$$\dot{x} = F(x,y;c,\varepsilon)$$
$$\dot{y} = G(x,y;c,\varepsilon) \; ,$$

where the slow manifold and the fast foliation just have to be transversal to eact other almost everywhere to ment the conditions for canard solutions.

The small amplitude periodic solutions are generated in a Hopf bifurcation [8] at $c = 1$ circling around $Z = (+1, -\frac{2}{3})$ with period $T = 2\pi \cdot \sqrt{\varepsilon}$. Thus the Hopf bifurcation is degenerate at $\varepsilon = 0$.

The large amplitude periodic solutions are relaxation oscillations. This is because (*) is a singular perturbed system [9].

The transition from small to large oscillations need not go via canards in regular perturbed systems This is seen in the Rayleigh oscillator (which is intimately connected with van der Pol's equation), cf. [10].

In this context, the presence of periodic solutions is only secondary. The essential feature of the canards are the invariant manifolds. What happens when a solution leaves the unstable manifold is another matter. From this point of views, the term "canard" has lost its metaphorical character.

We have examined a system with 2 equations. The concept of manifolds can be generalized to systems with more equations. With 3 equations we may have one or two fast equations. If we have one fast equation we may get chaotic behaviour [7]. If we have two fast equations with flow toward the slow manifold, we may expect to find canards similar to the ones obtained here.

When ε is very small we may not be able to compute anything. Analytical work is in progress to determine the curves $\tilde{C}$ and $\tilde{R}$ in Fig. 2a near the corner $0 < \varepsilon \ll 1$, $0.5 < c < 1.0$.

The work presented here was initiated by one of us (Morten Brøns), who considered surge in compressor systems [2].

REFERENCES

[1] E. Benoit, J.L. Callor, F. Diener and M. Diener: Chasse au Canard, Publ. de l'Institut de Recherche Mathématique Avancée, Strasbourg, 1980.
[2] M. Brøns: Bifurcation analysis of surge in compressor systems, submitted to J. Eng. Gas Turbines and Power, 1985.
[3] J. Carr: Applications of center manifold theory, Springer-Verlag, 1983.
[4] M. Diener and T. Poston: On the perfect delay convention or The revolt of the slaved variables, in H. Haken (ed.): Chaos and order in nature, Springer-Verlag, 1981.
[5] W. Eckhaus: Relaxation oscillations including a standard chase on french ducks in G. Verhulst (ed.):

Asymptotic analysis, Springer Lecture notes in mathematics No. 985, Springer-Verlag, 1983.
[6] N. Fenichel: Geometric singular perturbation theory for odrinary differential equations, J. Diff. Eq., Vol. 31, 1979,,p. 53.
[7] L. Garrido (ed.): Dynamical systems and chaos, Lecture notes in Physics No. 179, Springer-Verlag, 1983.
[8] B.D. Hassard, N.D. Kazarinoff and Y.-H. Wan: Theory and applications of Hopf bifurcation, Cambridge University Press, Cambridge, 1981.
[9] J. Kevorkian and J.D. Cole: Perturbation methods in applied mathematics, Springer-Verlag, 1981
[10] A.H. Nayfeh and D.T. Mook: Nonlinear oscillations, Wiley, 1979.
[11] L. Petzold: Automatic selection of methods for solving stiff and nonstiff system of ordinary differential equations, SIAM J. Sci. Stat. Comput., Vol. 4, No. 1, 1983, pp. 136-148.
[12] J.J. Stoker: Nonlinear vibrations, Wiley Interscience, 1950.

Complex and Distributed Systems: Analysis, Simulation and Control
S.G. Tzafestas and P. Borne (editors)
Elsevier Science Publishers B.V. (North-Holland)

SURVEY OF STRANGE ATTRACTORS AND CHAOTICALLY TRANSITIONAL PHENOMENA IN THE SYSTEM GOVERNED BY DUFFING'S EQUATION

Yoshisuke UEDA

Department of Electrical Engineering
Kyoto University
Kyoto 606, Japan

1. INTRODUCTION

In nonlinear systems on some conditions, chaotic motions can occur in spite of perfectly deterministic nature of equations describing the physical systems. An appearance of such chaotic motions is attributed to both global structure of solutions for deterministic differential equations of the systems and microfluctuations in the systems. The author has long been studying on these problems and has called this type of motions 'chaotically transitional phenomena'. The phenomena are represented by strange attractors on phase planes or in state spaces.

This paper summarizes strange attractors and their related properties which occur in the system governed by Duffing's equation [1-8].

2. MICROFLUCTUATIONS IN ACTUAL ELECTRIC CIRCUITS AND LAWS IN THE THEORY OF ELECTRO-MAGNETISM

Before entering into the main subject, let us arouse attention as to the treatment of microfluctuations in actual electric circuits.

Physical quantities exhibiting voltages, currents, etc., are always subject to noises acting on actual electric circuits. Constants of the circuits are also affected by small perturbations. Accordingly, these variables and parameters are regarded as representative points of narrow intervals covering these small uncertainties. In fact, we can get no more than such values by measurements. Therefore, it must be noted that all laws in the theory of electro-magnetism have been obtained empirically by neglecting such microfluctuations in actual systems. Because of these circumstances, various kinds of stability concepts have been recognized.

3. DETERMINISTIC STEADY MOTIONS OF DYNAMICAL SYSTEM

This report deals with the system governed by Duffing's equation,

$$\frac{d^2x}{dt^2} + k\frac{dx}{dt} + x^3 = B\cos t \tag{1}$$

or

$$\frac{dx}{dt} = y, \quad \frac{dy}{dt} = -ky - x^3 + B\cos t \tag{2}$$

The above equation is a mathematical model of a series-resonance circuit containing a saturable inductor. Various physical systems are also described by the equation.

A discrete dynamical system on R^2, or a C^∞-diffeomorphism of the xy plane into itself is introduced by using the solution $(x(t, x_0, y_0), y(t, x_0, y_0))$, which, when $t = 0$, is at the point (x_0, y_0), of Eq. (2). That is,

$$f_\lambda : R^2 \to R^2$$
$$p_0 \mapsto p_1 \tag{3}$$

where

$$\lambda = (k, B)$$
$$p_0 = (x_0, y_0) = (x(0, x_0, y_0), y(0, x_0, y_0))$$
$$p_1 = (x_1, y_1) = (x(2\pi, x_0, y_0), y(2\pi, x_0, y_0))$$

A motion which starts from an arbitrary point $p_0 = (x_0, y_0)$ at $t = 0$ is represented by a stroboscopic point sequence on R^2 generated by successive iterations of f_λ, i.e.,

$$\{p_n\} : p_0, p_1, p_2, \ldots \tag{4}$$

where

$$p_n = f_\lambda^{\,n}(p_0) = (x(2n\pi, x_0, y_0), y(2n\pi, x_0, y_0)) \quad (n \in Z^+)$$

The simplest type of steady motions described by Eq. (2) are represented by the fixed or the periodic points of f_λ. Simple fixed points are classified as follows:

(i) Completely stable points (Sink) or S
(ii) Directly unstable points (Saddle) or D
(iii) Inversely unstable points (Saddle) or I
(iv) Completely unstable points (Source) or U

There issue from D (or I) four invariant curves or branches, i.e., two α-branches (unstable manifolds) whose points converge toward D (or I) on indefinite iteration of f_λ^{-1} and two ω-branches (stable manifolds) whose points converge toward D (or I) on indefinite iteration of f_λ.

A deterministic steady motion is represented by an exact solution emanating from a minimal set of the discrete dynamical system (3).* Every motion of a minimal set has the property of recurrence.** Contrary to steady motions, transient states are represented by orbits composed of nonwandering points.***

It is to be noted that, however complicatedly a motion may behave, it is anyway reproducible except for microfluctuations. Accordingly, the realizability of the above introduced deterministic steady motions in the actual circuits depends on both the stability of minimal sets and the magnitude of microfluctuations.

4. ACTUAL STEADY MOTIONS OF THE ELECTRIC CIRCUIT

In the preceding section, the deterministic motion $\{p_n\}(n \in Z^+)$ of the discrete dynamical system (3) has been introduced. Minimal sets and wandering points have also been explained in relation to the steady and the transient motion, respectively. In the present section, we represent an actual motion by a stroboscopic sequence $\{q_n\}(n \in Z^+)$ in order to relate it to $\{p_n\}$. The sequence $\{q_n\}$ is obtained by experiments; and that, practically, we can get no more than such $\{q_n\}$ instead of $\{p_n\}$.

In the following, an attractor is defined as an ω-limit set of $\{q_n\}$ and a steady state as an observational motion taking place in the attractor. As an ω-limit set of $\{q_n\}$ cannot be derived from a finite term observation or a finite number of points, we are obliged to regard a point set $\{q_n | N_1 < n \leq N_1 + N_2\}$ (N_1, N_2 : sufficiently large natural numbers) as an attractor, where N_1 is determined experimentally by the acknowledgment of a vanishing transient state.

In order to infer the relation between $\{p_n\}$ and $\{q_n\}$, let us here introduce the (exterior) diameter e(U) and the interior diameter i(U) of a connected region U on R^2 as follows:

$$e(U) = \sup_{r,s \in U} d(r, s) = d(r_0, s_0) \tag{5}$$

where r_0 and s_0 indicate limit points defining the (exterior) diameter. See Fig. 1.

$$i(U) = \inf \{\text{Set of curve length connecting } r_0 \text{ and } s_0 \text{ via the inside of the set } U\} \tag{6}$$

From these definitions, it is easily seen that the following relation holds.

$$i(U) \geq e(U) \tag{7}$$

Consider an infinite sequence of connected re-

---

* A set $M \subset R^2$ is called minimal if it is nonempty, closed and f_λ-invariant, and has no proper subset possessing these three properties.

** A motion is called recurrent if for any $\varepsilon > 0$ there can be found a $N(\varepsilon) > 0$ such that any neighborhood of successive $N(\varepsilon)$ points of this motion approximates the entire orbit with a precision to within ε.

***A point $p \in R^2$ is called wandering if there exists a neighborhood $\sigma \subset R^2$ of it and a positive number N such that

$$\sigma \cap f_\lambda^{\,n}(\sigma) = \phi \quad \text{for all } n \geq N$$

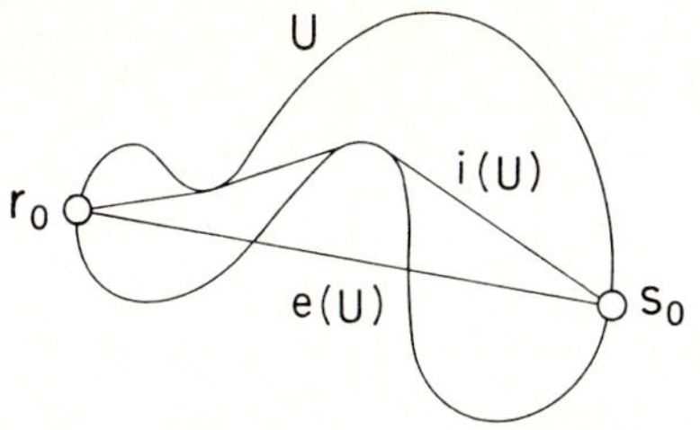

Fig. 1: The exterior and the interior diameter of a connected region U.

gions on R^2 of the form

$$\{U_n\}: U_0, U_1, U_2, \ldots \tag{8}$$

where $U_n = f_\lambda^{\,n}(U_0)(n \in Z^+)$ and U_0 is a neighborhood of the initial point p_0. The point p_n is contained in U_n. Let $\{q_n\}$ be the orbit which starts from the initial point q_0 close to p_0. Suppose next that the diameter of U_0 is assigned so far large in which U_0 covers the effects of small uncertain factors. Then finite number of images of $\{q_n\}$ can be contained in $\{U_n\}$, or in other words, for assigned diameter of U_0 there exists an N' such that

$$\{q_n | 0 \leqq n \leqq N'\} \subset \bigcup_{n=0}^{N'} U_n \tag{9}$$

Based on this property, asymptotic movements of $\{q_n\}$ to the attractors will be inferred from the asymtotic behavior of $\{U_n\}$. The asymptotic behavior of $\{U_n\}$ may be divided into two broad classes:

(1) In the case of regular attractor

$$\lim_{n\to\infty} e(U_n) = 0 \quad \text{and} \quad \lim_{n\to\infty} i(U_n) = 0 \tag{10}$$

(2) In the case of strange attractor

$$\lim_{n\to\infty} e(U_n) = C \quad \text{and} \quad \lim_{n\to\infty} i(U_n) = \infty \tag{11}$$

where C denotes a finite value determined uniquely from the discrete dynamical system (3), or the (external) diameter of the strange attractor.

It is supposed that the effects of microfluctuations will be neglected for the regular attractors. On the other hand, however small these factors may be, their effects cannot be neglected for the strange attractors. This situation will be seen in the following illustration.

Figure 2 shows an example of asymptotic behavior of $\{V_n\}$, where $V_n = f_\lambda^{\,n}(V_0)$ and V_0 is a neighborhood of q_0. We will see in the figure that, as n increases, V_n becomes longer, more complicated and is folded many times. In this way, $\{V_n\}$ tends to the whole attractor. Hence the effects of microfluctuations cannot be ignored, and stochastic property arizes and reproducibility disappears in the actual phenomena.

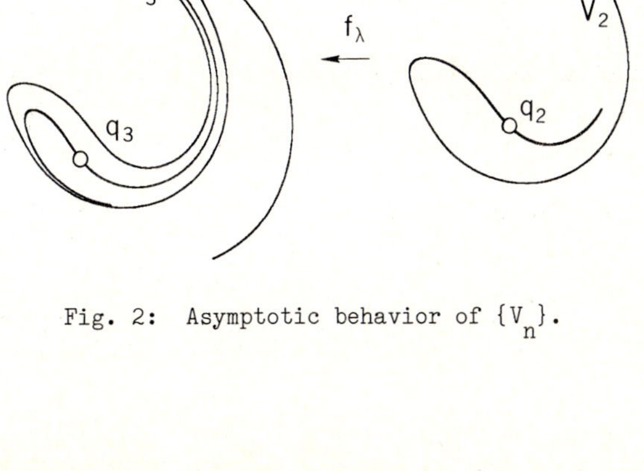

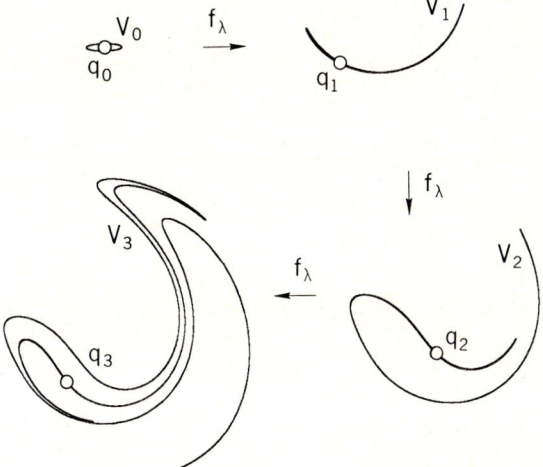

Fig. 2: Asymptotic behavior of $\{V_n\}$.

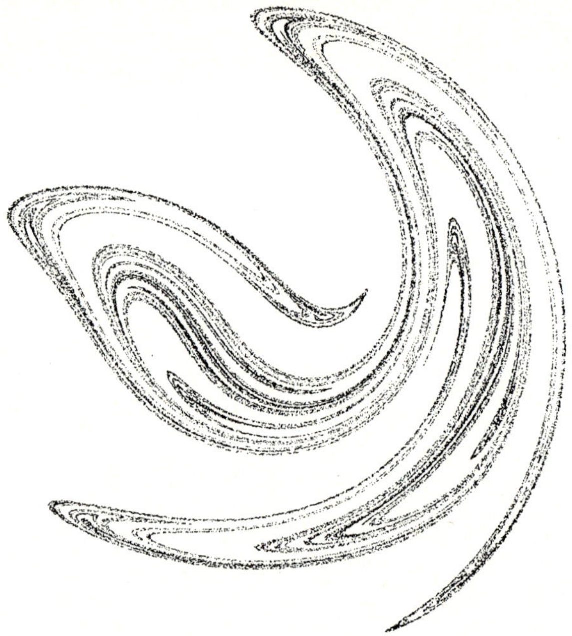

Fig. 3: Representative example of the strange attractors described by Duffing's Equation.

5. STRUCTURES OF STRANGE ATTRACTORS AND THEIR RELATED PROPERTIES

Before giving the details, a representative example of strange attractors in the system described by Eq. (2) is demonstrated. Figure 3 shows the strange attractor which occurs in the computer-simulated system for the parameter values

$$\lambda = (k, B) = (0.05, 7.5) \qquad (12)$$

It is obtained by stroboscopic sampling and plotting the computer solution $(x(t), y(t))$ ($t = 2n\pi$, $N_1 < n \leqq N_1 + N_2$) of Eq. (2) after the transient has vanished ($N_1 = 1{,}000$, $N_2 = 50{,}000$).

In the following, structures of the strange attractors and properties of the steady motions are summarized for the system governed by Duffing's equation. Examine the following items by referring to Fig. 3.

(1) Strange attractors are invariant closed sets with respect to f_λ of R^2 into itself. They are determined uniquely from the diffeomorphism (3). There exists at least one saddle type fixed point in the neighborhood of a center of gravitation of the attractor. Figure 4 shows the directly unstable fixed point D and the outlines of α- and ω-branches of it for the same system parameters as given by Eq. (12). If the α-branches of Fig. 4 are prolonged infinitely, they tend to themselves and are folded infinitely many times tracing out an appearance of the strange attractor. The α-branches have a property of self-similarity. That is, infinitely many branches go side by

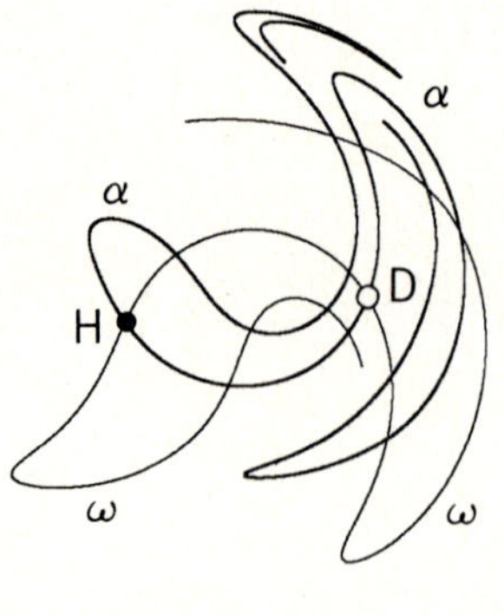

Fig. 4: Outline of α- and ω-branches of the saddle point D.

side in the neighborhood of an arbitrary point on them, and further magnification of the neighborhood would again yield a similar situation. Thus, strange attractors can be regarded as closures of α-branches of the saddle type unstable fixed points of the discrete dynamical system (3).

(2) The α-branches of the saddle point intersect the ω-braches of the same point, and form a doubly asymptotic structure. Existence of homoclinic points (e.g., the point H in Fig. 4) indicates that strange attractors contain infinitely many periodic groups. Every periodic group is regarded as unstable and the closer a periodic group gets to the homoclinic points, the longer its period becomes. Homoclinic points are accumulating points of the periodic points. This indicates that these periodic points cannot be isolated one another, and the existence of microfluctuations makes it impossible to discriminate one from the other. Periodic groups are the simplest minimal sets. However, the decomposition of the strange attractors, namely, numbers and characters of minimal sets contained in them, are not yet settled.

(3) Motions of representative points on the strange attractors depend sensitively on the initial conditions. That is, suppose that $\{q_n\}$ and $\{q'_n\}$ are the motions which start from the initial points q_0 and q'_0 close to each other, then the distance $d(q_n, q'_n)$ generally increases with n. At least this is the case as long as this distance remains small. When $d(q_n, q'_n)$ becomes of the order of the total size of the attractor, or the (exterior) diameter, it cannot increase any more. For the strange attractor of Fig. 3, we have

$$d(q_n, q'_n) = d(q_0, q'_0)\, \rho_u{}^n, \quad \rho_u = 1.8 \tag{13}$$

Since $\rho_u > 1$, the factor $\rho_u{}^n$ increases exponentially with n. The rate of exponential increase is given by an exponentlike quantity e_u:

$$e_u = \frac{1}{2\pi} \log \rho_u = \frac{1}{2\pi} \log 1.8 = 0.09 \tag{14}$$

It represents the average velocity of mixing or diffusion of a motion in the attractor.

(4) Solutions of the differential equation (2) emanating from the strange attractor at $t = 0$ constitute a solution bundle in the txy space. The solution bundle has the following properties:

(i) It is 2π-periodic with respect to t.

(ii) It is asymptotically orbitally stable, or it attracts its neighboring orbits as t increases.

(iii) It contains infinitely many unstable 2nπ-periodic solutions ($n \in Z^+$).

(iv) The distance of neighboring two solution curves belonging to it generally increases exponentially with respect to t.

The continuous-time evolution of the strange attractor is nothing less than that of the cross section of the solution bundle. It repeats stretching and folding movement periodically.

(5) A representative point exhibiting an actual motion continues to transit chaotically among the infinitely many solutions of the bundle due to the effects of microfluctuations. Thus a random process X(t) is composed of the totality of observational motions or computer solutions x(t). It is easily conjectured that, the smaller microfluctuations, the longer the representative point moves along the exact solution curve of the differential equation.

(6) The ensemble average $M(t) = \langle X(t) \rangle$ of the random process X(t) is a 2π-periodic function with respect to t. The random process X(t) has the properties of a periodic stationary random process whose probability distribution is invariant under periodic translations: $t \to t + 2n\pi$ ($n \in Z$).

Figure 5 shows the average power spectrum $\Phi_X(\omega)$ of the random process correlated with the strange attractor of Fig. 3. It is obtained by regarding the random process X(t) as the periodic random process $X_T(t)$ with a sufficiently long period T, where T is a multiple of 2π:

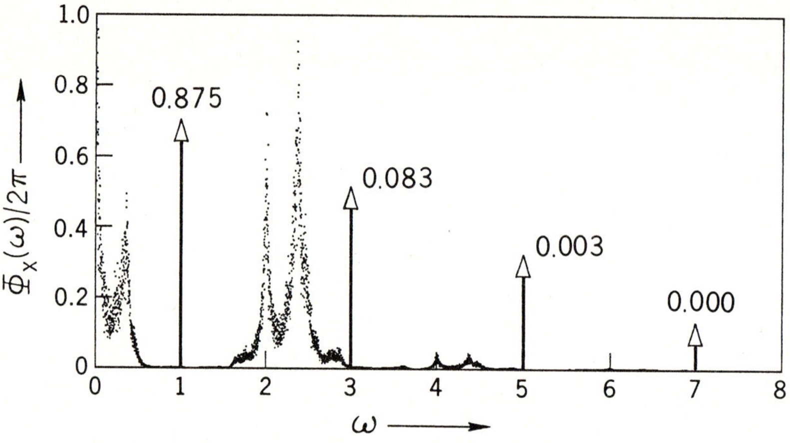

Fig. 5: Average power spectrum of the chaotically transitional process X(t).

$$\Phi_X(\omega) = \lim_{T\to\infty} \langle \frac{1}{T} \left| \int_{-T/2}^{T/2} x_T(t) e^{-i\omega t} dt \right|^2 \rangle \qquad (15)$$

In the figure, line spectra indicate the periodic components and continuous finitenesses show the statistical components. Numerical values in the figure represent the power focused on line spectra. Average power spectrum scarcely depends on the properties of microfluctuations but is determined from the global structures of the solution bundles emanating from the strange attractors. This implies that the properties of macrofluctuations (chaotically transitional processes) are little dependent on that of microfluctuations, although the appearance of macrofluctuations is caused by the existence of microfluctuations. However, further examinations will be required for the recognition of this statement, because it is merely a proposition of the author through his long-term computer experiments for chaotic phenomena in various deterministic systems.

(7) Every minimal set contained in the strange attractors is supposed to be unstable in the Lyapunov sense. Similarly, the structure of a strange attractor seems to be unstable in the Andronov-Pontryagin sense. Those account for the appearance of stochasticity in the actual phenomenon. It is conjectured that the strange attractors as well as the solution bundles will have structural stability in somewhat mild or coarse sense. However, no attempt has been made to investigate what that concept is.

6. PARAMETER REGIONS FOR DIFFERENT TYPES OF STEADY MOTIONS

In the system described by Eq. (1), various types of steady motions occur depending on the system parameters $\lambda = (k, B)$ as well as on the initial conditions. Figure 6 shows the regions on the kB plane in which different steady motions are observed. These regions are obtained by using analog and digital computers. Accordingly, remember that details are doubtful and further examinations will be required.

Strange attractors take place in the shaded regions. In the area hatched by full lines, strange attractor appears uniquely, while in the area hatched by dotted lines, two different steady states occur, i.e., chaotic and regular motions. Which one appears depends on the initial conditions.

The roman numerals I, II,... characterize 2π-periodic motions. The fractions p/q indicate the regions in which $2q\pi$-periodic motions occur. Representative trajectories have been collected in Ref. [4] at twenty-one sets of the parameters indicated by alphabets from 'a' to 'u' in the figure. The strange attractor of

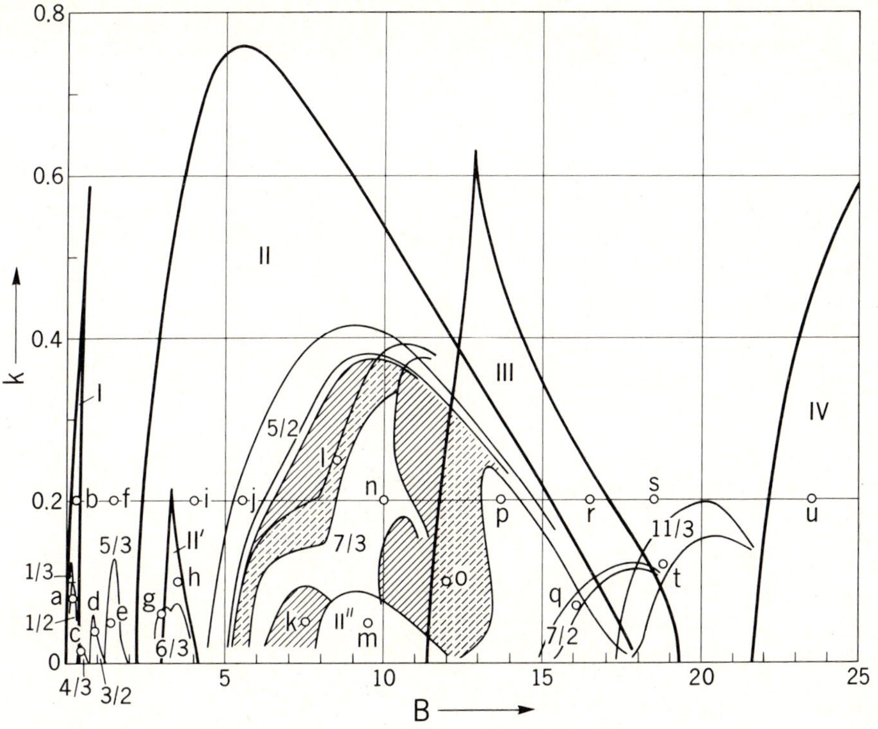

Fig. 6: Regions of different steady states for the system governed by Eq. (1). Reproduced with the courtesy of the Society for Industrial and Applied Mathematics [4].

Fig. 3 occurs at the point 'k' in Fig. 6. Note that Figures 2, 3 and 4 are obtained for the same system parameters, and the unit length of the x-axis is five times that of the y-axis.

7. TRANSITION OF THE STRANGE ATTRACTORS

When the system parameters are varied from the deterministic into the stochastic region, strange attractors usually develop from periodic points, producing periodic points of successive twice orders of the original order. This transition is known as pitchfork or period doubling bifurcation. It is frequently observed that stable periodic points disappear through coalescence with directly unstable periodic points, and then strange attractor takes their place. Strange attractors commonly change into periodic points by tracing pitchfork bifurcation inversely. It sometimes occurs that strange attractors disappear at the spot where the transition chains are formed. That is, as soon as the α-branches composing a strange attractor touch another ω-branches, the strange attractor cannot generally exist.

8. CONCLUSION

By summarizing our previous reports, structures of strange attractors and properties of chaotically transitional phenomena described by Duffing's equation (1) have been surveyed. Further, the whole aspect of the parameter regions and the transition of the strange attractors are also mentioned briefly.

The progress of computer techniques makes it possible to study chaotic phenomena in deterministic systems. It may be one of the remarkable impacts of technology on the sciences. However, it must be noted that computer solutions are sometimes accompanied by the essential and the difficult problems.

REFERENCES

[1] Y. Ueda et al., Trans. IECE Japan (1973) 56-A, 218.
[2] Y. Ueda, Trans. IEE Japan (1978) 98-A, 167.
[3] Y. Ueda, Jour. Statistical Physics (1979) 20-2, 181.
[4] Y. Ueda, Steady Motions Exhibited by Duffing's Equation, in: Holmes, P. J. (ed.), New Approaches to Nonlinear Problems in Dynamics (SIAM, Philadelphia, 1980) pp. 311-322.
[5] Y. Ueda, Explosion of Strange Attractors Exhibited by Duffing's Equation, in: Helleman, R. H. G. (ed.), Annals of the New York Academy of Sciences Vol. 357 (New York, 1980) pp. 422-434.
[6] Y. Ueda and N. Akamatsu, IEEE Trans. Circuits and Systems (1981) CAS-28, 217.
[7] H. Ogura et al., Prog. Theor. Phys. (1981) 66, 2280.
[8] Y. Ueda, Int. Jour. Nonlinear Mechanics (1985) 20, 481.

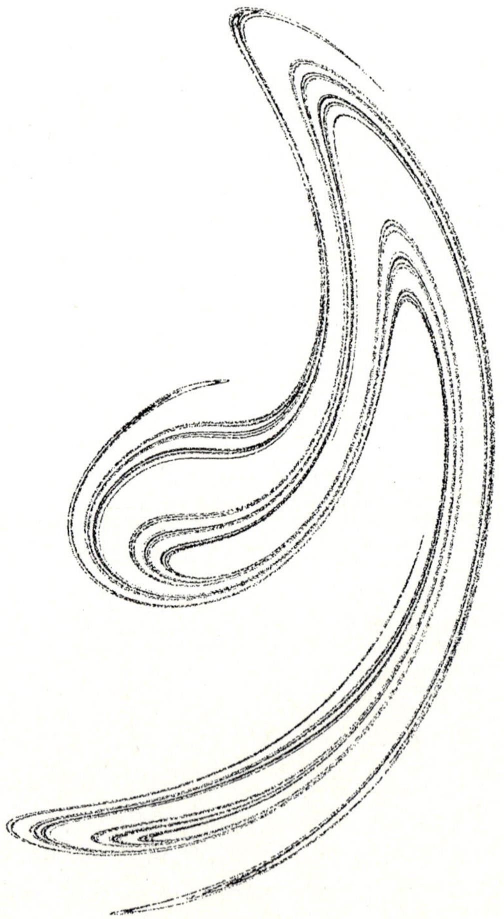

JAPANESE ATTRACTOR

The christener: Professor David Ruelle, IHES, France.
Differential equation: Equation (2).
Parameter: $\lambda = (k, B) = (0.1, 12.0)$, Point 'o' in Fig. 6.
Bounds: $2.36 < x < 3.96$, $-7.67 < y < 6.66$.
Interior fixed points: D(3.548, 0.726),
I(3.797, -1.535),
I(3.288, -0.853).

Complex and Distributed Systems: Analysis, Simulation and Control
S.G. Tzafestas and P. Borne (editors)
Elsevier Science Publishers B.V. (North-Holland)

ON COHERENCE AND CHAOS IN A PHYSICAL SYSTEM

Peter L. Christiansen

Laboratory of Applied Mathematical Physics
The Technical University of Denmark
DK-2800 Lyngby
Denmark

1. INTRODUCTION

In nonlinear physical systems af many degrees of freedom coherence and chaos will often be competing tendencies. The coherence manifests itself through pattern formation. Often the patterns depend strongly on the parameters of the system. The patterns may also be unstable in time resulting in chaotic behaviour. Examples of such systems from different branches of physics and numerous recent references are found in the contributions to the Proceedings of the 59th Nobel Symposium presented under Pattern formation [1]. The perturbed sine-Gordon equation [2] modelling the long Josephson junction is an excellent testing ground for these phenomena. In the present paper results obtained at the Laboratory of Applied Mathematical Physics and Physics Laboratory I, the Technical University of Denmark, in cooperation from this specific field of nonlinear dynamics will be presented. Experimental measurements are compared with direct computational solutions of boundary value problems at the 1+1 dimensional sine-Gordon equation. This equation is almost integrable and possesses soliton-like solutions. Approximative perturbation theory for the soliton dynamics on the longer junctions and modal expansions of quasi-linear type for the shorter junctions predict the experimental and direct computational results - at low computational cost - and provide insight in the nonlinear phenomena. For expansions in terms of genuinely nonlinear modes (soliton modes) we refer to Reference [3]. Our research is an example of the computational synergetics pioneered and described by Zabusky in Reference [4].

The paper first describes how the Josephson junction is modelled by the perturbed sine-Gordon equation with certain boundary conditions. Then the simpler dynamic states (which are the patterns in the present case) of the junction are described. Finally, we show examples of instability, hysteresis, and chaotic intermittency between the dynamic states.

MODELLING THE JOSEPHSON JUNCTION

A Josephson tunnel junction consists of two superconducting metal layers separated by a thin insulating oxide layer of uniform thickness (t_{ox}) that is small enough to permit quantum-mechanical tunnelling of electrons. Figure 1 shows a Josephson tunnel oscillator in which the two layers overlap in a region of length L in the X-direction and width W in the Y-direction. When L >> W the junction is essentially one-dimensional in space. The tunnelling supercurrent is

$$j(X,T) = j_o \sin\phi \ , \qquad (1)$$

where $\phi = \phi(X,T)$ is the difference between the phases of the order parameter of the two superconductors, j_o is the maximum current, and L is laboratory time. The voltage drop across the insulating layer is

$$V(X,T) = \frac{\hbar}{2e}\frac{\partial\phi}{\partial T} \qquad (2)$$

where $\hbar$ is Planck's constant divided by 2π and e is the electronic charge. Combination of Eqs. (1 - 2) and Maxwell's equations yields the <u>perturbed sine-Gordon equation</u> [2]

$$\phi_{xx} - \phi_{tt} - \sin\phi = \alpha\phi_t - \beta\phi_{xxt} - \gamma \qquad (3)$$

in normalized coordinates $x = X/\lambda_J$ and $t = T\omega_o$.

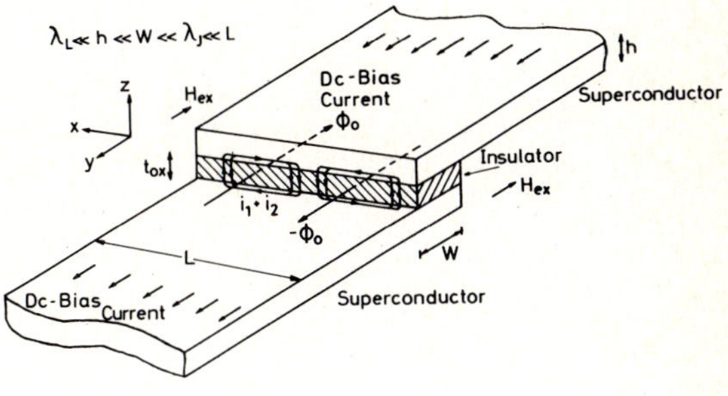

Figure 1. Josephson tunnel junction oscillator of overlap geometry [5].

The Josephson length is given by $\lambda_J = (\Phi_o/2\pi\, j_o\, L_p)^{\frac{1}{2}}$ and the Josephsonplasmafrequency $\omega_o = (2\pi\, j_o/\Phi_o\, \bar{C})^{\frac{1}{2}}$. Here $\Phi_o = h/2e$ is the magnetic flux quantum and L_p and C are inductance and capacitance per length unit of the junction respectively. The perturbation terms, $\alpha\phi_t$ and $-\beta\phi_{xxt}$, represent quasi-particle tunnelling loss and surface impedance loss respectively. Thus the coefficients $\alpha = G/\omega_o C$ and $\beta = \omega_o L_p/R_p$ where G^{-1} and R_p are the normal resistance and the surface resistance per length unit of the junction respectively. In the term $\gamma = j_B/j_o$, j_B is the externally applied bias current per length unit. Typical values of γ_J and ω_o are $\gamma_J = 1.56 \times 10^{-4}$ m and $\omega_o = 5.8 \times 10^{10}$ s^{-1} such that the propagation velocity for electromagnetic signals (i.e. solutions of the linear wave equation, $\phi_{xx} - \phi_{tt} = 0$) becomes $c = \lambda_J \omega_o = 9.05 \times 10^6$ m/s in laboratory coordinates along the junction. In the normalized coordinates this velocity is of course equal to unity.

At the ends of the junction, X = 0 and X = L, we apply the following sets of boundary conditions:

(i) Homogeneous open-end conditions

$$\phi_x(0,t) = \phi_x(\ell,t) = 0 \qquad (4a)$$

corresponding to zero current on the junction at the ends since the current is proportional to ϕ_X. Here $\ell = L/\lambda_J$ is the normalized length of the junctions. Eq. (4a) models the physical situation with no external magnetic field applied ($H_{ex} = 0$ in Fig. 1). The conditions neglect the coupling between the junction and the surrounding microwave circuit. Nevertheless, computational results obtained for this boundary condition agree well with experimental measurements [5].

(ii) Inhomogeneous open-end conditions

$$\phi_x(0,t) = \phi_x(\ell,t) = \eta \qquad (4b)$$

corresponding to the physical situation where an external magnetic field, H_{ex}, is applied ($H_{ex} \neq 0$ in Fig. 1). Here $\eta = (- w/j_o \lambda_J)H_{ex}$.

The initial conditions for the computational modelling of the oscillator are

$$\phi(x,0) = F(x) \quad \text{and} \quad \phi_t(x,0) = G(x) \qquad (5)$$

where the functions F and G are chosen such that stationary states of $\phi(x,t)$ are obtained in the numerical computations without too long transients. In practice the final ϕ and ϕ_t distributions from a nearby stationary state in parameter space $(\alpha,\beta,\gamma,\ell,\eta)$ are often used as the functions F and G respectively.

DYNAMIC STATES OF THE JOSEPHSON JUNCTION

The classical sine-Gordon equation, $\phi_{xx}-\phi_{tt}-\sin\phi = 0$, has 2π-kinks and anti-kinks

$$\phi(x,t) = 4\tan^{-1}[\exp(\pm(x-ut\,x_o)/\sqrt{1-u^2}] \qquad (6)$$

as soliton solutions. Here u is the constant velocity of the soliton, and x_o is the position of the soliton at t = 0.

The perturbed sine-Gordon equation (3) has similar soliton solutions in the looser sense. Each soliton carries a magnetic flux quantum. The dynamics of these solitons is investigated by means of perturbation theory in Reference [2]. As a result a first-order differential equation for the variable soliton velocity for a single soliton, u(t), is derived

$$\frac{du}{dt} = \pm\frac{1}{4}\pi\gamma(1-u^2)^{3/2} - \alpha u(1-u^2) - \frac{1}{3}\beta u\,. \qquad (7)$$

Eq. (7) expresses the balance between energy input in the system due to the γ-term and dissipation due to the loss terms, $\alpha\phi_t$ and $-\beta\phi_{xxt}$. The stationary velocity, u_∞, is determined from Eq. (7) by letting du/dt = 0 and solving the resulting equation with respect to u. In typical computer experiments u(t) rapidly adjusts towards the stationary velocity, u_∞. Applications of perturbation theory to Josephson junctions of finite length are surveyed in [6].

For a finite junction with open-end boundary conditions (4a) it is easy to show that solitons are reflected into antisolitons and vice versa at the boundaries. The bias current, γ, drives the soliton in the negative x-direction until it is reflected into an antisoliton at $x = \ell$. The antisoliton is driven in the positive x-direction and reflected into a soliton at x = 0, and a new cycle of this stationary state is initiated. We shall designate such a stationary state a soliton dynamic state. The periodic motion of the soliton on the junction is responsible for the emission of electromagnetic radiation, typically in the GHz-range, from the junction, which then acts as an oscillator. Figure 2 shows a computer picture of part of the oscillation cycle in the soliton dynamic state with one soliton.

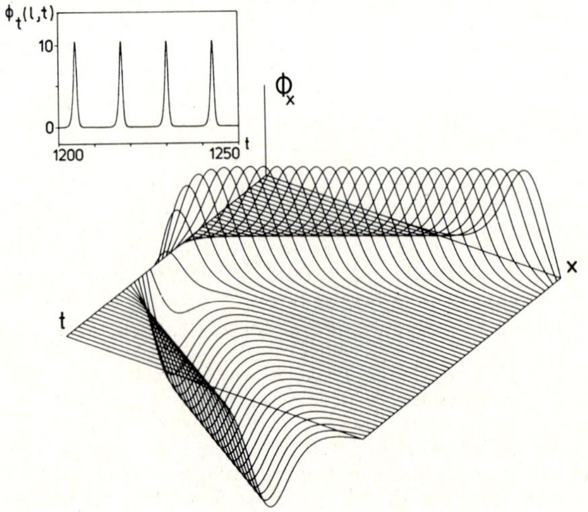

Figure 2. Computer solution of the perturbed sine-Gordon equation (3) with $\alpha = 0.05$, $\beta = 0.02$, $\gamma = 0.35$. Boundary conditions (4a) with $\ell = 6$. Initial conditions (5) with one soliton. The inset shows $\phi_t(\ell,t)$ [5].

In the inset, $\phi_t(\ell,t)$ is shown as a function of t for 50 time units. This quantity is proportional to the voltage on the oscillator at the end at $x = \ell$ according to (2). The DC-component of the voltage $\phi_t(\ell,t)$ has been computed for different values of the applied bias current γ in Eq. (3). The resulting curve shows agreement with experimentally measured IV-curves for the junction. Each soliton dynamic state corresponds to a branch of the IV-characteristic. These branches are designated zero field steps (ZFS) in the physical literature. Figure 3 shows experimental measurements of ZFS 1 and ZFS 2.

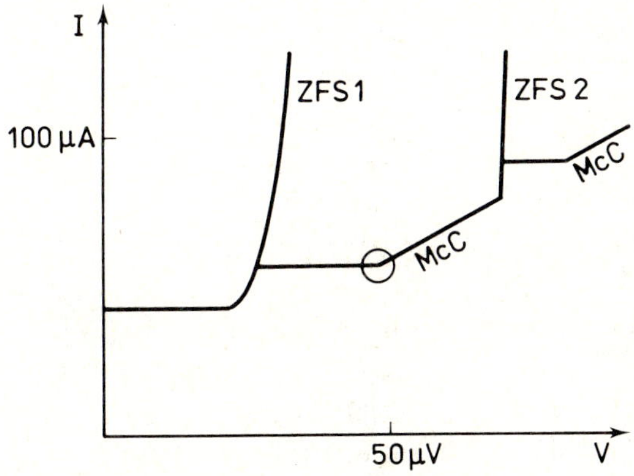

Figure 3. Experimental measurement of IV-characteristic for Josephson junction (Sample S6-7/4, temperature 6.7°K). Transitions from McCumber branch (McC) to ZFS 1 and ZFS 2 are observed [7].

Computational Fourier analysis of $\phi_t(\ell,t)$ provides the power spectrum for the radiation from the oscillator. The basic frequency f is given by

$$f = u/2\ell \ . \tag{8}$$

Also the computational power spectrum shows agreement with experimental measurements of the power spectrum.

The Josephson junction can also be excited in a state $\phi = \phi_o(t)$ without spatial structure. The corresponding solution to (3) and (4a) with $\alpha = \beta = \gamma = 0$ is [8]

$$\phi_o(t) = 2 \text{ am } [t/k;k] \ , \tag{9}$$

where am is the Jacobian elliptic amplitude function and the parameter k ($0 < k < 1$) is the modulus. In the IV-characteristic the uniform solution (9) corresponds to the McCumber branch (McC), which is also observed in Figure 3.

In order to tune the frequency of the electromagnetic radiation from the oscillator a constant external magnetic field may be applied. The corresponding mathematical model then consists of the sine-Gordon equation (3) with boundary condition (4b) and initial condition (5).

For a relatively weak magnetic field ($\eta = 0.75$) we get the computer picture shown in Figure 4.

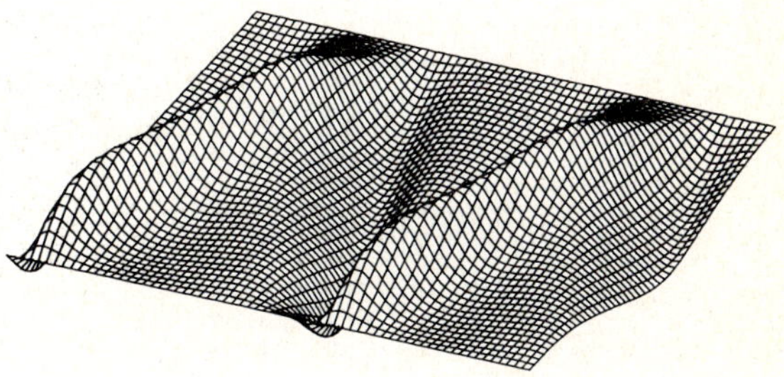

Figure 4. Computer solution ($\phi_x(x,t)$) of the perturbed sine-Gordon equation (3) with $\alpha = 0.252$, $\beta = 0$, $\gamma = 0.54$. Boundary conditions (4b) with $\ell = 5$, $\eta = 0.75$. Initial conditions (5) with one soliton.

A soliton travels towards the left and reacts with the boundary at $x = 0$. As a result of the boundary condition $\phi_x = \eta$ energy is absorbed from the incident soliton and the minimum energy for the sine-Gordon soliton (= 8 in normalized units) is no longer available. Therefore no antisoliton is reflected. Instead we observe the reflection of plasma oscillations which contain less energy. The plasma oscillations travel towards the other end in the figure and excite a soliton at the right end of the junction where the boundary condition $\phi_x(\ell,t) = \eta$ pumps energy into the system and makes this creation possible. This constitutes the first cycle of a stationary soliton dynamic state under the influence of a constant magnetic field. The state gives rise to a branch of the IV-characteristic for the oscillator designated the first Fiske step (FS 1). For higher values of η a soliton and the plasma oscillation travel in opposite directions at the same time in a symmetric configuration giving rise to the second Fiske step (FS 2) in the IV-characteristic.

INSTABILITY OF DYNAMIC STATE

Figure 3 shows examples of instability of the dynamic state represented by the McCumber branch. As the bias current γ is lowered the system jumps (at the circle) from this branch (McC) to the first zero field step (ZFS 1), i.e. the uniform excitation becomes unstable and is replaced by the single soliton excitation illustrated in Figure 2. The experiment corresponds to the normalized parameter values $\alpha \simeq 0.03$ and $\ell \simeq 3$. The instability occurs for $\gamma \simeq 0.14$. Direct computational solution of (3) and (4a) for $\alpha = 0.05$, $\beta = 0.02$, and $\ell \simeq 2$ yields the instability value $\gamma = 0.17$ [7], where the discrepancy can be explained by the slightly different parameter values of α and ℓ (β is unimportant). Using the modal expansion

$$\phi(x,t) = \phi_0(t) + \phi_1(t)\, e^{ibx} \qquad (10)$$

in (3) and (4a) we get the damped Hill's equation for $\phi_1(t)$

$$\ddot{\phi}_1(t) + (\alpha + \beta b^2)\,\dot{\phi}_1(t) + (b^2 + \cos(\phi_0(t)))y = 0 \qquad (11)$$

with

$$b = n\pi/\ell \quad , \quad n = 0,1,\ldots .$$

Linear instability analysis predicts the critical value of γ to be $\gamma = 0.17$ for $\alpha = 0.05$, $\beta = 0.02$, and $\ell \simeq 2$ [7] as in the direct computational solution.

HYSTERESIS BETWEEN DYNAMIC STATES

Figure 5 shows the power levels of the first two Fourier components of the solution to (3) and (4b) as η is increased. For small values of η the junction operates on the first zero field step (soliton trajectory to the left in Figure 5). For larger values of η the junction operates on the second Fiske step (soliton trajectory to the right in Figure 5). In the former mode the first Fourier component is dominant while the second Fourier dominates the latter mode. These observations have been verified experimentally [9]. A hysteresis phenomenon (indicated by arrows) is observed as η is raised or lowered.

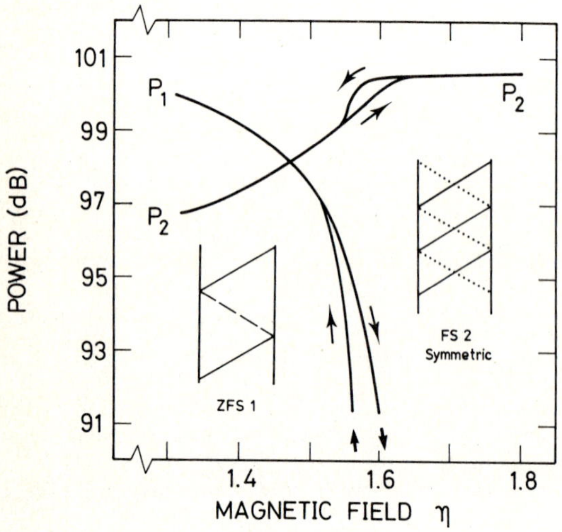

Figure 5. Power levels of first two Fourier voltage components at x = 0, normalized to $\phi_t^2 = 1.0 \times 10^{-10}$. Solution to (3) and (4b) with $\alpha = 0.05$, $\beta = 0.02$, $\gamma = 0.21$, and $\ell = 0.21$. Insets show corresponding soliton trajectories in the XT-plane.

CHAOTIC INTERMITTENCY BETWEEN DYNAMIC STATES

For certain parameter values in a narrow portion of parameter space ($\alpha,\beta,\gamma,\ell,\eta$) the oscillator switches back and forth between the first two Fiske steps (FS 1 and FS 2) as demonstrated in [10].

The switching appears in the solution curve, $\phi(0,t)$, as a shift in the average slope of curve shown in Figure 6. Each little wriggle of the curve marks the passage of a soliton through x = 0. The computational solution of (3), (4b), and (5) was continued for a very long time (t = 10,000). The resulting statistics are summarized in Figure 7.

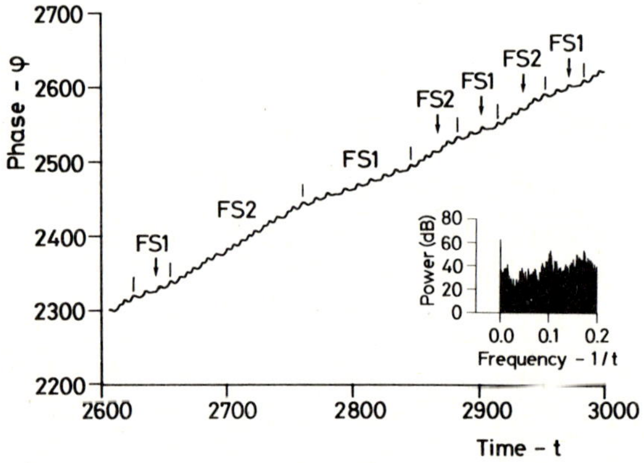

Figure 6. Computer solution, $\phi(0,1)$, and corresponding power spectrum normalized to $\phi_t^2 = 4 \times 10^{-7}$ of perturbed sine-Gordon equation (3) with $\alpha = 0.252$, $\beta = 0$, $\gamma = 0.480$. Boundary conditions (4b) with $\ell = 5$, $\eta = 1.25$. Initial conditions (5) with one or two solitons [10].

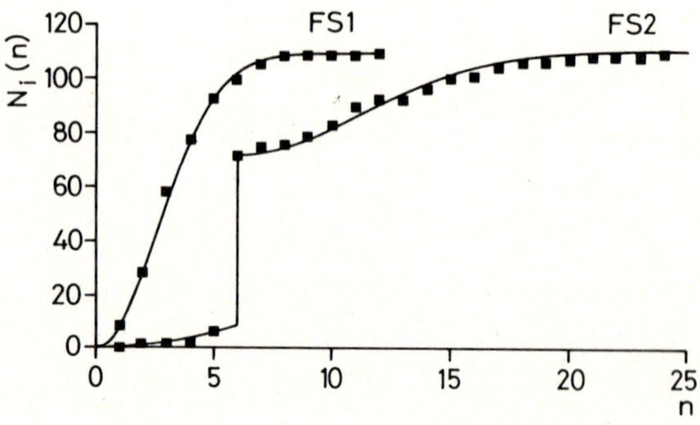

Figure 7. Abscissa: Length of time interval on FS 1 and FS 2 measured in terms of number of cycles, n, spent in the two soliton dynamic states respectively. Ordinate: Dots show number of intervals, $N_i(n)$, with i = 1,2, shorter than or equal to n. Full curves result from theoretical fits [10].

We observe a continuously increasing probability that the oscillator switches from FS 1 to FS 2 as a function of time spent on FS 1 and a discontinuously increasing probability (at n = 6) for the opposite switch from FS 2 to FS 1 as a function of time spent on FS 2. The reason for this asymmetry in the switching statistics is not known. The full curves in Figure 7 are obtained by means of an arbitrary probability model where the switching is assumed to be a Poisson process and the parameters in the model have been fitted in the best possible manner. The chaotic intermittent switching between the two soliton dynamic states gives rise to a special branch in the computed IV-characteristic for the oscillator. Recent experimental measurements [11] have perhaps also revealed such structures.

At present no exact or approximate theory for the chaotic intermittency between soliton dynamic states exists. Qualitatively we may argue that the infinitely dimensional nonlinear system of the perturbed sine-Gordon equation is almost integrable in some sense. Therefore there is a tendency in the system to form soliton-like solutions as we have seen in the computer experiments. The soliton formation effectively requires infinitely many degrees of freedom and thus projects the system down to a nonlinear low-dimensional system. This low-dimensional system exhibits chaotic intermittency between two critical points (corresponding to the two soliton dynamic states).

ACKNOWLEDGEMENTS

The financial support of the European Research Office of the United States Army through contract No. DAJA 37-82-C 0057 is gratefully acknowledged.

REFERENCES

[1] Proceedings of the 59th Nobel Symposium held at Gräftåvallen, Sweden, June 11-16, 1984 (ed.: S. Lundqvist), Pattern Formation; published in Physica Scripta T9 (1985) 93-136.

[2] McLaughlin, D.W. and Scott, A.C., Perturbation analysis of fluxon dynamics, Phys. Rev. A18 (1978) 1652-1680.

[3] Overman II, E.A., McLaughlin, D.W., and Bishop, A.R., Coherence and Chaos in the Driven Damped Sine-Gordon Equation: Measurement of the Soliton Spectrum, Physica D (to appear).

[4] Zabusky, N.J., Computational Synergetics and mathematical Innovation, J. Comp. Phys. 43 (1981) 195-249.

[5] Lomdahl, P.S., Soerensen, O.H., and Christiansen, P.L., Soliton excitations in Josephson tunnel junctions, Phys. Rev. B25 (1982) 5337-5748.

[6] Pedersen, N.F., Solitons in Josephson transmission lines, to appear in "Modern Problems in Condensed Matter Sciences" (eds.: A.A. Maradudin and V.M. Agranovich), North-Holland, Amsterdam.

[7] Pagano, S., Soerensen, M.P., Parmentier, R.D., Christiansen, P.L., Skovgaard, O., Mygind, J., Pedersen, N.F., and Samuelsen, M.R., Switching between dynamic states of the intermediate-length Josephson junction, Phys. Rev. B (to appear).

[8] Parmentier, R.D., Fluxons in Long Josephson Junctions, Solitons in Action (eds.: K. Lonngren and A.C. Scott), Academic Press, New York (1978) 173-199.

[9] Soerensen, M.P., Parmentier, R.D., Christiansen, P.L., Skovgaard, O., Dueholm, B., Joergensen, E., Koshelets, V.P., Levring, O.A., Monaco, R., Mygind, J., Pedersen, N.F., and Samuelsen, M.R., Magnetic field dependance of microwave radiation in intermediate length Josephson junctions, Phys. Rev. B30 (1984) 2640-2648.

[10] Soerensen, M.P., Arley, N., Christiansen, P.L., Parmentier, R.D., and Skovgaard, O., Intermittent Switching between Soliton Dynamic States in a Perturbed Sine-Gordon Model, Phys. Rev. Lett. 51 (1983) 1919-1922.

[11] Cirillo, M., Costabile, G., Pace, S., Parmentier, R.D., and Savo, B. Possible observation of chaotic intermittency effect in a long Josephson junction in magnetic field, Proceedings LT-17, Part II - Contributed Papers (eds.: U. Eckern, A. Schmid, H. Weber, and H. Wühl), North-Holland, Amsterdam (1984) 1131-1132.

Complex and Distributed Systems: Analysis, Simulation and Control
S.G. Tzafestas and P. Borne (editors)
Elsevier Science Publishers B.V. (North-Holland)

INFORMATION FLOW IN TRANSIENT AND NON-TRANSIENT CHAOS

P. Grassberger

Physics Department
University of Wuppertal
Gauss-Str. 20, D-5600 Wuppertal 1

ABSTRACT: The unpredictability of chaotic motion is related to an information flow associated with it. Topics discussed in the present talk are:
i) the relation between the information flow rate (=Kolmogorov entropy), Lyapunov exponents, and dimensions of the attractor;
ii) the information flow in strange repellers leading to metastable chaos.

1. INFORMATION FLOW AND DIMENSION

Physical systems involve typically a huge number of degrees of freedom ($\gtrsim 10^{20}$, say). Since it is impossible to treat all of them explicitely, one performs some kind of "coarse-graining". After this is done, one deals explicitely with few variables only, but these variables evolve in general non-deterministically with time.

Until the "chaos revolution" of the last years, bringing with it the concept of deterministic chaos, this was widely considered as the only source of randomness in nature.

Deterministic chaos is also related to coarse-graining, but of an essentially different kind. It results from the fact that we not only cannot deal with too many variables, but we also cannot deal with infinitely precise numbers. If we accept that space-time is continuous, this means that we must cut-off the digital expansions for all coordinates somewhere.

In the text-book examples of classical mechanics, this cut-off has no effect. But there exist simple and formally deterministic systems where trajectories emerging from near-by initial conditions diverge exponentially. Due to this "sensitive dependence on initial conditions", any ignorance about seemingly insignificant (and thus cutt off) digits, affects ultimately also significant digits leading to an essentially unpredictable and "chaotic" behaviour.

In this talk, I shall concentrate on dissipative chaotic systems. The sets of points in state space towards which nearly all trajectories are converging are called strange attractors in that case.

Let me stress again that chaotic motion is fundamentally different from ordered motion. The unpredictability cannot be avoided by just making more precise measurements of the initial conditions. Assume that we want to measure the initial conditions very precisely, say with some error $\pm\varepsilon$ in each variable of state space. Then we have first to build a suitable measuring device. But the same device can be used later to measure the state again with the same precision, while the equations of motion are unable to predict it with the same accuracy.

If the distance between near-by trajectories increases exponentially -- and this is expected by scale invariance, as long as this distance is much smaller than all typical length scales in the problem --, the lack of information is independent of ε and proportional to the elapsed time.

Consider now a piece of trajectory between times t_1 and t_2. Assume that t_1 is sufficiently large that transients have already died out, and that the system is on its attractor with some time-invariant probability distribution. The information $\hat{S}_\varepsilon$ needed to specify this trajectory to an error $\pm\varepsilon$ during the whole interval $[t_1,t_2]$ consists then of two parts:

(a) the information S_ε needed to specify it at time t_1, and
(b) the information needed to fix up the ignorance leaking in with constant rate due to the ignorance about originally insignificant digits.

Thus, we expect

$$\hat{S}_\varepsilon\ [t_1,t_2] \approx S_\varepsilon + (t_2-t_1)\ K \qquad (1.1)$$

for $\varepsilon \to 0$ and $t_2-t_1 \to \infty$.

The constant K is called the Kolmogorow-Sinai or "metric" entropy[1-3]. It is essentially the information flow rate in the limit of nearly error-free measurements[4].

For predictable systems, one would have $K = 0$. For the other extreme of Brownian motion (or, rather, its mathematical idealization as a Wiener process) one has $K = \infty$: even perfect knowledge of the state at some instant would not be sufficient to predict it even in the near future.

Let us now look at the dependence of S_ε on the error ε. If we want to specify a point x on a fixed interval with accuray $\pm\varepsilon$, the needed information (i.e., the number of significant bits) behaves for $\varepsilon \to 0$ like $\log_2 (1/\varepsilon)$. In the following, we shall always use natural logs instead of $\log_2$, implying that we measure information in "nats" and not in "bits". For a point in D-dimensional space, the information is D times as big. Thus, we expect that

$$S_\varepsilon \sim D \log (1/\varepsilon) \quad \text{for } \varepsilon \to 0, \tag{1.2}$$

where D is called the information dimension of the attractor[5-8].

We have to be somewhat more precise. The estimate $\log(1/\varepsilon)$ for the information stored in a point on an interval is true only if we know a priori that the point is indeed on the interval, and nothing more. Analogously, eq. (1.2) assumes a priori that the state is on the known attractor (i.e., any transients have died out), and distributed according to some invariant measure (= distribution), which is also assumed to be known from previous observations.

A typical case, assumed throughout the following, is that the system is ergodic and mixing. In that case, nearly all initial conditions within some suitable basin of attraction lead to the same invariant distribution, called the"natural measure"[9]. But in all chaotic systems there exist also other invariant distributions, reached from unitial conditions of measure zero. Quantities like D and K actually refer to one particular distribution. If nothing else is said, it will be the natural measure.

The most striking feature of strange attractors is that D is in general non-integer. Since D is closely related to the Hausdorff dimension (see sec. 2), the attractor is a "fractal" in the sense of Mandelbrot[10]. The observation of non-integer D in Couette-Taylor flow[11] (see Fig. 1) and in other hydrodynamic systems[12] is indeed a beautiful proof that strange attractors occur in real physical systems.

Measuring dimensions of strange attractors is thus a very active field of research.

As we have already stressed, the motion on a strange sttractor is unpredictable due to the divergence of nearby trajectories. This divergence is measured by the so-called Lyapunov exponents[13,8]. In order to define them, consider an infinitesimally small sphere in state space around some point $\vec{x}$ with radius ε. After a time $t \gg 0$, this sphere is transformed into an ellipsoid with semi-axes

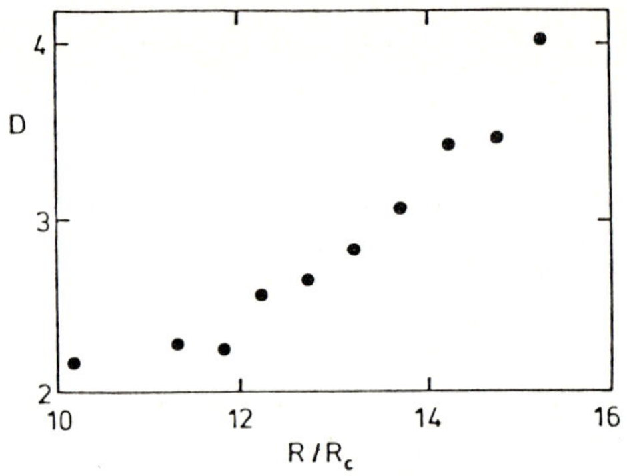

Fig. 1: Information dimension for Taylor-Couette flow, as a function of Reynolds number (adapted from ref. 11).

$$\varepsilon_i(t) \sim \varepsilon\ e^{\lambda_i t} \quad (i = 1,2, \ldots f; \; f = \#\text{ of degrees of freedom}) \tag{1.3}$$

for nearly all $\vec{x}$. In the following, we shall always assume that the λ_i are ordered by magnitude, $\lambda_1 \geq \lambda_2 \geq \ldots \geq \lambda_f$. Since the system is dissipative by assumption, their sum is < 0. But since it is also chaotic, at least $\lambda_1 > 0$. The distance between two arbitrary near-by points will increase then like $|\Delta\vec{x}(t)| \sim e^{\lambda_1 t}$.

At nearly every point in the basin of attraction, the semi-axes λ_i define the axes of a local coordinate system, up to signs. Those directions with $\lambda_i > 0$ are called instable, and those with $\lambda_i < 0$ are called stable. In addition, one can have "central" directions with $\lambda_i = 0$. In particular, the direction of the velocity vector in continuous-time systems has $\lambda_i = 0$.

In the next section, we shall see that we can attribute a "partial dimension" D_i to each of the stable, instable, and central directions. Just as D measures how the information depends on an uncertainty common to all state variables, D_i measures its dependence on the uncertainty of the i-th local coordinate only. Thus is is clear that

$$D = \sum_i D_i \,. \tag{1.4}$$

Stated differently, D_i is the density of information per digit of the i-th coordinate. Since the Lyapunov exponent λ_i is just the speed of the information flow along that coordinate, the rate of information flowing from the "insignificant digits" of x_i into the system is $D_i \cdot \lambda_i$. The Kolmogorov entropy, being the total rate of information flow, will then be given by the standard expression "flow rate = density x velocity", or[14-17]

$$K = \sum_i{}' D_i \cdot \lambda_i \quad (1.5)$$

where the sum extends over positive λ_i only.

The natural measure is characterized by maximal information density along the unstable directions: along these directions, each incoming bit is unpredictable, leading to $D_i = 1$. The resulting relation $K = \Sigma_i \lambda_i$ has been proposed by Ruelle[18]. Measures with $D_i \neq 1$ along unstable directions arise e.g. in the flow on repellers leading to transient chaos (see sec. 3). They provide non-trivial tests of eq. (1.5).

If the motion is invertible (which it is always for continuous time systems), information is leaving the system by convergence of points along the stable directions. The rate is again given by eq. (1.5), but this time the sum extends over all negative λ_i. Conservation of information (the average knowledge about the system stays constant with time) leads then to the simple relation

$$K_{in} - K_{out} = \sum_i D_i \cdot \lambda_i = 0. \quad (1.6)$$

A formula for D proposed in refs. 20,21 follows immediately if we assume that D is the maximum allowed by eqs. (1.4) and (1.6).

For 1-dimensional maps $x_{n+1} = f(x_n)$, information stored in x_1 is lost (i.e. is leaving the system) due to the non-invertibility of the map. Equation (1.6) does not make sense then, but eq. (1.5) can be simply tested numerically by iterating the reversed map $x_{n-1} = f^{-1}(x_n)$, and choosing the branches of f^{-1} at random. In general, this leads to fractal invariant measures, and eq.(1.5) was verified in all examples studied in ref. 22.

In section 2, we shall discuss technical details of the above concepts and relations.

In addition to Kolmogorov entropy and information dimension, often-studied quantities are the topological entropy[23] and the fractal (Hausdorff) dimension[10] of the attractor. They are related to the flow of Renyi informations[24] on the one hand, and to deviations from $\varepsilon_i(t) = \varepsilon e^{\lambda_i t}$ on the other[15, 14].

2. FORMAL DEVELOPMENTS

Technically, the information S_ε is defined via a partitioning of the attractor into (hyper-) cubes of size ε. Let us call p_i the probability that an arbitrary point $\vec{x}(t)$ falls into cube i. That is, p_i is the "mass" of cube i with respect to the natural measure μ:

$$p_i = \int_{\text{cube } i} d\mu(\vec{x}). \quad (2.1)$$

Then, S_ε is defined as $S = -\sum_i p_i \log p_i$.

Analogously, we can define $\hat{S}_\varepsilon[t_1,t_2]$ by partitioning not space but space-time (see fig. 2). Assume that the interval $[t_1,t_2]$ has been devided into n bins of length τ each. Let $p_{\{i_1 \ldots i_n\}}$ be the joint probability that $\vec{x}(t_1+\tau) \in$ cube i, $\vec{x}(t_1+2\tau) \in$ cube $i_2,\ldots,\vec{x}(t_2) \in$ cube i_n.
Then

$$S[t_1,t_2] = -\sum_{\{i\}} p_{\{i_1\ldots i_n\}} \log p_{\{i_1\ldots i_n\}} \quad (2.2)$$

The information dimension is[5,8]

$$D = \lim_{\varepsilon\to 0} \frac{S_\varepsilon}{\log 1/\varepsilon}, \quad (2.3)$$

and the metric entropy is[4]

$$K = \lim_{\substack{\varepsilon\to 0\\ \tau\to 0}} \lim_{t_2\to\infty} \frac{1}{t_2 - t_1} S_\varepsilon[t_1,t_2]. \quad (2.4)$$

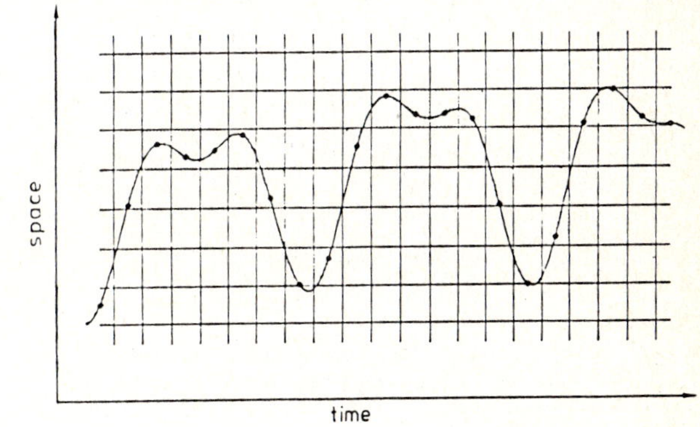

Fig. 2: Partitioning of space-time.

The equation defining S_ε can be interpreted as follows: S_ε is the average, weighted according to the natural measure, of the quantity $-\log p_i = -\log \int d\mu(\vec{x})$, where the integral goes over a cube of size ε in a fixed mesh, but with its center distributed randomly according to μ:

$$S_\varepsilon \approx -\int d\mu(\vec{x}) \quad \log M_\varepsilon(\vec{x}) = -\langle \log M_\varepsilon \rangle \quad (2.5)$$

with

$$M_\varepsilon(\vec{x}) = \int_{|\vec{y}-\vec{x}|<\varepsilon} d\mu\,(\vec{y}) \quad (2.6)$$

Equation (2.3) states then that the mass in a ball of size ε increases, in geometric average, like ε^D.

A completely analogous argument shows that

$$\hat{S}_\varepsilon\,[t_1,t_2] \approx -\,\langle\log \hat{M}_\varepsilon\,[t_1,t_2]\rangle \quad (2.7)$$

where $\hat{M}_\varepsilon(\vec{x},[t_1,t_2])$ is the mass of that domain around $\vec{x}(t_1)$, the trajectories emerging from which stay within a distance ε from $\vec{x}(t)$ for all $t_1 < t < t_2$ (see fig. 3):
$M_\varepsilon(\vec{x},[t_1,t_2]) = \mu(B_\varepsilon(\vec{x}))$ with $B_\varepsilon(\vec{x}) = \{\vec{y}(t_1) : |\vec{y}(t)-\vec{x}(t)|<\varepsilon \text{ for all } t[t_1,t_2]\}$

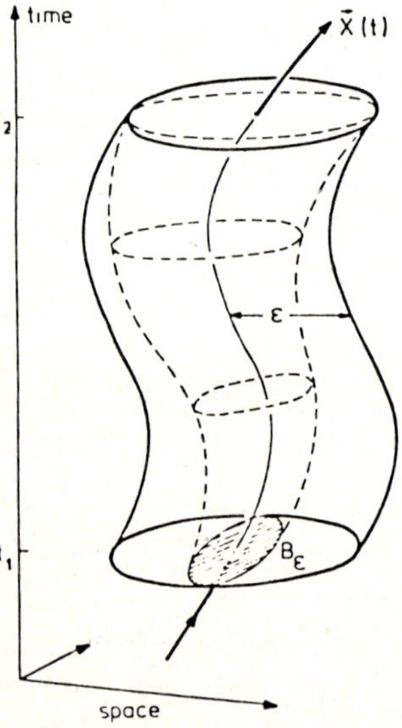

Fig. 3: The domain $B_\varepsilon(\vec{x})$ (shaded region) consists of those points, the trajectories emerging of which stay in an ε-sausage around $\vec{x}(t)$.

As we have said in sect. 1, the stable and instable directions define in (nearly) any point a foliation. We can thus generalize the above and consider, instead of balls, ellipsoids with $\varepsilon_1, \varepsilon_2 \ldots$ along the thereby induced directions. The definitions of S_ε, $\hat{S}_\varepsilon$, M_ε, $\hat{M}_\varepsilon$, and B_ε are generalized in an obvious way to $S_{\varepsilon_1\varepsilon_2}\ldots$ etc. During time evolution, their semi-axes will change as

$$\varepsilon_i \to \varepsilon_i \cdot e^{\Lambda_i(\vec{x},t)} \quad . \quad (2.8)$$

The average dilatation factors are by definition the Lyapunov exponents:

$$\lambda_i = \frac{1}{t}\int d\mu(\vec{x})\,\Lambda_i(\vec{x},t). \quad (2.9)$$

Let us now consider the domain $B_{\varepsilon_1\varepsilon_2\ldots}(\vec{x})$. It will be an ellipsoid with semi-axes

$$\approx \begin{cases} \varepsilon_i & \text{along the stable directions,} \\ \varepsilon_i\; e^{-\Lambda_i(\vec{x},t)} & \text{along the instable directions.} \end{cases} \quad (2.8a)$$

Scale invariance suggests that it scales with each ε_i,

$$\langle\log M_{\varepsilon_1\varepsilon_2\ldots}\rangle \underset{\varepsilon_i\to 0}{\sim} \sum_i D_i \log \varepsilon_i, \quad (2.10)$$

where the constants D_i are the partial dimensions discussed above. They satisfy

$$0 \le D_i \le 1 \quad\text{and}\quad D = \sum_i D_i \,. \quad (2.11)$$

From eqs. (2.4)-(2.11), one obtains finally eq.(1.5).

Equation (2.10) shows that the attractor factorizes essentially in a direct product of continua (corresponding to $D_i = 1$), discrete points ($D_i = 0$), and Cantor sets ($D_i \neq 0, \neq 1$), with orientation according to the (in-)stable directions. As an example, we show in fig. 4 the Hénon[25] attractor, for which $D_1 = 1$ and $D_2 \approx 0.25$[14]. For a different example, see ref. 26.

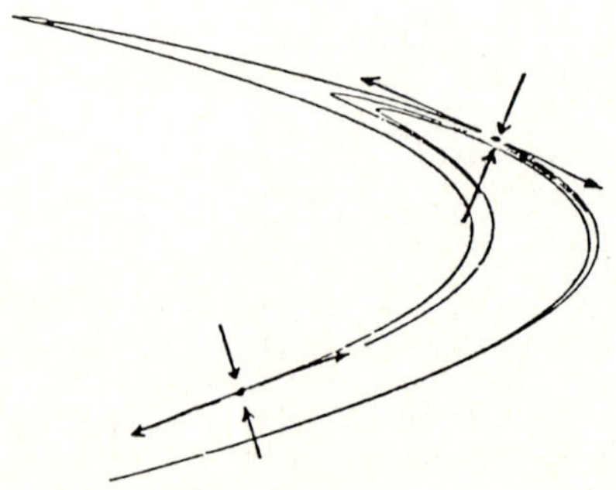

Fig. 4: Attractor of the Hénon map $(x,y) \to (1+y-ax^2,\ bx)$ with $a = 1.4$, $b = .3$. The arrows indicate the stable resp. instable directions.

As pointed out in sect. 1, we can apply the same argument to the time-reversed motion if it is also deterministic, and obtain eq. (1.6)[14,16]. An upper bound on D compatible with that relation is obtained if $D_i = 1$ for all $i < j$, $D_i = 0$ for all $i > j+1$ and $0 \le D_{j+1} < 1$.

Here, the integer j is uniquely determined by eq. (1.6). The bound is[27]

$$D \le D_{Lyap} \equiv j + \frac{\sum_{i=1}^{j} \lambda_i}{|\lambda_{j+1}|} \tag{2.12}$$

It was conjectured by Kaplan and Yorke[20,21] that indeed $D = D_{Lyap}$ for the natural measure in all typicla cases. One knows several examples[28,29], where $D < D_{Lyap}$, but all are "untypical" (i.e., correspond to a set of measure zero in some control parameter). For non-natural measures, the Kaplan-Yorke conjecture does not apply in general, but eqs. (1.5) and (1.6) are still correct[15,16,19].

A heuristic explanation why a typical attractor should fill mostly the directions of least contraction, as required by the Kaplan-Yorke conjecture, is found in ref. 15.

For an analogous discussion of fractal dimensions[10] and topological entropies[23], I refer to refs. 14,15,22.

3. TRANSIENT CHAOS

Even if one observes chaotic behaviour for very long times, one has no guarantee that the system will not fall sooner or later on a regular attractor. An example involving the logistic map $x_{n+1} = 1-ax_n^2$ is shown in fig. 5.

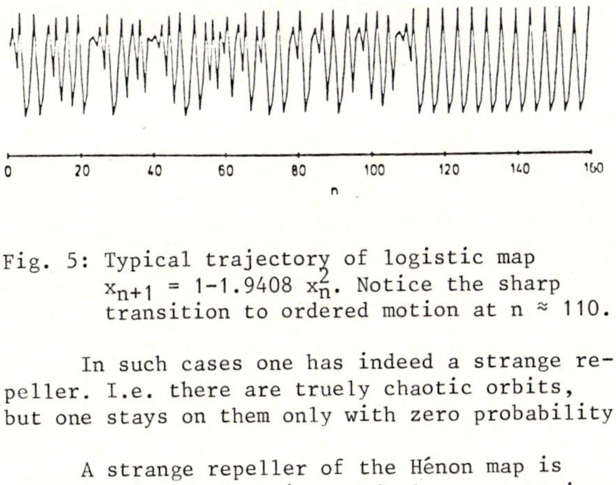

Fig. 5: Typical trajectory of logistic map $x_{n+1} = 1-1.9408\, x_n^2$. Notice the sharp transition to ordered motion at $n \approx 110$.

In such cases one has indeed a strange repeller. I.e. there are truely chaotic orbits, but one stays on them only with zero probability.

A strange repeller of the Hénon map is shown in fig. 6. Starting randomly, most trajectories are first attracted close to the repeller, the repulsion occuring with a much larger time scale. In such cases, we[19] proposed to speak of a semi-attractor instead of a repeller.

As long as one stays on the repeller, the motion is truely chaotic in the sense that there is (at least) one unstable direction, along which information is flowing in.

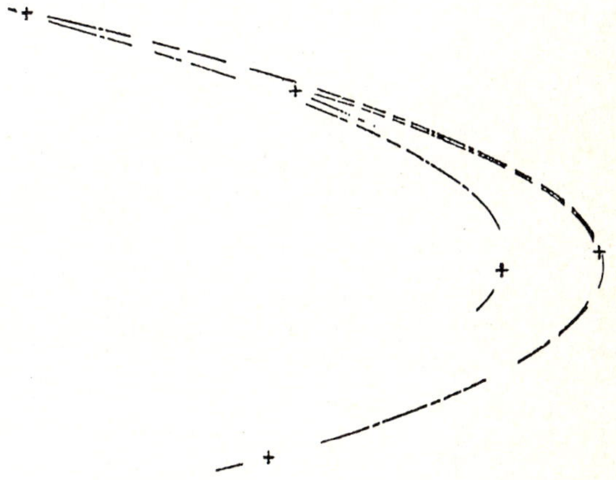

Fig. 6: Ω-limit set of the Hénon map with $a = 1.45$, $b = 0.2$. The attractor is a period - 5 cycle (crosses) The broken lines are a repeller (or rather "semi-attractor") on which trajectories stay ~ 20 iterations in average. They actually form a product of two Cantor sets, with $D_1 = 0.89$ and $D_2 = 0.19$.

Thus, eqs. (1.5) and (1.6) hold unchanged.

But now the initial conditions have to be very special if the trajectory is to stay on the repeller. If one starts with random initial conditions instead, the "wrong" incoming digits will push the trajectory off the repeller. Clearly, these will be also digits of the coordinates along the unstable directions. Thus, the flow $\sum_i{}' \lambda_i$ of incoming digits consists of 2 üarts: one is the Kolmogorow entropy of the flow on the repeller, the other is the rate α at which the trajectory is pushed away from it:

$$\sum_i{}' \lambda_i = K + \alpha \, . \tag{3.1}$$

The probability P_t to stay in some small neighbourhood of the repeller decays with time as

$$P_t \sim e^{-\alpha t} \, . \tag{3.2}$$

Otherwise said, the decay of transient chaos happens since the repeller is Cantorian along the unstable directions. Except for a set of initial conditions of measure zero, the trajectory will sooner or later fall into one of the holes of this Cantor set. Thus, we expect α to depend on the partial codimensions along the unstable directions. Indeed, combining eqs. (1.5) and (3.1), one finds

$$\alpha = \sum_i{}' (1-D_i)\lambda_i . \quad (3.3)$$

For a restricted class of models, eqs.(3.1) and (3.3) were proven rigorously in ref. 19. For other cases, including those shown in figs. 5 and 6, they were verified numerically.

Experimentally, semi-attractors should not only be seen in transients but also in some noisy systems. Consider e.g. the right-most point of the period-5 cycle in fig. 6. The semi-attractor approaches it to a distance of $\sim 10^{-2}$. If one has now noise with an amplitude of that order of magnitude, the orbit will be pushed onto the semi-attractor again and again. Such a behaviour has been seen in ref. 30.

4. CONCLUSIONS

We have seen that the unpredictability of chaotic motion is due to an information flow through chaotic systems. This information flow rate was found to be a sum (over all instable directions) of the product of the flow velocities (the Lyapunov exponents) times the information densities (the partial dimensions). The information balance equation leads then to a connection between dimension and Lyapunov exponents (a weakened form of the Kaplan-Yorke conjecture), which - unlike the Kaplan-Yorke conjecture - is true in all cases.

Throughout the present paper, we have assumed that some coarse-graining is done, and have looked at the limit behaviour when this coarse-graining is made finer and finer. The most clear-cut sign of deterministic chaos is that effective information flow rates and dimensions are non-zero and stay finite in this limit.

REFERENCES

1. A.N. Kolmogovov, Dokl. Akad. Nauk. SSSR 124, 754 (1959)
2. Ya.G. Sinai, Dokl. Akad. Nauk. SSSR 124, 768 (1959)
3. P. Billingsley, Ergodic theory and information (Wiley, N.Y. 1965)
4. R. Shaw, Z. Naturforsch. 36a, 80 (1981)
5. J. Balatoni and A. Renyi, Publ. Math. Inst. Hung. Acad. Sci 1, 9 (1956)
6. D. Farmer, Z. Naturforsch. 37a, 1304 (1982)
7. D. Farmer, Physica 4D, 366 (1982)
8. D. Farmer, E. Ott and J.A. Yorke, Physica 7D, 153 (1983)
9. R. Bowen and D. Ruelle, Invent. Math. 29, 181 (1975)
10. B.B. Mandelstam, The fractal geometry of nature (Freeman, San Francisco 1982)
11. A. Brandstaeter et al., Phys. Rev. Lett. 51, 1442 (1983)
12. B. Malraison et al., J. Physique-LETTRES 44, L897 (1983)
13. V.I. Oseledec, Trans. Mosc. Math. Soc. 19, 197 (1968)
14. P. Grassberger and I. Procaccia, Physica 13D, 34 (1984)
15. P. Grassberger, to appear in "Chaos - an Introduction"; A.V. Holden, ed. (Manchester Univ. Press, 1985)
16. F. Ledrappier and L.-S. Young, Berkeley preprint (1984)
17. Ya.B. Pesin, Russ. Math. Surveys 32, 55 (1977)
18. D. Ruelle, New York Acad. Sci. 136, 229 (1981)
19. H. Kantz, P. Grassberger, WU B 84-18 (Wuppertal 1984)
20. J.L. Kaplan, J.A. Yorke, in Lecture Notes in Math., vol. 730, p. 204 (Springer, Berlin 1978)
21. P. Frederickson et al., J. Diff. Eqns. 49, 185 (1983)
22. P. Grassberger, to appear in Physica D
23. R. Adler et al., Trans. Amer. Math. Soc. 114, 309 (1965)
24. A. Renyi, Probability Theory (North-Holland, N.Y. 1970)
25. M. Hénon, Commun. Math. Phys. 50, 69 (1976)
26. E.N. Lorenz, Physica 13D, 90 (1984)
27. F. Ledrappier, Commun. Math. Phys. 81, 229 (1981)
28. J.C. Alexander, J.A. Yorke, Univ. Maryland prepr. (1982)
29. P. Grassberger, I. Procaccia, Physica 9D, 189 (1983)
30. M. Dörfle and R. Graham, to be published.

Section IV

DISTRIBUTED PARAMETER SYSTEMS

Complex and Distributed Systems: Analysis, Simulation and Control
S.G. Tzafestas and P. Borne (editors)
Elsevier Science Publishers B.V. (North-Holland)

DISTRIBUTED-PARAMETER AND LARGE SCALE SYSTEMS: A LITERATURE OVERVIEW

Spyros G. Tzafestas
Control and Automation Group
Computer Engineering Division
National Technical University
Zografou 15773, Athens, Greece

ABSTRACT

This paper presents an overview of the literature on distributed-parameter and large scale systems extended over the last two decades. Due to space limitation details on the methods and results are not included. The material of the paper was used as an introduction to the session on modelling and control of distributed and large scale systems of the 11th IMACS World Congress.

1. INTRODUCTION

The theory of distributed-parameter (DPS) and large-scale systems (LSS) has reached a very advanced level of maturity and sophistication in the last twenty years.

The origin of the DPS field may be dated from the first papers of the Russian scientists Butkovsky and Lerner back in 1960. The results of DPS theory have already been applied to a large number of eingineering fields. The key areas of the theory that have found an impressive number of applications are stability, control, estimation, identification, optimal design, and simulation. Application areas in the last ten years include chemical engineering, petroleum and metallurgical industries, nuclear reactor control, plasma control, mechanical structure design problems (bridges, platforms, large antennas), resource recovery (oil, water, coal), environmetal problems (environmetal quality modeling, control, water quality management, etc), physiological systems (distribution and effect of drugs, etc) and sociological systems (dynamic modelling and control of behavior of groups of people) to name a few.

The two main avenues in the study of LSS are the hierarchical and the decentralized approach. Actually the two main difficulties in solving problems of LSS using standard multivariable techniques are:i) the numerical difficulty (computational effort required increases by a cubic or quartic order with a linear increase in the system dimension) and ii) the lack of centrality in LSS (all available information and computations about the system are not brought to a single central location as is explicitly or implicitly assumed in standard multivariable methods and tools). The decentralized control problem arises whenever a system is acted upon by a number of controllers which cannot communicate to each other on-line, even though they may have a set of a priori rules which provide them with some structural knowledge about each other.

Our purpose in this brief paper is to provide a list (nonexhaustive) of important contributions on DPS and LSS (hierarchical and decentralized) with short comments illustrating the problems and techniques covered by them. The material on LSS was drawn mainly from the joint EEC reports of M.G. Singh, A. Titli, G. Schmidt and S.G. Tzafestas.

2. DISTRIBUTED-PARAMETER SYSTEMS

The references on DPS modelling and control are given in ten main sections concerning Stability, Controllability, Observability, Optimal Control, Filtering, Smoothing and Prediction, Identification, Sensitivity, Computational Methods, Simulation, and Applications. We shall present a general mathematical model of DPS and then lead the reader through the sections by refering to this model in trying to illustrate the concepts contained in each section.

It is important to make the observation right from the beginning that there exists no unique mathematical model to describe all physical processes which belong to distributed parameter systems. Fortunately, most of the problems of interest to engineers and scientists involved in the study of distributed parameter systems can be described in state space form by the mathematical model

$$\frac{\partial \underline{y}(t,\underline{x})}{\partial t} = F(\underline{y}(t,x),\ \underline{u}_{\Omega}(t,\underline{x}))$$

where $\underline{y}(t,\underline{x})$ is the state vector and F(.) is a matrix partial differential operator which in general can be nonlinear. The spatial coordinate vector $\underline{x}=(x_1,x_2,\ldots,x_m)^T$ varies over a fixed m-dimensional domain Ω with boundary $\partial\Omega$. The vector $\underline{u}_{\Omega}(t,\underline{x})$ is the control vector distributed over the same fixed spatial domain. The above equation only describes the local behavior of the system at any point $\underline{x}\varepsilon\Omega$. Starting from a given point $\underline{x}$ and a set of initial data at $\underline{x}$, i.e. $\underline{y}(t_0,\underline{x})=y_0(\underline{x})$ the differential equation generally permits the construction of many possible solutions. In order to choose the solution which is appropriate to the physical situation, additional constraints or boundary conditions are introduced. They may be represented by a vector equation

$$G(\underline{y}(t,\underline{x}'),\underline{u}_{\partial\Omega}(t,\underline{x}'))=0,\quad \underline{x}'\varepsilon\partial\Omega$$

where G(.) is in general a nonlinear spatial differential matrix operator whose parameters may depend on $\underline{x}'$ and t, and $\underline{u}_{\partial\Omega}(t,\underline{x}')$ is the boundary control function.

Many physical systems in the neighbourhood of certain prescribed motions can be approximated by linear mathematical models. For a general class of distributed parameter systems defined in a fixed spatial domain Ω, the governing partial differential equations (PDE) can be expressed in the following general form:

$$\frac{\partial \underline{y}(t,\underline{x})}{\partial t} = \mathcal{L}_{\Omega}\{\underline{y}(t,\underline{x})\}+B(t,\underline{x})\underline{u}_{\Omega}(t,\underline{x})$$

with boundary conditions:

$$\mathcal{L}_{\partial\Omega}\{\underline{y}(t,\underline{x}')\}=u_{\partial\Omega}(t,\underline{x}'),\quad x'\varepsilon\,\partial\Omega$$

where $\mathcal{L}_{\Omega}$, and $\mathcal{L}_{\partial\Omega}$ are matrix, linear spatial differential or integrodifferential operators whose parameters may depend upon $\underline{x}$ and/or t.

In a stochastic environment, the linear DPS can be found by a stochastic vector process $\underline{w}(t,\underline{x})$ to represent the effect of distrubances where the model then becomes:

$$\frac{\partial \underline{y}(t,\underline{x})}{\partial t} = \mathcal{L}_\Omega\{\underline{y}(t,\underline{x})\} + B_1(t,\underline{x})\underline{u}_\Omega(t,\underline{x}) + B_2(t,\underline{x})\underline{w}(t,\underline{x})$$

In the following sections, we shall survey the ten major concepts (mentioned above) based on this mathematical model.

Stability

The problem of stability of DPS has been studied using Lyapunov method, semi-group theory, and the method of comparison of functions [1-5]. In [6] the stability of systems governed by a coupled nonlinear DPS in considered. A stability theorem is proved using the method of functional analysis and semi-groups. In [7] open loop stability in the sense of bounded outputs for bounded inputs is studied. A closed-loop stability criterion similar to Popov's stability criterion for lumped-parameter systems is also derived. In [8] the input-output stability criterion is extended to cases where the system parameters depend on the space variable. The method of comparison theorem is used in [9] to study nonlinear parabolic systems and applied to a system with temperature control. In [10] the stabilization problem of unstable DPS using feedback control is studied. The results are extensions of corresponding lumped parameter ones [11]. Finally in [12] a functional observer of the Luenberger type is used for the stabilization of a DPS with boundary control.

Controllability

Some basic work on DPS controllability for distributed and boundary control is obtained in [1-3]. The concept of exact controllability, i.e. achieving a given terminal state exaclty, for DPS is studied in [4]. In [5] the reachable states of a DPS are determined for cases when the control energy is constrained but can be either distributed or at the boundary. Existence and uniqueness questions regarding control that can drive the system to a desired state are examined in [6]. The modal approach to the study of DPS controllability is studied in [7], and the controllability of 1st and 2nd-order evolution equations is discussed in [8]. The duality between DPS controllability and observability is considered in [9-10].

Optimal Control

The initial papers on DPS optimal control are attributed to Butkovsky and Lerner [1]. The first comprehensive survey was given by Wang [2]. That paper still provides an excellent introduction to the subject, and the point of view developed is consistent with that used by a majority of subsequent authors. In [3] a bibliography was published covering both stability and control. In [4] a survey of soviet work in the field is presented. Some tutorial material on the same subject was published in [5]. A more mathematical foundation of the subject of control of DPS is provided in [6] and [7]. These books are aimed towards an audience of theorists, particularly those with a strong background in functional analysis. An extensive survey of the subject (by 1970) with more than 250 items: papers, reports and books, is given in[8]. As it is seen in the references [9-29], many results which were obtained in the lumped parameter control systems as well as the mathematical procedures implemented for their derivation are also applicable to DPS.

Observability

Observability for DPS has been defined in [1]. In [2] the determination of initial conditions of DPS on the basis of observed measurement is studied as an observability problem. In[3] observability is defined as the ability to establish the uniqueness of a solution of the system under study. Necessary and sufficient conditions for observability of a class of hyperbolic systems is derived in [4]. In[5] the concepts of observability are studied in conjuction with filter convergence of a class of stochastic DPS. The questions that are examined are:a) The effect of measurement locations on observability and 2) the optimal location of measurement for state estimation. It is shown that for systems whose solutions can be expressed as eigenfunction expansions only a few measurements need suffice for observability. Finally, in[6], observability is examined on the basis of the observed measurement data from a finite number of sensors over a finite period of time.

Filtering, Smoothing and Prediction

In the last 20 years filtering problems of DPS have received increased attention. Several statistical information processing techniques which have been applied to lumped parameter systems have also been extended to DPS. Among these stand out:Wiener-Hopf equation, orthogonal projection in Hilbert space [1,2], Maximum likelihood Bayesianapproach[3], Fokker-Plank equation[4], adaptive estimamation method via the partitioning estimation algorithms [5-6] and Monte Carlo approach [7]. The papers [8-34] present some of the basic contributions in this direction. The problem of estimating the state of DPS and/or an optimal control using only output measurements has also been studied for DPS. In lumped parameter systems, the auxiliary system that is used for the reconstruction of the state or control in addition to the output available data is referred to an observer [34]. A distributed parameter observer has been derived in [35]. Observers for bilinear DPS and for linear DPS using finite spatial measurements are treated in[36]and [37] respectively.

Identification

The concept of identification is a very wide one and different people use it in different ways[1]. In [2] of this section a comprehensive survey of methods for DPS identification is presented regarding basically structural and parameter identification. In[3], a method is given for estimating unknown parameters for DPS by invoking the Bayesian theoretic approach. The basic notion of this method is the separation principle of the identification scheme from the state estimation with considerable computational advantages. In [4] the identification of a parameter function of a single variable is made using the Galerkin approximation method. Methods dealing with determination of initial state are covered in[5-7]. Reference [8] deals with the problem of determination of unknown functions which give the input distributions for a class of DPS. State estimation as an identification problem is covered in the book of Phillipson [9].

Finally, reference [10] deals with parameter determination of a process described by a PDE. Further surveys on DPS idendification works may be found in [11-13].

Sensitivity

In many physical systems, the parameter values can vary from their assumed nominal values (fatigue, wearout, environmental changes). These perturbations may alter the performance of a system from the desired performance. In some cases feedback control can be utilized to insure that parameter variations alter the system performance with feedback to a lesser degree than an equivalent open-loop system subject to the same parameter variations.

Original work on this subject is included in [1] and [2]. In [3] of this section the sensitivity problem has been formulated so that the sensitivity results for lumped parameter systems could then be applied (see [4]). In [5] a nonlinear DPS is studied in which either the initial states or a vector of constant plant parameters are unknown. The result is the synthesis of a feedback control structure that adapts the nominal open-loop control in such a manner as to compensate for the errors introduced by model uncertainties. In [6] a performance index which includes the trajectory sensitivity functions and the sensitivity of a cost function is sought to be minimized. In [7] it is shown that an operator which corresponds to an extension of the Fisher information matrix plays an important role in measurement controls. The maximization of a scalar measure of the information operator for suboptimal measurement strategies implies that sensors be designed such that the collected data be as sensitive as possible to changes in unknown parameters. It is seen that a measurement optimization problem may be converted into a sensitivity problem which in many practical cases has considerable computational advantages.

Computational Techniques

The problem of determining the optimum control for DPS is generally very difficult (computationally) to solve. In most cases approximation in one form or another is imperative in the implementation and/or determination of the control law. A variety of computational methods unders the general heading of mathematical programming have been developed over the years. In [1] the general mathematical programming approach is surveyed as has been applied to DPS.

The dynamic programming approach which is included by some authors in mathematical programming is discussed in [2] and [3]. In [5] an example of applying the quadratic programming aproach to a discrete-time model for the process of one-sided heating of a metal is presented. In [6] the penalty method is applied to a distributed parameter control system. In [7] and [8] the conjugate gradient method is applied to both discrete and continuous time models. In [9] the conjugate method is used in conjuction with the penalty method. The penalty terms are added to the performance criterion to effectively reduce the constrained optimization problem to a series of uncostrained minimizations. In [10] and [11] optimal control problems under norm constraints on the control function are studied. In [12] an efficient computational technique is introduced which is based on the order expansion of the performance index. This technique is equivaluent to that derived for lumped systems in [13]. Further techniques and results are provided in [14-28].

Simulation

Regarding the computational solution and simulation of DPS there have been developed a variety of techniques by both numerical analysts and engineers [1-12] Actually, most simulation methods of DPS employ some kind of finite difference approximation for the time and/or space derivatives involved in the system PDEs.

A classification of DPS simulation methods is the following:

i) Discrete-time discrete-space (DTDS) of finite-difference methods [1-2].
ii) Continous-time discrete-space (CTDS) methods.
iii) Discrete-time continuous space (DTCS) methods [4].
iv) Function expansion (modal) methods [5].
v) Other methods (method of characteristics, Monte Carlo [9], finite elements [10], etc).

Applications

DPS theory encompasses a large number of fields. The purpose of this section is to report some of the significant applications of the theory. A comprehensive survey of applications of DPS theory is included in [1].In [2], the results of an experimental study are given in which real time state estimation, deterministic optimal and suboptimal feedback control, as well as stochastic feedback control are included. In [3], using a state-space representation, a stochastic model of air pollution is developed.

In [4] ten theoretical and five application chapters are involved. In the first application chapter, by Lausterer and Eitelberg, some further experimental results obtained with practical implementation DP algorithms are presented. The next chapter, by Seinfeld and Kravaris, explores the DP identification problem arising in the modelling of petroleum reservoirs and subsurface aquifers. The third chapter, by Ramirez, reviews the scope of DP identification and filtering applications to chemical engineering processes, and considers the class of fixed bed tubular reactors in detail. The fourth chapter, by Kanoh, presents various DP dynamic models of heat exhangers, studies their properties, and provides some lumped-parameter simulation results of them. The final chapter of the book, by Wiberg, deals with the application of DP methods to nuclear reactor systems. Other applications may be found in [5-9].

3. HIERARCHICAL CONTROL SYSTEMS

The main topics of research in the field of LSS are (i) model simplification/reduction/decomposition, (ii) structural properties of LSS (e.g. controllability, stability, etc), (iii) computational aspects for off-line design problems or for on-line control, and (iv) LSS control structures and algorithms.

The first two aspects are discussed in [1]. Regarding the computational procedures, one can easily verify that they fall in two categories;hierarchical and decentralized structures.

The literature on LSS is vast. Here we shall present an overview of the hierarchical approaches [2-103]. The basic concepts will be discussed first, then the references on hierarchical optimization and control methods will be given, and finally some application references will be discussed.

Hierarchical Optimization and Control Concepts

All procedures of attacking LS problems are based upon decomposing the overall problem into a number of simpler problems which are amenable to standard techniques. A first "hierarchical"configuration is the so called "multistrata hierarchy" used for example in the California Water Project for multihorizon planning. The subproblems are solved sequentially starting from subproblem K. Based on the Kth subproblem solution, the K-1 subproblem is solved and so on. This configuration can also manage the lack of centrality if some parallel decomposition is used.

In [3] a general theory of hierarchical structures of the above type is developed (decomposition-coordination). Most works on hierarchical computation and decision making were formulated within an optimization framework. For example [3-5,19,36,67].

Hierarchical Control Structures

Two different situations can be distinguished. The first occurs when the controller needs to solve a problem P within the desired time and this problem is solved by means of a hierarchical computing or optimization scheme instead of a standard one. In this case only the costs of communications and computations or of other controller aspects are affected. The second occurs when within the control structure one can distinguish some hierarchy which affects the actual control actions as happens in an "multilayer" control structure, where the control function is partitioned into several algorithms operating at different time scales (adaptation layer, optimization layer, regulation layer) [7-10,19].

There exist also hierarchical structures involving multiple decision makers [11,14]. Usually the overall dynamic optimization problem (long term global problem) is solved via a hierarchical method at long time periods and the subproblems (short-term problems) at shorter time periods with a frequency depending on how frequently new data becomes available [12,19].

Hierarchical Optimization Techniques

The general (overall) optimization problem (OP) has the form

$$\min_{d^o} J^o(d^o) \text{ subject to } d^o \varepsilon D^o \subset D$$

where

$J^o: D \to R$ (a real valued criterion function)

D is a suitable space of decision variables

D^o is a feasible space

This formulation includes numerous decision problems encountered in resource allocation, state estimation, parameter identification, network optimization, optimal control etc. The OPs are distinguished in dynamic (DOP) and static (SOP) problems.

The DOP formulation is [5,15,16,36,76-82,85] . Given the dynamic system

$$\dot{x}=f^{o}(x,m,t),x(t_o)=x_o, m\varepsilon R^{m}, x\varepsilon R^{nx}$$

choose the control variable m(t) restricted by

$$m(t)\varepsilon M, \quad \ell(x(t), m(t))\leq 0,$$

such that to minimize the performance index

$$J_d^{o}=\int_{t_o}^{t_f} q^{o}(x,m,t)dt+J_f^{o}(x(t_f))$$

The SOP is the steady-state form of the DOP, namely determine an m such that $q^{o}(x,m)$ is minimized subject to $f^{o}(x,m)=0$ $m\varepsilon M$, $\ell^{o}(x,m)\leq 0$.

In LSS theory the DOP and SOP are appropriately partitioned via interaction (coordination) variables and additive costs. See e.g. the references [3,5,15, 17,19,74].

The decomposition involves the choise of the coordination problem.
The existing decomposition methods are classified as:

1. Direct decomposition method [18,19, 20,22,24].
   This method is also known as primal method [18] or feasible method [20].
2. Interaction prediction decomposition method [3].
3. Price decomposition (interaction balance) method [3,4,5,19,25,26,27,60].
4. Mixed decomposition method [3,5,19,20 21,32].

The choice of the coordination strategy (i.e the algorithm used by the coordinator for updating the coordination variables ξ) is the next step. This strategy and its convergence depend upon the original problem properties, the decomposition adopted, the choice of the coordination mode and the coordination problem (see e.g. [5,19,26,27,36])

The future of hierarchical optimization and control will depend on new ideas of process decomposition and cordination strategies.

Applications

Here only a representative set of application references are given. A comprehensive list is far from the purpose of the present introductory paper. The applications deal with the state estimation problem [5,47], power dispatching (e.g. [63]), river pollution problem [5], production planning, water system planning and control, traffic control, transportation systems, proces control etc, [57,61,62,63,73,79]. We close by mentioning some applications which were studied in the framework of a European Economic Community project (UMIST, LAAS, Munich Techn. Univ, Patras Univ.). These concern the hierarchical control of sulphur production plant, the multilayer control of a freeway traffic system, and the decoupling real-time control of complex systems.

4. DECENTRALIZED CONTROL SYSTEMS

Although decentralized controllers have been designed over two decades, their design was based on ad hoc methods (e.g. assuming weak interacting subsystems etc). Some of the fundamental well known principles of centralized control fail in the case of decentralized control (e.g. the separation principle of LGQ control theory). Hence much of the work on decentralized control has been devoted on justifying the ad hoc design procedures used in current industrial practice and developing new issues on the basis of them. Decentralized control schemes are popular in both engineering and managerial/O.R. problems. Decentralized control is classified in stochastic and deterministc decentralized control. In the former case one starts from team theory concepts, and considers the value of information and the optimality of the control schemes upon the assumption of several types of information constraints. A practical way to design the stochastic control scheme is to consider controllers of given fixed structure.

The key concept in the analysis and design of deterministic decentralized controllers arises from the presence or absence of decentralized "fixed modes" Closely connected with decentralized control is the decentralized observation/filtering.

Stochastic Decentralized Control

Consider the following stochastic model

$$\dot{x}=Ax+\sum_{i=1}^{N} B_i u_i+\xi(t)$$

$$y=Cx+\sum_{i=1}^{N} D_i u_i+n(t)$$

where $\xi(t)$ and $n(t)$ are independent Gaussian white noise processes.

The information pattern determines what measurements are available to the controller. For the decentralized control problem, the information pattern is

$$\{z_i(t)=H_i(t)y(t),\ i=1,2,..,N\}$$

and is determined by the set of matrices

$$H=\{H_1,H_2,\dots,H_N\}$$

In the classical information pattern B, D and H are block diagonal matrices, and H=I, i.e. all controllers have identical information. The cost function is assumed of the following quadratic type:

$$J=\lim_{T\to\infty} \frac{1}{T}\int_0^T [x^TQx+\sum_{i=1}^{N} u_i^T R_i u_i]dt$$

with $Q\geqq 0$ and R=block diag R_i =0
A non classical information pattern occurs when

$$u_i(t)=\gamma_i(z_i^t),\quad z_i^t=\{z_i(\tau),\ 0\leqq\tau\leqq t\}$$

In the case of nonclassical information patterns, the optimal control is not linear and the optimal estimator may be infinite dimensional [2]. Much of the work has been concentrated on finding special cases where linear solutions are possible. Such cases are the so called "partially nested information structures" [3-6] and the " one step delay sharing patterns" [7-9]. Sufficient conditions for separation of estimation and control have been derived in [10] (see also [11]).

Particular fixed structure decentralized controllers (linear controller as a function of its local measurements, compensator of the same order as the system) were studied in [12]. It is noted that informational complexity poses additional computational problems not encountered in centralized or hierarchical procedures [13, 14] . The case of decentralized control which is periodically updated to take into account new information is studied in [15] .

Deterministic Decentralized Control (Stabilization)

The general problem is to choose local feedback control laws such that the overall interconnected system is stable. The theory is based on the concept of fixed modes [16-17], and structural fixed modes [18] (see also [19], [25], for the concept of structural controllability under the decentralization constraint). The calculation of the fixed modes can be done by the algorithm of [20] or by the more systematic method of [21]. The absence of unstable fixed modes in decentralized system with two control stations can be checked as shown in [22]. An iterative characterization of fixed modes was given in [23]. A characterization of fixed modes using the zeros of the system and of certain subsystems can be found in [24].
The problem of getting around the difficulty, that a system subject to the decentralized constraint cannot have its unstable modes stabilized using static decentralized feedback or its poles assigned by dynamic decentralized compensators, is discussed and solved in [26]. The treatment of fixed modes through block diagonally dominant matrices is made in [50]. Wang in [51] produces a new viable feedback matrix structure. Anderson and Moore in [28] showed that in the case of some decentralized fixed modes, these can be controlled using time varying feedback control laws. The synthesis problem of decentralized control in the absence of unstable fixed modes is studied in [30] (Ch. 3). A parametric optimization problem together with a numerical algorithm are studied in [31] for the special case where the decentralization constraint is imposed on the regulation structure of an interconnected linear system. Gradient methods for treating the non - linear parametric form of the problem are provided in [32](based on the Rosen projection operator) and [33]. Stability tests are provided in [34].

Other approaches for the synthesis of decentralized control can be found for example in [36] (prespecified degree of stability), [46,49] (model following method) and in [64] (decentralized control using overlapping information sets)

Decentralized State Estimation

The problem of estimating missing variables by means of observers [52-55] has been studied for LSS in [56]. Centralized observers are now well known [8,53]. A method for constructing an exact subobserver based on a direct computation of the interaction variables from the available measurements was proposed in [59]. An alternative scheme for decentralized observation is based on the unknown input observer of Basile and Marro [60,61]. An other way of solving the problem of observing a system with unknown inputs is by observing the unknown input itself,via a distrubance observer [62]. Detectability conditions are given in [63]. The information exchange between subobservers is studied in [64,65] where an appropriate subobserver scheme is developed. The selection of decentralized observation schemes is illustrated in [66] for the case of a linear mathematical model of a rear-axle test stand.

Two important applications of decentralized control (real time control of a telephone network and of the energy cycle of a ship) are studied in [39] and [47].

REFERENCES

REFERENCES ON DISTRIBUTED-PARAMETER SYSTEMS

Stability

1. J.C. Williams, "A Survey of Stability of distributed parameter systems", in Control of Distributed Parameter Systems, pp. 63-102, Proc of AACC-JACC Symp.,1969.
2. P.K.C. Wang, "Control of distributed parameter systems" in Advances in Control Systems, Vol. 1, pp. 75-172, (C. Leondes, Ed.) 1964.
3. P.K.C. Wang, "Theory of stability and control for distributed parameter systems" (A bibliography) Int. J. Control, Vol. 17, pp.101-116, 1968.
4. A.J.Berger and L.Lapidus, "Introduction to the stability of distributed systems via Lyapunov functions", A.I. Ch. E.I., Vol. 14, pp. 558-566, 1968.
5. W.E. Kastenberg, "On the stability of equilibrium of distributed parameter feedback control systems", Int. J. Control, Vol.6, pp. 523-532, 1967.
6. R.M. Crawford and W.E. Kastenberg, Stability analysis of distributed parameter systems in a Banach space, Int.J. Control, Vol.12, pp. 929-943, No. 6, 1970.
7. H.C. Khatri, "Stability conditions for a class of distributed parameter systems", Journal of the Franklin Institute, Vol. 29, pp. 43-56, Jan. 1971.
8. G. Jumarie, "A practical criterion for the input-output stability of distributed-parameter systems with space dependent coefficients", Int. J. Control, Vol. 21, pp. 285-292, No.2, 1975.
9. W.E. Kastenberg, "On asymtotic stability of nonlinear distributed parameter energy systems", Int. J. Control, Vol. 19, pp. 73-79.
10. Y. Sakawa and T.Mutsushita, "Feedback stabilization of a class of distributed systems and construction of a state estimator", IEEE Trans. Autom. Control, Vol. AC-20, pp. 748-753, No.6, Dec. 1975.
11. W.M. Wonham, "On pole assignment in multi-input controllable linear systems", IEEE Trans. Autom.Control, Vol. AC-12, pp. 660-665. Dec. 1967.
12. N.Fujii, "Feedback stabilization of distributed parameter systems by a functional observer", SIAM. J.Con - trol and Optimization, Vol.28, pp. 108-119, No.2. March 1980.

Controllability

1. H.O. Fattorini, "Some remarks on complete controllability, Siam,J.Control, Vol. 4, pp. 391-402, 1966.
2. H.O. Fattorini, "A remark on the bang-bang for linear control systems in infinite dimensional space",Siam J. Control, Vol. 6, pp. 349-352, 1968.
3. M. Vidyasagar and T.J. Higgins, "A controllability criterion for stable linear systems", IEEE Trans.Autom. Control, AC-15, pp. 391-395, 1970.
4. H.O. Fattorini and D.L. Russel, "Exact controllability theorems for linear equations in one space dimension", Arch. Rational Mech.Anal. Vol.43,pp. 275-280, 1971.
5. C.J. Herget, "On the controllability of distributed parameter systems",Int. J. Control, Vol. 11, pp.827-833, No.5., 1970.
6. T. Kobayashi, "Some remarks on controllability for distributed parameter systems", SIAM J.Control and Optimization, Vol. 18, pp. 733-742, 1978.
7. G.E. Mc Glothin, "A modal control for distributed systems with applications to boundary controllability", Int.J.Control, Vol. 20, pp. 417-432, No.3, 1974.
8. T. Kobayashi, "Discrete time controllability for distributed parameter systems",Int.J.Systems Sci.,Vol. 11, pp. 1063-1074, No.9, 1980.
9. G.E. Mc.Glothin, Controllability, observability and duality in a distributed parameter system with continuous and point spectrum", IEEE Trans.Autom. Control, Vol. AC-23, pp. 627-690, No.4, 1978.
10. R.Triggiani, "Controllability and observability in Banach space with bounded operators", SIAM J.Control, Vol. 13, pp. 462-465, No.3, Feb. 1975.

Optimal Control

1. A.G. Butkovsky and A.Y. Lerner, "The optimal control of systems with distributed parameters, Autom. Remote Control, Vol. 21, pp.472-477, 1960.
2. P.K.C. Wang, Control of distributed parameter systems, in Advances in Control Systems: Theory and Application, Vol. 1, pp. 75-172, Academic Press, 1964.
3. P.K.C. Wang, Theory of stability and control for distributed parameter systems (a Bibliography), Int.J. Control Vol. 17, pp. 101-120, Feb. 1968.
4. A.G. Butkovsky, A.I. Egorov and K.A. Lurie, "Optimal control of distributed systems" (a survey of soviet publications) Siam J.Control, Vol. 6, pp. 437-476,1968.
5. W.L. Brogan, "Optimal control theory applied to systems described by partial differential equations", in Advances of Control Systems, Academic Press, 1968.
6. J.L. Lions, "Optimal control of systems governed by partial differential equations", Springer, 1971.
7. N.U. Ahmed and K.L. Teo", Optimal control of distributed parameter systems, North-Holland-Elsevier, 1981.
8. A.C. Robinson, "A survey of optimal control of distributed parameter systems", Automatica, Vol. 7, pp. 371-388, 1971.
9. S.G. Tzafestas, "Optimal distributed parameter control using classical variational theory", Int.J.Control, Vol. 12, No.4, pp. 593-608, 1970.
10. T.L. Johnson and M.Athans, "A minimum principle for smooth first-order distributed systems", IEEE Trans. Autom. Control, Vol. AC-19 pp. 136-139, April 1974.
11. P.Stavroulakis and S.Tzafestas," Matrix Minimum Principle for distributed parameter control systems", Int. J. Systems Sci., Vol.11, pp. 793-801, No.7, 1980.
12. S. Fond, "A dynamic programming approach to the maximum principle of distributed parameter systems", J.

Optim. Theory Applic., Vol. 27, pp. 583-601, No.4, April 1979.

13. S.G. Tzafestas, "Multiplayer distributed parameter differential Games with sampled state feedback control", Int. J. Systems Sci., Vol. 10, pp. 989-1005, 1979.
14. S.G. Tzafestas, "Hierarchical control of distributed parameter systems using Stackelberg policies", Int.J. Systems Sci., Vol. 12, pp. 719-744, No.6, 1981.
15. T.K. Yu and J.H. Seinfeld, "Suboptimal control of stochastic distributed parameter systems" , AIChe Journal, Vol. 19, pp. 389-392, 1973.
16. J.G. Vermeychuk and L. Lapidus, "Suboptimal feedback control of distributed systems" , Part I, AIChe Journal , Vol. 19, pp. 123-129, 1973.
17. J.G. Vermeychuk and L. Lapidus, "Suboptimal feedback control of distributed systems", Part II, AIChe Journal, Vol. 19, pp. 129-137. 1973.
18. S.P. Chaudhuri, "Optimal control computational techniques of a class of nonlinear distributed parameter systems", Int. J. Control, Vol. 15, pp. 419,432, No.3, 1972.
19. M.S. Sholar and D.M. Wiberg, "Canonical equations of boundary feedback control of stochastic distributed parameter systems", Automatica, Vol. 8, pp. 287-298, 1972.
20. S.G. Tzafestas, "Parameter adaptive control of stochastic distributed systems" , Int. J. Systems Sci., Vol. 11, No.12, pp. 1397-1433, No.12, 1980.
21. M. Vidyasayar, and T.J. Higgins, "A basic theorem of distributed control and point control, ASME Trans., Dyn. Syst., Meas. Control, pp. 64-67, March 1973.
22. P.K.C. Wang, "Optimal control of parabolic systems with boundary conditions having time delays", Siam J. Control Optimiz., Vo. 13, pp. 274-293, 1975.
23. G. Knowles, "Time optimal control of parabolic systems with boundary conditions involving time delays", J. Optim. Theory Appl., Vol.21, pp. 563-574, No.4, Aug. 1979.
24. R.F. Baum, "Necessary conditions for distributed parameter systems with controls of fewer variables than state variables", J. Optim. Theory Appl., Vol. 30, pp. 663-681, No.4, April 1980.
25. B. Porter and A. Bradshaw, "Modal control of a class of distributed parameter systems", Int. J. Control, Vol. 15, pp. 673-681, 1972.
26. C.R. Johnson, Jr. and R.C. Montgomery, "A distributed system adaptive control strategy", IEEE Trans. Aerospace and Electronic Systems, Vol. AES-15, pp. 601-612, No. 5, Sept. 1979.
27. J.C.E. Martin, "Dynamic selection of actuators for lumped and distributed parameter systems", IEEE Trans. Control, Vol. AC-24, No.1, pp. 70-77, Feb. 1979.
28. S.E. Aidarous, M.R. Gevers and M.J. Installe, "Optimal point-wise discrete control and controllers allocation strategies for stochastic distributed systems , Int. J.Control, Vol. 24, No.4, pp.493-508, 1976.
29. C.G. Malandrakis, "Optimal sensor and controller allocation for a class of distributed parameter systems", Int. J. Systems Sci., Vol. 10, No.11, pp. 1283-1299, 1979.

Observability

1. P.K.C. Wang, "Control of distributed Parameter Systems" in Advances in Control Systems, Vol. 1 (C. Leondes, Ed.), 1964.
2. Y. Sakawa, "Observability and related problems for partial differential equations of parabolic type". SIAM J. Control, Vol. 13, No. 1, pp. 14-27, Jan. 1975.
3. R.E. Goodson and R.E. Klein, "A definition and some results for distributed system observability", IEEE Trans. Autom. Control., Vol. AC-15, No.2, pp. 165-174, April 1970.
4. T.K. Yu and J.H. Seinfeld, "Observability of a class of distributed parameter systems", IEEE Trans. Aut. Control, Vol. AC-16, pp. 495-500, 1971.
5. T.K.Yu and J.H. Seinfeld, "Observability and optimal measurement location in linear distributed parameter systems", Int.J.Control, Vol. 18, No.4, pp. 785-789, 1973.
6. T. Kobayashi, "Discrete time observability for distributed parameter systems", Int. J. Control, Vol. 31, No. 1, pp. 181-193, 1980.

Filtering, Smoothing and Prediction

1. R.E. Kalman and R.S. Bucy, "New results in linear filtering and prediction Theory", J.Bas.Eng., Vol.83, pp. 95-108, 1961.
2. J. Meditch, "An optimal linear smoothing theory",Inform. Control, Vol. 10, pp. 598-605, 1967.
3. H. Cox, "On the estimation of state variables and parameters of noisy dynamic systems", IEEE Trans.Auto. Control, 9, 1, pp.5-12, 1964.
4. H.Kushner, "Nonlinear filtering: The exact dynamic equations satisfied by the conditional mode", IEEE Trans. Auto. Control, Vol. AC-12, pp. 262-276, 1967.
5. D.G. Lainiotis, "Optimal adaptive estimation Structure and Parameter Adaptation", IEEE Trans. Auto. Control, Vol. AC-16, No.2, pp. 160-170, April 1971.
6. D.G. Lainiotis, "Partitioning: A unifying framework for adaptive systems, I: Estimation", Proceedings of IEEE, Vol. 64, No.8, pp.1126-1143, August 1976.
7. J. Handschin, "Monte Carlo techniques for prediction and filtering of non-linear stochastic processes", Automatica, 6, pp. 555-563, 1970.
8. H.J. Kushner, "Filtering for linear distributed parameter systems", SIAM J. Control, Vol. 8, No.3, pp. 346-359, Aug. 1970.
9. Y.Sakawa, "Optimal filtering of distributed parameter systems", Int. J. Control, Vol. 16, No.1, pp. 115-127, 1972.

10. H. Nagamine, S. Omatu, and T. Soeda, "The optimal filtering problem for a discrete-time distributed parameter system", Int. J. Systems Sci., Vol. 10, pp. 735-749, 1979.

11. L. Padmanabhan and G. Calantuoni, "Sequential estimation in distributed systems", Int. J. Systems Science, Vol. 5, pp. 973-986, 1974.

12. K.E. Bencala and J.H. Seinfeld, "Distributed parameter filtering:Boundary noise and discrete observations, Int. J. Systems Sci., Vol. 10, pp. 493-512, 1979.

13. J.S. Meditch, "Least Squares estimation for distributed parameter systems", Automatica, Vol. 7, pp. 315-322, 1971.

14. R. Curtain, "A survey of infinite-dimensional filtering, SIAM Review,"Vol. 17, No.3, pp. 395-411, 1975.

15. J.H. Seinfeld, G.R. Gavalas and M. Hwang, "Nonlinear filtering in distributed parameter systems", Trans. of ASME, J.Dynamic Systems, Measurement, and Control, Vol. 93, pp. 157-163, 1971.

16. T.K. Yu, J.H. Seinfeld and W.H. Ray, "Filtering in nonlinear time delay systems", IEEE Trans. in Auto. Control, Vol. AC-19, No.4, pp. 327-335, Aug. 1974.

17. W.H. Ray and J.H. Seinfeld, "Filtering in distributed parameter systems with moving boundaries", Automatica, Vol. 11, pp. 509-515, 1975.

18. J.S. Meditch, "On state estimation for DPS J.of the Franklin Inst. Vol. 290, pp. 49-59, 1970.

19. S.G. Tzafestas, Bayesian approach to distributed parameter filtering and smoothing, Int.J.Control, Vol. 15, No.2, pp. 273-295, 1972.

20. S.G. Tzafestas, "On the distributed parameter least-square state estimation theory", Int. J. System Science, Vol. 4, pp. 833-858, 1973.

21. S.Omatu, T. Soeda and Y. Tomita, "Linear fixed-point smoothing by using functional Analysis, IEEE Trans. Auto. Control, Vol. AC-22, pp. 9-18, 1977.

22. S. Omatu, S. Ohe and T. Soeda, "Fixed interval smoothing for linear distributed parameter systems", Int. J. Systems, Sci., Vol. 10, pp. 1219-1233, 1979.

23. S.G. Tzafestas, "On optimum distributed parameter filtering and fixed-interval smoothing for colored noise", IEEE Trans. Auto.Control, Vol. AC-17, No.4, pp. 448-458, Aug. 1972.

24. K. Watanabe, "Optimal filtering and smoothing algorithms for linear distributed parameter systems with pointwise observations", Int. J. Systems Sci., Vol. 12, No. 3, pp. 325-349, 1981.

25. J.H. Seinfeld, G. Gavalas and M. Hwang, "Nonlinear filtering in distributed parameter systems", J.Dynamic Syst. Meas. Control, Vol. 93D, pp. 157-163, 1971.

26. M. Hwang, J.H. Seinfeld and G. Gavalas, "Optimal least-square filtering and interpolation in distributed parameter systems", J. Math. Anal. Appl., Vol. 39, pp. 49-74, 1972.

27. D.M. Detchmendy and R. Sridhar, "Sequential estimation of states and parameters in noisy nonlinear dynamical systems", J. Bas. Eng., Vol. 88D, pp. 362-366, 1966.

28. H. Kugiwada, et. al., "Invariant imbedding and sequential interpolating filters for nonlinear processes", J.Bas. Eng. Vol. 91D,pp. 195-200, 1969.

29. R. Sahgal and R.R. Webb, "Smoothing algorithms for nonlinear distributed parameter systems", Automatica, Vol. 12, pp. 239-244, 1976.

30. K. Watanabe, T. Yoshimura and T. Soeda, "Optimal nonlinear estimation for distributed parameter systems via the partition theorem", Int. J. Systems Sci., Vol. 11, No. 9, pp. 1113-1130, 1980.

31. S. Tzafestas and P. Stavroulakis, "Partitioning approach to adaptive distributed parameter filtering", Ricerche di Automatica, Vol. XI, No. 1, pp. 51-71, 1980.

32. S. Tzafestas and P. Stavroulakis, "Partitioned adaptive filtering and control of distributed systems with space-dependent unknown parameters", Mathematics and Computers in Simulation, IMACS, Vol. XXIII, pp. 206-212, 1981.

33. K. Watanabe, "An alternative approach to the derivation of distributed-type partitioned filters", Int. J. Syst. Science,Vol. 12, No.3, pp. 325-349, 1981.

34. D. Luenberger, An introduction to observers, IEEE Trans. on Control, Vol. AC-16, pp. 596-602, 1971.

35. Y.A. Liu and L. Lapidus," Observer theory of distributed parameter systems", Int. J. Systems Sci., Vol. 7, pp. 731-742, 1976.

36. J. Furusho, H. Kanoh and M. Masubuchi, "Observers for bilinear distributed-parameter processes", Proc. 2nd IFAC Symp. on Control of DPS , Pergamon Press, 1978.

37. P.Stavroulakis and S. Tzafestas, "Distributed-parameter observer-based control implementation using finite spatial measurements", Math. Computers in Simulation Proceedings of IMACS, Vol. XXII, pp. 373-379, 1980.

Identification

1. Special Issue on Identification and System Parameter Estimation, Automatica, Jan. 1981.

2. C.S. Kubrusly, "Distributed parameter systems identification: A Survey", Int. J. Control, Vol. 26, No.4, pp. 504-535, 1977.

3. Y. Sunahara, A. Ohsumi and M. Imamura, "A method of parameter identification for linear distributed parameter systems", Automatica, Vol. 12, pp. 245-256, 1976.

4. L. Carotenuto and G. Raiconi, "Identifiability and identification of a Galerkin approximation for a class of distributed parameter systems" , Int. J. Systems Sci., Vol. 11, No.3, pp. 1035-1049, 1980.

5. T. Kobayashi, "Initial state determination for distributed parameter systems" , SIAM J. Control and Optimization, Vol. 14, pp. 14, pp. 934-944, 1976.

6. T. Kobayashi, "A well-posed approximate method for initial state determination of discrete-time distributed parameter systems", SIAM J. Control and optimization, Vo. 15, pp. 947-958, 1977.

7. P. Caravani and M. Ciaffi, "Approximate State Recovery of a distributed parameter system", IEEE Trans. Auto. Control, Vol. AC-23, pp. 1067-1074, 1978.

8. T. Kobayashi, Determination of unknown function for a class of distributed parameter systems", SIAM J. Control and Optimization, Vol. 17, pp. 469-474, 1979.

9. G.A. Phillipson, "Identification of Distributed Systems", Elsevier, New York, 1971.

10. D.C. Saha and G.P. Rao, "Identification of distributed parameter system via multidimensional distributions, Proc. of IEE, Vo. 127, No. 2, pp. 45-50, March 1980.

11. S.G. Tzafestas, "Parameter estimation in distributed-parameter dynamic models", I. Chem. E. Symp. Series No.35, 5:43-5:50, London, 1972.

12. R.E. Goodson and M.P. Polis, "A Survey of parameter identification in distributed systems", Proc. IFAC 1975 World Congress, Boston Cambridge.

13. R.E. Goodson and M.P. Polis, "Parameter identification in distributed systems: A Synthesizing overview", ASME Monograph, 1974.

Sensitivity

1. W.A. Porter, "Sensitivity problems in distributive systems, Int. J. Control, Vol. 5, pp. 393-412,1967.

2. W.A. Porter, "Parameter sensitivity in distributed feedback systems, Int. J. Control, Vol. 5, pp. 413--423, 1967.

3. K.C. Pedersen and L.R. Nardizzi, "Optimally sensitive control for distributed parameter systems", Int. J. Control, Vol. 116, No.4, pp. 723-735, 1972.

4. J.B. Cruz, Jr. and W.R. Perkins, "A new approach to the sensitivity problem in multivariable feedback system design", IEEE Trans. Auto.Control, Vol. AC-9, pp. 216-223, July 1964.

5. J.M. Davis and W.R. Perkins. "Comparison sensitivity of distributed parameter systems", IEEE Trans. on Auto. Control, AC-17, pp.100-105, Feb. 1972.

6. M.C.Y. Kuo and M.N.B. Ayiku, "Synthesis of low sensitivity optimal control in distributed parameter systems", Proceedings of IEE, Vol.125, No.6, pp. 550-554, 1978.

7. Y. Nakamiri, S. Miyamoto, S. Ikeda and Y. Sawaragi, "Measurement Optimization with sensitivity criteria for distributed parameter systems", IEEE Trans. on Auto. Control,Vol. AC-25, No.5, pp. 889-901, Oct. 1980.

Computational Techniques

1. S. Tzafestas, "Distributed-Parameter Optimal Control via Mathematical Programming", Journal of the Franklin Institute , Vol. 309, No.6, pp. 399-438, June 1980.

2. P. Wang and F. Tung, "Optimum control of distributed parameter systems, ASME Trans. J. Bas. Eng., Vol.86D, pp. 67-79, 1964.

3. S. Tzafestas and J. Nightingale, "Differential -dynamic programming approach to optimal nonlinear systems, Proc.IEE,Vol. 116, pp. 1078-1084, 1969.

4. P.A. Orner, P.F. Salamon, and W.Yu, "Least squares simulation of distributed systems", IEEE Transactions on Auto. Control, Vol. AC-20, No.1, pp. 75-83, Feb. 1979.

5. M.A. Sheirah and M.H. Hamza, "Optimal Control of Distributed Parameter Systems", Int. J. Control, Vol. 19, pp. 891-902, 1974.

6. H. Sasai, "A note on the Penalty Method for Distributed Parameter Optimal Control Problem", SIAM J. Control, Vol. 10, No.4, pp. 730-736, Nov. 1972.

7. D.E. Cornick and A.N. Michel, "Numerical optimization of Distributed Parameter Systems by the Conjugate Gradient Method", IEEE Transactions on Auto. Control, AC-17, pp. 358-362, June 1972.

8. D.E. Cornick and A.N. Michel, "Mumerical Optimization of Linear Distributed Parameter Systems", Journal of Optimization Theory and Applications, Vol. 14, pp. 73-98, 1974.

9. D.J. Ball and J.R. Hewit, "An optimal Control Problem for a class of Distributed Parameter Systems, Automatica, Vol. 9, pp. 263-267, 1973.

10. Y. Yavin and Y. Rasis, "The bounded energy optimal control for a class of heat conduction systems", Int. J. Control, Vol. 11, pp. 153-164, 1970.

11. R.N.P. Singh and V.S. Rajamani, "Minimal-time control of distributed-parameter systems with multiple-norm constraints on the control function", Int. J. Control, Vol. 15, pp. 241-254, 1972.

12. G. Zone and K.S. Chang, "A successive Approximation Method for Nonlinear Distributed Parameter Control Systems, Int. J. Control, Vol. 15, pp. 255-272, 1972.

13. S.K. Mitter, "Successive approximation methods for the solution of optimal control problems", Automatica, Vol. 3, pp. 133-140, 1966.

14. S.G. Tzafestas, "Design of DP optimal Controllers and Filters via Walsh-Galerkin expansion", Proc. IFAC Symp. on Control of DPS, Warwick Univ. June 1977.

15. J.H. Holliday and C. Storey, "Numerical Solution of Certain Nonlinear Distributed Parameter Opti-

mal Control Problems, Int. J. Control, Vol. 18, pp. 812-825.

16. S.G. Tzafestas, "Distributed Parameter Control in Function Space", Journal of the Franklin Institute, Vol. 295, pp. 317-342, 1973
17. G. Kranc and P.E. Sarachik, "An application of functional analysis to the optimal control problem", ASME Trans. J. Basic Eng., Vol. 85, pp. 143-150, 1963.
18. A. Balakrishnan, "An operator theoretic formulation of a class of control problems and a steepest descent method of solution, SIAM J. Control, Vo. 1, pp. 109-127, 1963.
19. H. Sasai and E. Shimemura, "On the convergence of approximating solutions for linear distributed parameter control problems", SIAM J. Control, Vol. 9, pp. 263-273, 1977.
20. G. Knowles, "Some Problems in the Control of Distributed Systems and their Numerical Solution", SIAM J. Control and Optimization, Vol. 17, No.1, pp. 5-22, 1977.
21. S.G. Tzafestas, "Final-value Control of Nonlinear Composite Distributed and Lumped Parameter Systems", J. Franklin Institute, Vol. 290, No.5, pp. 439-450, Nov. 1970.
22. M. Friedman and Y. Yavin, "Computation of Optimal Controls for two classes of Nonlinear Distributed Parameter Systems", Int.J.Control, Vol. 18, No.4, pp. 705-712, 1973.
23. S.E. Aidarous and M.A.R. Ghonaimy, "A direct method for optimization of stochastic distributed system, Automatica, Vol. 11, pp. 203-207, 1977.
24. S.E. Aidarous, M.R. Gevers and M.J. Installé, "Optimal Sensors Allocation Strategies for a class of Stochastic Distributed System", Int. J. Control, Vol. 22, No.2, pp. 197-213, 1975.
25. W.H. Chen and J.H. Seinfeld, "Optimal Location of Process Measurement", Int. J. Control, Vol. 21, pp. 1003-1008, 1975.
26. J.R. Canon and R.E. Klein, "Optimal Selection of Measurement Location in a Conductor for Approximate Determination of Temperature Distribution", Proc. Joint. Aut. Contr., Conf., 1970.
27. S. Omatu, S. Koide, and T. Soeda, "Optimal Sensor Location Problems for a Linear Distributed Parameter System", IEEE Transactions on Auto. Control, Vol. AC-23, No.4, pp. 665-673, Aug. 1978.
28. S. Kumar and J.H. Seinfeld, "Optimal Location of Measurement for Distributed Parameter Estimation", IEEE Trans. on Auto. Control, Vol. AC-23, No.4, pp. 690-698, Aug. 1978.

Simulation

1. L.Fox, "Numerical solution of ordinary and partial differential equations", Pergamon Press, Oxford, 1962.
2. R. Vichnevetsky, "Generalized finite difference approximation for the parallel solution of Initial-Value Problems", Simulation, Vol. 13, pp. 233-237, 1969.
3. W. Karplus, "Software for distributed system simulation", Proc. IASTED Simulation Course Interlaken, Switzerland, June 23-24, (see F. Cellier, Ed., Progreess in Modeling and Simulation, Academic Press, 1982).
4. R. Vichnevetsky, "A new stable computing method for the serial hybrid computer integration of PDEs", Proc. 1968 SJCC, Vol. 32, pp. 143-150, 1968.
5. R. Vichnevetsky, "Use of functional approximation methods in the computer sqlution of initial-value PDE problems", IEEE Trans.Comput., Vol. C-18, pp. 499-512, 1969.
6. R. Vichnevetsky, "Physical criteria in computer methods for PDEs, AICA Annales, pp. 3-16, 1974.
7. R. Vichnevetsky, "Propagation characteristics of semidiscretization of hyperbolic equations, Mathematics and Computers in Simulation, Vol. XXII, pp. 98-105 , 1980.
8. S. Tzafestas, "Hybrid simulation of distributed-parameter systems: The-state-of-the-art , in F. Cellier, Ed., Progress in Modeling and Simulation, Academic Press, 1982.
9. S. Tzafestas, "Monte Carlo simulation and filtering in DPS", in Simulation of Distributed-Parameter and Large-Scale Systems, (S.Tzafestas, Ed.) pp. 13-31, North-Holland, 1980.
10. L. Sagerlind, "Applied finite element analysis", J. Wiley, 1976.
11. R.Cavin, III, and S. Tandon, "Distributed parameter optimal control design via finite elements", Automatica, Vol. 13, p. 611, 1977.
12. L. Carotenuto and G. Raiconi, "The finite element method in distributed parameter control systems", in Distributed-Parameter Control Systems:Theory and Application, Pergamon Press, Oxford, 1982.

Applications

1. W.H. Ray, "Sóme recent applications of distributed parameter systems theory-A survey", Automatica, Vol. 14, pp. 281-287, 1978.
2. G.K. Lausterer and W.H. Ray, "Distributed parameter state estimation and optimal feedback control-An experimental study in two space dimensions", IEEE Trans. Auto. Control, Vol. AC-24, No.2, pp.

179-190, April. 1979.

3. A.A. Desalu, L.A. Gould and F.C. Schweppe, "Dynamic estimation of air polution" , IEEE Trans. on Auto. Control, Vol. AC-19, No.6, pp.904-910, Dec. 1974.
4. S.G. Tzafestas, (Ed.), "Distributed-parameter control systems: Theory and application", Pergamon Press, Oxford, 1982.
5. S. Banks and A. Pritchard (Eds.), "Control of distributed-parameter systems", (Proc. IFAC. Symp.), Pergamon Press, Oxford, 1978.
6. S.G. Tzafestas, "Simulation of distributed- parameter and large-scale systems" (Proc. IMACS Symp.), North-Holland, Amsterdam, 1980.
7. W.H. Ray and D.G. Lainiotis (Eds.), "Distributed-parameter systems" , Marcel Dekker, New York, 1978.
8. S.G. Tzafestas (Ed.), "Special Issue on distributed-Parameter systems, J.Franklin Inst., Vol. 15, No. 5/6 , 1983.
9. S.G. Tzafestas, J.H.Seinfeld and C.Saguez (Eds.), Special Issue on distributed-parameter systems, Large Scale Systems, Vol. 6, No.3 1984.

References on Hierarchical Control Systems

1. Sandell, Jr. N.R., P. Varaiya, M. Athans and M.G. Safanov-Survey of Decentralised Control Methods for Large Scale Systems, IEEE Trans. on Aut. Control, vol. AC-23, no.2, April 1978, pp. 108-128.
2. Findeisen, W. - Decentralised and Hierarchical Control (Decision Making) Under Consistency or Disagreement of Interests, Report, Institute of Automatic Control, Techn. Univ. of Warsaw, 1980. Also presented as a plenary paper at the Second IFAC-LSSTA Symposium, Toulouse, France, 1980.
3. Mesarovic, M.D., D. Macko and Y. Takahara - Theory of Hierarchical,Multilevel Systems, Academic Press, New York, 1970.
4. Lasdon,L.S. - Optimisation Theory for Large Scale Systems, MacMillan, London, 1970.
5. Singh, M.G. - Dynamical Hierarchical Control, North Holland, Amsterdam, 1980 (second edition).
6. Malinowski, K., and F.N. Bailey - Price Coordination with Communication Constraints, Proc. CSC 1979,Fort Lauderdale, Florida, USA, 1979.
7. Lefkowitz, I. - Multilevel approach applied to control system design, Trans. ASME, Series B, J. Bas. Eng. 88 (2), 1966, pp. 392-398.
8. Lefkowitz, I. - System control of Chemical and related process systems,Proc. 6th IFAC World Congress, Pt.IID, Boston, Massachusetts, USA, 1975.
9. Donoughe, J.F., and I. Lefkowitz-Economic tradeoffs associated with a multilayer control strategy for a static systems, IEEE Trans. on Autom. Control, vol. AC-17, no. 1, February 1972, pp. 7-15.
10. Findeisen, W.-Wielopozromoue ulidady sterovania Multilevel Control Systems , PWN, Warsaw, 1974, (German translation Hierarchische Steuerungssysteme.Verlag Technik, Berlin, 1977)
11. Findeisen, W., M. Bryds, K. Malinowski, P. Tatjewski and A. Wozinak- On-line hierarchical control for steady-state systems, IEEE Trans. on Autom. Control, vol. AC-23, no. 2, April, 1978, pp. 189-209.
12. Findeisen, W., and K. Malinowski - A structure for on-line dynamic coordination, Proceedings, First IFAC Symposium on Large-Scale Systems Theory and Applications, Udine, Italy, 1976.
13. Findeisen, W., and K.Malinowski - Two-Level Control and Coordination for Dynamical Systems, Arch. Autom i Telemech, vol. XXIV, No. 1, 1979.
14. Bailey, F.N., and K. Malinowski - Problems in the Design of Multilayer, Multiechelon Control Structures, Proceedings, IFAC MUTS Symposium Fredericton, N.B., Canada, 1977.
15. Hassan, M.F., and M.G. Singh - The Optimisation of Non-Linear Systems Using a New Two-Level Method, Automatica, 12, 1976, pp. 359-363.
16. Hassan, M.F., R. Hurteau, M.G. Singh and A. Titli- A Three-Level Costate Prediction Method for Continuous Dynamical Systems, Proceedings, IFAC MUTS Symposium, Fredericton, N.B., Canada, 1977.
17. Hassan, M.F., and M.G. Singh - Hierarchical Successive Approximation Algorithms for Non-Linear Systems, Part I and Part II, Large Scale Systems,vol. 2, Issue 2, 1981.
18. Findeisen, W. - Parametric Optimisation by Primal Method in Multilevel Systems, IEEE Trans. on Syst. Science and Cyber., vol. SSC-4, 1968, pp. 155-164.
19. Findeisen, W., F.N. Bailey, M. Bryds, K. Malinowski, P. Tatjewski and A. Wozniak- Control and Coordination in Hierarchical Systems,Wiley (International Series on Applied Systems Analysis), 1980.
20. Grateloup, G., and A. Titli - Methode de decomposition mixte et convergence d'un algorithme coordinateur de type gradient pour systemes de commande a deux niveaux, Bull. Acad. Pol. Sci.,Ser. Tech.Sci., 19, no. 5, 1971.
21. Grateloup, G., and A. Titli-- Two-level dynamic optimisation methods, J. Optimis. Theory Appl.,vol. 15, no. 3, 1975.
22. Dantzig, G.B. - Linear control processes and mathematical programming,SIAM J. Contr.,vol. 4, no. 1, 1960.
23. Geoffrion, A.M., Primal resource-directives approaches for optimising nonlinear decomposable systems,

Oper. Res., vol. 18, no. 3, 1970.

24. Geoffrion, A.M. - Elements of large-scale mathematical programming: Part I, concepts, Part II, synthesis of algorithms and bibliography, Management Sci., vol. 16, 1970.

25. Malinowski, K. - Properties of two balance methods of coordination, Bull. Pol. Acad. Sci. Ser. Tech. Sci., vol. 23, no. 9, 1975.

26. Tamura, H. - Application of duality and decomposition in high order multi-stage decision processes, Cambridge Univ. Rep. no. CUEB/B-Controll/TR49.

27. Tamura, H. - Decentralised optimisation for distributed-lag models of discrete systems, Automatica, vol. 11, 1975, pp. 593-602.

28. Tamura, H. - Discrete minimum principle and multistage decomposition algorithms for optimising discrete-time dynamic systems with distributed lags, in Handbook of Large Scale Systems Engineering Applications (M.G. Singh and A. Titli,eds.), North Holland,Amsterdam, 1979, pp. 81-95.

29. Takahara, Y., and M.D. Mesarovic - Coordinability of dynamic systems, IEEE Trans. on Autom. Control, vol.. AC-14, no. 6, December 1969, pp. 688-698.

30. Hurwicz, L. - Optimality and informational efficiency in resource allocation processes, in Mathematical Methods in the Social Sciences (K.Arrow, S. Karlin and P. Suppes, eds.), Stanford University, 1960, pp. 27-46.

31. Bailey, F.N., and A.J. Laud - An iterative procedure of decentralised estimation and control, in Proc. of the Workshop Discussion on Multilevel Control (W. Findeisen, ed.), Inst. of Automat. Control, Technical University of Warsaw, Warsaw, Poland, 1975.

32. Benveniste, A., P. Bernhard and A. Cohen - On the decomposition of stochastic control problems, INRIA, France, Rapport de Recherche, no. 187.1976.

33. Singh, M.G., S.A.W. Drew and J.F. Coales - Comparisons of practical hierarchical control methods for interconnected dynamical systems, Automatica, vol. 11, 1975.

34. Wismer, D.A. (editor) - Optimisation Methods for Large Scale Systems with Applications, McGraw-Hill, New York, 1971.

35. Findeisen, W. - A survey of problems in hierarchical control, in Proc. Workshop Discussion on Multilevel Control, Inst. of Automat.Control, Technical University of Warsaw, Warsaw, Poland,1975.

36. Singh, M.G., and A. Titli - Systems: Decomposition, Control and Optimisation, Pergamon Press, 1978.

37. Mahmoud, M.S. - Multilevel Systems Control and Applications: A Survey, IEEE Trans. Syst. Man and Cybern., vol. 7, 1977, pp. 125-143.

38. Calvet, J.L., and A. Titli - Hierarchical Optimisation and Control of Large Scale Systems with Dynamical Interconnection System, Proc. 2nd. IFAC Symposium on Large Scale Systems Theory and Applications, Toulouse, France, 1980, pp. 117-126.

39. Findeisen, W. - Control and coordination in multilevel systems, Proc. 2nd. Polish-Italian Conf. on Applications of Syst. Theory, Pugnochiuso, Italy, 1974.

40. Bryds, M. - Hierarchical control of steady-state, in Proc. Second Workshop on Hierarchical Control, Inst. of Automat. Control. Technical University of Warsaw, 1978, pp. 19-68.

41. Tatjewski, P.-Multilevel optimisation methods, in Proc. Second Workshop on Hierarchical Control, Inst. of Automat. Control, Technical University of Warsaw, 1978, pp. 241-266.

42. Tatjewski, P., and M. Cygler - Completely decentralised output control based on an approximate mathematical model,Large Scale Systems, vol. 2, no.4, 1981, pp. 243-255.

43. Malinowski, K. and A. Rusiaynski - Application of interaction balance method to real process coordination, Control and Cybernetics, vol. 4, no. 2, 1975.

44. Roberts, P.D. - Multilevel approaches to the combined problem of system optimisation and parameter identification, Int. J. Systems Sci., vol. 3, no.3, 1977, pp. 273-299.

45. Ellis, J., and P.D. Roberts - Simple models for integrated optimisation and parameter estimation,Int. J. Systems Sci., vol. 12, no.4, 1981, pp. 455-472.

46. Chong, C., and M. Athans - On the periodic coordination of linear stochastic systems, Proc., 6th IFAC Congress, Part IVA, Boston, Massachusetts, USA, 1975.

47. Arafeh, S., and A.P. Sage- Multilevel discrete time system identification in large scale systems, Int. J. Systems Sci., vol. 5, 1974, pp. 753-791.

48. Arafeh, S., and A.P. Sage - Hierarchical system identification of states and parameters in interconnected power systems, Int. J.Systems Sci., vol. 5, 1974, pp. 817-846.

49. Hassan, M., G. Salut, M.G. Singh and A. Titli - A decentralised computational algorithm for the global Kalman Filter, IEEE Trans.on Automat. Control, vol. AC-23, no.2, 1978, pp. 262-268.

50. Hassan, M., M.S. Mahmoud and M.G. Singh - A multiple projection technique for parameter estimation in large scale systems, Proc.2nd IFAC LSSTA Symposium, Toulouse, France, 1980.

51. Bryds, M., Machalak P. and B. Ulanicki - Optimising Control of Time Varying Systems by price Mechanism with Local Feedback, in A. Titli and M.G.

Singh (editors) Large Scale Systems - Theory and Applications , Pergamon Press, 1980.

52. Bryds, M., W. Findeisen and P. Tatjewski - Hierarchical control for systems operating in the steady state, Large Scale Systems, Vol. 1, 3, 1980, pp. 193-214.

53. Cohen, G., and G. Joalland - Coordination methods by the prediction principle in large dynamic constrained optimisation problems, Proc. IFAC LSSTA Symposium, Udine, Italy, 1976.

54. Cohen, G. - Optimisation by decomposition and coordination: A unified approach, IEEE Trans. on AC, vol. AC-23, no.2, 1978, pp. 108-129.

55. Mesarovic, M.D., J.D. Pearson, D. Macko and Y. Takahara - On the synthesis of dynamic multilevel systems, Proc. 3rd IFAC Congress, London, U.K., 1966, pp.408.1 -408.9.

56. Wozniack, A. - Parametric method of coordination using feedback from the real process, Proc. IFAC LSSTA Symposium, Udine, Italy, 1976.

57. Singh, M.G. and A. Titli (editors) - Handbook of Large Scale Systems Engineering Applications, North Holland, Amsterdam, 1979.

58. Singh, M.G. - Multilevel state estimation, Int. J. Systems Sci., vol. 6, 1975, pp. 533-555.

59. Bailey, F.N., and A.J. Laub - An iterative procedure of decentralised estimation and control, Proc. Workshop Discussion on Hierarchical Control, (W.Findeisen, ed.), Inst. of Automat. Control, Technical University of Warsaw, Warsaw, Poland, 1975.

60. Malinowski, K. - Applicability of Lagrange Multiplier Method to Optimisation of Quadratic Systems, IEEE Trans. on Automat. Control, vol. AC-22, no.3, 1977.

61. Haimes, Y.Y. - Hierarchical Analyses of Water Resources System: Modelling and Optimisation of Large Scale Systems, McGraw-Hill, New York, 1977.

62. Singh, M.G. and A. Titli (Editors) - Control and Management of Integrated Industrial Complexes, Proc.IFAC Workshop, Toulouse, France, 1977.

63. Irving, M.R., H. Nicholson and M.J.H. Sterling - Hierarchical control of electric power systems, in Handbook of Large Scale Systems Engineering Applications (M.G. Singh and A. Titli, eds.), North Holland, Amsterdam, 1975, pp. 496-515.

64. Findeisen, W., and K. Malinowski - Hierarchical approach to realtime control of multireservoir systems,Int. Symposium on Real-Time Operation of Hydrosystems, Waterloo, Canada, 1981.

65. Bernussou, J., and A. Titli - Interconnected dynamical systems:Stability, decomposition and decentralisation, North Holland, 1982.

66. Basar, T., and G. Olsder - Dynamic non-cooperative game theory, Academic Press, London, 1982.

67. Titli, A. - Structures de commande hierarchisees en vue de l'optimisation des processes complexes, These d'état, Toulouse, 1972. Published by Dunod, Paris,1975

68. Malinowski, K., and F.N. Bailey - Coordination with Information Constraints, Part I and Part II, in preparation, Warsaw Tech. University.

69. Titli,A., M.G. Singh and M.F. Hassan - Hierarchical optimisation of dynamical systems using multiprocessors, Journal of Computers and Electrical Engg.,1978.

70. Mahmoud, M.S., M.F. Hassan and M.G. Singh - A new hierarchical approach to the joint problem of identification and optimisation, Large Scale Systems 1, 2, 1980, pp. 159-166.

71. Li, R.,and M.G. Singh - An on -line hierarchical approach to the joint problem of identification and optimisation, Large Scale Systems, 2,2, 1981, pp. 205-216.

72. Hassan, M.F., A. Titli, M.G. Singh and R. Hurteau-Stability, stabilisation and performance of multilevel controllers subject to structural perturbations, Proc. IEE, 127, 5, 1980, pp. 207-213.

73. Hassan,M.F., and M.G. Singh - Stability, stabilisation and performance of large scale systems subject to structural perturbations - Part II, Proc. IEE, 127, 5, 1980, pp. 214-219.

74. Titli A. and M.G. Singh (editors) - Large Scale Systems:Theory and Applications, Pergamon Press, 1980.

75. Stephanopoulos, G., and A.W. Westerberg - The use of Hestenes' Method of Multipliers to Resolve Dual Gaps in Engineering System Optimisation, Journal of Optimisation Theory and Applications, 15, 1975, pp. 285-309.

76. Singh, M.G. and M.F. Hassan - Hierarchical optimisation of non-linear dynamical systems with non-separable cost functions, Automatica, 14, 1978.

77. Titli, A., and Singh M.G. - Commande hierarchises des systems lineaires, and - commande decentralisee des systems non-lineaires, in A. Titli et al "Analyse et Commande des Systems Complexes", Collins, France, 1978.

78. Singh, M.G., and A Titli - Practical hierarchical optimisation and control algorithms, Invited paper for the 7th IFAC World Congress, Helsinki, 1978.

79. Hassan, M.F., R. Hurteau, M.G. Singh and A. Titli-Stochastic optimal control for a large scale river system, Proc. 7th IFAC World Congress, Helsinki, 1978.

80. Singh, M.G. and M.F. Hassan - A two level prediction algorithm for non-linear systems, Automatica, 13, 1, January, 1977.

81. Hassan, M.F., and M.G. Singh - A two level costate prediction algorithm for non-linear systems, Automatica, 13, November, 1977.

82. Singh, M.G., M.F. Hassan and A. Titli - A feedback solution for large scale interconnected dynamical systems using the prediction principle, IEEE Trans. SMC, 6, 1976, pp. 233-239.

83. Chemouil, P., M.R. Katebi, D.Sastry and M.G. Singh - Parameter Estimation in large scale systems using the Maximum A Posteriori Approach, CSC Report 484, 1980. To appear in Automatica, 1981. Also Proc. 8th IFAC World Congress, Kyoto, 1981.

84. Mahmoud, M.S. and M.G. Singh - Large Scale Systems Modelling, Pergamon Press, Oxford, 1981.

85. Hassan, M.F. and M.G. Singh - The control of a synchronous machine using a hierarchical model follower, Automatica, March, 1977.

86. Mahmoud, M.S. and M.G. Singh - Decentralised state reconstruction of interconnected discrete systems, CSC Report No. 488, 1980. Large Scale Systems, 2, 1981.

87. Blandin, P. - Commande hierarchise e d'un complexe de production de soufre. These de doctorat de specialite . V.P.S. Toulouse, 1976.

88. Payne, H.J. - Models of Freeway Traffic and Control. Simulation Council Proc. 1, 51-61, 1971.

89. Cremer M., Papageorgiou, M. - Parameter Identification for a Traffic Flow Model, Automatica, 17, 837-843, 1981.

90. Papageorgiou, M. - Applications of Automatic Control concepts to Traffic Flow Modelling and Control. Springer-Verlag, Berlin, Heidelberg, New York, Tokyo, 1983.

91. Cremer, M., Papageorgiou, M., Schmidt, G. - Use of Technical Control Equipment for the Improvement of Traffic Operations on Highspeed Highways. Forschung Straßenbau and Straßenverkehrtstechnik, 307 (in German), 1-44, 1980.

92. Van Maarseveen, M.F.A.M. - Application of Martingales in Stochastic Systems Theory - Surveillance and Control of Freeway Traffic Flow. Dissertation, Technische Hogeschool Twente, Netherlands, 1982.

93. Nahi, N.E. - Freeway Traffic Data Processing. Proc. IEEE 61, 537-541, 1973.

94. Ghosh, D., Knapp, C.H. - Estimation of Traffic Variables Using a Linear Model of Traffic Flow. Transportation Research, 12, 395-402, 1978.

95. Kurkjian, A., Gershwin, S.B., Houpt, P.K., Willsky, A.S., Chow, E.Y. - Estimation of Roadway Traffic Density on Freeways Using Presence Detector Data, Transportation Science, 14, 232-261, 1980.

96. Payne, H.J. - Calibration of Freeway Incident Detection Algorithms. Proc. 3rd IFAC/IFIP/IFORS Symp. on Control in Transportation Systems, Columbus, Ohio, 377-378, 1976.

97. Levin, M., Krause, G.M. - A Probabilistic Approach to Incident Detection on Urban Freeways, Traffic Enging. and control, 19, 107-109, 1979.

98. Willsky, A.S., Chow, E.Y., Gershwin, S.B., Greene, C.S., Houpt, P.K., Kurkjian, A.L. - Dynamic Model-Based Techniques for the Detection of Incidents on Freeways. IEEE Trans. on Automatic Control, AC-25, 347-359, 1980.

99. Cremer, M. - Incident Detection on Freeways by Filtering Techniques, Prepr. 8th IFAC World Congress, Kyoto, Japan, XVII 96-101, 1981.

100. Findeisen, W., Lefkowitz, I. - Design and Applications of Multi-layer Control, 4th IFAC World Congress, Warsaw, Poland, 3-22, 1969.

101. Lefkowitz, C. - Vertical decomposition and multilayer control, Encyclopedia on Systems and Control, Pergammon Press (to appear 1985).

102. Papageorgiou, M. - Multilayer Control System Design Applied to Freeway Traffic, IEEE Trans. on Automatic Control, AC-28, (to appear 1983).

103. Papageorgiou, M., Posch, B., Schmidt, G. - Comparison of Macroscopic Models for Control of Freeway Traffic, Transportation Research, 17B, 107-116, 1983.

References on Decentralized Control Systems

1. Singh, M. "Decentralised control", North Holland, Amsterdam, 1981.

2. Witsenbausen, H. "A counter-example in stochastic optimal control", SIAM J. Control, 6, 1, 131-147, 1968.

3. Ho, Y.C. and Chu, K.C. "In formation structures in dynamic multiperson control problems", Automatica 10, 341-351, 1974.

4. Ho, Y.C. "Team decision theory and information structures", Proc. IEEE, June 1980.

5. Ho, Y.C. and Chu, K. "Team decision theory and information structures in optimal control, problems - Part I", IEEE Trans. AC-17, 15-22, 1972.

6. Chu, K.C. "Team decision theory and information structures in dynamic multiperson control problems", IEEE Trans. AC-17, 15-22, 1972.

7. Kurtaran, B. and Sivan, R. "L-Q-G control with one-step-delay sharing pattern", IEEE Trans. AC-571-574, 1974.

8. Kwakernaak, H. and Sivan, R."Linear optimal control systems", John Wiley, New York, 1972.

9. Kailath, T. "An innovations approach to least squares estimation - Part 1: linear filtering with additive white noise", IEEE Trans. AC-13, 646-655, 1968.

10. Yoshikawa, T. and Hirodki, K. "Separation of estimation and control for decentralised stochastic control systems", Automatica 14,623-628, 1978.

11. Streibel, C. "Sufficient statistics in the optimum control of stochastic systems", J. Math. Analysis Applic. 12, 576-592, 1965.

12. Chong, C. and Athans, M. "On the stochastic control of linear systems with different information sets", IEEE Trans. AC-16, 5, 423-430, 1971.

13. Singh, M.G. "Dynamical hierarchical control", North Holland, 1977.

14. Singh, M. and A. Titli "Systems : decomposition, optimisation and control", Pergamon Press, Oxford, 1978.

15. Chong, C. and Athans, M. "On the periodic coordination of linear stochastic systems", Automatica 12, 321-335, 1978.

16. Wang, S.H., Dawson, E.J. "On the stabilisation of decentralised control systems. IEEE AC-18, 473, 1973.

17. Corfmat, J.P., Morse, A.S. "Decentralised control of linear multivariable systems. Automatica 12, 479, 1976.

18. Sezer, N.E., Siljak, D.O. "Structurally fixed modes", Systems and Control Letters, Vol. 1, No. 1, p. 60, July 1981.

19. Lin, C.T. "Structural controllability", IEEE AC-19, 201, 1974.

20. Dawson, E.J. "Decentralised stabilisation and regulation in multivariable systems", in Directions in decentralised control, many person optimization and large scale systems (Editors, Y.C. Ho, S. Mitter), Plenum Press, pp. 303-323.

21. Anderson, B.D.O., Clements, D.J. "Algebraic characterisation of fixed modes in decentralised control", Automatica, 17, 703, 1981.

22. Willems, J.L. "Decentralised stabilisation", Paper proposed for publication in the Encyclopedia of Systems and Control, Pergamon Press. To appear.

23. Dawson, E.J., Ozguner, V. "Characterisation of decentralised fixed modes for interconnected systems", Session 41, Decentralised Control, 8th Triennal IFAC World Congress, Kyoto, 1981.

24. Seraji, H. "On fixed modes in decentralised control systems. SIAM Journal of Control, 1984, Vol. 35, No. 5, 775-784.

25. Lin, C.T. "Structural controllability", IEEE AC-19, 1974, 201-208.

26. Armentano, V.A., Singh, M.G. "A procedure to eliminate decentralised fixed modes with reduced information exchange", IEEE AC 1982, AC-27, No.1.

27. Wang, S.H. "Stabilisation of decentralised control systems via time varying controllers", IEEE AC 1982, vol. 27, No.3.

28. Anderson, B.D.O., Morse, J.B. "Time varying feedback laws for decentralised control" , IEEE AC 1981, vol. 26, No.5.

29. Purviance, J.E., Tylee, J.L. "Scalar sinusoidal feedback laws in decentralised control", IEEE, CDC Conference, Orlando USA, Dec. 1982.

30. Bernussou, J., Titli, A. "Interconnected dynamical systems: stability, decomposition and decentralisation", North Holland, Amsterdam, 1982.

31. Elgerd, O.I. "Electric energy systems theory: an introduction", McGraw Hill Company, THM Edition, 1973.

32. Rosen, J.B. "The gradient projection method for nonlinear programming- Part I - Linear Constraints", SIAM 8(1), 1960.

33. Armentano, V.A. and Singh, M.G. "A new approach to the decentralised controller initialisation problem", 8th Triennal IFAC World Congress, Kyoto, 24-28 August, 1981.

34. Geromel, J.C. and Bernussou, J. "An algorith for optimal decentralised regulation of linear quadratic interconnected systems", Automatica, vol.15, 489-491, 1979.

35. Levine, W.S. and Athans, M. "On the determination of the optimal output feedback gains for linear multivariable systems", Automatic Control, vol. AC-15, No.1, 1970.

36. Pietras, J.V. and Looze, D.P. "Decomposition of. decentralised gain computation for interconnected systems", JACC Conference, San Francisco, August 1980.

37. Hassa M., Singh, M. and Titli, A. "A near optimal decentralised controller with a pre-specified degree of stability", Automatica, July 1979.

38. Kleinrock, L. " Queueing Systems", vol. 1: Theory, John Wiley, 1975.

39. Garcia, J.M. "Problemes lies a la modelisation du trafic et a l' acheminement des appels dans un reseau telephonique" , These de Docteur-Ingenieur, Universite Paul Sabatier, 733, Toulouse, 1980.

40. Le Gall, F. and Bernussou, J. "Reseaux de Telecommunications: leur modelisation", Note Interne ASCI, No. 80.I.22, LAAS Toulouse, 1981.

41. Forestier, J.P. and Varaiya, P. "Multilayer control of large Markov chains" , IEEE Trans. Aut. Control, AC-23, 2, 1978.

42. Garcia, J.M., Copinah, B. and Varaiya, P. "Routing in telephone networks", Conference Decision and Control, San Diego, Dec., 1981.

43. Narendra, K.S., Wright, E.A. and Mason, L.G. "Application of learning automata to telephone network traffic routing problems", IEEE Trans. System Man and Cybernetics, SMC-7, 1977.

44. Srikantakumas, P.R. and Narendra, I.S. "A learning model for routing in telephone networks", SIAM Journal of Control and Optimization, vol. 20, 1, 1982.

45. Bonatti, I.S. "Gestion de reseaux de service: appli-

cation au reseau telephonique interurbain", These de Docteur-Ingenieur, Universite Paul Sabatier, No. 767, Toulouse, 1981.

46. Hassan, M. and Singh, M. "A decentralised controller with trajectory improvement", Proc. IEE, May 1980.

47. Chen, Yu Li and Singh, M. "Certain practical considerations in the model following method of decentralised control", Proc. IEE, 128, 4, July 1981.

48. Bardiaud, Ingenieur thesis, Nantes, 1978.

49. Markland, C.A. "Optimal model-following control-system synthesis techniques", Proc. IEE, 1970, 117 (3), pp. 623-627.

50. Feingold, D.G. and Varga, R.S. "Block diagonally dominant matrices and generalisation of the Gershgorin circle theorem", Pacific Journal Mathematics, 12,pp. 1241-1250, 1962.

51. Wang, S.H. "An example in decentralised control systems", IEEE Trans. on Automatic Control, vol. AC-23, p. 938. October 1978.

52. Luenberger, D.G. "Observing the state of a linear system", IEEE Trans. on Military Electronics, vol.MIL-8 (1964), 74-80.

53. Luenberger, D.G. "An Introduction to Observers", IEEE Trans, on Automatic Control, vol. AC-16 (1971), 596-602.

54. Weihrich, G. "Drehzahlregelung von Gleichstromantrieben unter Verwendung eines Zustands - und Storgrobenbeobachters - Regelungstechnik 26 (1978), 349-354,392-397.

55. Eckelmann, W. "Erfahrungen mit einem Zustandsbeobachter fur die Regelung von Destillationskolonnen", Regelungstechnische Praxis 22 (1980), 120-126.

56. Lappus, G. and Schmidt, G. "Supervision and control of gas transportation and distributed systems", 6th IFAC IFIP Conference on Digital Computer Applications to Process Control, Dusseldorf, October 1980.

57. Astrom, K. "Introduction to stochastic control theory", Academic Press, 1970.

58. Follinger, O. "Reduktion der Systemordnung", Regelungstechnik 30 (1982), 367-377.

59. Sundareshan, M.K. "Decentralised observation in large-scale systems", IEEE Trans. on Systems, Man and Cybernetics, vol. SMC-7 (1977), 863-867.

60. Basilo, G. and Marro, G. "On the observability of linear time-invariant systems with unknown inputs", J. Optim. Theory Appl. 3 (1969), 410-415.

61. Wiswanadham, N. and Ramakrishna, A. "Decentralised estimation and control for interconnected systems", Large Scale Systems 3 (1982), 255-266.

62. Muller, P.C. and Luckel, J. "Zur Theorie der StorgroBenaufschaltung in linearen MehrgroBenregelsystemen", Regelungstechnik 25 (1977), 54-59.

63. Hautus, M.L.J. "Controllability and observability conditions of linear autonomous systems", Indagationes Mathematicae 31 (1969), 443-448.

64. Siljak, D.D. and Vukcevic, M.B. "On decentralised estimation", Int. J. Control, 27 (1978), 113-131.

65. Siljak, D.D. "Large scale dynamic systems", North Holland, New York, Amsterdam, Oxford 1978.

66. Litz, L. and Roth, H. "State decomposition for singular perturbation order reduction - a model approach", Int. J. Control, 34 (1981), 937-954.

67. Kuhn, U. "Verfahren zur desentralen Zustandsbeobachtung linearer Systeme mit komplexer Struktur", Regelungstechnik, to appear.

Complex and Distributed Systems: Analysis, Simulation and Control
S.G. Tzafestas and P. Borne (editors)
Elsevier Science Publishers B.V. (North-Holland)

ANALYSIS OF NONLINEAR DISCRETE STOCHASTIC DISTRIBUTED SYSTEMS BY USING AN EXTENSION OF THE NORMAL FORM METHOD*

Guy JUMARIE

Department of Mathematics and Computer Sciences,
Université du Québec à Montréal; P.O. Box 8888, St. A;
Montréal; QUE; H3C 3P8; CANADA
(514) 670-9981

KEYWORDS

Stability, bifurcation, distributed systems, stochastic systems, the normal form approach, finite elements.

ABSTRACT

Briuno extended the Poincaré's theory in a method referred to as the *normal form method* which is very powerful to analyze the stability and the bifurcations of deterministic systems governed by nonlinear ordinary differential equations. The purpose of the present paper is two-fold: first to modify this method in order that it apply to discrete systems, and second to use the latter to investigate nonlinear distributed dynamics as they occur in control systems. The method yields approximate results for stochastic systems. Due to the constraint on the length of the paper, the main result is state for lumped parameter systems, and then the approach to distributed systems is outlined via finite approximation.

1. SUMMARY OF THE NORMAL FORM

1.1 Introduction to the Normal Form

A useful technique to analyze the qualitative behaviour of a dynamical system, say $S : T \to X$ where T denotes the time interval and S refers to the state space, is to consider the new system $\Sigma : T \to Y$ derived from S by means of a diffeomorphisme $\xi : X \to Y$, that is to say a transformation which is C as well as its inverse. The interest of the approach is to obtain Σ in a simpler form than that of S, and all the problem is to define such a mapping. This simplest possible form for Σ is referred to as the *normal form*.

Since Poincaré, the problem of findind the normal form of dynamical systems has been considered by several authors, and the most recent related results are those of Briuno (1971, [1]) which are summarized below.

1.2 Briuno's Theorem

Let $x = (x_1,\dots,x_m)^T \in C^n$ denote the state vector of the dynamical system, and let $\Psi := (\Psi_1,\dots,\Psi_m)$ denote a vector the components $\Psi_i(x)$'s of which are power series in $x_1,\dots,x_m$; converging in a neighbourhood of zero, and such that $\Psi_i(0) = 0$. As a result, zero is a singular point for the dynamical system described by the differential equation

$$\dot{x} = \Psi(x),\ x(t_o) = x_o \qquad (1.1)$$

the fundamental result is now the following

Theorem 1.1 (Briuno, 1971, [1]). Given the dynamical system defined by the equation (1.1), there exists a formal invertible transformation $(x \to y)$,

$$x_i = y_i \sum_{q \in N_i} b_{iq} u^q,\ i = 1,\dots,m \qquad (1.2)$$

$$=: \xi(y)$$

which brings this system into the normal form

$$\dot{y}_i = y_i \sum_{\lambda^T.q = o} c_{iq} y^q \qquad (1.3)$$

with the following notations:
$q := (q_1,q_2,\dots,q_m) \in N_i$; $N_i = \{q$ integer vector: $q_i \geq -1$, $q_k \geq 0$ when $k \neq i$; $q_1+q_2+\dots+q_m \geq 0\}$; $i = 1,\dots,m$; $y^q := y_1^{q_1}\dots y_m^{q_m}$; b_{iq} and c_{iq} are constant complex valued numbers; $\lambda := (\lambda_1,\dots,\lambda_m)^T$ is the eigenvalue vector of the linear part of the system (1.1); $\lambda^T.q$ holds for the inner product

$$\lambda^T.q := \lambda_1 q_1 + \dots + \lambda_m q_m. \quad \square \qquad (1.4)$$

The condition $\lambda^T.q = 0$ defines the *resonant forms* which are only those which appear in the

*Research supported by the National Research Council of Canada.

right side of equation (1.3); and in addition, the linear part of (1.3) is the Jordan's canonical form.

When the linear part of system (1.1) is diagonalizable, that is to say when one has

$$\dot{x}_i = \lambda_i x_i + \sum_{|\gamma|=2}^{\infty} b_{i\gamma} y^\gamma, \quad i = 1,\dots,m \tag{1.5}$$

$|\gamma| := (\gamma_1 + \dots + \gamma_n)$; then it is possible to take the transformation in the form

$$x_i = y_i + \sum_{|\gamma|=2}^{\infty} b_{i\gamma} y^\gamma, \quad i = 1,\dots,m \tag{1.6}$$

and the normal form can be written as

$$\dot{y}_i = \lambda_i y_i + \sum_{\lambda_{\gamma i}=0} c_{i\gamma} y^\gamma; \quad i = 1,\dots,m \tag{1.7}$$

where $\gamma = (\gamma_1,\dots,\gamma_m)$ denotes and integer vector with $\gamma_i \geq 0$ and $\lambda_{\gamma i}$ holds for

$$\lambda_{\gamma i} := \lambda^T \gamma - \lambda_i . \tag{1.8}$$

Once more the condition $\lambda_{\gamma i} = 0$ defines the resonant terms in the right side term of equation (1.7).

Briuno stated conditions for the transformation $\xi(y)$ to be convergent, and they are rather restrictive, so that it may happen that the series diverge, in which case we shall consider the transformation on the formal standpoint only. A case in which the convergence is ensured corresponds to the following theorem.

Theorem 1.2 (Poincaré, 1879). Refer to the diagonal case (1.5) and assume (i) that $\lambda_{\gamma i} \neq 0$ $(i = 1,\dots,n;\ |\gamma| = 2;\ \gamma_i \geq 0)$ and (ii) that, in the complex plane, there exists a straight-line $\mathcal{D}$ intersecting the origin, such that all the points $\lambda_1,\dots,\lambda_n$ remain on the same side of $\mathcal{D}$; then there exists a unique invertible transformation $\xi(.)$, analytic at the origin, which brings system (1.5) into the linear normal form

$$\dot{y}_i = \lambda_i y_i, \quad (i = 1,\dots,m). \ \Box \tag{1.9}$$

As a result, locally, it is possible to derive the state x in the form

$$x_i(t) = \xi_i(y(t)) \tag{1.10}$$

with $y_i(t) = y_i(t_o) \exp[\lambda_i (t - t_o)]$. $\Box$

2. NORMAL FORM FOR DISCRETE SYSTEMS

The most direct generalization of Briuno theory to discrete systems is given by the following result.

Theorem 2.1 Consider the discrete systems

$$\mathbf{x}_i(n+1) = \lambda_i x_i(n) + \sum_{|\gamma|=2}^{\infty} a_{i\gamma} x^\gamma(n), \quad i = 1,\dots,m \tag{2.1}$$

with $\lambda_1 \neq \lambda_2 \neq \dots \neq \lambda_n$; there exists a transformation

$$x_i(n) = y_i(n) + \sum_{|\gamma|=2}^{\infty} b_{i\gamma} y^\gamma(n), \quad i = 1,\dots,m \tag{2.2}$$

which converts it into the normal form

$$y_i(n+1) = \lambda_i y_i(n) + \sum_{\lambda'_{\gamma_i}=0} c_{i\gamma} y^\gamma(n), \quad i = 1,\dots,m \tag{2.3}$$

where λ'_{γ_i} denotes

$$\lambda'_{\gamma_i} := \lambda_1^{\gamma_1} \lambda_2^{\gamma_2} \cdots \lambda_m^{\gamma_m} - \lambda_i . \ \Box \tag{2.4}$$

In the next section, we shall proove this result, and we shall simultaneously derive the normalization formulae.

3. NORMALIZATION FORMULAE

Our purpose, in this section, is to derive explicit normalization formulae for nonlinear discrete systems.

3.1 Class of Considered Systems

With the notations of Section 3, we consider the system

$$x_i(n+1) = \Psi_i(\mathbf{x}(n)), \quad i = 1,\dots,m. \tag{3.1}$$

We assume that the linear part of this system has ℓ critical eigenvalues (that is to say with zero real parts) and that the remaining $(m-\ell)$ eigenvalues have negative real parts; and we shall denote by x_j and $\mathbf{x}_k$ the corresponding components. Clearly, $j = 1,2,\dots,\ell$; $k = \ell, \ell+1,\dots,m$ while $i = 1,2,\dots,m$ is the standard subscript. In addition, we shall denote by $\{n_k\}$ the vector whose the components are x_k.

By using a linear transformation, it is possible to convert the system (3.1) into the new one

$$x_j(n+1) = \lambda_j x_j(n) + \sum_{|\gamma|=2}^{\infty} a_{i\gamma} x^\gamma(n) \tag{3.2a}$$

$$\{x_k(n+1)\} = A\{x_k\} + \{\sum_{|\gamma|=2}^{\infty} a_{k\gamma} x^\gamma(n)\}. \tag{3.2b}$$

where the same notation $x_i(n)$ is conserved for convenience only.

3.2 Definition of the Problem

In substance, we are looking for a transformation in the form

$$x_i = y_i + \sum_{|\mu|=K} b_{i\mu}^{(K)} y^{\mu}; \ \mu = (\mu_1,\ldots,\mu_\ell,0,\ldots,0) \tag{3.3}$$

which brings the system (3.2) into the form

$$y_j(n+1) = \lambda_j y_j(n) + \sum_{|\gamma|=2}^{\infty} c_{j\gamma} y^{\gamma} \tag{3.4a}$$

$$\{y_k(n+1)\} = A\{y_k(n)\} + \sum_{|\gamma|=2}^{\infty} c_{k\gamma} y^{\gamma} \tag{3.4b}$$

Assume that the system has been normalized for $K = 2,3,\ldots,N$; then the normalization with $K = N+1$, will not modify the coefficients of y^{γ}, for $|\gamma| \le N$; and will affect only the other coefficients whose the order is larger or equal to $(N+1)$. By this way, one can determine the normalized form of order N, what is sufficient for practical studies.

3.3 Derivation of the Normalization Formulae

(i) We take $x_i(n)$ and $x_i(n+1)$ in equation (3.3) and we substitute into equations (3.2) to obtain (we drop the superscript K of $b^{(K)}$ to simplify the notations)

$$y_j(n+1) + \sum_{|\mu|=K} b_{j\mu} y^{\mu}(n+1) - \lambda_j[y_j(n) + \sum_{|\mu|=K} b_{j\mu} y^{\mu}(n)] = \sum_{|\gamma|=2}^{\infty} a_{j\gamma} x^{\gamma} \tag{3.5a}$$

$$\{y_k(n+1)\} + \{\sum_{|\mu|=K} b_{k\mu} y^{\mu}(n+1)\} - A\{y_k(n) + \sum_{|\mu|=K} b_{k\mu} y^{\mu}(n)\} = \sum_{|\gamma|=2}^{\infty} a_{k\gamma} x^{\gamma} \tag{3.5b}$$

In these equations, x^{γ} should be expressed in terms of y^{γ} but this is without consequences for the following.

(ii) We now account for $y_i(n+1)$ as expressed by equations (3.4) and we re-write equations (3.5) in the form

$$\sum_{|\gamma|=2}^{\infty} c_{j\gamma} y^{\gamma}(n) + \sum_{|\mu|=K} b_{j\mu} \prod_{i=1}^{\ell} [\lambda_i y_i(n) + \sum_{|\gamma|=2}^{\infty} c_{i\gamma} y^{\gamma}(n)]^{\mu_i}\} - \lambda_j \sum_{|\mu|=K} b_{j\mu} y^{\mu}(n) - \sum_{|\gamma|=2}^{\infty} a_{j\gamma} x^{\gamma} = 0 \tag{3.6a}$$

$$\{\sum_{|\gamma|=2}^{\infty} c_{k\gamma} y^{\gamma}(n)\} + \{\sum_{|\mu|=K} b_{k\mu} \prod_{i=1}^{\ell} [\lambda_i y(n) + \sum_{|\gamma|=2}^{\infty} c_{i\gamma} y^{\gamma}]^{\mu_i}\} - A\{\sum_{|\mu|=K} b_{k\mu} y^{\mu}\} - \{\sum_{|\gamma|=2}^{\infty} a_{k\gamma} x^{\gamma}\} = 0 \tag{3.6b}$$

(iii) We now equate to zero the coefficient of y^{μ}, $|\mu| = K$, in equation (3.6b) to obtain

$$c_{j\mu} = a_{j\mu} \qquad , |\mu| \le K-1 \tag{3.7a}$$

$$c_{j\mu} = a_{j\mu} - \lambda'_{\mu j} b_{j\mu}^{(K)}, \ |\mu| = K \tag{3.7b}$$

$$\{c_{k\mu}\} = \{a_{k\mu}\} + \{A - (\prod_{j=1}^{\ell} \lambda^{\mu_j}) I\} \{b_{k\mu}^{(K)}\}, \ |\mu|=K \tag{3.7c}$$

One then has the following results:

First, when $\lambda'_{\mu j} = 0$, $c_{j\mu} = a_{j\mu}$ and $b_{j\mu}^{(K)}$ is arbitrary, for instance $b_{j\mu}^{(K)} = 1$.

Second, when $\lambda'_{\mu j} \neq 0$, then

$$c_{j\mu} = 0; \ b_{j\mu}^{(K)} = a_{j\mu}/\lambda'_{\mu j}. \tag{3.8}$$

Third, $\{b_{k\mu}^{(k)}\}$ is given by the equation

$$\{b_{k\mu}^{(K)}\} = \{(\prod_{j=1}^{\ell} \lambda^{\mu j}) I - A\}^{-1} \{a_{k\mu}\}. \tag{3.9}$$

(iv) There now remains to determine the coefficients $c_{j\gamma}$, when $\gamma \ge K+1$, and to this end, it is necessary to express the explicit form ot the term $a_{j\gamma} x^{\gamma}$ by means of y. For instance, when $K = 2$ and $\gamma = 3$, one has

$$c_{j\gamma} = a_{j\gamma} - \sum_{i=1}^{\ell} \sum_{|\mu|=2} (\gamma_i - \mu_i + 1) c_{i\mu} b_{j,\gamma-\mu+\delta} + \sum_{m,k} c_{jmk} (b_{m,\gamma-\delta_k} + b_{k,\gamma-\delta_m}) \tag{3.10}$$

where the symbol δ_m holds for $(0,0,\ldots,1,0,0,0)$ in which 1 is the m-th component of the vector. □

4. STOCHASTIC SYSTEMS

4.1 Systems with Parametric Excitations

In order to illustrate the main idea of the approach, we consider the discrete equation $(x_n, \alpha, \beta, \gamma \in \mathbb{R})$

$$x(n+2) = (\alpha \cos \omega n + \beta \sin \omega n)\, x(n) - \gamma^2 x(n). \tag{4.1}$$

At first glance, this system is time varying and consequently, the normal form approach does not apply. So, to circumrent this difficulty, we shall introduce the new state y_n defined as $y(n) = \alpha \cos \omega n + \beta \sin \omega n$, so that system (4.1) is now written in the form

$$x(n+2) = y(n)x(n) - \gamma^2 x(n) \tag{4.2a}$$

$$y(n+2) + \omega^2 y(n) = 0 \tag{4.2b}$$

where the coefficients α and β are defined by means of the initial conditions on $y(n)$.

By using the standard matrix transformation, one may bring this system (4.2) into the form (3.2), but here it is easier to define the new state

$$X_1 := x_1 + (j/\gamma)x_2;\ X_2 := x_1 - (j/\gamma)x_2 \tag{4.3a}$$

$$X_3 := y_1 + (j/\omega)y_2;\ X_4 := y_1 - (j/\omega)y_2 \tag{4.3b}$$

where $(x_1(n), x_2(n)) := (x(n),x(n+1))$ and $(y_1(n), y_2(n) := (y(n), y(n+1))$; so that the dynamic system is

$$X_1(n+1) = -j\gamma X_1(n) + \frac{j}{4\gamma}[X_3(n) + X_4(n)]$$

$$[X_1(n) + X_2(n)] \tag{4.4a}$$

$$X_2(n+1) = j\gamma X_2(n) - \frac{j}{4\gamma}[X_3(n) + X_4(n)]$$

$$[X_1(n) + X_2(n)] \tag{4.4b}$$

$$X_3(n+1) = -j\omega X_3(n) \tag{4.4c}$$

$$X_4(n+1) = j\omega X_4(n). \tag{4.4d}$$

4.2 Discrete Systems with White Noises

We now consider a discrete system defined by the equations

$$x_j(n+1) = \lambda_j x_j(n) + \sum_{|\gamma|=2}^{\infty} a_{i\mu} x^\gamma(n) +$$

$$+ [\mu_j x_j(n) + \sum_{|\gamma|=2}^{\infty} b_{i\mu} x^\gamma(n)]v(n) \tag{4.5a}$$

$$\{x_k(n+1)\} = A\{x_k\} + \{\sum_{|\gamma|=2}^{\infty} a_{k\gamma} x^\gamma(n)\} +$$

$$+ [B\{x_k\} + \{\sum_{|\gamma|=2}^{\infty} c_{k\gamma} x^\gamma(n)\}]w(n) \tag{4.5b}$$

where $v(n)$ and $w(n)$ denote Gaussian white noises with zero mean and unit variance, while the other terms have significances similar to those in equations (3.2a,b).

The normal approach may be used by taking an approximation for the white noise. Explicitly, let ω denote a small frequency considered as a sampling period, then a scalar valued white noise $n(t)$ may be represented in the form

$$n(t) := \sum_k \alpha_k \cos k\omega t + \beta_k \sin k\omega t$$

$$= \sum_k y_k$$

where y_k satisfies the equation

$$y_k(n+2) + k^2\omega^2 y_k(n) = 0$$

and α_k and β_k are Gaussian variables; so that the initial stochastic differential equation is converted into a random equation.

5. DISTRIBUTED SYSTEMS

5.1 General Model

The considered system is defined by the equation

$$x(n+1,z) = B(z,\partial/\partial z)x(n,z) + \sum_{|\gamma|=2}^{\infty} a_{i\gamma}(z)x^\gamma(n) \tag{5.1}$$

$$x(0,z) - x_0(z)$$

together with time independent conditions (BC) given on the boundary $\mathcal{B}(\Omega)$ of the admissible domain Ω for the distributed m'-vector z. n is the time, $x(t,z)$ is the state vector, $B(z,\partial/\partial z)$ is a mxm-matrix whose the coefficients $p_{ij}(z,\partial/\partial z)$ are polynomials $\sum_k b_{ij,k}\partial^{n_{ij,k}}/\partial z^{n_{ij,k}}$ where $b_{ij,k}(z)$ depends explicitly upon z. We shall assume that the function $a_{i\gamma}(z)$ and $b_{ij}(z)$ are continuous in $\bar{\Omega}$ closure of Ω : $\bar{\Omega} = \Omega + \mathcal{B}(\Omega)$.

5.2 Theoretical Consideration

The analysis of the stability, the bifurcations and the periodic solutions of distributed systems is considerably much more difficult than for lumped parameter systems mainly because the formers are infinite dimensional systems. So basically, the approach consists of approximating the latter by a finite dimensional system.

A technique which appears to be useful and needs theoretical investigations is the following.

Let $\phi(z)\epsilon C^n$, $\phi(z) > 0$, and consider the function $x(n/\phi)$ defined as

$$x(n/\phi) := \int_\Omega x(n,z)\phi(z)dz \tag{5.2}$$

On multiplying both sides of (5.1) by $\phi(z)$ and integrating over Ω it is possible to derive a finite differential equation for $x(n/\phi)$. One can then apply the normal form approach to this system on analyze the dependence of the result so obtained upon $\phi(z)$.

6. AN APPROACH VIA FINITE ELEMENTS

The problem of determining the best finite approximation to an infinite dimensional system is an up-to-date one presently in control theory, and to the best of our knowledge, we have not yet general satisfactory solution. It seems that each case is especial and requires a special answer.

The techniques which are mostly used on a practical standpoint are finite difference schemes and expansions by using a complete set of orthogonal functions.

Another approach which seems to be of interest is the finite element technique so fruitfully utilized in structural mechanics. Shortly, this methods works as follows.

Let $\phi_i(z)$, $i = 1,2,\ldots,N$ denote a set of trial functions defined on their respective domains Ω_i, $\Omega_i \cap \Omega_j = 0$, $\cup\Omega_i = \Omega$; and let the estimate $\hat{x}(n,z)$ be

$$\hat{x}(n,z) = \xi(n)\phi_1(z)+\ldots+\xi_N(n)\phi_N(z) \qquad (5.3)$$

On substituting (5.3) into (5.1), we do not yet 0, but rather a residual function $R(n,\phi_1,\phi_2,\ldots,\phi_N)$, and the components $\xi_i(n)$ are defined by the N equations

$$\int_{\Omega_i} R(n,\phi_1,\phi_2,\ldots,\phi_N)\phi_i d_z = 0.$$

For more details, see for instance HUEBNER (1975).

We so obtain a nonlinear discrete system and it is then possible to use the normal form technique.

7. CONCLUDING REMARKS

The conclusion is mainly related to the numerical aspects of the question.

(i) The normalization formulae are a bit cumbersome, but it is possible to set up a tableau of such equations, as done for instance by Starzhinskii (1980) for continuous systems.

(ii) Despite its efficiency, the use of the finite element technique should be subject to some care. Shortly, when the problem is sufficiently regular, a Ritz-Gallerkin method is required, while the finite element technique is convenient mainly for problems with discontinuity in their structural definition.

REFERENCES

[1] BRIUNO, A.D.; Analytical Form of Differential Equations, *Transaction of the Moscow Mathematical Society*, Vol. 25, pp. 132-198, 1971 (English translation by the American Mathematical Society).

[2] BRIUNO, A.D.; Analytical Form of Differential Equations, *Transaction of the Moscow Mathematical Society*, Vol. 26, pp. 199-299 (English translation by the American Mathematical Society), 1971.

[3] BRIUNO, A.D.; Normal Form and Averaging Method, *Soviet. Math. Dokl.*, Vol. 17, pp. 1268-1269, 1976.

[4] HUEBNER, K.H.; *The Finite Element Method for Engineers*, John Wiley, 1975.

[5] STARZHINSKII, V.M.; *Applied Methods in the Theory of Nonlinear Oscillations*, Mir Publishers, Moscow, 1980.

[6] VIDAL, P.; *Systèmes échantillonnés non linéaires* (Nonlinear Sampled Data Systems), Gordon and Breach, 1968.

Complex and Distributed Systems: Analysis, Simulation and Control
S.G. Tzafestas and P. Borne (editors)
Elsevier Science Publishers B.V. (North-Holland)

STABILITY ANALYSIS VIA EIGENVALUE ENCLOSURE IN DISTRIBUTED PARAMETER CONTROL SYSTEMS

D. Franke

Universität der Bundeswehr Hamburg
Holstenhofweg 85
D-2000 Hamburg 70

The paper provides a generalization of Gershgorin's theorems to certain infinite-dimensional matrices, which arise in series representation of the state of linear distributed parameter control systems. Modal expansions are covered as a special case, but also the more general method of weighted residuals (MWR). The paper gives an answer to the question: How are the eigenvalues of the actually infinite-dimensional control system related to those of its finite Galerkin approximation? Gershgorin disks will be constructed for all eigenvalues of the closed-loop system thus providing a sufficient stability criterion. The basic results will be illustrated by means of an example.

1. INTRODUCTION

The method of weighted residuals, MWR [1], is a well-known tool for the approximate numerical solution of boundary value problems in applied mathematics. Such problems arise also in control engineering when feedback control is to be designed for certain distributed parameter systems (DPS). Only linear and time-invariant systems will be regarded in this paper. Due to the infinite dimension of DPS series truncation is an inherent feature of the MWR which results in a reduced model with a finite set of eigenvalues $\rho_1,\ldots,\rho_n$.

Now let the reduced model be stable. Then the question arises: Can one establish conditions such that the original infinite-dimensional system will also be stable? The answer is, of course, related to the spectrum of the DPS. Under specific conditions, namely if the spectrum of the DPS consists only of eigenvalues, stability of such a control system depends only on the locations of the eigenvalues of the closed-loop operator [2]. In this case the known stability condition for finite-dimensional systems can directly be generalized to DPS.

The present paper gives an answer to the question: How are the eigenvalues of the actually infinite-dimensional control system related to those of its finite MWR-model?

This question will be investigated by means of a generalization of Gershgorin's theorems [3] to certain infinite matrix representations. As a result, a new sufficient stability criterion will be obtained.

To fix ideas, consider the following linear, time-invariant distributed parameter plant:

$$\partial x(t,z)/\partial t = Dx(t,z) + \underline{b}^T(z)\underline{u}(t), \quad 0<z<1, \tag{1}$$

with boundary conditions

$$R_o x(t,0) = R_1 x(t,1) = 0. \tag{2}$$

Here, $x(t,z)\in L_2[0,1]$ is the time and space dependent state, $\underline{u}(t)\in \mathbb{R}^p$ is the control, D, R_o and R_1 are spatial differential operators such that the associated eigenvalue problem has a discrete spectrum of (infinitely many) eigenvalues. However, these eigenvalues are not required to be known. $\underline{b}^T(z)$ is a given row vector. Assume that

$$\underline{y}(t) = \int_0^1 \underline{c}(z)x(t,z)dz \in \mathbb{R}^q \tag{3}$$

is the output to be controlled via feedback law

$$\underline{u}(t) = -\underline{K}\,\underline{y}(t), \tag{4}$$

where $\underline{K}$ is a constant gain matrix to be determined, and $\underline{c}(z)$ is given. In a more general case the feedback law can also be admitted to be dynamical.

Following the MWR-procedure, we approximate $x(t,z)$ by

$$x^{(n)}(t,z) = \sum_{i=1}^{n} x_i(t)\phi_i(z) = \underline{\phi}^T(z)\underline{x}(t), \tag{5}$$

where $\phi_1(z),\ldots,\phi_n(z)$ are elements of an arbitrarily selected sufficiently smooth complete orthonormal set $\{\phi_i(z)\}$ in $[0,1]$, where

$$R_o\phi_i(0) = R_1\phi_i(1) = 0, \quad i = 1,2,3,\ldots \tag{6}$$

Insertion into Eq. (1) yields the residual

$$R(t,z) = \underline{\phi}^T(z)\dot{\underline{x}}(t) - D\underline{\phi}^T(z)\underline{x}(t) - \underline{b}^T(z)\underline{u}(t). \tag{7}$$

In this paper, we use as weighting functions the same $\phi_i(z)$ as above and hence the Galerkin version of the MWR. This has the advantage that by $n\to\infty$ the approximate system will be sure to tend to the original one [1].

Therefore, by evaluating

$$\int_0^1 \underline{R}(t,z)\underline{\phi}(z)dz = \underline{0}, \quad (8)$$

one finally arrives at the state representation

$$\underline{\dot{x}}(t) = \underline{A}^{(n)}\underline{x}(t) \quad (9)$$

for the n-th order reduced overall control system, where

$$\underline{A}^{(n)} = \int_0^1 \underline{\phi}(z)D\underline{\phi}^T(z)dz - \int_0^1 \underline{\phi}(z)\underline{b}^T(z)dz \cdot \underline{K} \cdot \int_0^1 \underline{c}(z)\underline{\phi}^T(z)dz. \quad (10)$$

Since $\underline{K}$ is the only unknown in Eq. (10), known pole placement methods may be applied to determine favourable eigenvalue locations $\rho_1,\ldots,\rho_n$ for system (9).

It is stressed again that by increasing n the approximate system will tend more and more to the original one, and therefore the eigenvalues ρ_i will tend to those of the original control loop. This motivates the procedure of eigenvalue enclosure derived in the following section.

Galerkin's method covers, of course, the modal representation as a special case. Here, the $\phi_i(z)$ are the eigenfunctions of operator D along with boundary operators R_0 and R_1. Eq. (10) then takes the special form

$$\underline{A}^{(n)} = \underline{\Lambda} - \underline{B}^* \underline{K} \underline{C}^*,$$

where $\underline{\Lambda} = \mathrm{diag}(\lambda_i)$ is constituted by the first n eigenvalues λ_i of operator D, and the elements of $\underline{B}^*$ and $\underline{C}^*$ are Fourier coefficients of $\underline{b}(z)$ and $\underline{c}(z)$, respectively.

This special case has been investigated by Franke [4], [5], [6].

2. BASIC RESULTS

The eigenvalues of an n-dimensional square matrix $\underline{A}^{(n)} = (a_{ik})^{(n)}$ can be enclosed in Gershgorin disks [3], centred at a_{kk}, with radii

$$r_k = \sum_{\substack{i=1 \\ i\neq k}}^{n} |a_{ik}|, \quad k = 1,\ldots,n. \quad (11)$$

All elements a_{ik} are assumed to be real-valued for the present.

The basic idea to be proposed is the following: We consider a sequence $\{\underline{A}^{(n)}\}$ of finite Galerkin approximations $\underline{A}^{(n)}$ with increasing dimension n and ask: Can one impose conditions on the a_{ik} such that

$$\lim_{n\to\infty} \sum_{\substack{i=1 \\ i\neq k}}^{n} |a_{ik}| \text{ exists for each } k? \quad (12)$$

This is essentially a question of diagonal dominance. Two theorems will be given below, based on different assumptions concerning diagonal dominance.

Theorem 1:

Suppose that for the sequence $\{\underline{A}^{(n)}\}$ the following inequalities hold for each integer n:

$$\left.\begin{aligned} a_{kk} &\leq - M_1 \cdot k^{\alpha}, \\ |a_{ik}| &\leq M_2 \cdot i^{\beta} \cdot k^{\gamma}, \quad i\neq k \end{aligned}\right\} i,\ k=1,\ldots,n, \quad (13)$$

where M_1 and M_2 are positive coefficients, and α, β, γ are real numbers with properties

$$\alpha \geqq 0 \quad \text{and} \quad \beta+\gamma+1<\alpha. \quad (14)$$

Then the eigenvalues of $\underline{A}^{(n)}$ as $n\to\infty$, and hence the eigenvalues of control system (1)-(4), will be enclosed in Gershgorin disks centred at a_{kk}, k=1,2,3,..., with radii

$$r_k = M_2 \cdot k^{\gamma+\delta} \cdot [\zeta(\delta-\beta) - k^{-(\delta-\beta)}] \qquad k=1,2,3,\ldots, \quad (15)$$

where δ is an arbitrarily selected real number in the interval

$$\beta+1<\delta\leqq\alpha-\gamma, \quad (16)$$

and ζ denotes Riemann's ζ-function [7]

$$\zeta(x) = \sum_{i=1}^{\infty} i^{-x}. \quad (17)$$

For example,

$$\zeta(2) = \frac{\pi^2}{6}, \quad \zeta(4) = \frac{\pi^4}{90}.$$

Proof of Theorem 1:

For finite n, apply the similarity transformation

$$\underline{T}^{(n)} = \mathrm{diag}(i^{\delta}) \quad (18)$$

to matrix $\underline{A}^{(n)}$ to obtain

$$\underline{A}^{*(n)} = \underline{T}^{-1}\underline{A}^{(n)}\underline{T} = (a_{ik}^*)^{(n)}, \quad (19)$$

where δ is a real number yet unspecified. This yields

$$a_{ik}^* = \begin{cases} a_{ik}, & i=k \\ a_{ik} \cdot k^{\delta} \cdot i^{-\delta}, & i\neq k \end{cases} \quad i,k=1,\ldots,n, \quad (20)$$

and hence

$$\left.\begin{aligned} a_{kk}^* &\leq - M_1 \cdot k^{\alpha}, \\ |a_{ik}^*| &\leq M_2 \cdot i^{\beta-\delta} \cdot k^{\gamma+\delta}, \quad i\neq k \end{aligned}\right\} i,k=1,\ldots,n. \quad (21)$$

Therefore, the Gershgorin disks for the eigenvalues of $\underline{A}^{*(n)}$ are centred at a_{kk}^*, k=1,...,n, with radii

$$r^*_k = \sum_{\substack{i=1\\ i\neq k}}^{n} |a^*_{ik}| \leq M_2 \cdot k^{\gamma+\delta} \cdot \sum_{\substack{i=1\\ i\neq k}}^{n} i^{-(\delta-\beta)}, \quad k=1,\ldots,n. \tag{22}$$

Now

$$\lim_{n\to\infty} \sum_{\substack{i=1\\ i\neq k}}^{n} i^{-(\delta-\beta)} = \zeta(\delta-\beta) - k^{-(\delta-\beta)} \tag{23}$$

takes a finite value for any $\delta-\beta>1$, which holds according to Eq. (16). Since the eigenvalues of $\underline{A}^{*(n)}$ coincide with those of $\underline{A}^{(n)}$, Eqs. (22), (23) readily result in the upper bound (15) for the disks of the original control system.□

Remarks:

a) Due to assumptions (13) and (14), $\alpha\geq 0$, the centres a_{kk} of the Gershgorin disks will not accumulate at the origin of the complex plane.

b) The obtained result guarantees a finite Gershgorin disk for each fixed k. The radii may increase with growing k, depending on the sign of $\gamma+\delta$. However, due to Eq. (16) we have $\gamma+\delta\leq\alpha$, and therefore the radii r_k will not grow faster than $|a_{kk}|$. Moreover, if M_2 is small enough,

$$M_2 \cdot \zeta(\delta-\beta) < M_1, \tag{24}$$

all disks will lie completely in the open left hand side of the complex plane, thus assuring stability of the original control system (1)-(4).

c) The assumption (13) are somewhat restrictive in so far as *all* diagonal elements a_{kk} are required to be negative. It should be noted that this assumption will be relaxed in the next section.

Theorem 2:

Suppose that for the sequence $\{\underline{A}^{(n)}\}$ the following inequalities hold for each integer n:

$$\left.\begin{aligned} &a_{kk} \leq -M_1 \cdot k^{\alpha}, \\ &|a_{ik}| \leq M_2 \cdot i^{\beta} \cdot k^{\gamma} \cdot |i-k|^{-\delta}, \quad i\neq k \end{aligned}\right\} i,k=1,\ldots,n, \tag{25}$$

where again M_1 and M_2 are positive coefficients, and $\alpha,\beta,\gamma,\delta$ are real numbers with properties

$$\alpha \geq 0, \quad \beta+\gamma \leq \alpha \text{ and } \delta > 1. \tag{26}$$

Then the eigenvalues of $\underline{A}^{(n)}$ as $n\to\infty$, and hence the eigenvalues of control system (1)-(4), will be enclosed in Gershgorin disks centred at a_{kk}, $k=1,2,3,\ldots$, with radii

$$r_k = 2M_2 \cdot k^{\beta+\gamma} \cdot \zeta(\delta), \quad k=1,2,3,\ldots, \tag{27}$$

where again ζ denotes Riemann's ζ-function.

Proof of Theorem 2:

For finite n, apply the similarity transformation

$$\underline{T}^{(n)} = \operatorname{diag}(i^{\beta}) \tag{28}$$

to matrix $\underline{A}^{(n)}$ to obtain

$$\underline{A}^{*(n)} = \underline{T}^{-1}\underline{A}^{(n)}\underline{T} = (a^*_{ik})^{(n)}, \tag{29}$$

where

$$a^*_{ik} = \begin{cases} a_{ik}, & i=k \\ a_{ik} \cdot k^{\beta} \cdot i^{-\beta}, & i\neq k \end{cases} \quad i,k=1,\ldots,n, \tag{30}$$

and hence

$$\left.\begin{aligned} &a^*_{kk} \leq -M_1 \cdot k^{\alpha}, \\ &|a^*_{ik}| \leq M_2 \cdot k^{\beta+\gamma} \cdot |i-k|^{-\delta}, \quad i\neq k \end{aligned}\right\} i,k=1,\ldots,n. \tag{31}$$

Therefore, the Gershgorin disks for the eigenvalues of $\underline{A}^{*(n)}$ are centred at a^*_{kk}, $k=1,\ldots,n$, with radii

$$r^*_k = \sum_{\substack{i=1\\ i\neq k}}^{n} |a^*_{ik}| \leq M_2 \cdot k^{\beta+\gamma} \cdot \left[\sum_{i=1}^{k-1} (k-i)^{-\delta} + \sum_{i=k+1}^{n} (i-k)^{-\delta}\right]. \tag{32}$$

Now,

$$\sum_{i=1}^{k-1} (k-i)^{-\delta} = \sum_{\nu=1}^{k-1} \nu^{-\delta} \leq \sum_{\nu=1}^{\infty} \nu^{-\delta} = \zeta(\delta) \tag{33}$$

for each $k=1,2,3,\ldots$, and

$$\sum_{i=k+1}^{n} (i-k)^{-\delta} = \sum_{\nu=1}^{n-k} \nu^{-\delta} \leq \sum_{\nu=1}^{\infty} \nu^{-\delta} = \zeta(\delta). \tag{34}$$

Therefore, for $k=1,2,3,\ldots$,

$$\lim_{n\to\infty} \sum_{\substack{i=1\\ i\neq k}}^{n} |i-k|^{-\delta} \leq 2 \cdot \zeta(\delta) \tag{35}$$

takes a finite value for any $\delta>1$, which holds according to Eq. (26). Since the eigenvalues of $\underline{A}^{*(n)}$ coincide with those of $\underline{A}^{(n)}$, Eqs. (32)-(35) readily result in the upper bound (27) for the disks of the original control system.□

3. CONSTRUCTION OF SHARPER ENCLOSURES BASED ON RELAXED ASSUMPTIONS

We suppose that for some N-th order Galerkin model (9) the matrix $\underline{A}^{(N)}$ according to Eq. (10) is such that it can be stabilized via $\underline{K}$. Let $\rho_1,\ldots,\rho_N$ be the corresponding eigenvalues of $\underline{A}^{(N)}$ and $\underline{V}^{(N)} = (v_{ik})^{(N)}$ the associated modal matrix. Both, N and $\underline{K}$ will be kept in the subsequence, and the question is: Which is the effect of $\underline{K}$ when applied to the actual infinite-dimensional control system (1)-(4)?

To answer this question we consider again the sequence $\{\underline{A}^{(n)}\}$, n>N, and rewrite $\underline{A}^{(n)}$ in the partitioned form

$$\underline{A}^{(n)} = \begin{bmatrix} \underline{A}^{(N)} & \underline{B} \\ \underline{C} & \underline{D} \end{bmatrix} \quad (36)$$

where the meanings of $\underline{B}$, $\underline{C}$, $\underline{D}$ are self-evident. Before applying theorem 1 or 2, $\underline{A}^{(n)}$ will be subject to an additional similarity transformation:

$$\tilde{\underline{A}}^{(n)} := \underline{T}_1^{-1}\underline{A}^{(n)}\underline{T}_1 , \quad (37)$$

where

$$\underline{T}_1 = \begin{bmatrix} \underline{V}^{(N)}\cdot \operatorname{diag}(\alpha_i) & \underline{0} \\ \underline{0} & \underline{I} \end{bmatrix}. \quad (38)$$

For the moment, the diagonal elements $\alpha_1,\ldots,\alpha_N$ are arbitrary positive real numbers, and $\underline{I}$ denotes identity matrix. Hence we have

$$\tilde{\underline{A}}^{(n)} = \begin{bmatrix} \operatorname{diag}(\rho_i) & \operatorname{diag}(\alpha_i^{-1})\cdot\underline{V}^{(N)-1}\cdot\underline{B} \\ \underline{C}\cdot\underline{V}^{(N)}\cdot\operatorname{diag}(\alpha_i) & \underline{D} \end{bmatrix}. \quad (39)$$

Some (or all) of ρ_i may have been selected conjugate complex, which makes $\underline{V}^{(N)}$ and hence $\tilde{\underline{A}}^{(n)}$ complex-valued.

Transformation $\underline{V}^{(N)}$ has been applied to make the upper left hand corner of $\tilde{\underline{A}}^{(n)}$ diagonal, which results in sharper enclosures for the dominant eigenvalues.

Transformation $\operatorname{diag}(\alpha_i)$ has been introduced to obtain free parameters for radius adjustment. Due to Eq. (39), small α_i, i=1,...,N, will result in sharp enclosures for the first N eigenvalues, but will enlarge the other disks. Therefore, the α_i should be selected well-balanced. However, as Klickow [8] pointed out, the sharpness can considerably be improved by constructing overlapping domains via different selection of the α_i. Sharpness can, of course, always be improved by enlarging the dimension N of the finite Galerkin model.

With regard to $\{\tilde{\underline{A}}^{(n)}\}$ the assumptions made for theorems 1 and 2 can now be relaxed. Since

$$\operatorname{Re}(\rho_i)<0, \quad i=1,\ldots,N, \quad (40)$$

inequalities (13) and (25), when applied to the $\tilde{a}_{ik}$, are *not* required to hold for the first N rows of $\tilde{\underline{A}}^{(n)}$.

After these preparations, theorem 1 and 2 can be extended in a straight-forward manner to $\{\tilde{\underline{A}}^{(n)}\}$.

Suppose that the conditions of one of the above theorems are satisfied. This does, of course, not necessarily imply stability of the given control systems, because some of the disks may not lie completely left hand side of the imaginary axis. However, the following *sufficient stability criterion* holds:

The distributed parameter system described by Eqs. (1)-(4) will be stable if positive numbers $\alpha_1,\ldots,\alpha_N$ can be selected such that

$$\text{(i)} \quad r_k<|\operatorname{Re}(\rho_k)|, \quad k=1,\ldots,N, \quad (41)$$

$$\text{(ii)} \quad r_k<|a_{kk}|, \quad k=N+1,\ldots . \quad (42)$$

Then all Gershgorin disks and hence all eigenvalues will be located left hand of the imaginary axis in the complex plane.

4. EXAMPLE

The calculation of many examples has shown that theorem 1 fits especially to those cases where the $\phi_i(z)$ are the eigenfunctions of operator D, whereas theorem 2 fits to the more general Galerkin case. The following example makes use of both theorems.

Let the plant be given by

$$\frac{\partial x}{\partial t} = \frac{\partial}{\partial z}\left[a(z)\frac{\partial x}{\partial z}\right] + b(z)u(t), \quad 0\leq z\leq 1,$$

with boundary conditions

$$x(0,t) - x(1,t) = 0,$$

and output equation

$$y(t) = \int_0^1 c(z)x(t,z)\,dz.$$

Let $a(z)$ be given by

$$a(z) = a_0 + a_1\cdot z > 0,$$

whereas $b(z)$ and $c(z)$ will be specified later.

The controller equation is

$$u(t) = -Ky(t).$$

We select the orthonormal basis

$$\phi_i(z) = \sqrt{2}\cdot\sin i\pi z, \quad i-1,2,3,\ldots,$$

to arrive, after some elementary calculations, at the following expressions for the elements a_{ik}, according to Eq. (10):

$$a_{kk} = -\left(a_0+\frac{a_1}{2}\right)k^2\pi^2 - Kb_kc_k,$$

$$a_{ik} = -2a_1\cdot[(-1)^{i+k}-1]\cdot ik\frac{i^2+k^2}{(i-k)^2(i+k)^2} -$$

$$- Kb_ic_k, \quad i\neq k, \qquad i,k=1,2,3,\ldots,$$

where

$$b_i = \int_0^1\phi_i(z)b(z)\,dz, \qquad c_k = \int_0^1\phi_k(z)c(z)\,dz.$$

It can be observed that

$$|a_{ik}|<4\cdot|a_1|\cdot ik\cdot(i-k)^{-2}+|Kb_i\cdot c_k|, \qquad i\neq k.$$

With respect to the first right hand expression we apply transformation $\underline{T}^{(n)}=\mathrm{diag}(i)$, according to theorem 2, to obtain

$$|a^*_{ik}|<4\cdot|a_1|\cdot k^2\cdot(i-k)^{-2}+k\cdot|K\cdot c_k|\cdot|b_i|\cdot i^{-1}, \qquad i\neq k,$$

and therefore

$$\sum_{\substack{i=1\\ i\neq k}}^{\infty}|a^*_{ik}|<8\cdot|a_1|\cdot k^2\cdot\zeta(2)+k\cdot|K\cdot c_k|\cdot\sum_{\substack{i=1\\ i\neq k}}^{\infty}\left|\frac{b_i}{i}\right|, \qquad k=1,2,\ldots .$$

The last sum requires $|b_i|\leq \mathrm{const}\cdot i^{-\varepsilon}$, $\varepsilon>0$, for convergence.

Therefore, pointwise control at a point $z_1\in[0,1]$, where $b(z)=\delta(z-z_1)$, cannot be admitted here. $b(z)\equiv$ const. meets the above requirement. However, pointwise measurement at any point $z_2\in[0,1]$ can be admitted.

By summarizing it can be stated that

$$a_{kk}\leq -M_1\cdot k^2,$$

$$r_k < M_2\cdot k^2$$

for large k, whenever the control action is smoother than pointwise.

The numerical calculation has been carried out for

$$a(z) = 0.8+0.4z,$$

$$b(z) = \begin{cases} 2z, & 0\leq z\leq 0.5 \\ 2\cdot(1-z), & 0.5<z\leq 1, \end{cases}$$

$$c(z) = \delta(z-0.5).$$

N=2 has been chosen for the reduced control system design, which means a drastic order reduction. K=36.5 yields the dominant eigenvalues $\rho_1=-37.7$ and $\rho_2=-43.3$, and the associated modal matrix

$$\underline{V}^{(2)} = \begin{bmatrix} 1 & 1 \\ 1 & -1 \end{bmatrix}.$$

Following the procedure described in the foregoing section, one has now free parameters α_1 and α_2 for radius adjustment. In a first step, $\alpha_1=2.39$ and $\alpha_2=1.5$ have been selected such that the dominant disks K_1, K_2 and K_3 are located as far as possible left of the imaginary axis. In a second step, $\alpha_1=10$ and $\alpha_2=5.3$ have been selected aiming at a smaller sector for inclusion of the non-dominant disks. Then, by overlapping the results from these two procedures [8], one obtains the final region outlined in Fig. 1. As can be seen, not only stability but also favourable damping is guaranteed. It can be shown that all disks K_i, $i\geq 5$, are located within a 2x38°-sector as outlined in Fig. 1.

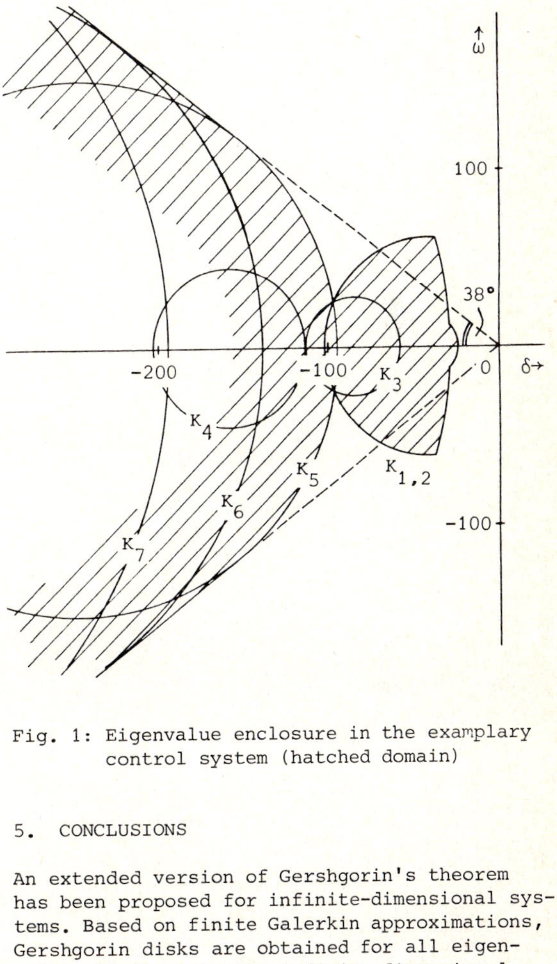

Fig. 1: Eigenvalue enclosure in the examplary control system (hatched domain)

5. CONCLUSIONS

An extended version of Gershgorin's theorem has been proposed for infinite-dimensional systems. Based on finite Galerkin approximations, Gershgorin disks are obtained for all eigenvalues of the actually infinite-dimensional feedback control system, thus providing a sufficient stability criterion. The procedure applies to numerous control problems of distributed parameter systems and therefore deserves further interest.

REFERENCES

[1] Finlayson, B.A.: The Method of Weighted Residuals and Variational Principles. Academic Press, New York, London 1972

[2] Curtain, R.F.: Compensator Design for Distributed Parameter Systems (Survey Paper). Preprint of the 3rd IFAC-Symposium on Control of Distributed Parameter Systems,

Toulouse 1982, pp. S.P.13-S.P.19

[3] Zurmühl, R.: Matrizen und ihre technischen Anwendungen.
4. Auflage, Springer Verlag Berlin, New York 1964

[4] Franke, D.: Eigenvalue Estimation by Means of Gershgorin Disks for Infinite-Dimensional Linear Feedback Control Systems.
Preprints of the 3rd IFAC-Symposium on Control of Distributed Parameter Systems, Toulouse 1982, pp. X.17-X.20

[5] Franke, D.: On the Influence of Neglected Modes in Distributed Parameter Control Systems.
Regelungstechnik 32(1984), pp. 151-156

[6] Franke, D.: Eigenvalue Estimation in Non-Ideal Modal Profile Control of Distributed Parameter Systems.
Preprints of the Fourth IMA International Conference on Control Theory.
Univ. of Cambridge, England 1984, pp. 163-168

[7] Jahnke, Emde and Lösch: Tafeln höherer Funktionen.
Teubner Verlag, Stuttgart 1966

[8] Klickow, H.-H.: Analyse von Regelungen mit verteilten Parametern durch Einschließung der Eigenwerte.
Regelungstechnik 32(1984), pp. 165-167.

Complex and Distributed Systems: Analysis, Simulation and Control
S.G. Tzafestas and P. Borne (editors)
Elsevier Science Publishers B.V. (North-Holland)

LOW ORDER NON-LINEAR MODELS FOR DISTRIBUTED PARAMETER PROCESSES IN INDUSTRIAL PRODUCTION

Mats Molander, Bengt Lennartson & Birger Qvarnström
Chalmers university of technology
Control Engineering Laboratory
S-412 96 GOTHENBURG, Sweden

A programmed method is presented which models industrial production systems with interconnected subprocesses of distributed and lumped parameter types (e.g. distillation columns, furnaces, heat exchangers). In the modelling, partial differential equations are transformed into ordinary differential equations by polynomial approximations and the use of the collocation method.

1. INTRODUCTION

In the process industry, automatic control contributes to improved product quality, higher efficiency and better energy conservation. When developing control systems it is important to have a suitable mathematical model of the plant to be controlled, and the model must have the lowest possible dimension and a form that matches standard computational methods. It is difficult to satisfy this requirement for processes of the distributed parameter type, e.g. distillation columns, rotary and tunnel kilns, heat exchangers, etc.

The aim of the work presented here is a computer aided technique to develop, for control purposes, models of production systems, comprising both distributed and lumped parameter processes.

The partial differential equations (PDE) related to distributed processes are transformed into ordinary differential equations (ODE) by polynomial approximation [8], a method that has been successfully used by several authors [1], [4], [9], [10] for distillation columns, tunnel kilns and similar processes.

Our work aims specifically at creating a tool with the widest possible applicability while remaining easy to use. The distributed processes are generalized to be parallel, interacting flows of material or energy, within a class represented by our basic mathematical PDE-model. The model allows non-linearities of the kind that might occur in practice. A new idea, inspired by Cho and Joseph [1], is that the conventional state variables are replaced by functions (g, h) of these variables. This will allow cases in which flow rates are dependent on state variables and will also improve the accuracy of stationary and dynamic solutions.

Another feature of the work is that structures of coupled PDE- and ODE-processes are allowed. The result is a total model in state space form. To handle the boundary conditions it is then necessary to explicitly indicate each flow direction (denoted by superscripts +, -), in a similar way to that suggested by Wysocki [10].

The input file of the program comprises material and energy parameters, definitions about the structure and the couplings and user-written expressions for the nonlinearities. Given control variables, dynamic simulations are possible, stationary solutions can be computed and linearized control models can be transferred to other programs.

2. THE BASIC MATHEMATICAL MODEL

The permissible form of the partial differential equations describing the distributed subsystems (PDE) is:

$$\begin{aligned}\frac{\partial x_i(z,t)}{\partial t} &= a_i^+(x,u,z,t)\,\frac{\partial}{\partial z}\,g_i^+(x,u,z,t)\\ &+ a_i^-(x,u,z,t)\,\frac{\partial}{\partial z}\,g_i^-(x,u,z,t)\\ &+ b_i^+(x,u,z,t)\,\frac{\partial^2}{\partial z^2}\,h_i^+(x,u,z,t)\\ &+ b_i^-(x,u,z,t)\,\frac{\partial^2}{\partial z^2}\,h_i^-(x,u,z,t)\;+\\ &+ f_i(x,u,z,t)\end{aligned}$$

where

$0 < z < \ell$

$x = (x_1, \ldots, x_i, \ldots, x_n)$ are the state variables.

$u = (u_1, \ldots, u_m)$ are the controls.

The left-hand term may be zero for some values of i (no accumulation).

The a_i and g_i functions describe "transport" flows, the b_i and h_i functions "diffusion" or "conduction" flows and f_i "transfer" or "production/loss" of material or energy (see Gould [3] for details of flow process models). The plus signed terms describe flows in the positive direction and the minus signed, flows in the negative direction. The equations are formulated with the space-derivatives acting upon the functions g and h , which gives good generality but makes it necessary to have the boundary-conditions stated with the same functions. The form of the boundary conditions is (separate for plus and minus signed functions):

$$G_{10}^{i+} g_i^+(0,t) + G_{11}^{i} g_i^+(\ell,t) = C_1^{i+}$$

and two independent conditions of the type:

$$G_{20}^{i+} g_i^+(0,t) + G_{21}^{i+} g_i^+(\ell,t) + H_{20}^{i+} h_i^+(0,t) +$$

$$+ H_{21}^{i+} h_i^+(\ell,t) +$$

$$+ (dH)_{20}^{i+} \frac{\partial h_i^+}{\partial z}\bigg|_{z=0} + (dH)_{21}^{i+} \frac{\partial h_i^+}{\partial z}\bigg|_{z=\ell} = C_3^{i+}$$

The permissible form of the lumped system equations (ODE) is

$\frac{d}{dt} x = f(x,u)$; $y = g(x,u)$ where y is the output.

3. DEVELOPMENT OF A CONTROL MODEL

For control purposes, the most convenient model is a so-called state space model, that is, a system of first order, coupled ordinary differential equations. A simple and straightforward way to obtain models of the sought type is a finite-difference method, but this often (as exemplified below) leads to rather large dimensions. We have instead chosen the method of orthogonal collocation (see Villadsen [8]).

In this method the variables are approximated by interpolation polynomials in the following way:

$$x_i \approx x_i^* = \sum_{j=0}^{p+1} \ell_j^{p+2}(z)\, x_i(z_j)$$

$$g_i \approx g_i^* = \sum_{j=0}^{p} \ell_j^{p+1}(z)\, g_i(z_j)$$

$$h_i \approx h_i^* = \sum_{j=0}^{p+1} \ell_j^{p+2}(z)\, h_i(z_j)$$

$$0 = z_0 < z_1 < z_2 \ldots < z_{p+1} = \ell$$

where p is the number of interior interpolation (or collocation) points.

$\ell_j^N(z)$ are the so-called Lagrange polynomials which possess the following properties:

$$\ell_j^N(z_j) = 1$$

$$\ell_j^N(z_i) = 0 \qquad i \neq j$$

where z_i is a collocation point and N the total number of points.

For a collocation point this gives:

$$x_i^*(z_j) = x_i(z_j)$$

and

$$\frac{d}{dt} x_i^*\bigg|_{z=z_j} = \frac{d}{dt} x_i\bigg|_{z=z_j}$$

By eliminating $g_i^*(0) = g_i(0)$, $h_i^*(0) = h_i(0)$ and $h_i^*(\ell) = h_i(\ell)$, through the boundary-conditions, and by demanding that the polynomials satisfy the partial differential equations at the collocation points, a model of the form:

$$\frac{d}{dt} x_i(z_j) = f(x(z_1), \ldots, x(z_p), u)$$

is obtained.

The method is called orthogonal if the collocation points are chosen as zeroes to a class of orthogonal polynomials. In the examples presented in this paper we have used zeroes to the orthogonal polynomials $P_n(z)$ which are defined by the following relations:

$$\int_0^{\ell} P_n(z) \cdot P_m(z) \cdot z(z-1)\, dz = 0 \qquad n \neq m$$

n, m = degrees of the polynomials .

The introduction of the functions g and h allows nonlinearities and coupling between the equations in the derivative terms. One situation when this occurs is when the flow in a flow process is dependent on the composition of the flowing medium. In addition to this advantage, it can be shown that if most of the z-dependence is limited to those functions, a more accurate solution is obtained (see Appendix A).

Positive and negative flow terms are split in order to handle situations with boundary conditions at both boundaries. When, in these cases, several subsystems are connected through their boundary conditions, this splitting makes it possible to eliminate the boundary values separately for each g and h function.

As an example of the dimension reduction that this method gives compared with a simple finite difference method we study models of a pure time delay. A delay can be described by the following partial differential equation:

$$\frac{\partial x(z,t)}{\partial t} = -\frac{\partial x(z,t)}{\partial z} \quad ; \quad 0 < z < 1$$

$$x(0,t) = u(t)$$

A finite difference model would be:

$$\frac{dx_i}{dt} = -n \cdot (x_i - x_{i-1})$$

$$x_0 = u$$

where n is the number of discrete states.

To compare the methods we look at the frequency-response at z = 1 (delay of 1). The results are shown below.

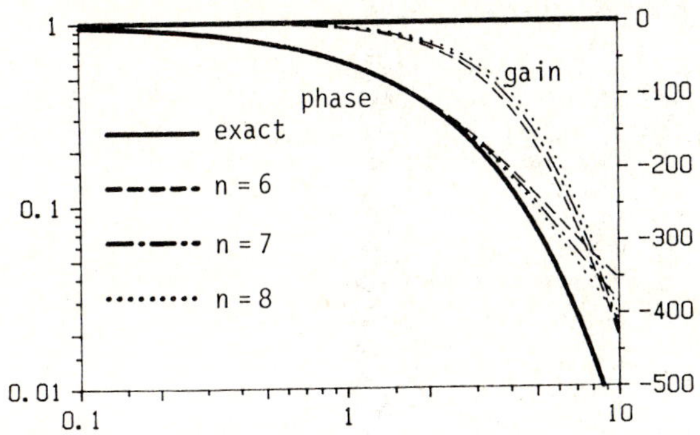

Fig. 1 Frequency curves for a finite difference model of pure transport delay. n = number of discrete states.

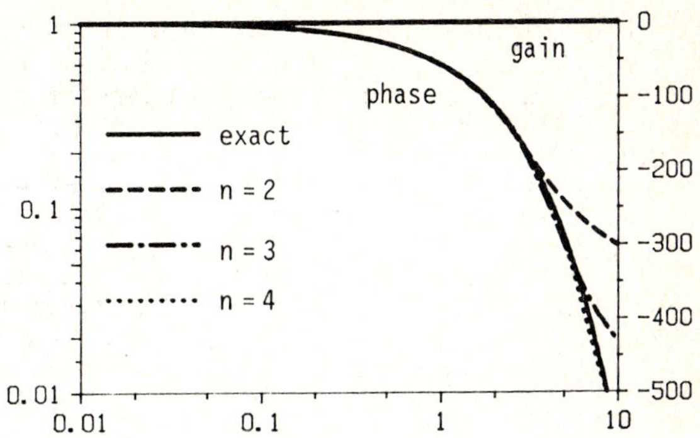

Fig. 2 Frequency curves for a collocation model of pure transport delay. n = number of collocation points.

In fig. 1 the exact frequency curves are compared with the curves for finite-difference models with n = 6, 7 and 8. Fig. 2 shows the frequency response for orthogonal collocation models with 2, 3 and 4 collocation points. The accurate frequency curves in the collocation case are explained by the fact that the model gives zeroes in the right half-plane. The finite-difference model gives no zeroes at all.

4. PROGRAMMING THE MODEL

The definition of the total system to be modelled is made in a special user-dedicated data-file where all functions and coefficients of the differential equations and boundary conditions must be programmed. The structure of the file is as below:

* GLOBAL SYSTEM DATA *
Definitions of the total system.

* INITIAL SYSTEM DATA *
Some initial definitions for each subsystem.

* PARTIAL DIFF. EQ. *
Expressions for the functions and coefficients.

* ORDINARY DIFF. EQ. *
Expressions for the functions.

* BOUNDARY CONDITIONS *
Expressions for the coefficients.

To exemplify the procedure we look at a model for batch-distillation of a binary-mixture. The column is assumed to be of packed-bed type. A simplified model of the system will contain two lumped subsystems (the reboiler and the condenser), and one of distributed type (the column). Under certain assumptions the equations can be written as follows.

The reboiler:

O.D.E.

$$\frac{d}{dt} x_B = \frac{L}{H_B} x(0,t) + \frac{(V - L)}{H_B} x_B - \frac{V}{H_B} f(x_B)$$

$$\frac{d}{dt} H_B = L - V$$

The column:

P.D.E.

$$\frac{\partial}{\partial t} x_1 = \frac{L}{h_L} \frac{\partial}{\partial z} x_1 - \frac{V}{h_L} \frac{\partial}{\partial z} x_2$$

$$0 = V \frac{\partial}{\partial z} x_2 - \psi \left[f(x_1) - x_2 \right]$$

$$0 < z < \ell$$

B.C.

$$x_1(\ell,t) = x_c$$

$$x_2(0,t) = f(x_B)$$

The condenser:

O.D.E.

$$\frac{d}{dt} x_c = \frac{V}{H_c} (x_2(\ell,t) - x_c)$$

$$\frac{d}{dt} H_c = V - L$$

Only concentrations of the more volatile component are considered. x_1 is the concentration in the liquid and x_2 the concentration in the vapour. L and V are the liquid and vapour flows. f is a known function (for the assumptions made see Gould [3]). The definition of this system is done as below. First some initial data are needed:

* GLOBAL DATA *

NDS = 1 ! Number of Distributed Systems

NLS = 2 ! Number of Lumped Systems

* INITIAL SYSTEM DATA *

SYS1 = LPS ! Subsystem 1 is Lumped

NXD = 2 ! Number of dynamic relations

SYS2 = DPS ! Subsystem 2 is Distributed

NXD = 1

NXA = 1 ! Number of algebraic relations
! (no time derivatives)

NP = <integer> ! Number of collocation points

ℓ = <real> ! Height of the column

SYS3 = LPS

NXD = 2

A definition of the column equations (partial differential) would be:

* PARTIAL DIFF. EQ. *

* SYS2 (The column) *

$$a_1^- = L/h_L$$

$$a_1^+ = - V/h_L$$

$$g_1^- = x_1$$

$$g_1^+ = x_2$$

$$a_2^+ = V$$

$$g_2^+ = x_2$$

$$f_2 = - \psi \left[f(x_1) - x_2 \right]$$

The subscript indices of a, g and f refer to the value of i in the basic PDE. The algebraic equations (zero left hand sides) should come last.

In the ordinary differential equations it is allowed to use the boundary values of the functions g and h . Definitions for the reboiler and condenser would be:

* ORDINARY DIFF. EQ. *

* SYS1 (The reboiler) *

$$f_1 = (L \cdot g_{12}^-(0) + (V - L) \cdot x_B - V \cdot f(x_B))/H_B$$

$$f_2 = V - L$$

* SYS3 (The condenser) *

$$f_1 = V \cdot (g_{12}^{+}(\ell) - x_c)/H_c$$

$$f_2 = L - V$$

The first subscript index of g refers to the value of i in the basic PDE and the second gives the number of the subsystem.

In the section for boundary conditions, definitions for every equation (separate for plus and minus flows) must be written. When a boundary condition is defined for some certain g and h functions, the boundary values of these functions can be used in the conditions which follow it, but not in those which precede it. The outputs from the lumped systems are treated as boundary conditions. For our example we would get:

* BOUNDARY CONDITIONS *

* SYS1 (Output) *

$$y_{11} = x_B$$

* SYS2 , EQ.1 *

! Makes $g_{12}^{+}(0)$ and

$$G_{10}^{+} = 1$$

! $g_{12}^{+}(\ell)$ available

$$c_1^{+} = f(y_{11})$$

* SYS2, EQ.2 *

! Makes $g_{22}^{+}(0)$ and

$$G_{10}^{+} = 1$$

! $g_{22}^{+}(\ell)$ available

$$c_1^{+} = f(y_{11})$$

* SYS3 (Output) *

$$y_{13} = x_c$$

* SYS2 EQ.1 *

! Makes $g_{12}^{-}(0)$ and

$$G_{11}^{-} = 1$$

! $g_{12}^{-}(\ell)$ available

$$c_1^{-} = y_{13}$$

For a given total state vector (all the state-variables at all collocation points) the program has the following scheme in evaluating the state derivatives.

1. All the functions and coefficients in the partial-differential equations are evaluated at all collocation points (using the expressions in the user-written system definition file).

2. Boundary values for the g and h functions, and the outputs from the lumped systems, are evaluated in the order given in the system definition file. This order must be consistent.

3. The functions (derivatives of the states) in the ordinary differential equations are evaluated.

4. The derivative of the total state-vector is computed.

5. EXAMPLES OF APPLICATIONS

5.1 Multi-component distillation

The method is used to model a 3-component packed-bed distillation column. Only two of the components are considered since the third concentration can always be obtained using the fact that the sum of the concentrations (in mole-ratios) must equal unity.

The data used are taken from a plate-column described by Toijala, Gustafson [7]. The data are translated to approximate a packed-bed column, following Gould [3].

To determine the necessary dimension, the non-linear model is linearized for different numbers of collocation points, and the frequency response from one of the controls to one of the product concentrations is studied. The results are shown in Fig. 3. As another check, non-linear simulations of the step response (same input and output) are made for two different model dimensions (Fig. 4).

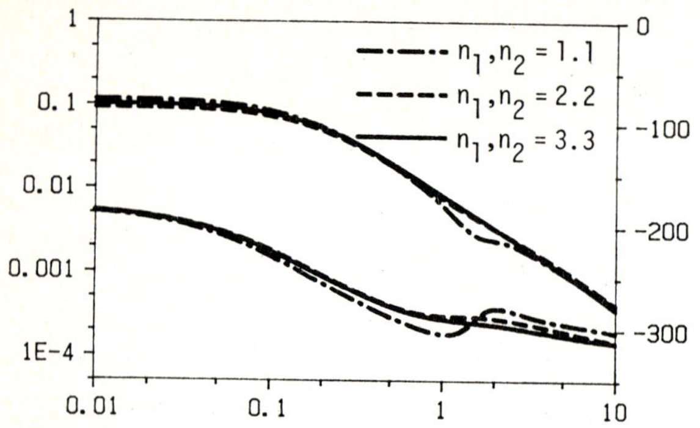

Fig. 3 Bode plots of $x_{c2}(j\omega)/L(j\omega)$ of a 3-component distillation column (see ref. [6]). n_1 and n_2 are the number of collocation points in each section.

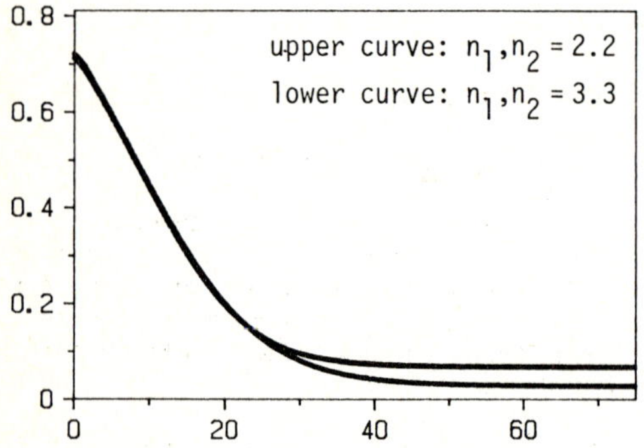

Fig 4. Step responses for a 3-component distillation column (see ref. [6]). L input and x_{c2} output.

The figures show that two points in each section are sufficient to describe the dynamic response. This should be enough for control purposes.

5.2 Tunnel furnace

As a simplified model of a tunnel furnace we take a counter-current heat-exchanger where one of the streams is the material for burning (the charge) and the other is a gas-flow. In the middle of the furnace hot gas is injected, corresponding to burners. Only the charge is assumed to have heat capacity. The gas is assumed to have no radiation.

The basis for this model is a furnace described by El Hajjar [2] but, as our main purpose is to present a modelling procedure, it has been simplified to a more manageable dimension.

Two simulations are made. The first is a step response where at equilibrium, the flow of injected hot gas is raised by 10 %. The charge temperature at the injection point is taken as output. Curves are presented for three different dimensions, and they show (Fig. 5) that 4 collocation points in the heating section and 2 in the cooling section seems to desribe the dynamics accurately. In the other simulation (Fig. 6) the charge has a uniform temperature at the beginning and the temperature profile of the charge during the recovery to equilibrium is registered. This gives a better picture of the furnace behaviour than looking at a single point.

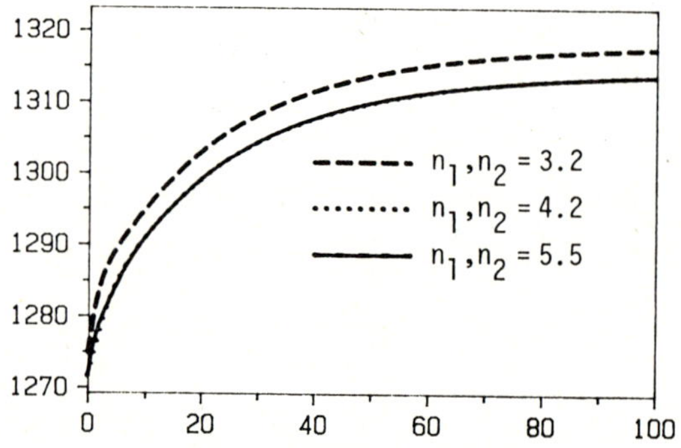

Fig. 5 Step responses for models of a tunnel furnace. n_1 and n_2 are the numbers of points in each section.

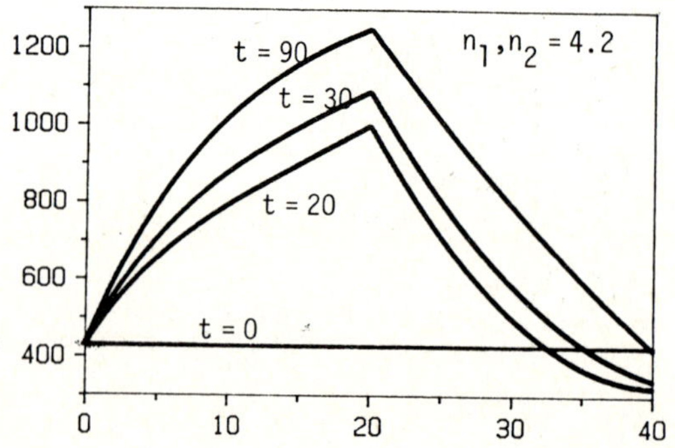

Fig. 6 Simulated temperature profiles along a tunnel furnace at different times during recovery to equilibrium.

Equations and parameter values used in the examples above are given in Molander et al. [6].

6. CONCLUSION

The user's experience of the first version of the program, completed February 1985, is that processes very different in physical nature can be handled without a deep understanding of the mathematics involved. To compare models of different dimension and to examine the effect of non-linearities is very easy. Often the dimension can be reduced more than one would expect. It has been observed that further model reduction can be carried out if only moderate performance of the control system is required. Another observation is that the polynomial approximation tends to conserve the properties that determines the stability margins of the resulting control system. In other words, the controller based on the reduced model seems to be reliable when coupled to the real process.

REFERENCES

[1] Cho, Y.S. and B. Joseph. "Reduced-Order Steady-state and Dynamic Models for Separation Processes". AIChE J 29, No. 2, pp. 261 - 276 (1983).

[2] El Hajjar, H.. "Contrôle et Conduite Numériques d'un Four Tunnel d'Industrie Céramique". Thèse de Docteur Ingenieur n° 833, Université Paul Sabatier, Toulouse (1983).

[3] Gould, L.A.. "Chemical Process Control: Theory and Applications". Addison-Wesley (1969).

[4] Hariri, A. and J.P. Babary. "Modelling and Simulation of Tunnel Kilns in the Ceramic Industry". To appear in Int. J. of Modelling and Simulation.

[5] Luyben, W.L.. "Process Modelling, Simulation and Control for Chemical Engineers". McGraw-Hill (1973).

[6] Molander, M., B. Qvarnström and B. Lennartson. "Modelling of Coupled Flow-Processes by Polynomial Approximations". Report R 85-02, Control Eng. Lab., Chalmers Univ. of Technology, Gothenburg, Sweden

[7] Toijala, K. and S. Gustafsson. "Process Dynamics of Multi-Component Distillation. Comparison of Theoretical Model with Published Experimental Results". Kemian Teollisuus 29 (1972) 2:95-103.

[8] Villadsen, J. and M.L. Michelsen. "Solution of Differential Equation Models by Polynomial Approximation". Prenctice-Hall (1978).

[9] Wong, K.T. and R. Luus. "Model Reduction of High-Order Multistage Systems by the Method of Orthogonal Collocation". Can. J. Chem. Eng. 58, 382 (1980).

[10] Wysocki, M.. "Application of Orthogonal Collocation to Simulation and Control of First Order Hyperbolic Systems". Math. & Comput. Simulation (Netherlands), Vol. 25, No. 4, pp. 335 - 345 (Aug. 1983).

APPENDIX A

INTEGRATON ACCURACY

To show how different polynomial approximations affect the accuracy of the solutions we look at the following rather academic but illustrative example

P.D.E. $$\frac{\partial x}{\partial t} = -\frac{\partial}{\partial z}[(1+z)x] - x$$

B.C. $$x(0,t) = x(z,0) = 1$$

With our formulation, where the polynomials approximate functions of x (g and h) instead of x itself, we get

$g(x,z) = -(1+z)x$
$f(x,z) = -x$.

To approximate x directly by a polynomial we have to rewrite the equations as

P.D.E. $$\frac{\partial x}{\partial t} = -(1+z)\frac{\partial x}{\partial z} - 2x$$

Below are pictures showing the profiles obtained with the two different approximations. The curves marked description 1 show the solutions obtained with the first formulation (polynomial approximating $g=-(z+1)x$) and curves marked description 2 show solutions obtained with the second formulation (polynomial approximating x). The profiles are evaluated at two times (t=0.6 and t=3) and with two different numbers of collocation points (1 and 3).

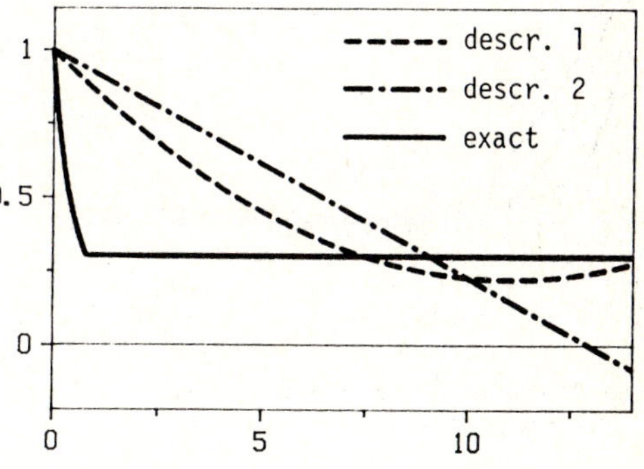

Fig. 7 Profiles at $t = 0.6$. One collocation point.

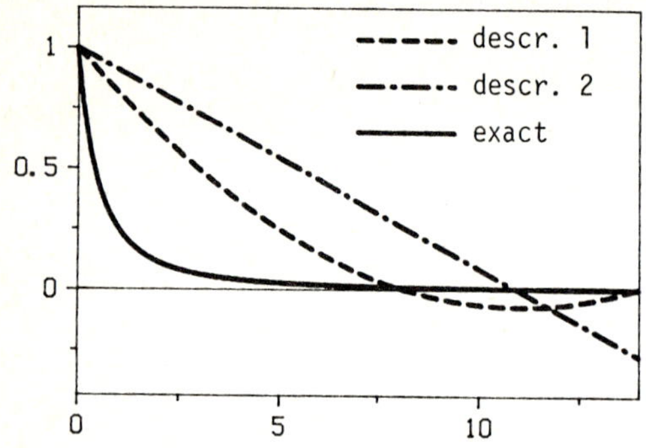

Fig. 8 Profiles at t = 3.0. One collocation point.

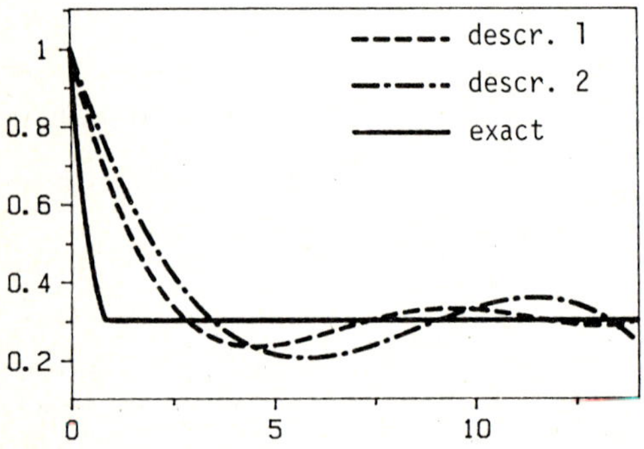

Fig. 9 Profiles at t = 0.6 . Three collocation points.

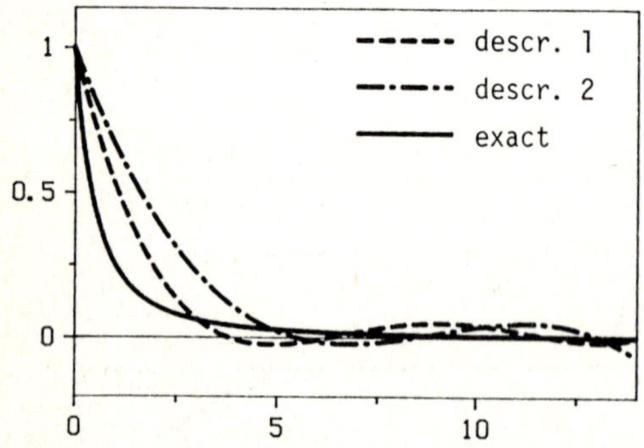

Fig. 10 Profiles at t = 3.0 . Three collocation points.

In this example the introduction of the g function evidently gives more accurate solutions. To explain this we look at the following, more general, equation

$$\frac{\partial x}{\partial f} = a \frac{\partial g}{\partial z} + b \frac{\partial^2 h}{\partial z^2} + f$$

The solution is obtained by integrating this equation. The accuracy then obviously depends on how well we can represent the right-hand side. Suppose that, at some instant, the profile of the time derivative can be accurately described by a polynomial of degree n . Suppose further that x is known at $n+1$ discrete (collocation) points. If a and b are independant of the space coordinate z then for the whole expression to be of degree n, g must be a polynomial of degree $n+1$, h a polynomial of degree $n+2$ and f a polynomial of degree n . The known values of x at the collocation points is enough to determine f , the extra coefficients in g and h are determined through the boundary conditions. Thus f,g and h can be determined exactly and thereby also the time derivative of x . Future values of x can now be evaluated with an accuracy dependent only on the integration method. If instead we try to approximate x by a polynomial this might not be the case. The polynomial will be of degree $n+2$ ($n+1$ collocation points plus two boundary conditions) but we now have to rewrite the equation in a form where the derivatives act directly upon x . The first term on the left-hand side will, for instance, be

$$a \left[\frac{\partial g}{\partial x} \frac{\partial x}{\partial z} + \frac{\partial g}{\partial z} \right]$$

We cannot be sure that this is a polynomial of degree n ($\partial x/\partial z$ is a polynomial of degree $n+1$ but $\partial g/\partial x$ and $\partial g/\partial z$ can be anything) so the exact time derivative might not be achieved even if it is a true polynomial and we use enough collocation points to fit a polynomial of that degree. The conclusion is therefore that the best accuracy is achieved using the equations formulated with g and h functions containing as much as possible of the space dependence.

Complex and Distributed Systems: Analysis, Simulation and Control
S.G. Tzafestas and P. Borne (editors)
Elsevier Science Publishers B.V. (North-Holland)

SIMULATION OF A NON LINEAR ULTRAFILTRATION MODEL USING A MIXED-FINITE ELEMENTS METHOD

Y. Jarny, J.D. Picot

Laboratoire d'Automatique, Ecole Nationale Supérieure de Mécanique,
1 rue de la Noë, 44072 Nantes Cédex, France

The ultrafiltration process uses membranes for the separation of macromolecules (B) from solutions (A+B) in which they are dissolved. Concentration of (B) near the membrane is governed by a non linear convective diffusion equation, the flow velocity decreases when the concentration grows on the boundary, the diffusion coefficient is a non linear function of the concentration. A numerical model is obtained by using mixed-finite elements. Results of runs on laboratory data are shown.

1. INTRODUCTION

Ultrafiltration is a membrane separation process. The rate of separation is strongly reduced by the concentration polarization of the membrane. This phenomenon may be described by a distributed parameter model. V.L. Vilker et al. (1981) solved the convective diffusion equation of this model by searching for asymptotics solutions, assuming the diffusivity to be constant.

The aim of this paper is to apply a mixed-finite elements method to the resolution of these equations. The flow velocity and the diffusion coefficient are non linear functions of concentration. This numerical method was studied by C. Chavent et al. (1982), G. Cohen (1978), for convective diffusion equations occuring in the modelling of oil reservoirs.

The numerical model which is obtained, allows to simulate the unsteady concentration profile in an unstirred cell, near the membrane. The resulting ultrafiltrate flux given by the model is compared to laboratory data provided by S. Poyen (1984).

2. MODEL EQUATIONS

A description of an ultrafiltration cell is given in figure 1. Macromolecules (B) are dissolved in a solvent (A), a transmembrane pressure differential ΔP is applied to the solution to force (A) through the membrane. ΔP must be sufficient to overcome the hydraulic resistance of the membrane and the osmotic pressure created by the accumulation of (B) near the membrane. The model equations describe the evolution of the concentration of (B) as a function of time t and of a space variable x, in a cylindrical cell.

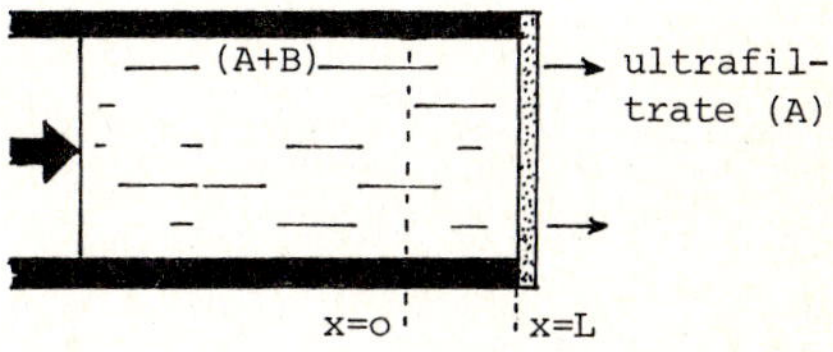

Figure 1. Schematic diagram of the ultrafiltration process

Hypotheses and notations

The following are assumed :
- the membrane is empermeable to (B),
- the solution is incompressible,
- concentration of (B), noted C, in a cross section of the cylinder is homogeneous,
- at the entrance, x=o, the concentration stays unchanged, C = Ci,
- when pressure ΔP is applied, t=o, concentration C is constant in the cell, C = Ci,
- the flux of (B) due to the diffusion is proportional to the concentration gradient (Fick's law),
- the diffusion coefficient D is a positive function of C, figure 2,
- the flow velocity V through the membrane is described by $V = Lp\,[\Delta P - \Delta\eta]$ (1)

where Lp is the hydraulic permeability of the membrane,
- the osmotic pressure $\Delta\eta$ is a monotone positive increasing function of C at the boundary x = L, as in figure 2.

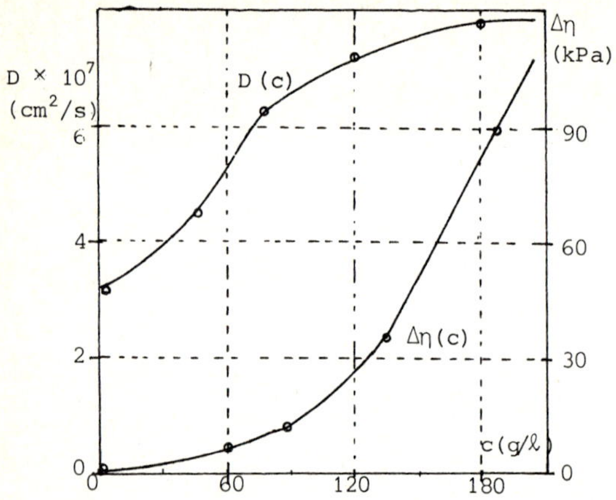

Figure 2. The osmotic pressure and the diffusion coefficient as a function of concentration

with these hypotheses, a mass balance leads to the following evolution equations

$$\frac{\partial c}{\partial t}(x,t) = \frac{\partial}{\partial x}\left[D(c)\frac{\partial c}{\partial x}(x,t)\right] - V\frac{\partial c}{\partial x}(x,t) \quad (2)$$

$$(x,t) \in]0,L[x]0,T[$$

$$C(x,0) = C_i, \; x \in]0,L[\quad (3)$$

and to the boundary conditions

$$-D\frac{\partial c}{\partial x}(0,t) + V.C\,(0,t) = V.C_i \quad (4)$$

$$D\frac{\partial c}{\partial x}(L,t) - V.C\,(L,t) = 0 \quad (5)$$

It is observed experimentally that e, the thickness of the boundary layer, is very small compared to the cell length : a typical ratio is 1 %, so it is useless to take L equal to the cell length. Simulation runs, J.D. Picot (1984), showed that if L is taken large enough compared to e, then the concentration gradient at x=o is nul so, equation (4), can be approximated by

$$C(0,t) = C_i \quad (6)$$

3. NUMERICAL APPROXIMATION USING A MIXED-FINITE ELEMENTS METHOD

Reduced equations. Let C_M be the concentration which satisfies

$$\Delta\eta\,(C_M) = \Delta P \quad (7)$$

then the reduced variable y is introduced

$$y(x,t) = [C(x,t) - C_i]/[C_M - C_i] \quad (8)$$

and we set $C_L(t) = C(L,t)$, $y_L(t) = y(L,t)$ (9)

$$a(y) = D(C) \quad (10)$$

$$b(y_L) = V(C_L) = Lp[\Delta P - \Delta\eta(C_L)] \quad (11)$$

$$r(x,t) = -\,a(y)\,.\,\frac{\partial y}{\partial x}(x,t) \quad (12)$$

$$K = C_i/[C_M - C_i] \quad (13)$$

Equations (2) to (6) become

$$\frac{\partial y}{\partial t}(x,t) + \frac{\partial r}{\partial x}(x,t) + b(Y_L)\frac{\partial y}{\partial x}(x,t) = 0 \quad (14)$$

$$y(x,0) = 0 \quad (15)$$

$$y(0,t) = 0 \quad (16)$$

$$r(L,t) + b(y_L).[y(L,t) + K] = 0 \quad (17)$$

We set

$$\alpha(y) = \int_0^y a(z)\,dz \quad (18)$$

Discretization. The interval)0,L(is divided in N equal intervals $I_i =]x_i, x_{i+1}[$ of length h,

$i = 1, \ldots, N.$

The mixed-finite elements method consists in introducing the following spaces

Y_h^m, R_h^m, $m \in N$

$$Y_h^m = \{v \in L^2(\Omega),\; v|I_j \in P^m(I_j),\; j = 1,\ldots N\} \quad (19)$$

$$R_h^m = \{\mu \in C^\circ(\Omega),\; \mu|I_j \in P^{m+1}(I_j),\; j = 1,\ldots N\} \quad (20)$$

where P^m is the set of polynomials of degree less than or equal to m, and in searching for $(y,r) \in Y_h^m \times R_h^m$ such that, for j = 1, ..., N,

$$\frac{\partial}{\partial t}\int_{I_j} y\,vdx + \int_{I_j}\frac{\partial r}{\partial x}\,vdx + b(y_L)\int_{I_j}\frac{\partial y}{\partial x}\,vdx$$

$$+\,b(Y_L)\,[y_j^+ - y_j^-]\,v(x_j^+) = 0,\; \forall v \in P^m(I_j) \quad (21)$$

$$\int_0^L r\,udx = \int_0^L \alpha(y)\frac{du}{dx}\,dx - \Big[\alpha(y)\,u\Big]_{x=0}^{x=L}$$

$$\forall u \in R_h^m \quad (22)$$

At each point x_i, functions $y(t) \in Y_h^m$ are discontinuous, the notations are

$$y_j^+(t) = y(x_j^+,t) \text{ and } y_j^-(t) = y(x_j^-,t)$$

Resulting approximation of transport equation (21) is due to Raviart-Lesaint (1974). It corresponds to a conservative scheme : on each interval I_i, the accumulation rate of y is equal to the difference between incoming and out-going flux. The flow velocity V is a function of the boundary value $y_L(t)$, the approximation of y beeing discontinuous at the boundary, equation (21) is solved by taking

$$y_L(t) = y^+_{N+1}(t) \tag{23}$$

It can be observed that the convective flux are approximated by an upwind scheme.

The interval $)0,t($ is divided in M intervals

$]t_k, t_{k+1}[$ of length τ, for k=1 to M+1, the applications $r(t_k) \in R^m_k$ and $y(t_k) \in y^m_k$ are noted r^k and y^k.

Time discretization is obtained by taking the explicit approximation

$$\frac{\partial}{\partial t} y(t_k) = (y^{k+1} - y^k)/\tau \tag{24}$$

4. NUMERICAL RESOLUTION

Discrete equations. With m=o, the dimensions of the spaces Y^m_n and R^m_h are respectively N and N+1.

Let r^k_i be the value $r^k(x_i)$, $i = 1, \ldots, N+1$ and $\{u_{ih}, i = 1, \ldots, N+1\}$ the basis functions of R^m_h such that

$$r^k(x) = \sum_{i=1}^{N+1} r^k_i \, u_{ih}(x), \ \forall x \in (0,L) \tag{25}$$

Let y^{+k}_i and y^{-k}_i be the value of $y^k(x^+_i)$ and $y^k(x^-_i)$, $i = 1, \ldots, N+1$, for m=o we get $y^{+k}_i = y^{-k}_{i+1}$ this value is noted y^k_i.

Let $\{v_{ih}, i = 1, \ldots, N\}$ be the basis function of Y^m_h such that

$$y^k(x) = \sum_{i=1}^{N} y^k_i \, v_{ih}(x), \ \forall x \in \bigcup_{i=1}^{N} I_i \tag{26}$$

Hence, replacing v by v_{ih} in (21), $i = 1, \ldots, N$ and u by u_{ih} in (22), $i = 1, \ldots, N+1$,

y^k_L, $[y^k_i]^N_{i=1} \in R^N$ and $[r^k_i]^{N+1}_{i=1} \in R^{N+1}$ are solutions

of the recurring equations, for k=1 to M

$$\frac{h}{\tau}(y^{k+1}_i - y^k_i) + r^k_{i+1} - r^k_i + b(y^k_L)[y^k_i - y^k_{i-1}] = 0 \tag{27}$$

$i = 1, \ldots, N$

$$\sum_{j=i-1}^{i+1} q_{ij} \, r^k_j = \alpha^k_{i-1} - \alpha^k_i, \text{ with} \tag{28}$$

$$\alpha^k_i = \int_o^{y^k_i} a(z)\, dz, \ i = 1, \ldots, N+1 \tag{29}$$

$$y^k_o = 0 \tag{30}$$

$$r^k_{N+1} + b(y^k_L)\,[y^k_N + K] = 0 \tag{31}$$

The function a is assumed to be continuous and piecewise linear, then the values

α^k_i, $i = 1, \ldots, N+1$ are exactly computed.

The matrix $Q = q_{ij}$ is a band matrix, computing Q with the Simpson rule leads to exact values and Q is symetric but tridiagonal, the approximation of Q by the trapezoidal rule gives a diagonal matrix and the resolution is simplified.

Resolution of the non-linear equation (31)

When the matrix Q is diagonal, the non-linearity of equations (27), (31) is reduced to just one scalar equation which may be written at each step k

$$F^k(\omega) = b(\omega)\,[K + y^k_N] + \frac{2}{h}[\alpha^k_N - \alpha(\omega)] = 0 \tag{32}$$

We get the following result concerning the solution of equation (32)

If $y^k_N \in [0,1[$, $\forall k > 0$

then $\exists\, \omega \in\,]y^k_N, 1]$, ω unique solution of (32).

This solution is noted y^k_L.

Proof. For k=1 to M+1 and for every ω in $[0,1]$, (omitting k), the functions a, b and F are such that

F is continuous and derivable,

$b'(\omega) < 0$ and $a(\omega) > 0$

hence $F'(\omega) \ll 0$

and F is monotone decreasing ; from equation (32), we get $F(y_N) = b(y_N)\,[y_N + K] > 0$

and $F(1) = r_{N+1} = 2\,[\alpha_N - \alpha(1)]/h$,

the function α is monotone increasing,

therefore $F(1) \leqslant 0$

and equation (32) admits a unique solution in $]y_N, 1]$.

This solution is computed using the secant method.

Stability. The explicit approximation in equation (24) leads to a numerical recurring scheme which is conditionnally stable. A simplified study assuming a and b to be constant, ends in the sufficient condition on the choice of time step

$$\tau < \tau_c = h^2/[2a(1 + Pe/2)] \tag{33}$$

where Pe, Peclet number, is equal to b.h/a. Since Pe is time warying, equations (27) to (32) are solved with a variable time step.

An implicit approximation of equation (24) would allow to avoid the constraint (33) on the time step, but it would be necessary to resolve at each time step a system of non-linear equations.

Algorithm. Suppose that y_i^{k-1} (i=1, ..., N+1), $b(y_L^{k-1})$ and r_i^{k-1} (i=1, ..., N+1) are known, then the solution at time t_k is calculated as follows,

Step 1. Calculate a time step τ which ensures stability, equation (33).

Step 2. Calculate y_j^k, j=1, ..., N using explicit equation (27).

Step 3. Calculate α_j^k using equation (29).

Step 4. Calculate y_L^k and r_j^k, (j=1, ..., N+1) solution of the non linear equations (28) to (31).

Step 5. If $b(y_L^k) > 0$ go to step 1, else stop.

For k=o, the algorithm is initialized at step 3, y_i^o, (i=1,...,N), are known, r_j^o, (j = 1,N+1) and y_L^o are calculated to satisfy equation (31).

5. NUMERICAL RESULTS

Simulations are run with laboratory data from V.L. Vilker (1) concerning ultrafiltration of BSA solutions (bovine serum albumin) and from S. Poyen (4) concerning ultrafiltration of water dextran solutions.

The following results have been obtained with the parameter values of tables 1 and 2. For each experiment, the evolution of the ultrafiltrate volume which has been measured is compared to the values computed with the model.

$$v(t^k) = S\tau . \sum_{j=1}^{k} b(y_L^j) \tag{34}$$

S is the section cell.

Ultrafiltration of BSA

Table 1. Experimental data

Lp	L	S	ΔP	C_i	D
cm / kPa.mn	cm	cm^2	kPa	g/l	cm^2/s
(9.1)10^{-4}	0.25	1.524	70	101	(1.8)10^{-7}

Figure 3 shows the variations of the osmotic pressure with concentration. To get a good adjustement between measured and predicted volumes, Vilker took a diffusion coefficient value five times as great as experimental value of table 1 ; results of figures 4 and 5 have been obtained with D equals to (3.6) 10^{-7} cm^2/s.

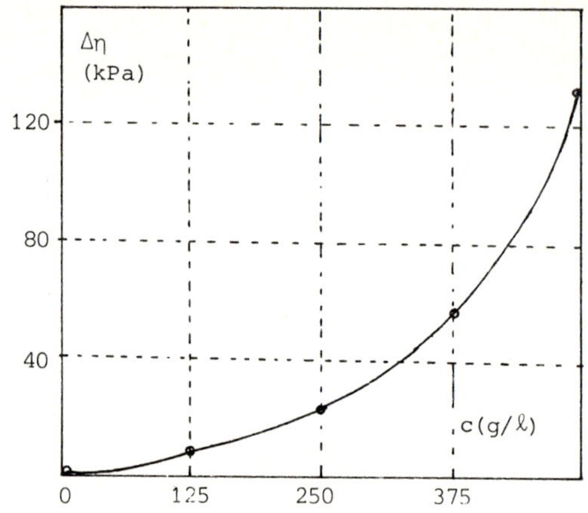

Figure 3. Osmotic pressure as a function of concentration (BSA)

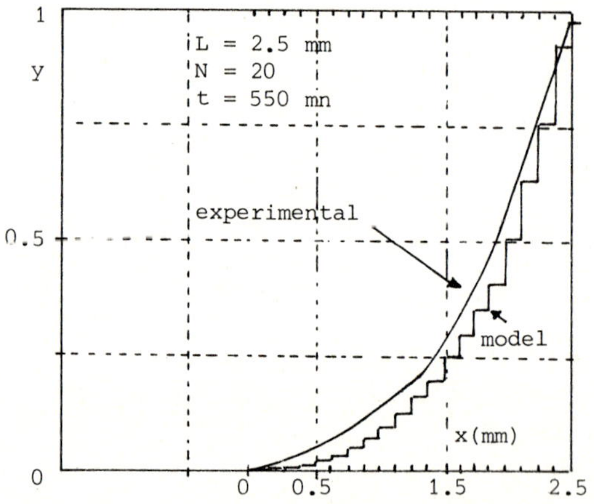

Figure 4. Experimental and model concentration profiles in polarized layer (BSA)

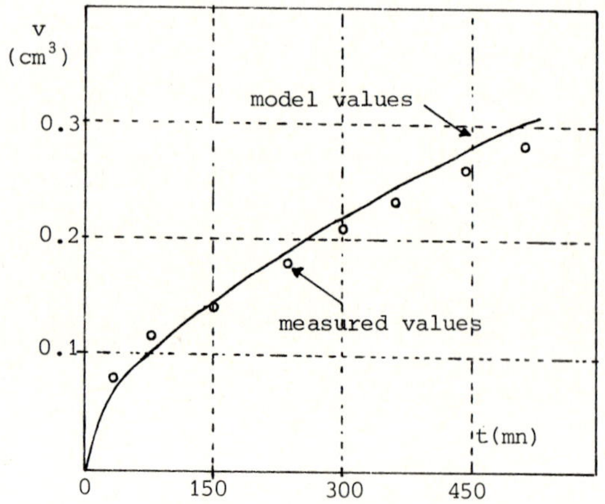

Figure 5. Ultrafiltrate volume as a function of elapsed time (BSA)

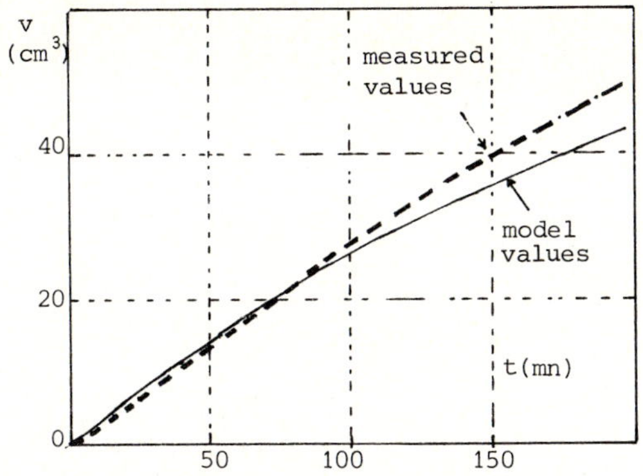

Figure 6. Ultrafiltrate volume as a function of elapsed time (Dextran 70)
N = 50 ; L = 0.8 mm

Ultrafiltration of Dextran 7Ø

Table 2. Experimental data.

Lp	L	S	ΔP	C_i	T
0.885 cm / kPa.mn	0.08cm	41.8cm^2	100KPa	1g/1	200 mn

The osmotic pressure and the diffusion coefficient are given on figure 2. The comparison of measured and computed volumes of ultrafiltrate is shown on figure 6. The influence of variations of parameters values on the evolution of the volume of ultrafiltrate has been studied in simulation, it results a great dependance with the law $\Delta\eta$ (c) ; now this law like the law D(c) are determined by experiments which are subject to high uncertainty.

However numerical results are in good agreement with experiments.

6. CONCLUSION

The concentration polarization occuring in ultrafiltration is described by a non linear convective diffusion equation. An approximation using mixed-finite elements is presented. A numerical difficulty for this type of equation is due to the simultaneous variations of flow velocity and diffusion coefficient during a simulation. These variations may be characterized by the Peclet number. The interest of this approximation method is to get a numerical conservative scheme which is adapted to variations of Peclet number.

Works with this model and various laboratory data are in progress, they will allow to contribute to a better knowledge of phenomena occuring in ultrafiltration.

REFERENCES

[1] Vilker, V.L., Colton, C.K., Smith, K.A., Concentration polarization in protein ultrafiltration, AICHE Journal, vol. 27, n°4, 1981.

[2] Chavent, G., Salzano, G., A finite element method for the 1-d water flooding problem with gravity, Journal of computational physics, vol. 45, n° 3, 1982.

[3] Cohen, G., Résolution numérique et identification pour une équation quasi-parabolique non linéaire dégénérée de diffusion et transport en dimension 1, Thèse, Paris, 1978.

[4] Poyen, S., Contribution à l'étude de l'ultrafiltration, ultrafiltration dynamique d'huiles de moteur, promotion de turbulence, étude théorique en statique et dynamique, Thèse, Rennes, 1984.

[5] Picot, J.D., Etude en simulation d'un modèle d'ultrafiltration, DEA, Nantes, 1984.
Raviart, P.A., Les méthodes d'éléments finis mixtes en mécanique des fluides, Editions Eyrolles, 1981.

[6] Lesaint, P., Raviart, P.A., On a finite element method for solving the neutron transport equation, Mathematical aspects of finite elements in partial differential equations, Ed. Carl de Boor, Academic Press, 1974.

Complex and Distributed Systems: Analysis, Simulation and Control
S.G. Tzafestas and P. Borne (editors)
Elsevier Science Publishers B.V. (North-Holland)

NUMERICAL RESULTS BY ADI METHODS ON PERIODIC REACTION-DIFFUSION SYSTEMS FOR EPIDEMIC MODELS

Luciano Galeone - Carmela Mastroserio
Dipartimento di Matematica
Università di Bari - Italy

Abstract. In this paper we study the numerical solution of reaction-diffusion systems with periodic in time nonlinear term and in particular we are interested to a system which models the evolution of a class of bacterial diseases. In order to solve the system we use a class of alternating direction methods. In particular we study the existence and the symptotic stability of the numerical periodic solutions. To simulate the behaviour of the theoretical solution, we impose that the numerical scheme is of positive type. Some numerical experiments are presented.

1. INTRODUCTION

The aim of this paper is to study the numerical solution of reaction-diffusion systems with periodic in the time non linear term F, obtained by means of a class of alternating direction methods.
This kind of systems have been introduced to model the spread of some viral and bacterial diseases that may have seasonal fluctuation, such as typhoid fever, viral hepatitis in some Mediterranean regions.
The system that here we consider has the following general form:

$$(1.1)\quad \begin{aligned} \frac{\partial U_1(x,t)}{\partial Ut} &= d_1 \Delta U_1(x,t) + F_1(U_1(x,t), U_2(x,t), t) \\ & \qquad , (x,t) \in \Omega \times R_+ \\ \frac{\partial U_2(x,t)}{\partial Ut} &= d_2 \Delta U_2(x,t) + F_2(U_1(x,t), U_2(x,t), t) \end{aligned}$$

with initial and boundary conditions

(1.1a) $U_i(x,0) = U_i^o(x) > 0$, for $x \in \Omega$, $i=1,2$

(1.1b) $U_i(x,t) = 0$, for $(x,t) \in \partial\Omega \times R$, $i=1,2$,

where $U_1(x,t)$ denotes the spatial density of the bacterial population and $U_2(x,t)$ that one of the infective human population in Ω .

The functions $F_1(U,t)$ with $U=(U_1,U_2)$ are periodic in the time with period . Systems of the type (1.1) have been studied in [3] where the θ-method was used.

In this paper we consider $\Omega=[0,1]^2$ and d_1, d_2 positive constants. Moreover the reaction terms F_1 and F_2 have the following particular form

$$(1.2)\quad \begin{aligned} F_1(U,t) &= -a_{11}U_1(x,t) + a_{12}U_2(x,t) \\ F_2(U,t) &= -a_{22}U_2(x,t) + p(t)h(U_1(x,t)), \end{aligned}$$

where the coefficients are nonnegative and the function p(t) is a periodic function given by:

$$p(t) = p_0 + p_1(t) \quad , \quad t \in [0,\infty[.$$

The first term p_0 denotes a constant in the time average livel, while the second term $p_1(t)$ may be thought of as a periodic perturbation, with some period $\omega > 0$, that corresponds to seasonal variations around the average level p_0. Furthermore we require that for the function h in (1.2), the following properties hold:

i) h is an increasing monotone function;
ii) h is a sufficiently regular function such that $h''(z) < 0, \forall z > 0$;
iii) $h(0) = 0$.

The behaviour of the theoretical solution of this system has been studied in [2].

In order to find the numerical solution of (1.1), we introduce a class of alternating direction methods [5], that presents the practical advantage of the reduction of computational

---

* Work performed in the context of the Special Program "Infectious Diseases" of the C.N.R.

effort. Furthermore we wish to analyze the behaviour of the obtained numerical solution over long intervals of time and to this end we introduce two threshold parameters to study the stability of the trivial solution. Finally, we find some sufficient conditions such that a positive periodic asymptotically stable solution exists.

To prove this results we require the positivity of the numerical scheme. We observe that this property signifies a very strong stability form and it is closely related to the positivity of the matrices which appear in the scheme. We reach this form of stability thanks to some results on M-matrices and to the quasi-monotonicity property of the non linear term $F=(F_1,F_2)$.

2. NUMERICAL SCHEME

In this section we consider a uniform mesh on Ω given by a set of lines parallel to the axes x and y, with $\Delta x = \Delta y = h$, where $h = \frac{1}{N+1}$, N positive integer. Moreover we denote by $u^n = (u_1^n, u_2^n)' \in R^{2N^2}$ the numerical solution at the time $t_n = n\Delta t$, n=0,1..., where Δt is the time increment.

The coordinates of u_i^n i=1,2, that correspond to the interior points of the spatial mesh, are arranged according to the columns of the grid. Moreover we set

$$f(u^n, t_n) = (f_1(u^n,t_n), f_2(u^n,t_n))'$$

where $f_i(u^n,t_n) \in R^{N^2}$ is the vector whose elements are given by the values of F_i evaluated in the mesh points.

When we discretize the equations (1.1), with respect to the spatial variables, using, for example, second order contered differences, we can write the resulting system in the following way:

$$U_t = (L_x + L_y)U + f(U,t) \tag{2.1}$$

where U(t) denotes a R^{2N^2}-vector corresponding to the mesh.

The $2N^2$-matrices L_x and L_y are given by

$$L_x = \frac{1}{h^2}\begin{pmatrix} d_1 A\otimes I & 0 \\ 0 & d_2 A\otimes I \end{pmatrix}, \quad L_y = \frac{1}{h^2}\begin{pmatrix} d_1 I\otimes A & 0 \\ 0 & d_2 I\otimes A \end{pmatrix} \tag{2.2}$$

where A is the tridiagonal N-matrix obtained when we discretize the $-\Delta$ operator in one direction, and I denotes the identicity matrix of the same dimension. The simbol $\otimes$ denotes the Kronecker product between two matrices. Thanks to the form of the system, we discretize the equations respect to the time, using a class of alternating direction methods [5] and we obtain the following numerical scheme:

$$\begin{aligned} (I+\theta\Delta t L_x)u^* &= (I-\psi\Delta t L_y)u^n + \Delta t f(u^n,t_n), \\ (I+\theta\Delta t L_y)u^{n+1} &= (I-\psi\Delta t L_y)u^* \\ u^o &= U(to) \end{aligned} \tag{2.3}$$

where θ and ψ are two non negative parameters such that $\theta+\psi = 1$. The scheme (2.3) shows that the numerical solution at every time-step is obtained resolving 2N tridiagonal systems of order N; in this way the computational effort and the dimension of the problem is considerably reduced.

In order to study the behaviour of the periodic solutions given by the numerical scheme (2.3), we need to see that the following result holds:

THEOREM 2.1. Let $d=\max\{d_1,d_2\}$ and $c=\max\{a_{11}, a_{22}\}$, if Δt and h are taken to satisfy:

$$\psi \frac{\Delta t}{h^2} \le \frac{1-\alpha}{2d}, \quad \Delta t < \frac{\alpha}{c}, \tag{2.4}$$

with $0 < \alpha < 1$, then we have the following properties:

a) $(I+\theta\Delta t Lx)$ and $(I-\theta\Delta t Ly)$ are M-matrices;
b) $(I-\psi\Delta t Lx)$ and $(I-\psi\Delta t Ly)$ are positive matrices;
c) if $u^o \ge 0$, then $u^n \ge 0$ for every $n \in N$, and $u^n > 0$ for n sufficiently large;
d) if $0 \le u^o \le v^o$, then $u^n \le v^n$ for every $n \in N$.

PROOF. The proof of a) and b) can be found in [5]. To prove the other results, we introduce the Jacobian matrix of the nonlinear term, given by

$$J(U,t) = \begin{pmatrix} -a_{11} & a_{12} \\ p(t)h'(U_1) & -a_{22} \end{pmatrix} \tag{2.5}$$

Moreover we need also to introduce the numerical Jacobian matrix:

$$(2.6)\qquad \tilde{J}^n = \begin{pmatrix} \tilde{J}^n_{11} & \tilde{J}^n_{12} \\ \tilde{J}^n_{21} & \tilde{J}^n_{22} \end{pmatrix},$$

where the bloks $\tilde{J}_{ij}$ are diagonal N^2-matrices so defined:

$$\tilde{J}^n_{11} = -a_{11}I,\quad \tilde{J}^n_{12} = a_{12}I,\quad \tilde{J}^n_{22} = -a_{22}I$$

$$J^n_{21} = p(t_n)\begin{pmatrix} h'(\xi^n_{1,1}) & & 0 \\ & \ddots & \\ 0 & & h'(\xi^n_{1,N^2}) \end{pmatrix},$$

with $\xi^n_1 \in R^{N2}$.

This expression of the matrix $\tilde{J}^n$ follows from the particular term F given in (1.2). For a general non linear term, the matrix $\tilde{J}^n$ is defined in [3]

Now, thanks to the hypotheses (2.4), we deduce that the matrix

$$(2.7)\qquad (I-\Psi\Delta t\, Ly + \Delta t\tilde{J}n) \text{ is non-negative}$$

and then, in an analogous way to the proof of the theorem 2.1 in [3], we can see that u^*, in the numerical scheme (2.3), is non-negative when $u^n > 0$. So the proof of c) follows immediately. Finally, by the positivity of the matrix $(I-\Psi\Delta t Ly + \Delta t\tilde{J}^n)$, we can see in an analogous way that also the property d) holds.

3. PERIODIC SOLUTIONS

In this section we introduce some conditions about the existence and stability of periodic solutions of the numerical scheme (2.3). To this end we study the problem of existence and stability of equilibrium points for over and upper solutions.

Now we define the operator of monodromy associated to the system (2.3), as

$$(3.1)\qquad H_\omega : R^{2N^2} \to R^{2N^2},\ \ni' \forall\, u^o \in R^{2N^2} : H_\omega(u^o) = u^p,$$

where p is an integer such that $p\Delta t = \omega$ and u^p is the solution of the numerical scheme at time ω, with initial condition u^o.

We will say that u^n is an ω-periodic solution when

$$u^{kp+r} = u^r,\ \text{for } k \in N,\ r \in N \qquad 0 \le r \le p.$$

REMARK 1. It follows immediately that u^n is an ω-periodic solution if u^o is a fixed point for H_ω and that H_ω is a monotone non-decreasing operator, thanks to the property d) of Theorem 2.1. Now we give the following

DEFINITION 1. A function $\underline{U}(x,t)$ is called a lower solution of system (1.1) if:

$$(3.2)\qquad \begin{aligned} &\frac{\partial \underline{U}}{\partial t} \le D\Delta\underline{U}(x,t) + F(\underline{U},t), && (x,t) \in \Omega \times R_+ \\ &\underline{U}(x,t) < 0 && (x,t) \in \partial\Omega \times R_+ . \end{aligned}$$

In an analogous way we can give the definition of an upper solution $\overline{U}(x,t)$. Now we introduce the following numerical systems:

$$(3.3)\qquad \begin{aligned} &(I+\theta\Delta t Lx)V^* = (I-\psi\Delta t Ly)V^n + \Delta t g_1(V^n) \\ &(I+\theta\Delta t Ly)V^{n+1} = (I-\psi\Delta t Lx)V^* \end{aligned}$$

and

$$(3.4)\qquad \begin{aligned} &(I+\theta\Delta t Lx)W^* = (I-\psi\Delta t Ly)W^n + \Delta t g_2(W^n) \\ &(I+\theta\Delta t Ly)W^{n+1} = (I-\psi\Delta t Lx)W^* \end{aligned}$$

where g_1 and g_2 are such that

$$g_1(V^n) < f(u^n, t^n) < g_2(W^n)$$

and we observe that V^n and W^n, $n=0,\ldots,$ are approximations of the over and upper solutions as defined in (3.2).

For the problem that we are studing we can consider the matrix C given by:

$$(3.5)\qquad C = \begin{pmatrix} -a_{11} & a_{12} \\ p_{max}h'(0) & -a_{22} \end{pmatrix}$$

where $p_{max} = \max_{0<t<\omega} p(t)$.

It is easy to see that, by the property ii) in section 1, $J(U,t) < C$, where J is the Jacobian matrix given in (2.5), and that, by the iii)

$$F(U,t) \le C \cdot U$$

Then the system (3.4) becomes

$$(I+\theta\Delta tLx)W^* = (I-\psi\Delta tLy + \Delta t\tilde{C})W^n \tag{3.6}$$
$$(I+\theta\Delta tLy)W^{n+1} = (I-\psi\Delta tLx)W^*$$

where $\tilde{C} = C \otimes I$.

Moreover the matrix

$$B^{\epsilon} = \begin{pmatrix} -a_{11} & a_{12} \\ p_{min}h'(0)-\epsilon & -a_{22} \end{pmatrix} \tag{3.7}$$

with $p_{min} = \min_{0<t<\omega} p(t)$ and $\epsilon > 0$, is such that

$$F(U,t) \geq B^{\epsilon}U$$

when U is in a neighbourhood of the origin.

Then for system (3.3) we can take

$$(I+\theta\Delta tLx)V^* = (I-\psi\Delta tLy + \Delta t\tilde{B}^{\epsilon})V^n$$
$$(I+\theta\Delta tLy)V^{n+1} = (I-\psi\Delta tLx)V^*$$

with $\tilde{B}^{\epsilon} = B^{\epsilon} \otimes I$.

We can prove the following results:

THEOREM 3.1. Under the hypotheses of Theorem 2.1 we have:

1) if u^o, v^o and w^o are the initial conditions of the systems (2.3), (3.3) and (3.4) respectively, such that $v^o \leq u^o \leq w^o$, then it is $v^n \leq u^n \leq w^n$ for n=1,2...

2) If the sequences v^n and w^n are convergent to the points v and w, then it is $v \leq w$.

3) The monodromy sequence $(H^n_\omega(w))_{n\in N}$ $((H^n_\omega(v))_{n\in N})$ is not increasing (not decreasing) monotone and convergent. The fixed point $\xi_+ = \lim_n H^n_\omega(w)$ $(\xi_- = \lim_n H^n_\omega(v))$ is the initial value of an ω-periodic solution of system (2.3).

4) If w is asymptotically stable for (3.4), the fixed point ξ_+ is independent of the particular upper-solution w^n.

PROOF:

1) Let us consider u^o, v^o as initial condition of the systems (2.3) and (3.3). If we subtract the first equations of these systems and then we add and substract the quantity $f(v^o, t_o)$, in the right hand side of the obtained equation, we have

$$(I+\theta\Delta tLx)(u^*-v^*) =$$
$$= (I-\psi\Delta tLy)(u^o-v^o) + \Delta t(f(u^o,t^o) - f(u^o,t^o)) + $$
$$+ \Delta t(f(v^o,t^o) - g_1(v^o)) ,$$

that is:

$$(I+\theta\Delta tLx)(u^*-v^*) = (I-\psi\Delta tLy + \Delta t\tilde{J}_o)(u^o-v^o) +$$
$$+ \Delta t(f(v^o,t^o) - g_1(v^o)).$$

For the positivity of the matrices which appear and for the definition of g_1, it follows that $u^*-v^* \geq 0$ and then that $u-v \geq 0$. In the same way we can see that $u^n - v^n$ for $n > 1$ and analogously that $u^n - w^n \leq 0$.

2) The Proof follows from the property 1).

3) First of all see that $H_\omega(w) \leq w$. Let $\bar{\xi}^n$ be the numerical solution of (2.3) with $\bar{\xi}^o_o = w$, then we can prove, by induction that ξ^o + $- \bar{\xi}^n \geq 0$. We have that:

$$(I+\Delta t\theta L_x)\bar{\xi}^o = (I-\Delta t\psi L_y)\bar{\xi}^o + \Delta t\, g_2(\bar{\xi}^o).$$
$$(I+\Delta t\theta L_y)\bar{\xi}^o = (I-\Delta t\psi L_x)\bar{\xi}^o$$

Then it's immediate to verify that $\bar{\xi}^o - \bar{\xi}^1 \geq 0$. Now, let us suppose that $\bar{\xi}^o - \bar{\xi}^{n-1} \geq 0$. In a way similar to that used in the proof of 1) we obtain that

$$(I+\Delta t\theta L_x)(\bar{\xi}^o-\bar{\xi}^*)=(I-\psi L_1 + t\, J^{n-1}_1)(\bar{\xi}^o - \bar{\xi}^{n-1}) +$$
$$+ \Delta t(g_2(\bar{\xi}^o)-f(\bar{\xi}^o,t_{n-1}))$$
$$(I+\Delta t\theta L_y)(\bar{\xi}^o - \bar{\xi}^n) = (I-\psi\Delta tL_x)(\bar{\xi}^o - \bar{\xi}^*)$$

from which it follows that $\bar{\xi}^o - \bar{\xi}^n \geq 0$. This result, for n=p, proves that $H_\omega(w) \leq w$, and then thanks to the property of monotonicity of H_ω, we have that $(H^n_\omega(w))_{n\in N}$ is a not--increasing sequence.

Moreover, if $u^o \geq 0$, the monodromy sequence is bounded from below and then it is a convergente sequence. If we put $\xi_+ = \lim_{n\to\infty} H^n_\omega(w)$ an ω-periodic solution of system (2.3) exists.

Analogously we can prove the result for the sequence $H^n_\omega(v)$, which results bounded from above by $(H_\omega^n(w))_{n\ N}$.

4) The proof follows for the used A.D.I. scheme in an analogous way to that in [3].

4. STABILITY FOR THE PERIODIC SOLUTIONS.

Thanks to those results, it is possible to introduce threshold parameters to study the stability and instability of the origin. We observe that if the trivial solution is globally asymptotically stable, the epidemic tends to extintion, if it is instable then a globally asymptotically stable positive periodic solution exists which corresponds to a seasonally fluctuation epidemic level.

We have the following results, the proof of which can be obtained in a similar way as in [3]. We suppose that the hypotheses of Theorem 2.1 are verified.

LEMMA 4.1 If the matrix $-C$ is an M-matrix, then, the matrix $Lx + Ly - \tilde{C}$ is an M-matrix.

PROOF. We observe that the matrix $Lx + Ly$ may be written in this way:

$$Lx + Ly = \begin{pmatrix} d_1 & 0 \\ 0 & d_2 \end{pmatrix} L$$

with $L = \frac{1}{h^2}(A \otimes I + I \otimes A)$.

In order to prove the result we may use the following equivalence:

(3.8) A is an M-matrix $\Longleftrightarrow \exists$ a positive diagonal matrix $D \ni D^{-1}AD$, is a diagonally dominant.

At the first, we apply the (3.8) to the matrix $-C$. Then, thanks to the structure of $Lx + Ly$ and of $\tilde{C}$, we can see that the matrix $(Lx+Ly-\tilde{C})$ satisfies (3.8) with the diangonal matrix given by $\begin{pmatrix} d_1 & 0 \\ 0 & d_2 \end{pmatrix} \otimes I$.

THEOREM 4.1. If the matrix $-C$ is an M-matrix, then the iteration matrix

$$T = (I+\theta\Delta tL_x)^{-1}(I-\psi\Delta tL_x)(I+\theta\Delta tLy)^{-1}(I-\psi\Delta tL_y+\Delta t\tilde{C})$$

has spectral radius $\rho(T) < 1$, for Δt sufficiently small.

PROOF. It is known that the spectral radius of a positive matrix A is minor of one iff $(I-A)^{-1} \geq 0$. We may apply this results to matrix T. To show that $(I-T)^{-1} \geq 0$, we shall see that the matrix $(I-T)$ is an M-matrix. To this end we can use the following result: if A and B are two matrices such that $A^{-1} > 0$ and $A^{-1}B > 0$, then the matrix $(A-B)$ is an M-matrix iff the matrix $I-A^{-1}B$ is an M-matrix.

If we apply this result to the matrices $A = (I+\theta\Delta tL_x)(I+\theta\Delta tL_y)$ and $B = (I-\Delta t\psi L_x)(I-\Delta t\psi L_y + \Delta t\tilde{C})$, by the above Lemma, the matrix $(A-B)$ is an M-matrix and then the thesis is reached.

COROLLARY. If $a_{11}\, a_{22} - a_{12}\, P_{max} h'(0) > 0$, then the trivial solution is asymptotically stable for the numerical scheme.

THEOREM 4.2. Let λ_1 be the minimum eigenvalue of the matrix A. If the matrix $2 \frac{\lambda_1}{h^2} D-C$ is an M-matrix, then the result of Theorem 4.1 holds.

COROLLARY. If it is:

$$\theta_{max} = \frac{a_{12} P_{max} h'(0)}{(a_{11} + 2 \frac{d_1\lambda_1}{h^2})(a_{22} + 2 \frac{d_2\lambda_1}{h^2})} < 1,$$

then the trivial solution is asymptotically stable for the numerical scheme.

By considering the matrix B defined in (3.7), we can see:

THEOREM 4.3. If

$$\theta_{min} = \frac{a_{12} P_{min} h'(0)}{(a_{11} + 2 \frac{d_1\lambda_1}{h^2})(a_{22} + 2 \frac{d_2\lambda_1}{h^2})} > 1,$$

then the trivial solution is unstable.

THEOREM 4.4. Let the trivial solution be unstable. If a quasi-monotone matrix C exists such that for some $R > 0$

$$F(U,t) \leq C \cdot U \qquad |U| > R$$

and such that Theorem 4.1 holds, then there is an unique positive asymptotically stable periodic solution for the numerical scheme.

COROLLARY. If it is:

$$\lim_{z\to\infty} \sup \frac{h(z)}{z} < a_{21},$$

and

$$\frac{a_{12}a_{21}\ p_{max}}{(a_{11} + 2\frac{d_1\lambda_1}{h^2})(a_{22} + \frac{d_2\lambda_1}{h^2})} < 1,$$

then the result of Theorem 4.4 holds.

5. A NUMERICAL EXAMPLE.

The results shown in the previous section have been tested for the A.D.I. method, given by (2.3) with $\theta=1$ on the following reaction-diffusion system:

$$\frac{\partial U_1}{\partial t} = d_1\Delta U_1 - U_1 + 4U_2$$

$$\frac{\partial U_2}{\partial t} = d_2\Delta U_2 - U_2 + (2+\text{sen}t)(U_1/(1+U_1))$$

with homogeneous Dirichlet boundary conditions and with the following initial conditions:

$$U_1(x,0)=\begin{cases} 1/10 & \text{for } d < 1/9 \\ 1/(1+d) - 4/5 & \text{for } 1/9 < d < 1/4 \\ 0 & \text{for } d > 1/4 \end{cases}$$

$$U_2(x,0)=\begin{cases} 1/(1+d) - 4/5 & \text{for } d < 1/4 \\ 0 & \text{for } d > 1/4 \end{cases}$$

where $d=(x_1 - \frac{1}{2})^2 + (x_2 - \frac{1}{2})^2$ and $x=(x_1,x_2)\in [0,1]^2$.

In the numerical experiments we have set $\Delta t=0.1$, $\Delta x = 0.1$.

The Fig. 1 shows the different behaviour of the numerical solution ω_1^n, evaluated for $n=0,1...200$ in a point of the mesh, in terms of different values of the parameters $d_1 = d_2$, as indicated in the picture.

We observe that when the constants $d_1 = d_2$ are greater then a fixed value close to 0.09, we have that $\theta_{max} > 1$ and the trivial solution is asymptotically stable, that is the epidemic tends to extinction.

In the other case a periodic epidemic level is reached by the concentration of bacteria.

This two different behaviour are better shown in Fig. 2 and Fig. 3.

We have obtained analogous pictures for the other solution u_2^n.

REFERENCES.

[1] A. Berman, R. Plemmons: Non negative matrices in the mathematical Sciences, Accademic Press, New York, 1979.

[2] V. Capasso, L. Maddalena: Periodic solutions for a reaction-diffusion system modelling the spread of a class of epidemics, SIAM J. APPL. MATH., Vol. 43, N.2, 1983, pp. 417-427.

[3] L. Galeone, C. Mastroserio: Stability of the numerical solution of the periodic reaction-diffusion system. Appl. Num. Math. (1985), N.4, Vol.1, pp. 315-324.

[4] M.A. Krasnosel'skii: The operator of traslation aloug the trajectories of Differential Equations, Trans, Math. Monographs, 19, A.M.S., Providence, 1968.

[5] C. Mastroserio, M. Montrone: Invariant regions and asymptotic behaviour for the numerical solution of reaction-diffusion systems by a class of alternating direction methods, Calcolo XXI, 1984, pp. 269-279.

[6] R.S. Varga: Matrix Iterative Analysis, Prentice-Hall, Englewood Cliffs, N.J. 1962.

Fig. 1

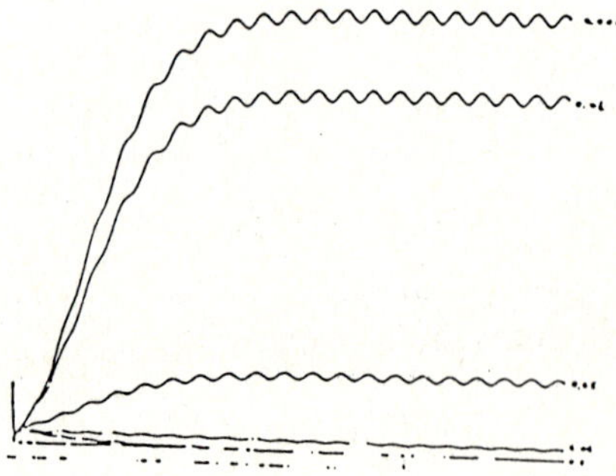

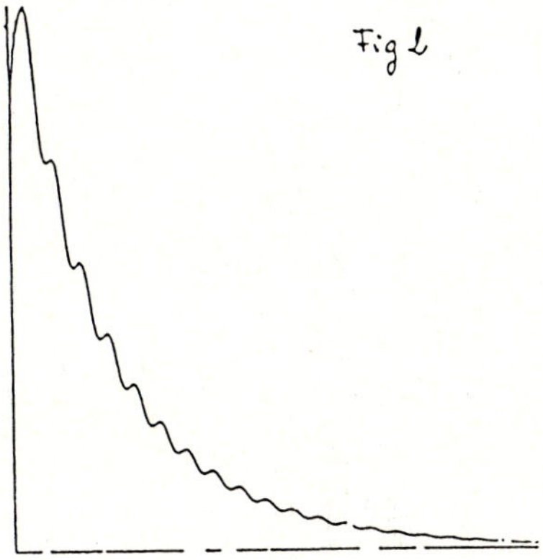
Fig 2

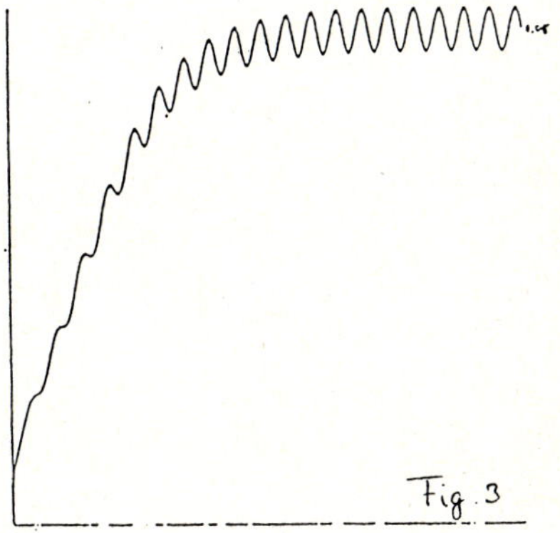
Fig. 3

Complex and Distributed Systems: Analysis, Simulation and Control
S.G. Tzafestas and P. Borne (editors)
Elsevier Science Publishers B.V. (North-Holland)

NEW RESULTS ON WALSH FUNCTION ANALYSIS AND IDENTIFICATION OF DISTRIBUTED-PARAMETER SYSTEMS

S. Tzafestas and J. Kalat*
Computer Science Division
National Technical University
Zografou 15773, Athens, Greece

ABSTRACT

This paper presents some new results on Walsh functions which are used for the analysis and identification of distributed parameter systems (DPS). Specifically a recursive Walsh function generation method is given and the Walsh transform of integral expressions of two-variable functions is derived. Then the identification of first-order time-varying DPS is studied and a necessary condition for simultaneous identification of boundary conditions and parameters is given. After that, relationships between Walsh functions and polynomials are established which make possible the estimation of boundary conditions directly from the state function Walsh series expansion. The paper closes with the application of the Walsh function technique for solving (simulating) second-order DPS of the parabolic form with three different types of boundary conditions.

1. INTRODUCTION

Walsh functions constitute a wonderful set of orthogonal functions which have received over the recent years increasing attetion in a variety of engineering areas such as communication, signal processing, control, etc. Due to the fact that Walsh functions and transforms are naturally more suited for digital computation, an effort was made to gradually replace the Fourier transform by Walsh-type transforms. Actually the Walsh function field has experienced a significant development and a large amount of theoretical and applied results are presently available.

The foundations of the Walsh functions field were made by Rademacher (1922) [1] Walsh (1923) [2], Fine (1950) [3-4], Paley (1952) [5], and Kaczmarz and Steinhaus (1951) [6]. The engineering approach to the study and utilization of these functions was originated by Harmuth [7,8] who has introduced the concept of "sequency" to represent the associated generalized frequency defined as one-half the mean rate of zero crossings. The variety of Walsh function definitions is due to the existence of different orderings. In the sequency ordering (or Walsh ordering), which is popular in communication engineering, Walsh functions are ordered according to the zero crossings (or sign changes), This ordering implies that ith Walsh function wal (i,t) has i zero crossings in the interval $t \varepsilon [0,1]$, and, obviously, is directly related to the sequency concept. The Paley ordering [5] is cha-

* On leave from Computer Centre, Techn. University of Bialystok, Bialystok, Poland.

racterized by the fact in this form Walsh functions are represented by products of Rademacher functions which lead to useful recursive Walsh signal generation algorithms. A third ordering was proposed by Henderson [9,10] and is merely Paley s ordering in reversed binary. This ordering is computationally attractive and arizes when one computes fast Walsh transforms (FWT) without sorting. Yuen [11] calls the index i, in wal(i,t), the "zequency" of wal(i,t) to show that it represents both the generalized frequency and the number of wal (i,t).

Surveys on Walsh transform theory and its application to system analysis and control can be found in Tzafestas [12] and Rao and Tzafestas [13]. A well-balanced set of important papers on the subject reflecting the state of art is provided in Tzafestas [14]. The key point for the analysis and synthesis of dynamic systems via the Walsh function expansion approach is the concept of operational matrices which reduces the computation of integral expressions to a set of linear algebraic equations in the Walsh domain.

In this paper the Walsh expansion approach is extended for simulating second-order DPS and identifying time varyings DPS. Two numerical examples are included.

2. SOME NEW RESULTS ON WALSH FUNCTIONS

2.1. Walsh Function Generation

One of the most popular ways of Walsh function definition in papers on Walsh function applications is the one that is based on Rademacher function products. Namely, to obtain the k-th Walsh function one has to express the number k in a binary system i.e.

$$k=d_{\ell+1}2^{\ell}+d_{\ell}2^{\ell-1}+\ldots+d_2 2+d_1 \tag{1}$$

and then to multiply the Rademacher functions as follows [1,15]:

$$\Phi_0(t)=r_0(t)=1$$

$$\Phi_k(t)=(r_{\ell+1}(t))^{d_{\ell+1}}(r_1(t))^{d_\ell}\ldots(r_1(t))^{d_1} \tag{2}$$

$$\text{for } k=1,2,\ldots.$$

where

$$r_\ell(t)=\text{sign}(\sin(\ell\pi t)),\ \text{for } \ell=1,2,\ldots \tag{3}$$

However it is usually necessery to generate a set of Walsh functions $\Phi_i(t)$, $i=0,1,\ldots,2^L-1$ and then a new recursive method, requiring only one multiplication for every Walsh function, can be derived as:

$$\Phi_0(t)=1$$

$$\Phi_{2^N+i}(t)=\Phi_i(t)r_N(t),\ \text{for } N=1,2,\ldots \tag{4}$$

$$i=0,1,\ldots,2^N-1$$

The proof of the formula (4) follows from properties of Walsh functions. Namely

$$r_N(t)\Phi_i(t)=\Phi_{2^N}(t)\Phi_i(t)=\Phi_{2^N\oplus i}(t)=\Phi_{2^N+i}(t) \tag{5}$$

$$\text{for } i=0,1,\ldots,2^N-1$$

where $i\oplus j$ denotes modulo-2 addition on the binary digits of i and j.

2.2 Walsh Transform of Space and Time Varying Functions

An important tool in Walsh function approach to the analysis and identification problems are operational matrices.

It has been shown by Paraskevopoulos and Bounas [5] that the followling relationship is true.

$$\underbrace{\int_0^x \cdots \int_0^x}_{m\text{-times}} \underbrace{\int_0^t \cdots \int_0^t}_{n\text{-times}} \Psi_M^T(x) Y_{MN} \Phi_N(t) \underbrace{dt \ldots dt}_{n\text{-times}} \underbrace{dx \ldots dx}_{m\text{-times}}$$

$$= \Psi_M^T(x) (P_M^T)^m Y_{MN} (P_N)^n \Phi_N(t) \qquad (6)$$

where $\Psi_n(x)$, $\Phi_N(t)$ denote vectors of Walsh functions with respect to x and t, P_M (or P_N) are the operational matrices [17] of order M (or N). Making use of the relationship (6) the identification of time-invariant lumped and distributed system has been possible [16], [18]-[20] To identify time or space varying distributed parameter systems, the relationship (6) has been extended as follows:

If the Walsh series of the functions y(x,t) and a(t) are given

$$y(x,t) = \Psi_M^T(x) Y_{MN} \Phi_N(t) \qquad (7)$$

$$a(t) = \Phi_\alpha^T(t) a_\alpha \qquad \alpha \leqq N \qquad (8)$$

Then the following relationships hold:

$$\underbrace{\int_0^x \cdots \int_0^x}_{m\text{-times}} \underbrace{\int_0^t \cdots \int_0^t}_{n\text{-times}} y(x,t) a(t) \underbrace{dt \ldots dt}_{n\text{-times}} \underbrace{dx \ldots dx}_{m\text{-times}}$$

$$= \Psi_M^T(x) \left[\sum_{\ell=0}^{\alpha-1} (P_M^T)^m Y_{0,\ell} (P_N)^n a_\ell \right] \Phi_N(t) \qquad (9)$$

$$\underbrace{\int_0^x \cdots \int_0^x}_{m\text{-times}} \underbrace{\int_0^t \cdots \int_0^t}_{n\text{-times}} y(x,t) \underbrace{\int_0^t \cdots \int_0^t}_{k\text{-times}} a(t) \underbrace{dt \ \ldots \ dt}_{(n+k)\text{-times}}$$

$$\underbrace{dx \ldots dx}_{m\text{-times}} = \Psi_M^T(x) \left[\sum_{\ell=0}^{\alpha-1} (P_M^T)^m Y_{0,\ell}^{(k)} (P_N)^n a_\ell \right] \Phi_N(t) \qquad (10)$$

where

$$Y_{k,\ell_{ij}} = Y_{i \oplus k, j \oplus \ell} \qquad (11)$$

for $i,k=0,1,\ldots,M-1$, $j,\ell=0,1,\ldots,N-1$

$$Y_{0,\ell}^{(s)} = \sum_{r=0}^{N-1} Y_{0,\ell} p_{\ell r}^s \qquad (12)$$

where p_{1r}^s denotes the (1,r)th element of $(P_N)^s$.

The last two relationships (9) and (10) are based on the property of mod-2 addition:

$$i \oplus j = k \equiv k \oplus j = i, \qquad (13)$$

for every $i,j=0,1,2,\ldots$

The form of these relationships is convenient since the matrices $Y_{0,\ell}$ and $Y_{0,\ell}^{(k)}$ corresponding to each Walsh coefficient a_ℓ are explicitly given. The concept of the matrix $Y_{k,\ell}$ defined by (11) is also very useful for describing properties of Walsh series. For example, the following relationship has been established for double Walsh series in [20]:

$$\Psi_M^T(x) A_{MN} \Phi_N(t) \Phi_N^T(t) (B_{MN})^T \Psi_M(x) = \Psi_M^T(x) C_{MN} \Phi_N(t) \qquad (14)$$

$$C_j = j\text{th column of } C_{MN} = \sum_{i=1}^{M} \Lambda_{i-1}^{(M)} A_{MN} \Lambda_{j-1}^{(N)} b_i \qquad (15)$$

where b_i is the ith row of matrix B_{MN} and $\Lambda_i^{(M)}$ and $\Lambda_j^{(N)}$ are (MxM) and (NxN) matrices, respectively, which can be obtained from

$$\Lambda_i^{(M)} = \left[\begin{array}{c|c} \Lambda_i^{(M/2)} & 0_{(M/2)} \\ \hline 0_{(M/2)} & \Lambda_i^{(M/2)} \end{array} \right] \quad \text{and}$$

$$\Lambda_{(i+M/2)}^{(M)} = \left[\begin{array}{c|c} 0_{(M/2)} & \Lambda_i^{(M/2)} \\ \hline \Lambda_i^{(M/2)} & 0_{(M/2)} \end{array} \right] \qquad (16)$$

$$\text{for } i=0,1,\ldots,(M/2-1) \text{ and } \Lambda_0^{(M/2)} = I_{(M/2)} \qquad (17)$$

Using the $Y_{k,\ell}$ notation, the relations (15) to (17) are found to be equivalent to

$$C_{MN} = \sum_{k=0}^{M-1} \sum_{1=0}^{N-1} A_{k,1} b_{k,1} \tag{18}$$

The method of the $Y_{k,\ell}$ matrix generation has been presented in [21]. This method is closely related to the concept of a similar matrix proposed by Chen and Shih [22].

3. IDENTIFICATION OF TIME VARYING FIRST ORDER DISTRIBUTED PARAMETER SYSTEMS

Consider linear time-varying distributed parameter system described by the following partial differential equation

$$a(t)\frac{\partial y(x,t)}{\partial t} + b(t)\frac{\partial y(x,t)}{\partial x} + c(t)y(x,t) = u(x,t) \tag{19}$$

where $a(t)$, $b(t)$ and $c(t)$ are unknown parameter functions to be identified; $u(x,t)$ and $y(x,t)$ are the two-variable input and output functions, with known double Walsh series expansions. Here it is also assumed that the first derivative of $a(t)$ exists and can be expanded in Walsh series. To identify the above system, eqn (19) is first converted to an integral one, namely

$$\int_0^x y(x,t)\int_0^t a'(t)dtdx - \int_0^x\int_0^t y(x,t)a'(t)dtdx + \int_0^t b(t)y(x,t)dt$$
$$+\int_0^x\int_0^t y(x,t)c(t)dtdx + \int_0^x y(x,t)a(0)dx - \int_0^x y(x,0)a(0)dx$$
$$-\int_0^t b(t)y(0,t)dt = \int_0^x\int_0^t u(x,t)dtdx \tag{20}$$

where $y(x,0)$ and $y(0,t)$ are the unknown initial and boundary conditions. To determine the unknown parameters, $y(x,0)$ and $y(0,t)$, we first approximate them by a finite Walsh series as follows:

$$a'(t) = \Phi_\alpha^T(t)\hat{a}_\alpha \qquad \alpha \leq N \tag{21}$$

$$b(t) = \Phi_\beta^T(t)b_\beta \qquad \beta \leq N \tag{22}$$

$$c(t) = \Phi_\gamma^T(t)c_\gamma \qquad \gamma \leq N \tag{23}$$

$$a(0)y(x,0) = \Psi_q^T(x)\hat{d}_q = \sum_{i=0}^{q-1} \hat{d}_i \Psi_M^T(x)E_{i+1,1}\Phi_N(t) \qquad q \leq M \tag{24}$$

$$b(t)y(0,t) = \Phi_p^T(t)\hat{f}_p = \sum_{i=0}^{p-1} \hat{f}_i \Psi_M^T(x)E_{1,i+1}\Phi_N(t) \qquad p \leq N \tag{25}$$

where E_{ij} is an MxN matrix having the (i,j)th element unity and the remainig elements zero.

Now, let us expand the input and output functions into double Walsh series as follows:

$$u(x,t) = \Psi_M^T(x)U_{MN}\Phi_N(t) \tag{26}$$

$$y(x,t) = \Psi_M^T(x)Y_{MN}\Phi_N(t) \tag{27}$$

Introducing (21)-(27) into (20), and making use of the transformations (9) and (10) one obtains:

$$\Psi_M^T(x)\Bigg[\sum_{\ell=0}^{\alpha-1} P_M^T(Y_{0,\ell}^{(1)} - Y_{0,\ell}P_N)a_\ell + \sum_{\ell=0}^{\beta-1} Y_{0,\ell}P_N b_\ell +$$
$$+\sum_{\ell=0}^{\gamma-1} P_M^T Y_{0,\ell}P_N c_\ell + P_M^T Y_{MN}a(0) - \sum_{i=0}^{q-1} P_M^T E_{i+1,1}\hat{d}_i -$$
$$\sum_{i=0}^{p-1} E_{1,i+1}P_N\hat{f}_i\Bigg]\Phi_N(t) = \Psi_M^T(x)P_M^T U_{MN}P_N\Phi_N(t) \tag{28}$$

Equating the coefficients of like Walsh function products $\Psi_i(x)\Phi_j(t)$, $i=0,1,\ldots M-1$ and $j=0,1,\ldots,N-1$ in (28), a system of MN linear algebraic equations, containing $\alpha+\beta+\gamma+1+p+q$ unknows, is obtained.

This system can be written in the standard form

$$A\theta = h \qquad (29)$$

where A is the $MN\times(\alpha+\beta+\gamma+1+p+q)$ matrix of known elements, θ is the vector of unknowns:

$$\theta^T = [\hat{a}_\alpha, b_\beta, c_\gamma, a(0), \hat{d}_q, \hat{f}_p]$$

and h is the MN vector of known elements. If there is enough input-output data equation (29) can be solved with the least squares technique to yield $\theta^* = (A^TA)^{-1}A^Th$, provided that A^TA is invertible. The solution θ^* gives the Walsh coefficients of the parameters b(t) and c(t) and also the value of $\alpha(0)$ directly, whereas the series coefficients of the parameter a(t) and of y(0,t) and y(x,0):

$$a(t) = \Phi_\alpha^T(t)a_\alpha \qquad (30)$$

$$y(x,0) = \Psi_q^T(x)d_q \qquad (31)$$

$$y(0,t) = \Phi_p^T(t)f_p \qquad (32)$$

one determined from the relations

$$a_\alpha = P_\alpha^T\hat{a}_\alpha + [a(0),0,\ldots,0]^T \qquad (33)$$

$$d_q = \frac{1}{a(0)}\hat{d}_q \qquad (34)$$

$$f_p = B_p^{-1}\hat{f}_p \qquad (35)$$

under the assumption, that $a(0)\neq 0$ and the matrix B is invertible where B_p is a pxp matrix corresponding to the vector b_p i.e.

$$B_{ij} = b_{i\oplus j} \quad \text{for } i,j=0,1,..,p-1 \qquad (36)$$

The assumption that A^TA must be invertible is not trivial as if the following condition are satisfied:

$$q>M/2 \quad \text{and} \quad p>N/2 \qquad (37)$$

the the final system of equations (29) is linearly depedent for every input output data. To obtain linear independence of it, both series (24) or (25) must be appropriately truncated. As a result of this, the estimates of the boundary conditions as well as of the parameters may be degraded [19].

To overcome this problem the following relationships between Walsh functions and polynomials have been established, which allow for the simultaneous estimation of boundary and initial conditions on the basis of the state variable Walsh function expansion.

<u>Lemma</u> For a given Walsh series

$$\sum_{i=0}^{N-1} c_i\Phi_i(t) = \Phi_N^T(t)c_N \qquad (38)$$

it is possible to obtain a polynomial of (L-1)th degree

$$b_0+\beta_1t+\ldots+\beta_{L-1}t^{L-1} = [1,t,..,t^{L-1}]\cdot b_L \qquad (39)$$

such that its mean values will be the same as those of the Walsh series at the first L subintervals $\tau_j = [j/N, j+1/N]$, $L\leq N$. The coefficients b_i of this polynomial can be calculated by

$$b_L = Z_{LN}c_N \qquad (40)$$

where

$$Z_{LN} = D_{LL}^{-1}V_{LL}^{-1}T_{LL}W_{LN} \qquad (41)$$

$$D_{ij}^{-1} = \begin{cases} iN^{i-1} & \text{for } i=j \\ 0 & \text{for } i\neq j \end{cases}$$

$$V_{ij} = i^j$$

$$T_{ij} = \begin{cases} 1 & \text{for } i\geq j \\ 0 & \text{for } i<j \end{cases}$$

and W_{LN} represents the first L rows of the Walsh matrix W_{NN} [23]. The proof of this Lemma follows directly from the properties of integrals of polynomials. As mentioned above this lemma can be used to estimate the boundary and initial conditions on the basis of the double Walsh series (27). Let us assume that the initial condition y(x,0) is to be estimated, i.e. the vector d_q in (31) is to be determined. For simplicity let q=M, where M is an integer power of 2. By the discrete Walsh transform, the calculation of the vector d_q is equivalent to the determination of the vector g_M of mean values of the function y(x,0) in subintervals $hx_i = [(i-1)/M, i/M)$, for i=1,...,M. In order to obtain the vector g_M, let us consider a set of functions $g_i(t)$ defined as follows

$$g_i(t)=\int_{(i-1)/M}^{i/M} y(x,t)dx, \quad i=1,2,..,M \tag{42}$$

Our task is then reduced to that of determining the values $g_i(0)$ for i=1,..,M. Making use of the above lemma, for each function $g_i(t)$ one can obtain a sequence of polynomials of the ℓ-th degree, where $\ell=0,1,...,N-1$. If the functions $g_i(t)$ are regular enough in the interval 0,L/N then the zero power coefficients of the polynomials of the degree from 0 to L will converge into $g_i(0)$ values. Thus the estimate of the initial condition may be achieved.

Using the above procedure, the problem of simultaneous parameter and boundary condition identification can be reduced to a pure parameter identification problem. There is no need to say, of course, that if the boundary and initial conditions are known, we don't need to use the above procedure. A solution to the identification problem of time-varying parameters of a second-order distributed parameter system can be found in [24]. Here we include a numerical example of space dependent parameter identification (see Example 2).

4. WALSH FUNCTION ANALYSIS OF PARABOLIC DISTRIBUTED PARAMETER SYSTEMS.

The Walsh functions have been applied successfuly in the analysis of various of systems such as lumped systems with constant and time-varying coefficients [17], [25], [26] delay systems [27] or nonlinear ones [28], [29]. This approach has been used also to solve first-order partial differitial equations [30].

The main problem in extending this technique to higher order distributed system analysis is that we have to transform the partial differential equation under consideration into an integral one containing only known initial and boundary conditions.

The analysis of distributed parameter system via Walsh functions will be illustrated in this section by treating three distinct sets of boundary conditions for parabolic systems with constant parameters.

Consider linear time-invariant distributed parameter system desribed by the following parabolic equation:

$$\frac{\partial y(x,t)}{\partial t}=\alpha\frac{\partial y^2(x,t)}{\partial x^2}+\beta\frac{\partial y(x,t)}{\partial x}+cy(x,t)+u(x,t) \tag{43}$$

with the initial condition

$$y(x,0)=y_0(x) \tag{44}$$

and three different boundary conditions

$$\left.\begin{aligned} &y(0,t)=y^0(t) \\ &\frac{\partial y(x,t)}{\partial x}\bigg|_{x=1}=y^1(t) \end{aligned}\right\} \qquad (45)$$

$$\left.\begin{aligned} &y(0,t)=y^0(t) \\ &y(1,t)=y^1(t) \end{aligned}\right\} \qquad (46)$$

$$\left.\begin{aligned} &\frac{\partial y(x,t)}{\partial x}\bigg|_{x=0}=y^0(t) \\ &\frac{\partial y(x,t)}{\partial x}\bigg|_{x=1}=-Bi\, y(1,t) \end{aligned}\right\} \qquad (47)$$

Integrating (43) with respect to t and making use of the initial condition (44) the following integral-partial equation is produced:

$$y(x,t)-y_0(x)=\int_0^t\left[\alpha\frac{\partial^2 y(x,t)}{\partial x^2}+\beta\frac{\partial y(x,t)}{\partial x}+cy(x,t)+u(x,t)\right]dt \qquad (48)$$

In the next step, we have to replace the state function and its first derivative by integral expressions containing the second partial derivative $\partial^2 y/\partial x^2$ and given boundary conditions.

The first boundary condition (45) is the simplest one. Integrating $\partial^2 y/\partial x^2$ twice from 1 to t and from 0 to t, one obtains

$$\frac{\partial y(x,t)}{\partial x}=\int_1^x\frac{\partial^2 y(x,t)}{\partial x^2}dx+y^1(t) \qquad (49)$$

$$y(x,t)=\int_0^x\int_1^x\frac{\partial^2 y(x,t)}{\partial x^2}dxdx+xy^1(t)+y^0(t) \qquad (50)$$

Thus the first parabolic boundary problem defined by (43), (44), (45) is equivalent to the integral equation

$$\int_0^x\int_1^x\frac{\partial^2 y(x,t)}{\partial x^2}\,dxdx-\int_0^t\{\alpha\frac{\partial^2 y(x,t)}{\partial x^2}+\beta\int_1^x\frac{\partial^2 y(x,t)}{\partial x^2}dx+c\int_0^x\int_1^x\frac{\partial^2 y(x,t)}{\partial x^2}dxdx\}dt=y_0(x)-y^0(t)+xy^1(t)+$$
$$+\int_0^t\{\beta y^1(t)+c\left[y^1(t)-y^0(t)\right]+\int_0^x\int_1^x u(x,t)dxdx\}dt \qquad (51)$$

The only unknown in this equation is the second partial derivative $\partial^2 y/\partial x^2$.

In order to transofrm the second boundary value problem, i.e. equation (43), (44) and (46) let consider the following integral expressions:

$$y(x,t)=\int_0^x\int_0^x\frac{\partial^2 y(x,t)}{\partial x^2}\,dxdx+y^0(t)+\int_0^x\frac{\partial y(x,t)}{\partial x}\bigg|_{x=0}dx \qquad (52)$$

$$y(x,t)=\int_1^x\int_0^x\frac{\partial^2 y(x,t)}{\partial x^2}\,dxdx+y^1(t)+\int_1^x\frac{\partial y(x,t)}{\partial x}\bigg|_{x=0}dx \qquad (53)$$

Combining equations (52) and (53), the unknown boundary condition $\partial y/\partial x|_{x=0}$ can be written as follows

$$\frac{\partial y(x,t)}{\partial x}\bigg|_{x=0}=y^1(t)-y^0(t)-\int_0^1\int_0^x\frac{\partial^2 y(x,t)}{\partial x^2}dxdx \qquad (54)$$

Then

$$\frac{\partial y(x,t)}{\partial x}=\int_0^x\frac{\partial y(x,t)}{\partial x}dx-\int_0^1\int_0^x\frac{\partial\, y(x,t)}{\partial x}dxdx+y^1(t)-y^0(t) \qquad (55)$$

$$y(x,t)=\int_0^x\int_0^x\frac{\partial^2 y(x,t)}{\partial x^2}dxdx-\int_0^x\int_0^1\int_0^x\frac{\partial^2 y(x,t)}{\partial x^2}dxdxdx+y^0(t)(1-x)+xy^1(t) \qquad (56)$$

and the second boundary-value problem can be reduced to the following integral one

$$\int_0^x\int_0^x \frac{\partial^2 y(x,t)}{\partial x^2}dx\,dx - \int_0^x\int_0^1\int_0^x \frac{\partial^2 y(x,t)}{\partial x^2}dx\,dx\,dx -$$

$$\int_0^t\{\alpha\frac{\partial^2 y(x,t)}{\partial x^2} + \beta\left[\int_0^x \frac{\partial^2 y(x,t)}{\partial x^2}dx - \int_0^1\int_0^x \frac{\partial^2 y(x,t)}{\partial x^2}dx\,dx\right] +$$

$$c\left[\int_0^x\int_0^x \frac{\partial^2 y(x,t)}{\partial x^2}dx\,dx - \int_0^x\int_0^1\int_0^x \frac{\partial^2 y(x,t)}{\partial x^2}dx\,dx\,dx\right]\}\,dt =$$

$$= y_0(x) - y^0(t) - x(y^1(t) - y^0(t)) + \int_0^t\{\beta\left[y^1(t) - y^0(t)\right] +$$

$$+\int_0^x\int_0^x u(x,t)\,dx\,dx\}\,dt \qquad (57)$$

Finally, integral expressions of the state function and its first derivative for the third boundary-value problem can be obtained from the following equality:

$$\left.\frac{\partial y(x,t)}{\partial x}\right|_{x=1} = \int_0^1 \frac{\partial^2 y(x,t)}{\partial x^2}dx + y_0(t) \qquad (58)$$

Namely:

$$\frac{\partial y(x,t)}{\partial x} = \int_0^1 \frac{\partial^2 y(x,t)}{\partial x^2}dx + y_0(t) \qquad (59)$$

$$y(x,t) = \int_1^x\int_0^x \frac{\partial^2 y(x,t)}{\partial x^2}dx\,dx - \frac{1}{Bi}\int_0^1 \frac{\partial^2 y(x,t)}{\partial x^2}dx +$$

$$+(x-1-\frac{1}{Bi})y^0(t) \qquad (60)$$

Making use of these relationships, the integral equation corresponding with the boundary problem (43), (44), (47) is derived.

$$\int_1^x\int_0^x \frac{\partial^2 y(x,t)}{\partial x^2}dx\,dx - \frac{1}{Bi}\int_0^1 \frac{\partial^2 y(x,t)}{\partial x^2}dx - \int_0^t\{\alpha\frac{\partial^2 y(x,t)}{\partial x^2} +$$

$$+\beta\int_0^x \frac{\partial^2 y(x,t)}{\partial x^2}dx + c\left[\int_1^x\int_0^x \frac{\partial^2 y(x,t)}{\partial x^2}dx\,dx -\right.$$

$$\left.\frac{1}{Bi}\int_0^1 \frac{\partial y(x,t)}{\partial x}dx\right]\}\,dt = y_0(x) - y^0(t)(x-1-\frac{1}{Bi})$$

$$+\int_0^t y^0(t)\left[\beta + c(x-1-\frac{1}{B_i})\right] + \int_1^x\int_0^x u(x,t)\,dx\,dx\}\,dt \qquad (61)$$

To solve the integral equations (51), (57) and (61) in the Walsh function domain, let assume the form of double series approximating the unknown second derivative

$$\frac{\partial^2 y(x,t)}{\partial x^2} = \Psi_M^T(x)\tilde{Y}_{MN}\Phi_N(t) \qquad (62)$$

and expand the known function u(x,t) and boundary-inital conditions as follows

$$u(x,t) = \Psi_M^T(x)U_{MN}\Phi_N(t) \qquad (63)$$

$$y_0(x) = \Psi_M^T(x)f_M \triangleq \Psi_M^T(x)F_{MN}\Phi_N(t) \qquad (64)$$

$$y^0(t) = g_N^T\Phi_N(t) \triangleq \Psi_M^T(x)G_{MN}\Phi_N(t) \qquad (65)$$

$$y^1(t) = h_N^T\Phi_N(t) \triangleq \Psi_M^T(x)H_{MN}\Phi_N(t) \qquad (66)$$

Then the integral equation can be reduced to the double Walsh identities using the operational matrix P, and the operational matrix for backward integration Q [12] :

$$\int_1^x \Psi_M(x)dx = Q_M^T\Psi_M(x) \qquad (67)$$

and an operational matrix $E_{1,1}$

$$\int_0^1 \Psi_M(x)dx = E_{1,1}\Psi_M(x) \qquad (68)$$

where E_{ij} is the MxM matrix having (i,j)th element unity and the remaining elements zero.

Namely

$$\Psi_M^T(x)\{P_M^TQ_M^T\tilde{Y}_{MN} - (\alpha I_M + \beta Q_M^T + cP_M^TQ_M^T)\tilde{Y}_{MN}P_N\}\Phi_N(t)$$

$$=\Psi_M^T(x)\{F_{MN} - G_{MN} + P_M^TH + \left[\beta H_{MN} + c(P_M^TH_{MN} - G_{MN}) +\right.$$

$$\left.P_M^TQ^TU_{MN}\right]P_N\}\Phi_N(t) \qquad (69)$$

$$\Psi_M^T(x)\{(P_M^T(I_M-E_{11})P_M^T\tilde{Y}_{MN}-\left[Q_M I_M+\beta(I_M-E_{11})P_M^T+cP_M^T(I-\right.$$

$$\left.-E_{11})P_M^T\right]\tilde{Y}_{MN}\cdot P_N\}\Phi_N(t)=\Psi_M^T(x)\{F_{MN}-G_{MN}-P_M^T(H_{MN}-G_{MN})$$

$$+\left[\beta(H_{MN}-G_{MN})+P_M^T P_M^T U_{MN}\right]P_N\}\Phi_N(t) \tag{70}$$

$$\Psi_M^T(x)\{(Q_M^T P_M^T-\frac{1}{B_i}I_M)Y_{MN}-\left[\alpha I_M+\beta P_M^T+c(Q_M^T P_M^T-\frac{1}{Bi}E_{11})\right.$$

$$\left.\tilde{Y}_{MN}\right]P_N\}\Phi_N(t)=\Psi_M^T(x)\{F_{MN}-P_M^T G_{MN}+(1+\frac{1}{B_i})G_{MN}+\left[cP_M^T G_{MN}+\right.$$

$$\left.+(\beta-1-\frac{1}{B_i})G_{MN}+Q_M^T P_M^T U_{MN}\right]P_N\}\Phi_N(t) \tag{71}$$

Equating the coefficients of like Walsh function products $\Psi_i(x)\Phi_j(t)$, $i=0,1,\ldots M-1$ and $j=0,1,\ldots N-1$ in (69)-(71) we finally obtain three systems of algebraic equations which are equivalent to the above three parabolic boundary-value problems.

The solution of each equation is represented by matrix of Walsh coefficients of the second derivative. The Walsh coefficients matrix Y_{MN} of the state functions of the first, second and third boundary value problems are then obtained from the formulas given below:

$$Y_{MN}=P_M^T Q_M^T Y_{MN}+P_M^T H_{MN}+G_{MN} \tag{72}$$

$$Y_{MN}=P_M^T(I_M-E_{11})P_M^T\tilde{Y}_{MN}+P_M^T(H_{MN}-G_{MN})+G_{MN} \tag{73}$$

$$Y_{MN}=(Q_M^T-\frac{1}{B_i}I_M)P_M^T\tilde{Y}_{MN}+P_M^T G_{MN}-(1+\frac{1}{B_i})G_{MN} \tag{74}$$

Making use of the Walsh transform for the integral expression (9) it is also possible to solve the same boundary value problems for time (or space) varying distributed parabolic systems.

5. NUMERICAL EXAMPLES

Example 1:

Let us consider the following parabolic boundary value problem

$$\frac{\partial y(x,t)}{\partial t}=0.9\frac{\partial^2 y(x,t)}{\partial x^2} \tag{75}$$

$$y(0,t)=0 \qquad y(x,0)=1$$

$$\left.\frac{\partial y(x,t)}{\partial x}\right|_{x=1}=0 \tag{76}$$

It is in the form of the first boundary-value problem (43)-(45) and the equivalent system of Walsh function equations is given by equation (69). For M,N=4 the following mean values of the state function have been obtained.

y(x,t) M,N=4	$0\le t<1/4$	$1/4\le t<1/2$	$1/2\le t<3/4$	$3/4\le t\le 1$
$0\le x<1/4$	0.29916	0.04546	0.10437	0.00315
$1/4\le x<1/2$	0.66955	0.25995	0.19732	0.08941
$1/2\le x<3/4$	0.83502	0.47303	0.25319	0.15444
$3/4\le x\le 1$	0.89817	0.58772	0.29399	0.17682

In the next two tables the mean values of the state function have been calculated by the proposed method for M,N=8 and for comparison purposes, by a series solution [20]:

$$y(x,t)=\frac{4}{\pi}\sum_{k=1}^{\infty}\frac{\exp\left[-0.9(2k-1)^2(\frac{\pi}{2})^2 t\right]\sin(\frac{2k-1}{2}\pi x)}{2k-1}$$

y(x,t) M,N=8	$0\le t<1/4$	$1/4\le t<1/2$	$1/2\le t<3/4$	$3/4\le t<1$
$0\le x<1/4$	0.26498	0.11420	0.06418	0.03647
$1/4\le x<1/2$	0.62198	0.31145	0.17637	0.10034
$1/2\le x<3/4$	0.82180	0.46187	0.26380	0.15063
$3/4\le x\le 1$	0.90119	0.54834	0.31273	0.17805

y(x,t) [20]	$0\le t<1/4$	$1/4\le t<1/2$	$1/2\le t<3/4$	$3/4\le t\le 1$
$0\le x<1/4$	0.27286	0.10901	0.06240	0.03581
$1/4\le x<1/2$	0.61162	0.31011	0.17769	0.10199
$1/2\le x<3/4$	0.81030	0.46341	0.26593	0.15269
$3/4\le x\le 1$	0.89374	0.54605	0.31368	0.18005

Example 2:

The identification method presented in this paper can be used not only for the identification of time-varying parameters of the first-order system, but also for higer-order ones.

Let us consider the second order hyperbolic distributed-parameter system

$$m(x)\frac{\partial^2 y(x,t)}{\partial t^2} - EF\,\frac{\partial^2 y(x,t)}{\partial x^2} = 0 \qquad (77)$$

with inital and boundary conditions:

$$y(x,0)=0 \qquad\qquad y(0,t)=0 \qquad (78)$$

$$\left.\frac{\partial y(x,t)}{\partial x}\right|_{x=1} = 0 \quad \left.\frac{\partial y(x,t)}{\partial t}\right|_{t=0} = V = \text{cont.} \qquad (79)$$

where the parameter m(x) is to be identified. The state function has been determined exactly by the characteristic method for

$$L=3.3, \qquad E=10^9$$
$$x\in[0,L] \qquad L=3.3$$
$$t\in[0,T] \qquad T=0.004$$

$$m(x)=\begin{cases}80.0 & \text{for } 0\leq x\leq L/2\\ 20.0 & \text{for } L/2<x\leq L\end{cases}$$

Expanding the state function in double Walsh series as:

$$y(xL,tT)=\Psi_M^T(x)\,Y_{MN}\,\Phi_N(t) \qquad (80)$$

and assuming the following form of the unknown parameter Walsh series

$$m(xL)=\Psi_q^T(x)\,m_q \qquad (81)$$

the following system of Walsh functions has been derived

$$\Psi_M^T(x)\left[\sum_{1=0}^{q-1} P_M^T Q_M^T \tilde{Y}_{1,0}\cdot m_1\right]\Phi_N(t)=\kappa\cdot\Psi_M^T(x)\left[Y_{MN}P_N^2\right]\Phi_N(t) \qquad (82)$$

where

$$\tilde{Y}_{MN}=Y_{MN}-VTE_{11}P_N$$

$$\kappa=EFT^2/L^2$$

Equation (82) was solved for M,N=8 and for q=4. In the table below mean values calculated from this solution are compared with the exact ones.

Values	$0\leq x<L/4$	$L/4\leq x<L/2$	$L/2\leq x<3L/4$	$3L/4\leq x\leq L$
Calculated	77.839	79.714	18.987	23.936
Exact	80.0	80.0	20.0	20.0

6. CONCLUSIONS

This paper presents an extension of the Walsh function approach for the analysis and identification of distributed parameter systems. The algorihms derived reduce the problem of identification of time-varying distributed systems and of second-order boundary value problems to that of solving linear algebraic equations.

The concept of the $Y_{k,1}$ matrix allows one to apply Walsh functions and operational matrices to the identification of time or space-varying systems. Making use of this concept the final system of equations involves much less unknowns than by say using Chebyshev function approach [32].

REFERENCES

1. H. Rademacher, "Einige Satze von Allgemeinen Orthogonal functionen", Math. Ann., Vol. 87, pp. 122-138, 1922.
2. J.L. Walsh, "A Closed Set of Normal Orthogonal Functions", Amer.J.Math., Vol. 45, pp. 5-24, 1923.
3. N.J.Fine "On the Walsh Functions" Trans. Amer. Math.Soc., Vol. 65,pp. 372-414, 1949.
4. N.J. Fine, "The Generalized Walsh Functions", Trans.Amer.Math.Soc., Vol. 69, pp. 66-77, 1950.
5. R.E.A.C. Paley, "A Remarkable Series of Orthogonal Functions",Proc. London Math. Soc., Vol. 2, No. 24, pp. 241-279, 1952.
6. S. Kaczmarz and H. Steinhaus, Theorie der Orthogonalreihen, New York, Chelsea, 1951.
7. H.F. Harmuth, "Transmission of Information by Orthogonal Functions, New York, Springer, 1969.
8. H.F. Harmuth, "Applications of Walsh Functions in Communications, IEEE Spectrum, Vol. 6, pp. 82-91, 1969.
9. K.W. Henderson, Comment on "Computation of the Fast Walsh Fourier Transform", IEEE Trans. Comp. (Corresp.), Vol. C-19,
10. K.W. Henderson, "Some Notes on the Walsh Functions", IEEE Trans. Comp. (Corresp.), Vol. EC-13, pp. 50-52, 1964.
11. C.K.Yen, "Remarks on the Ordering of Walsh Functions, IEEE Trans.Comp. (corresp.), pp. 1452, 1972.
12. S.G. Tzafestas "Walsh transform theory and its application to system analysis and control:an overview", Math. and Comp. in Simul., pp. 214-225, 1983.
13. G.P. Rao and S.G. Tzafestas "A decade of picewise constant orthogonal functions in systems and control Math. and Comp. in Simul., Vol. 27, no 5,6 pp. 389-407 1985.
14. S.G.Tzafestas "Walsh Functions in Signal and Systems Analysis and Design", Van Nostrand Reinhold Co., 1985.
15. R.E.A.C. Paley "A Remarkable Series of Orthogonal Functions", Proc.London Math.Soc., no. 24, pp. 241-279, 1932.
16. P.N. Paraskevopoulos and A.C. Bounas "Distributed parameter system identification via Walsh functions", Int. J. System Sci., Vol. 9, no. 1, pp. 75-83, 1978.
17. C.F. Chen and C.H.Hsiao, "A state space approach to Walsh series solution of linear systems", Int. J. System Sci., Vol. 6, no 9, pp 833-858, 1975.
18. G.P.Rao and L.Sivakumar, "System identification via Walsh functions", Proc, IEE, vol. 122, no. 10 pp 1160-1161, 1975.
19. S.G. Tzafestas,"Walsh series approach to lumped and distributed system identification", J. Franclin Inst., vol. 305, no. 4, pp. 199-220, 1978.
20. M.S.P. Sinha, V.S. Rajamani and A.K. Sinha, "Identification of non-linear distributed system using Walsh functions", Int. J.Control, vol. 32,no 4, pp 669-676, 1980.
21. J.Kalat, "Identification of varying parameter via Walsh function in distributed parameter system", 10th IMACS World Congress, vol. 3 pp.377-379, 1982.
22. W.L.Chen and Y.P.Shih, "Analysis and optimal control of time-varying linear system via Walsh functions",

Int.J.Control, vol. 27, no. 6, pp. 917-932, 1978.

23. C.F. Chen and C.H.Hsiao, "A state space approach to Walsh series solution of linear systems", Int. System Sci., vol. 6, no. 9, pp.833-858, 1975.

24. J. Kalat, "Identification of varying parameters in distributed parameter systems",(Ph. D.),IPPT, Polish Academy of Science, Warsaw, 1985.

25. M.S. Corrington, "Solution of differential and integral equations with Walsh functions", IEEE Trans. on Circuit Theory, vol. CT-20, pp.470-476, 1973.

26. C.H. Chen and C.F. Hsiao, "Solving integral equations via Walsh functions", Comput. and Elect. Engn., vol. 6, pp. 279-292, 1979.

27. W.J. Chen and Y.P. Shih, "Shift Walsh matrix and delay differential equations", IEEE Trans., vol. AC-23, no. 6, pp. 1023-1028, 1978.

28. M. Maqusi, "On the Walsh analysis of nonlinear systems",IEEE Trans., vol. EMC-20, pp. 519-523, 1978.

29. K.R. Palanisamy, "Analysis of nonlinear systems via single term Walsh series approach", Int.J.Syst.Sci., vol. 13, no.8, pp.929-935, 1982.

30. Y.P. Shih and J.Y.Han,"Double Walsh series solution of first order PDE" Int. J. Syst. Sci., vol. 9, no. 5, pp. 569-578, 1978.

31. M.P. Polis, R.E. Goodson and M.J. Wozny, "On parameter identification for distributed system using Galerkin's criterion", Automatica, no. 9, pp. 53-64, 1973.

32. P.N. Paraskevopoulos and G.Th. Kekkeris,"Identification of time-invariant and time-varying distributed parameter systems using Chebyshev functions", Proc. of Int. AMSE Conf. "Modelling and Simulation", Nice, vol. 1, pp. 51-69, 1983.

Complex and Distributed Systems: Analysis, Simulation and Control
S.G. Tzafestas and P. Borne (editors)
Elsevier Science Publishers B.V. (North-Holland)

IDENTIFICATION OF SPATIALLY VARYING PARAMETERS IN DISTRIBUTED PARAMETER SYSTEMS

A. EL BADIA - M. COURDESSES*

Laboratoire d'Automatique et d'Analyse des Systèmes du C.N.R.S.
7, avenue du Colonel Roche
31077 Toulouse Cédex, France

Identification of spatially varying parameters in distributed parameter systems is considered, within deterministic framework, as the direct resolution of an inverse problem. The observation is assumed to be given at one boundary.

1. INTRODUCTION

In a major body of work concerning parameter identification in "real physical systems" it is implicitely assumed that a distributed phenomenon can be approximated by using a lumped model. As a result, not much literature has been devoted to the problem of parameter identification in distributed system as compared to the classical problem.

If the goal to be reached is to simulate a process or to derive control strategies, the parameters are often of little interest and a lumped model is valid. If it is the determination of these parameters which is aimed at it is evident that a distributed model is then necessary. In the latter case, particular care should be taken to provide for a unique identification of the parameters. As in lumped systems, there are different ways of approaching the identifiability questions in distributed systems.

In addition to the numerous difficulties encountered in identifying parameters in lumped systems there are many other problems that appear in parameter identification in distributed systems : the mathematical formulation is complicated by the infinite dimensional of both the state space and the parameter space.

These obstacles can be overcome by using both the variational formulation and the infinite dimensional gradient of the cost function, but there are still many difficult problems, as for instance, the choice of approximation techniques and optimization schemes.

A number of approximate methods are available and the basic questions are as follows :

- At what point in the solution procedure should approximations be introduced?
- Does the approximate solution converge to the true solution? (most authors use numerical experiments to show that the result converge), but other questions concern the identifiability of the parameters in the approximate model and the convergence of the parameter estimates to the parameters in the identification. Few results are available when these questions are addressed simultaneously (Chavent [1] Banks [2]) and the relationship between identifiability and identification has received little attention in the literature.

Some authors completely ignore questions of identifiability or the effect of the approximate scheme on this property. However, since non identifiability can present a problem, it becomes necessary to use either a different approximate and/or optimization scheme, or to change the types of measurement in order to establish identifiability. A further question is wheter it is possible to study identifiability on the finite dimensional model approximation. In many approximation schemes the original problem is replaced by that of identifying the parameters of an approximated model and the identifiability of the parametrization is investigated following the definitions given for finite dimensional systems [3].

However the results do not provide for the identifiability of the parameter in the distributed model and can produce some rather stringent conditions on the number and location of the sensors.

In this paper the identification of a parameter function $a(x)$ in the one dimensional diffusion equation is considered within a deterministic framework. The model is based on a finite difference scheme and the identification problem can be considered as the direct resolution of an inverse problem.

* Université Paul Sabatier Toulouse.

2. IDENTIFIABILITY

In the present work we consider the problem of identifying $a(x)$ in the one dimensional parabolic equation :

$$\begin{aligned}
&\frac{\partial u}{\partial t} = \frac{\partial}{\partial x}\left(a(x)\frac{\partial u}{\partial x}\right) + f(x,t) \text{ in } (0,1)\times(0,T)\\
&u(x,0) = u_0(x) \qquad \text{in } (0,T)\\
&\frac{\partial u}{\partial x}(0,t) = \frac{G_0(t)}{a(0)} \qquad \text{in } (0,T)\\
&\frac{\partial u}{\partial x}(1,t) = G_1(t) \qquad \text{in } (0,T)
\end{aligned} \tag{1}$$

where $f(x,t)$; $G_0(t)$; $G_1(t)$; $a(0)$; and $u_0(x)$ are known.

Given the point measurement $Z(t) = u(0,t)$, the question is : can $a(x)$ be uniquely determined.

Pierce [5] considered a similar parabolic equation with null initial condition, with non homogeneous boundary condition of the third kind on $x = 0$ and with homogeneous condition of the same kind on $x = 1$. He showed that in such a case the values $u(t,xp)$; $xp = 0$ determine the spectral density function of the operator. Suzuki and Murayama [4] derive conditions under which $Z(t) = \tilde{Z}(t)$ implies $a(x) = \tilde{a}(x)$ ($\tilde{Z}(t) = \tilde{u}(0,t)$ where $\tilde{u}(x,t)$ is the solution of (1) with $\tilde{a}(x)$). All the results are inspired by Gel'fand Levitan Ideas, detailed discussion of these results is given by El Badia [6]. He shows that some alterations must be made in the proofs of their theorems and adds some extensions to other problems such as the determination of $a(x)$ in equation 1 with non homogeneous boundary conditions and null initial condition (or constant).

3. IDENTIFICATION PROBLEM

A wide range of optimization and approximation techniques have been proposed to solve the distributed system identification problem (D.S.I.P.) - Chavent [1] refers to the "jungle" of identification methods since no particular method yields satisfactory results for all distributed systems. In fact Polis [7] showed that the D.S.I.P. may be separated into a number of largely independent subproblems and each one must be considered in an attempt to treat the D.S.I.P..

- The only answer which seems to emerge concerns the approximation; early approximation seems to be better.

- In recent work, almost all authors use off line optimization schemes with gradient technique [9]. However, this technique requires the repeated solution of the system and adjoint equations and the simulation time is always important.

- The purpose of this paper is to study a direct solution of the identification problem without excessive computer time. We use a finite difference method at the first step of the D.S.I.P. and then the identification of $a(x)$ is replaced by that of identifying the point values of $a(x)$.

We shall discuss the identification problem both in the case where the measurements are available at $x = 0$, which is the identifiability condition of $a(x)$ [4] and in the case where measurements are assumed to be of the form $u_T(x) = u(x,T)$ where T is the final time and $z(t) = u(0,t)$.

The proofs of stability and convergence are not given here but can be found in [6].

4. FIRST APPROXIMATION METHOD [8]

In this section we are interested in observation processes where there is a discrete observation location on the boundary [9].

In order to give an explicit formulation for a at the discretization points we have to formulate differently equations (1)

Let us write (1) as :

$$\frac{\partial}{\partial x}\left((a(x))\frac{\partial u(x,t)}{\partial x}\right) = \frac{\partial u(x,t)}{\partial t} - f(x,t) \tag{2}$$

$$u(0,t) = z(t) \tag{3}$$

$$a_0\frac{\partial u}{\partial x}(0,t) = G_0(t) \quad \text{initial conditions} \tag{4}$$

$$u(x,0) = u_0(x) \quad \text{boundary condition} \tag{5}$$

By applying a finite difference scheme we obtain

$$a_{j+\frac{1}{2}}\frac{u^n_{j+1}-u^n_j}{h^2} - a_{j-\frac{1}{2}}\frac{u^n_j-u^n_{j-1}}{h^2} = \frac{u^{n+1}_j-u^n_j}{s} - f^n_j \tag{6}$$

$$j=o,\dots,J-1; \quad n=o,\dots,N-1$$

$$u^n_o = z^n \quad : \quad n=0,\dots,N \tag{7}$$

$$\frac{u^n_1-u^n_{-1}}{2h} = \frac{G^n_o}{a(0)}; \quad n=0,\dots,N \tag{8}$$

$$u^o_j = u_0(jk); \quad j=0,\dots,J \tag{9}$$

where $u^n_j \simeq u(jh,ns)$; $Jh = 1$, $Ns = T$ and $jh = x_j$ and $ns = t_n$.

Approximation of $a_{\frac{1}{2}} = a(\frac{h}{2})$

Assume that :

$$\left.\frac{\partial u(x,t)}{\partial x}\right|_{x=\frac{h}{2}} = \left.\frac{\partial u(x,t)}{\partial x}\right|_{x=0} + O(h) \qquad (10)$$

$$\int_0^{\frac{h}{2}} \frac{\partial u(x,t)}{\partial t}\, dx = \frac{h}{2} \left.\frac{\partial u(x,t)}{\partial t}\right|_{x=o} + O(h^2) \qquad (11)$$

Integrating equation (2) on $(o, \frac{h}{2})$ we obtain :

$$a(\tfrac{h}{2}) \frac{\partial u}{\partial x}(\tfrac{h}{2},t) = a(o) \frac{\partial u}{\partial x}(o,t) + \int_0^{\frac{h}{2}} \left[\frac{\partial u(x,t)}{\partial t} - f(x,t)\right] dx$$

$\forall t \; [0,T]$.

Using relations (10) and (11) we obtain :

$$a(\tfrac{h}{2}) \frac{\partial u}{\partial x}(o,t) \simeq a(o) \frac{\partial u}{\partial x}(o,t) + \frac{h}{2} \frac{\partial u}{\partial t}(o,t) - \int_0^{\frac{h}{2}} f(x,t)\, dx$$

so $a(\frac{h}{2})$ is given by :

$$a(\tfrac{h}{2}) \simeq a(o) + \frac{h}{2} \frac{\int_0^T \frac{\partial u}{\partial t}(o,t)dt}{\int_0^T \frac{\partial u}{\partial x}(o,t)dt} - \frac{\int_0^T \int_0^{\frac{h}{2}} f(x,t)\, dx\, dt}{\int_0^T \frac{\partial u}{\partial x}(o,t)\, dt} \qquad (12)$$

where $\frac{\partial u}{\partial x}(o,t) = \frac{G_o(t)}{a(o)}$ and $\frac{\partial u}{\partial t}(o,t) = \frac{\partial z(t)}{\partial t}$

are known.

If $\frac{\partial u}{\partial x}(o,t) = 0$, then $a(\frac{h}{2}) = a(o)$

Approximation of $\frac{u_1^n - u_o^n}{h}$

If we write equation (6) for j = o we obtain :

$$a_{\frac{1}{2}} \frac{u_1^n - u_o^n}{h^2} - a_{-\frac{1}{2}} \frac{u_o^n - u_{-1}^n}{h^2} = \frac{u_o^{n+1} - u_o^n}{s} - f_o^n; \qquad (13)$$

$n = o, \ldots, N-1$

With (7), (8) and assuming that :

$a_{-\frac{1}{2}} = a_{\frac{1}{2}} + O(h)$

equation (12) can now be written :

$$\frac{u_1^n - u_o^n}{h} = \frac{G_o^n}{a(o)} + \frac{h}{2\, a_{\frac{1}{2}}} \left(\frac{z^{n+1} - z^n}{s} - f_o^n\right) \qquad (16)$$

4.1 Algorithm

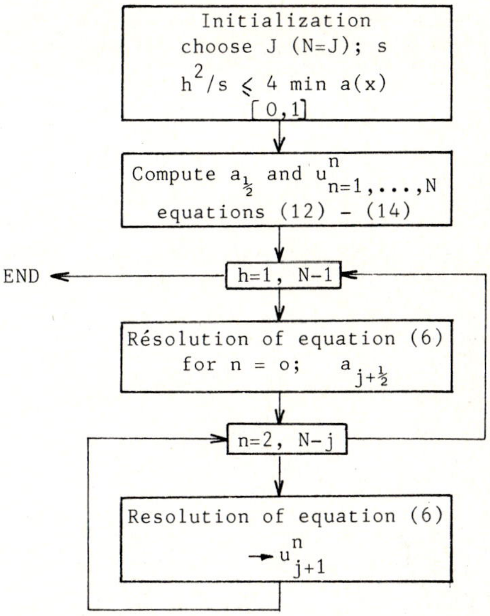

Remarks :

a) Let us define the matrix $U = (u(jh,ns))$ $j=o,\ldots,N-1$, $n=o,\ldots,N-1$

and $V = (u_j^n)$ $j=o,\ldots,N-1$, $n=o,\ldots,N-1$

as elements of a space U_h with the norm

$$\|V\|_{U_h} = \max_j \left\{ \max_n |u_j^n| + \max_n \left| \frac{u_{j+1}^n - u_j^n}{h} \right| \right\}$$

If $\|U-V\|_{U_h} \leqslant C\, h^{k+1}$ then $\|\Omega - \Gamma\|_a \leqslant C' h^k + |\varepsilon_o|$

where Ω and Γ are vectors defined respectively by (a_j) and $(a(jh))$.

$\varepsilon_o = a(\frac{h}{2}) - a_{\frac{1}{2}}$; $|\varepsilon_o| = o(h)$

and $\|\cdot\|_a = \max_j |a_j - a(jh)|$

The proof is given in [6].

b) Although the approach is deterministic the effect of random output noise is tested in the numerical applications using $z(t) = u(o,t) + V(t)$, where $V(t)$ is a white Gaussian zero mean noise. $V(t)$ is a perturbation of an initial condition of the equation (2), so this perturbation implies a perturbation of the same order in $u(x,t)$ and in $a(x)$ because of remark a).

4.2 Numerical applications [10]

We consider the D.P.S. (2....5) where

$a(x) = 0.5 + 0.5\, x - 0.5\, x^2$;

$f(x,t) = G_o(t) = G_1(t) = 0$;

$u(x,0) = 10 + 270\, x^2 - 180\, x^3.$

The data were generated by

$u(x,t) = (-45 + 270\, x^2 - 180\, x^3)\, e^{-6t} + 55$

which is the exact solution of the D.P.S.

The estimated a(x) and the true value are shown in fig. 1, with $h = s = 10^{-2}$ and in fig. 2 with $h = 10^{-2}$; $s = 10^{-3}$.

In the last case the data were generated with the same u(o,t) to which were added random numbers with zero mean and the standard deviation $\sigma = 0.2$ (fig. 3); $h = 10^{-2}$, $s = 10^{-2}$.

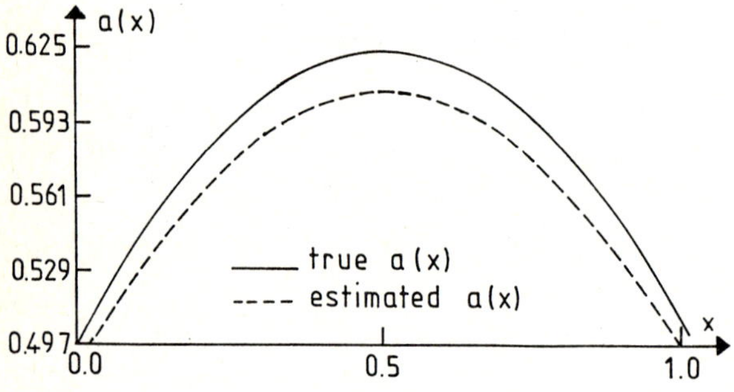

Fig. 1. a(x) obtained by the noise-free data $h = s = 10^{-2}$

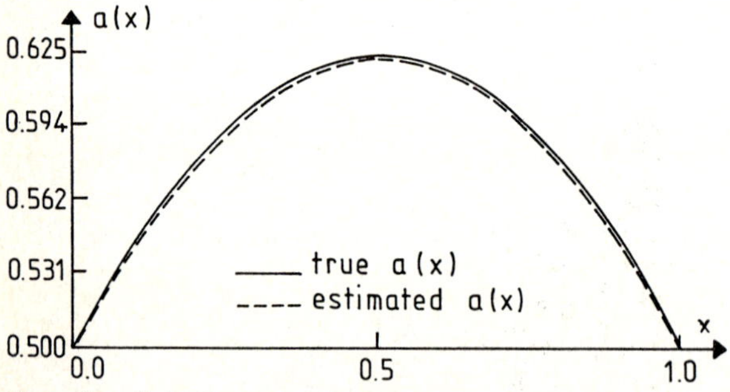

Fig. 2. a(x) obtained by the noise free data $h = 10^{-2}$. $s = 10^{-3}$

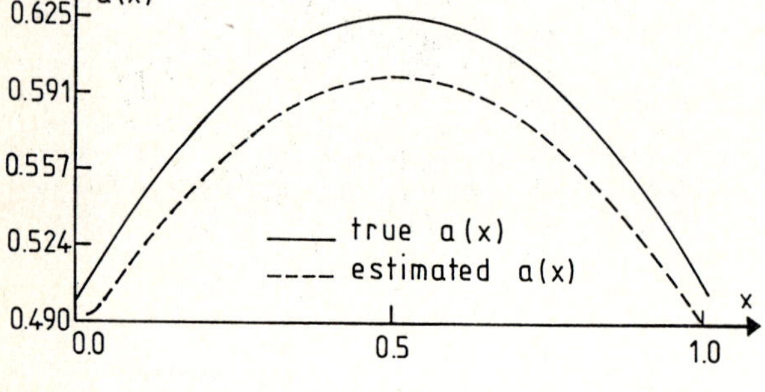

Fig. 3. a(x) obtained by the noiy data $h = s = 10^{-2}$.

5. SECOND APPROXIMATION METHOD [6]

It is clear that the boundary condition at x = 1 does not appear in the algorithm (however this condition is included in z(t).).

The goal of this second method is to avoid this problem.

It should be noted that equations (1) are equivalent to [6].

$$\begin{aligned} &\frac{\partial u}{\partial t} - \frac{\partial}{\partial x}\left(a(x)\frac{\partial u}{\partial x}\right) = f \\ &a(0)\left.\frac{\partial u}{\partial x}\right|_0 = g_0 \qquad t \in [0,T] \\ &u(0,.) = z \qquad t \in [0,T] \\ &u(.,0) = u_0 \qquad x \in [0,1] \\ &u(.,T) = u_T \qquad x \in [0,1] \end{aligned} \tag{17}$$

Let us define :

$$\begin{cases} v(x,t) = a(x)\dfrac{\partial u(x,t)}{\partial x} \\ \nabla_s u^n = \dfrac{1}{s}(u^n - u^{n-1}) \\ \nabla u_j = \dfrac{1}{h}(u_{j+1} - u_j) \end{cases}$$

Now we consider the finite difference scheme :

$$\begin{cases} u^n_{j+1} = u^n_j + \dfrac{h\, v^n_j}{a_j} & \begin{cases} j=0,\dots,J-1 \\ n=1,\dots,N \end{cases} \\ v^n_{j+1} = v^n_j + h\, \nabla_s u^n_{j+1} - h\, f^n_{j+1} & \begin{cases} j=0,\dots,J-1 \\ n=1,\dots,N \end{cases} \\ a_{j+1} = \dfrac{v^N_{j+1}}{\nabla u_{T,j}}\ ; & j=0,\dots,J-1 \end{cases}$$

with

$$\begin{cases} u^n_o = z(ns) \\ v^n_o = G_0(ns); & n=0,\dots,N \\ u^N_j = u_T(jh); & j=0,\dots,J. \end{cases}$$

5.1 Numerical applications

We consider the equations (17) where :

- $a(x) = x^2 - x + 1$
- $T = 50$
- $h = 10^{-2}$; $s=1$

the data were generated by :

$u(x,t) = x + t + xt$

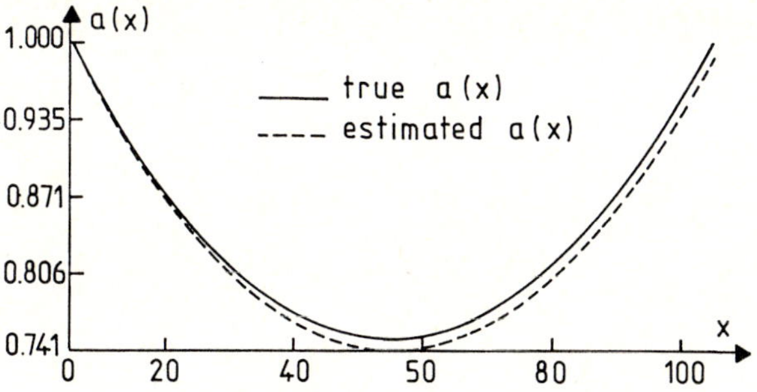

Fig. 4

6. CONCLUSION

In this paper two algorithms have been developed for identification of parameter of a linear distributed parameter system using finite differences schemes.

The advantages of these algorithms are that they do not require initializations of a(x) and do not lead to excessive computer time requirements; moreover, in the first case, we only have one measurement point which can lead to difficulties if Chavent's method is used.

REFERENCES

[1] Chavent, G., Identification of distributed parameter systems : about the output least square method its implementation and identifiability, Proc. 5th IFAC Symp. on Identification and System Parameter Estimation Darmstadt FRG, Pergamon Press, (1979).

[2] Banks, H.T. and Kunisch, K., Approximation theory for nonlinear partial differential equations with applications to identification and control, Lefschetz Center for Dynamical Systems, Report 81.7, Brown. Univ., (1981).

[3] Carotenuto, L. and Raiconi, G., Identifiability and identification of a Galerkin approximation for a class of distributed parameter systems, Int. J. Syst. Sci., vol. 11, n° 9, (1980).

[4] Suzuki, T. and Murayama, R., A uniqueness theorem in an identification problem for coefficients of parabolic equations., Proc. Japan Acad., 56, n° 6, Ser. A, (1980).

[5] Pierce, A., Unique identification of eigenvalues and coefficients in a parabolic problem., SIAM J. Control and Optimization, Vol. 17, n° 4, (1979).

[6] El Badia, A., Thesis, Toulouse, France (1985).

[7] Polis, M.P., The distributed system parameter identification problem : A survey of recent results, Proc. 3rd IFAC Symp. on Control of Distributed Parameter Systems, Toulouse, France (1982).

[8] Godounov, S. and Riabenki, V., Schemas aux différences, Editions de Moscou (MIR - 1977).

[9] El Badia, A. and Courdesses, M., Identification of spatially varying parameter in Distributed Parameter Systems, 11th IMACS World Congress System Simulation and scientific computation, Oslo, Norway (1985).

[10] Kravaris, C. and Seinfeld, J.H., Identification of parameters in Distributed Parameter Systems by regularization, SIAM J. Control and Optimization, vol. 23, n° 2, (1985).

Complex and Distributed Systems: Analysis, Simulation and Control
S.G. Tzafestas and P. Borne (editors)
Elsevier Science Publishers B.V. (North-Holland)

PARAMETER AND STATE IDENTIFICATION FOR THE DIFFUSION EQUATION VIA AUGMENTED LAGRANGIANS

Luciano CAROTENUTO and Giancarlo RAICONI

Dipartimento di Sistemi - Università della Calabria - 87030 Rende (CS) - Italy

The identification of the parameter function in the steady state diffusion equation with a finite number of observations is approached by minimizing a suitable regularization functional with respect both to the parameter and to the state, constrained by the system equation and the measurements. A discretized version of the problem is built up and is solved using an augmented Lagrangians method which ensures that the constraints are exactly satisfied. Numerical results are reported for problems in one and two dimensions.

1. INTRODUCTION

This paper deals with the identification of the parameter u(x) in the steady-state diffusion equation

$$\begin{aligned} &\nabla_x (u\, \nabla_x y) = f\,, \quad x \in \Omega \subset \mathbb{R}^n \\ &y(x) = y_b(x)\,, \quad x \in \partial\Omega \qquad (1) \\ &u(x) \geq \varepsilon > 0\,, \quad \text{a.e. } x \in \Omega \end{aligned}$$

given q exact measurements of the state y defined through a linear operator ℓ:

$$\underline{y}^{\circ} = \ell(y)\,,\ \underline{y}^{\circ} \in \mathbb{R}^q \qquad (2)$$

The parameter identification of (1) is relevant, for example, in modelling underground aquifers: in this case the parameter u(x) is the transmissivity of the aquifer, the state y(x) is the piezometric head, which is usually measured at some points (wells) in the spatial domain. The above problem has been considered by several authors: an extensive bibliography can be found in the surveys [1,2] . It is well known that a double indetermination arises from the problem formulation: obviously infinitely many state functions satisfy the observation constraint, and for each of them there exist infinitely many parameter functions satisfying the system equation. Moreover it can be seen, inverting equation (1), that the computation of u given y and the necessary initial data, is inherently unstable since u depends on the derivatives of y.

In this paper the above difficulties are overcome by selecting a pair of functions (u,y), belonging the Sobolev space $H^1(\Omega) \times H^1(\Omega)$, such that

a) equations (1) and (2) are satisfied,

b) the functional

$$I(u) = \frac{1}{2}\int_{\Omega} \{\rho(u-u_e)^2 + \sum_{i=1}^{n}\left(\frac{\partial u}{\partial x_i}\right)\}\partial x,\ \rho > 0$$

is minimized.

The minimization of I(u) may be regarded either as a mathematical device to obtain, at least in special cases, a unique, regularized, solution of the problem, or as a way to formalize the qualitative statement that the parameter should both be close to an a-priori estimate u_e and vary slowly with respect to x. Obviously the latter interpretation must be supported by the experience on the real system modelled by (1). The special case mentioned above is considered by Carotenuto et al.in [3], where it is assumed that y, ∇_x y, $\nabla_x \cdot \nabla_x$ y are known over Ω : equation (1) is then regarded as a linear equation in u and, under suitable assumptions, I(u) has a unique minimum in the manifold defined by (1). Here we consider a much more realistic set of data $\underline{y}^{\circ}$, and do not perform any a-priori interpolation of y(x). Rather the state itself is regarded as an unknown, constrained by the measurements.

A further point to be considered is that only an approximate finite dimensional solution of the identification problem can be computed. We follow the approach of the discretization at the beginning: the infinite dimensional constrained optimization problem, to which the identification is reduced, is approximated by a finite dimensional one, following the Ritz-Galerkin method which is of common use for the numerical solution of (1).

This procedure leads to the minimization of a quadratic function subject to bilinear constraints: such a problem can be handled profitably by the technique of Augmented

---

This work was supported by M.P.I. (60%)

Lagrangians [4].
A nice feature of the approach is that any solution thus obtained satisfies exactly both the observation data and a discrete model of the system, which is useful for simulation purposes. It should be noted that other approaches, such as those based on the discretization at the end, may fail to fulfil this property.

2. THE DISCRETIZED PROBLEM

Let us assume that Ω is a poligonal domain; let Π_Ω be a mesh with interior nodes $x_1, x_2, \ldots, x_N$, and boundary nodes $x_{N+1}, \ldots, x_M$; Π_Ω is chracterized by a mesh-size parameter h. Let S_h be the subspace of $H^1(\Omega)$ spanned by the piecewise polynomial basis functions $\phi_1(x), \ldots, \phi_M(x)$ with $\phi_i(x_j) = \delta_{ij}$ (lagrangian finite elements). The subspace spanned by $\phi_1(x), \ldots, \phi_N(x)$ is denoted by $S_h^\circ = S_h \subset H_0^1(\Omega)$. When $u(x), y(x)$ are approximated by

$$u_h(x) = \sum_{i=1}^{M} u_i \phi_i(x); \quad u_h \in S_h \tag{3}$$

$$y_h(x) = \sum_{i=1}^{N} y_i \phi_i(x) + \sum_{i=N+1}^{M} y(x_i) \phi_i(x); \tag{4}$$

$y_h \in S_h$

the Ritz-Galerkin approximate version of equation (1) takes the form:

$$\underline{A}(\underline{u})\underline{y} + B(\underline{u})\underline{y}_b = \underline{f}$$

where

$$\underline{u} = [u_1 \ldots u_M]', \quad \underline{y} = [y_1 \ldots y_N]',$$

$$\underline{y}_b = [y(x_{N+1}) \ldots y(x_M)]'$$

$$[A(\underline{u})]_{ij} = -\sum_{k=1}^{M} u_k \int_\Omega \phi_k \nabla_x \phi_i \cdot \nabla_x \phi_j \, dx;$$

$$i,j = 1, \ldots, N \tag{5}$$

$$[B(\underline{u})]_{ij} = -\sum_{k=1}^{M} u_k \int_\Omega \phi_k \nabla_x \phi_i \cdot \nabla_x \phi_{j+N} \, dx;$$

$i = 1, \ldots, N$; $j = 1, \ldots, M - N$

$$[\underline{f}]_i = \int_\Omega f \phi_i \, dx; \quad i = 1, \ldots N$$

The functional $I(u)$ is likewise approximated by

$$J(\underline{u}) = (1/2)\underline{u}'(Q + \rho T)\underline{u} - \rho \underline{u}'\underline{g} + 1/2\rho^2 \int_\Omega u_0^2 \, dx$$

where

$$[Q]_{ij} = \int_\Omega \nabla_x \phi_i \nabla_x \phi_j \, dx$$

$i = 1, \ldots, M$; $j = 1, \ldots, M$

$$[T]_{ij} = \int_\Omega \phi_i \phi_j \, dx$$

$i = 1, \ldots, M$; $j = 1, \ldots, M$

$$[\underline{g}]_i = \int_\Omega u_e \phi_i \, dx; \quad i = 1, \ldots, M$$

Where the observation operator ℓ is concerned we have

$$\ell(y_h) = C\underline{y}; \quad [C]_{ij} = \ell_i(\phi_j(x));$$

$i = 1, \ldots, q$; $j = 1, \ldots, N$

The previously defined matrices enjoy some properties which will be used later:

- $A(\underline{u})$ is symmetric, negative definite $\forall \underline{u} > 0$
- $A(\underline{u})$, $B(\underline{u})$ are linear with respect to $\underline{u}$:
  $$A(\underline{u}) = \sum_{i=1}^{M} A_i u_i, \quad B(\underline{u}) = \sum_{i=1}^{M} B_i u_i$$
- T is positive definite, Q is positive semidefinite, so that, $\forall \rho > 0$, $J(\underline{u})$ is strictly convex.

The identification problem in finite dimension is then stated as:
find $\hat{\underline{u}}$, $\hat{\underline{y}}$ such that

$$J(\hat{\underline{u}}) \le J(\underline{u})$$

in the subset of $\mathbb{R}^M \times \mathbb{R}^N$ defined by

$$A(\underline{u})\underline{y} + B(\underline{u})\underline{y}_b = \underline{f} \tag{6}$$

$$u_i \ge \varepsilon > 0, \quad i = 1, 2, \ldots M$$

$$C\underline{y} = \underline{y}^\circ. \tag{7}$$

Where the existence of the minimum is concerned we note that the constraints (6), (7) are equivalent, for $\underline{u} > \underline{0}$, to

$$\underline{\theta}(\underline{u}) = C A^{-1}(\underline{u})[\underline{f} - B(\underline{u})\underline{y}_b] - \underline{y}^\circ = \underline{0}.$$

Since A, B are continuous functions of $\underline{u}$, the set

$$\mathcal{U}_a = \{\underline{u} \in \mathbb{R}^M: u_i \ge \varepsilon > 0, \ i = 1, \ldots, M; \ \underline{\theta}(\underline{u}) = \underline{0}\}$$

is closed: hence the minimum of J on $\mathcal{U}_a$ exists, provided that $\mathcal{U}_a$ itself is nonvoid.

3. OPTIMIZATION

In the statement of the problem the inequality

constraint $u_i \geq \varepsilon$, with ε arbitrary positive constant, has been introduced to ensure the invertibility of $A(\underline{u})$, consistently with the continuous problem formulation. We shall use a minimization algorithm wich does not keep into account the inequality constraints: because the physical parameter $u(x)$ is intrinsically positive, results with non positive u_i's which would be obtained will be discarded.
Let us define the Lagrangian function

$$L_1(\underline{u},\underline{y},\underline{\lambda},\underline{\mu}) = J(\underline{u}) + \underline{\lambda}^T[A(\underline{u})\underline{y} + B(\underline{u})\underline{y}_b - \underline{f}] + \underline{\mu}^T[C\underline{y} - \underline{y}^o]$$

with multipliers $\underline{\lambda} \in \mathbb{R}^N$, $\underline{\mu} \in \mathbb{R}^q$; L_1 can be directly regarded as an approximation of the Lagrangian for the infinite dimensional problem:

$$L(u,y,\lambda,\underline{\mu}) = I(u) + \langle \nabla_x(u\nabla_x y) - f, \lambda \rangle_{L_2(\Omega)} +$$

$$+ \langle \ell(y) - \underline{y}^o, \underline{\mu} \rangle_{\mathbb{R}^q} \ . \ \lambda \in H_0^1(\Omega),\ \underline{\mu} \in \mathbb{R}^q$$

Indeed it can be verified that,

$$L_1(\underline{u},\underline{y},\underline{\lambda},\underline{\mu}) = L(u_h, y_h, \lambda_h, \underline{\mu})$$

for all $\underline{u},\underline{y},\underline{\lambda},\underline{\mu}$, with u_h, y_h, given by (3),(4) and $\lambda_h = \sum_{i=1}^{N} \lambda_i \phi_i(x) \in S_h^o$.
The first order optimality conditions are

$$\nabla_u L_1 = (Q+\rho T)\underline{u} - \rho\underline{g} + [A_1\underline{y} \cdots A_M\underline{y}]'\underline{\lambda} +$$

$$+ [B_1\underline{y}_b \cdots B_M\underline{y}_b]'\underline{\lambda} = \underline{0} \qquad (8)$$

$$\nabla_y L_1 = A(\underline{u})\underline{\lambda} + C'\underline{\mu} = \underline{0}$$

together with constraints (6),(7).
With a slight loss of generality, but consistently with several practical applications it is assumed in the following that pointwise observations are taken and that the finite element mesh is chosen in such a way that the observation points are nodes of the mesh, say $x_{i1}, \ldots, x_{iq}$. Then the constraint (7) can be written:

$$y_{i_j} = y_j^0 \ ; \ j = 1,\ldots,q$$

that is $C = [\underline{e}_{i_1} \cdots \underline{e}_{i_q}]'$

This enables to put the constraints into simpler form. Indeed equation (6) can be written as:

$$F(\underline{y},\underline{y}_b)\underline{u} = \underline{f} \qquad (9)$$

$$F(\underline{y},\underline{y}_b) = \sum_{i=1}^{N} F_i y_i + \sum_{i=1}^{M-N} F_{i+N}\, y_{bi}$$

where the expressions of F_i's come out directly from (5). Furthermore (7) is replaced into (6) by defining a vector $\underline{z} \in \mathbb{R}^p$, $p = N-q$, consisting of the unobserved components of $\underline{y}$: $\underline{z} = P\underline{y}$, where $P \in \mathbb{R}^{pxN}$ is obtained from the NxN identity by eliminating rows $i_1,\ldots,i_q$. Then equation (9) is rewritten as

$$[E(\underline{z}) + D]\ \underline{u} = \underline{f}$$

$$D = \sum_{k=1}^{q} F_{i_k}\, y_k^o + \sum_{i=1}^{M-N} F_{N+i}\, y_{bi}; \quad E(\underline{z}) = \sum_{i=1}^{p} E_i z_i;$$

each E_i is equal to the matrix F_j multiplying the element y_j of $\underline{y}$ indentified with z_i.
With the above notation the identification problem is finally stated as :

Pb : minimize $J(\underline{u})$
subject to
$[E(\underline{z}) + D]\ \underline{u} = \underline{f}$

$\underline{u} \in \mathbb{R}^M$, $\underline{z} \in \mathbb{R}^p$.

The corresponding Lagrangian is:

$$L_2(\underline{u},\underline{z},\underline{\lambda}) = J(\underline{u}) + \underline{\lambda}'[(E(\underline{z})+D)\underline{u} - \underline{f}]$$

The first order optimality conditions are

$$[E(\underline{z}) + D]\ \underline{u} = \underline{f}$$

$$\nabla_u L_2 = (Q+\rho T)\underline{u} - \rho\underline{g} + [E(\underline{z}) + D]'\lambda = \underline{0} \qquad (11)$$

$$\nabla_z L_2 = [E_1\underline{u}\ \ E_2\underline{u} \cdots E_p\underline{u}]'\lambda = \underline{0}$$

It can be verified by direct substitution that (8) and (11) are equivalent, in the sense that if $\hat{u},\hat{y},\hat{\lambda},\hat{\mu}$ satisfy (8) then $\hat{\underline{u}}$, $\hat{\underline{z}} = P\hat{\underline{y}}, \hat{\underline{\lambda}}$ satisfy (11); viceversa if $\hat{\underline{u}}$, $\hat{\underline{z}},\hat{\underline{\lambda}}$ is a solution of (11), then $\hat{\underline{\mu}}$ exists such that

$$\hat{\underline{u}},\hat{\underline{y}} = \begin{bmatrix} P \\ C \end{bmatrix}^{-1} \begin{bmatrix} \hat{\underline{z}} \\ \underline{y}^o \end{bmatrix} \ ,\ \hat{\underline{\lambda}},\ \hat{\underline{\mu}} \ \text{ satisfy (8)}$$

In order to obtain a well posed problem we make an "identifiability" assumption:
Hypothesis i - The data $\underline{y}^o$, $\underline{y}_b$ are such that there exists at least one $\underline{z} = \bar{\underline{z}}$ for which rank $[E(\bar{\underline{z}}) + D] = N$.
Due to the genericity of the property of a matrix to have full rank hypothesis i entails that rank $[E(\underline{z}) + D] = N$ for all $\underline{z} \in \mathbb{R}^p$, except at most on a manifold of dimension less than p. A consequence is that, $\forall\ \underline{f} \in \mathbb{R}^N$, the

set $U_b(z) = \{u \in \mathbb{R}^M : [E(z) + D] u = f\}$ is non void for almost all $z \in \mathbb{R}^p$. It cannot be guaranteed that problem Pb has a unique local, and then global, solution, but the problem:

Pbz - minimize $J(u)$ subject to $[E(z) + D]\ u = f$ has, for almost all $z \in \mathbb{R}^p$, a unique solution $\hat{u}(z)$ because $J(u)$ is quadratic strictly convex and the constraint is linear.

By defining

$$J^*(z) = \begin{cases} J[\hat{u}(z)] & \text{if } U_b(z) \neq \emptyset \\ +\infty & \text{otherwise} \end{cases}$$

a conceptually straightforward technique to solve the identification problem is to perform the unconstrained minimization of $J^*(z)$. But such an approach is unsatisfactory because each evaluation of $J^*(z)$ (and possibly of its gradient) requires the computation of a pseudo-inverse. Among the several method available to solve equality constrained problems, the method of augmented Lagrangians, as proposed by Di Pillo & Grippo in [4], seems to be well adapted to the present case.

Let us define the following augmented Lagrangian:

$$S(u,z,\lambda\,;c_1,c_2) = L_2(u,z,\lambda) + $$
$$+\ c_1/2\,\|[E(z)+D]\,u-f\,\|^2 + c_2/2\,\|\nabla_u L_2(u,z,\lambda)\|^2$$

The structure of the problem enables to prove the following results:

Proposition 1. Let $(\hat{u},\hat{z})$ be a local isolated minimum point of problem Pb with rank $[E(\hat{z}) + D] = N$: then, for any $c_2 > 0$ there exists $c^* > 0$ such that S a local isolated minimum point at $(\hat{u},\hat{z},\hat{\lambda})$ for any $c_1 > c^*$; moreover $\hat{\lambda}$ is the optimal multiplier of problem Pb.

Proposition 2. Let Σ be a compact subset of $\mathbb{R}^p$ and rank $[E(z) + D] = N$, $\forall\, z \in \Sigma$: then there exists $\hat{c}$ such that, if $(\hat{u},\hat{z},\hat{\lambda})$ is an isolated local minimum point of S with $c_1 > \hat{c}$, $\hat{z} \in \Sigma$, then $(\hat{u},\hat{z})$ is a local isolated minimum point of problem Pb.

These results are proved in [5]: they hold for any bilinearly constrained quadratic minimization problem and hold under assumptions weaker than those used in [4] for the general case.

Propositions 1 and 2 establish an equivalence between the local minima of the constrained problem and the unconstrained local minima of S, for a finite value of the penality coefficients. This fact enables, at least in principle, to solve Pb by a single unconstrained minimization of S with respect to $\hat{u},\hat{z},\hat{\lambda}$. On the contrary, in the case of a pure penality method, the above mentioned equivalence holds only asymptotically ($c_1 \to \infty$), while other lagrangian approaches [6] require a sequence of unconstrained minimizations.

The minimization of S can be performed by any standard algorithm for differentiable functions. Nevertheless, the dimension of meaningful d.p.s. identification probelms is large and our approach involves about 1.5 more variables than those based on penality: then the class of practically useable algorithms is greatly restricted.

4. APPLICATIONS

Let $\Omega = (0,L)$, $x_i = ih$, $i = 1,..,N$, $h = 1/(N+1)$. For convenience we slightly modify the previous notation for boundary data by putting: $x_0 = 0$, $x_{N+1} = L$, $y_0 = y(0)$, $y_{N+1} = y(L)$. The bases function, ϕ_i are chosen to be piecewise linear on $[0,L]$, with $\phi_i(x_j) = \sigma_{ij}$, $i = 0,1,..,N+1$, so that S_h is the subspace of $H^1(\Omega)$, consisting of continuous functions, linear on each subinterval $[x_i, x_{i+1}]$. In order to investigate this particular case the matrix $F \in \mathbb{R}^{N\times(N+2)}$ is written into the form

$$F = F_1 F_2,\quad F_1 \in \mathbb{R}^{N\times(N+1)},\quad F_2 \in \mathbb{R}^{(N+1)\times(N+2)}$$

$$F_1 = \begin{bmatrix} -d_0 & d_1 & 0 \dots 0 & 0 \\ 0 & -d_1 & d_2 \dots 0 & 0 \\ \dots & \dots & \dots & \dots \\ 0 & 0 & 0 \dots -d_{N-1} & d_N \end{bmatrix}$$

$$F_2 = \frac{1}{2}\begin{bmatrix} 1 & 1 & 0 \dots 0 & 0 \\ 0 & 1 & 1 \dots 0 & 0 \\ \dots & \dots & \dots & \dots \\ 0 & 0 & 0 \dots 1 & 1 \end{bmatrix}$$

where $d_i = (y_{i+1} - y_i)/h$, $i = 0,1,..,N$

Since rank $[F_2] = N+1$, it results rank $[F]$ = rank $[F_1]$: then by ispection, we verify that rank $[F] < N$ if and only if there exist i,j, $i \neq j$, such that $d_i = d_j = 0$. This implies that assumption- i - is satisfied unless the data are very particular: obviously the assumption can be always verified by modifying the definition of the mesh.

The features of the identification method are tested by a numerical experiment. The state function is assumed to be given by $y(x) = [(1+x)^5 - 1]/31$, $x \in [0,1]$, the forcing function by $f(x)= 0$. A family of parameter functions fitting with y,f is $u(x)= u(0)(1+x)^{-4}$. The data used for the identification are the boundary values $y(0)= 0$, $y(1)= 1$ and the interior values $y(k/7)$, $k= 2,4,6$.

The discrete model is built up taking $N= 20$. Since the augmented Lagrangians S is a function of $3N+2-q$ variables, a moderately large unconstrained minimization problem must be solved: however the special structure of F enables to compute very fastly both S and its gradient. A quasi-Newton algorithm has been used to carry out the minimization: very few trials has been necessary to select c_1 ,c_2 such that both the algorithm converges to a local minimum of S and the optimality conditions (11) are satisfied. This is practical test to check that the local minimizer of S is also a local solution of Pb. The identification is performed with $u_e= 1$, for several values of the weighting coefficient ρ : the results are shown in figures 1,2, where $\hat{u}_h(x)$ and $\hat{y}_h(x)$, respectively, are plotted. Note that only the graph of y_h for $\rho= 100$ can be distinguished from the others: substantially different parameter functions correspond to very close state functions. The numerical experiment also shows clearly the dependence of the identified parameter on ρ: the effect of increasing ρ is to make the parameter function to be closer, in the mean square sense, to the constant u_e, but less smooth because it is constrained by the system equation and the data.

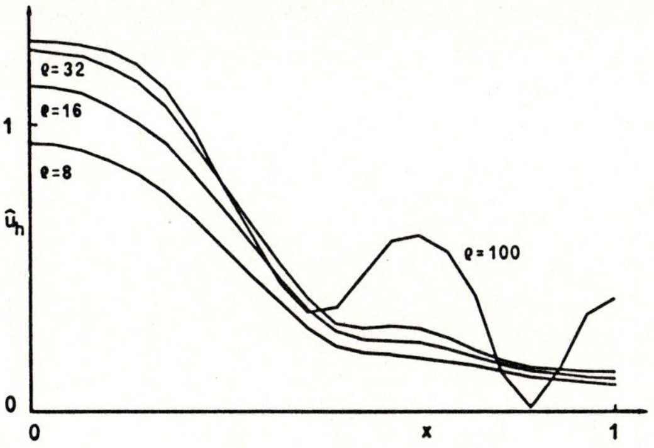

figure 1

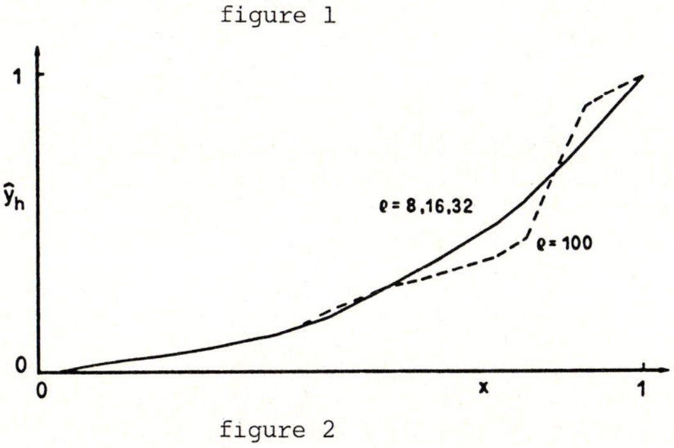

figure 2

The method has been applied also to the square domain $\Omega =(0,1)x(0,1)$, where the state is assumed to be given by

$y(x_1,x_2)= \exp\left[(x_1- x_2)/2\right]\sin(\pi x_2/4)$

and $f(x_1, x_2) = 0$. The $q= 11$ interior measurement points are shown in figure 5. A parameter u consistent with the given y and f is : $u(x_1, x_2) = \exp(\pi^2 x_1/8+x_2)$. The discretized problem is obtained by dividing the domain into triangular elements; both u and y are approximated by piecewise linear functions. A mesh with $N=41$, $M=61$ is used:

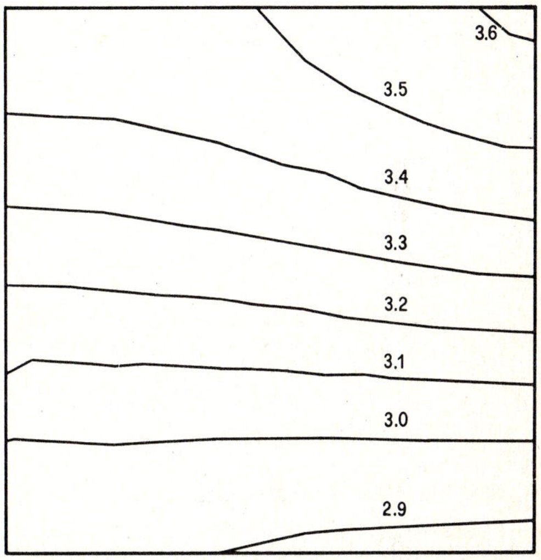

figure 3

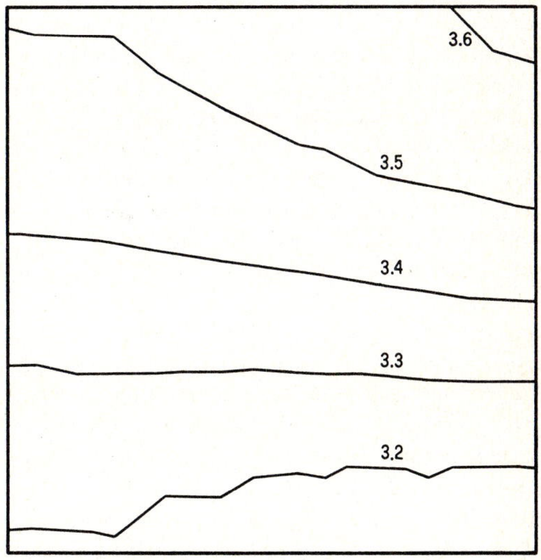

figure 4

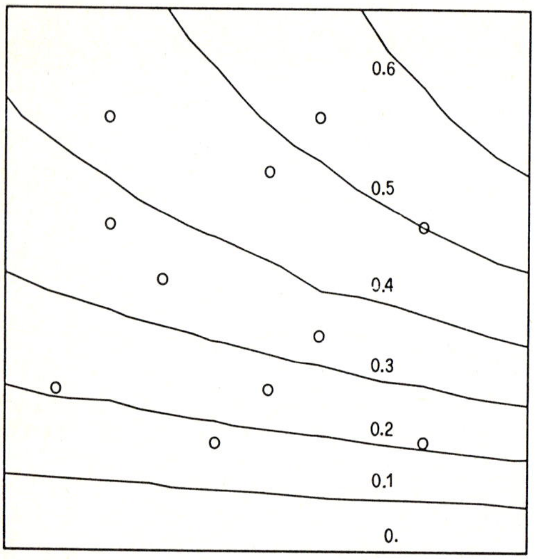

figure 5

since p= N-q= 30 the augmented Lagrangian S is a function of 132 variables. Both a quasi - Newton and a conjugate gradient algorithm where used for the minimization: the latter proved to be more satisfactory. Figures 3, 4 show the level curves of $\hat{u}_h(x)$ for $\rho = 2$, $\rho = 6$ respectively ($u_e = 3.5$); figure 5 shows the level curves of $\hat{y}_h(x)$: also in this case plots of $\hat{y}_h$ corresponding to different ρ are practically indistinguishable.

5. CONCLUSIONS

The problem of identifying the parameter function u in the diffusion equation is well known to be ill posed because the solution both is not unique and may be unstable with respect to data. The identification method here proposed uses as data a finite number of exact measurements of the state y and is based on the minimization of a quadratic regularization functional of u . The key feature of the method is that no a-priori interpolation of the state is introduced: so the state is a further unknown of the problem. After reducing the system equation to finite dimension by means of finite element approximation, the identification problem is stated as the minimization of a quadratic function subject to bilinear constraints. Since the regularization functional depends on two scalar parameters this approach leads to compute a family of pairs u, y which satisfy "exactly" both the discrete version of system equation and the measured data. Then it is not meaningful to speak about the "true" parameter: because all are consistent with the a-priori data, the selection of the parameter for further applications of the model must be left to the subjective judgement of the user.

REFERENCES

[1] Goodson R.E., Polis M.P., A survey of parameter identification in distributed systems , Proc. 6th IFAC World Congress, Boston (1975)

[2] Polis M.P., The distributed parameter identification problem: A survey of recent results , Proc. 3rd IFAC Symposium on Control of Distributed Parameter Systems, Tolouse (1982)

[3] Carotenuto L., Di Pillo G., Raiconi G., A regularized solution of the identification problem for the distributed parameter model of an underground aquifer, Proc. 7th IFAC World Congress, Helsinki (1979)

[4] Di Pillo G., Grippo L., A new class of augmented Lagrangians in nonlinear programmin, SIAM J. Contr. & Opt. vol.17 (1979)

[5] Carotenuto L., Raiconi G., On the minimization of quadratic functions with bilinear constraints via augmented Lagrangians , Tech. Rep. n.52 Dip. Sistemi Univ. Calabria (1985)

[6] Hestnes M.R., Optimization theory. The finite dimensional case. J. Wiley (1975)

Complex and Distributed Systems: Analysis, Simulation and Control
S.G. Tzafestas and P. Borne (editors)
Elsevier Science Publishers B.V. (North-Holland)

ADAPTIVE STATE ESTIMATION ALGORITHM FOR LINEAR DISCRETE-TIME DISTRIBUTED PARAMETER SYSTEMS WITH APPLICATION TO AIR POLLUTION PROCESSES

Józef KORBICZ+, Michael Z. Zgurovsky, Alexander N. Selin

Technical University of Kiev, prosp. Pobedy, 37
252056 Kiev, USSR

In this report, the adaptive state estimation problem (i.e. filtering and prediction) for linear discrete-time distributed parameter systems with unknown parameters is presented. The parameter identification algorithm is derived using the least-squares criterion and the adjoint method. Finally, the results obtained are applied to estimation of air pollutant concentrations levels in zone of a big town.

1. INTRODUCTION

The state estimation problem of distributed dynamical systems is one of important problems for modeling and optimal control in a practical field. A great deal of work has been carried on estimation problems for stochastic continuous-time [1] and discrete-time [2] distributed parameter systems (DPS) in the case when mathematical models and statistical characteristics of noises are known. On the other hand for DPS with unknown parameters or unknown noises some adaptive filter [3,4] may be used. However, these algorithms require much computations and therefore for DPS with unknown time and space-variant parameters the estimation problem must be solved by more effective algorithms. Some examples of such systems are system of estimation of air pollutant concentrations [5] or system of estimation of oil recovering processes parameters [6].

In this report, an adaptive filter and predictor for discrete-time DPS with unknown time and space-variant parameters is developed. The parameter identification algorithm is derived by unified approach based on the optimal control theory. In comparison with known adaptive estimation methods [3,8] our approach does not require solving of the Riccati equation because the estimation error covariance matrix is evaluated approximately [9,10]. This procedure allows the decreasing of the computation.

+On the leave from: Technical University of Zielona Góra, ul.Podgorna 50, 65-246 Zielona Góra, Poland

2. DESCRIPTION OF THE SYSTEM

Let a linear stochastic DPS be described in an open domain $\Omega \in R^r$ of an r-dimensional Euclidean space with smooth boundary $\partial\Omega$ by the following discrete-time equation:

$$X(k+1,z)=\mathcal{L}_z(\theta)X(k,z)+B(k,z)U(k,z)+C(k,z)W(k,z), \quad z\in\Omega \tag{1}$$

with boundary and initial conditions:

$$\beta_z(\theta)X(k+1,z)=C_b(k,z)W_b(k,z), \quad z\in\partial\Omega \tag{2}$$

$$X(k,z)|_{k=0}=X_0(z), \quad z\in\Omega \tag{3}$$

where $k=0,1,2,\dots,K$ is the discrete time ($t_k=k\Delta t$, Δt is a sampling time), $z=(z_1,z_2,\dots,z_r)\in\Omega$ is the spatial coordinate vector, $X(k,z)$ is the n-dimentional state vector, $U(k,z)$ is the p-dimensional control vector or deterministic input, $W(k,z)$ and $W_b(k,z)$ are q and q'-dimensional vector-valued stochastical inputs over Ω and $\partial\Omega$, respectively, $\theta(k,z)$ is the s-dimensional unknown vector of quasi-stationary parameters, $\mathcal{L}_z(\theta)$ and $\beta_z(\theta)$ are linear spatial matrix differential operators over Ω and $\partial\Omega$, respectively, $B(k,z)$, $C(k,z)$ and $C_b(k,z)$ are known matrix functions.

The state $X(k,z)$ is estimated from measurements which are taken at the selected discrete N points z^j, $j=\overline{1,N}$ of the coordinate space $\overline{\Omega}=\Omega\cup\partial\Omega$ and described by

$$Y(k)=H(k)X_N(k)+V(k) \quad (4)$$

where $Y(k)$ is the mN-dimensional observation vector, $V(k)$ is the mN-dimensional vector-valued measurement noise, $H(k)=\mathrm{diag}\{H(k,z^j),\ j=\overline{1,N}\}$ is the known $(mN\times mN)$ matrix and $X_N(k)=\mathrm{col}\{X(k,z^j),\ j=\overline{1,N}\}$ is the nN-dimensional column vector at the measured points. Assume, that the white gaussian processes $W(k,z)$, $W_b(k,z)$ and $V(k)$ are stochastically independent of each other and also independent of initial condition $X_0(z)$. Assume futher that the mean and covariance function are given by

$$E[W(k,z)]=0,\ E[V(k)]=0,\ E[X_0(z)]=\hat{X}_0(z)$$

$$E[W(k,z)W^T(l,y)]=Q(k,z,y)\delta_{kl},\quad z,y\in\Omega$$

$$E[W_b(k,z)W_b^T(l,y)]=Q_b(k,z,y)\delta_{kl},\quad z,y\in\partial\Omega \quad (5)$$

$$E[V(k)V^T(l)]=R(k)\delta_{kl},$$

$$E[X_0(z)X_0^T(y)]=P_0(z,y),\quad z,y\in\Omega$$

where $E[\cdot]$ and "T" denote the expectation and transpose operators, respectively, δ_{kl} is the Kronecker delta function and $Q(k,z,y)$, $Q_b(k,z,y)$, $R(k)$ and $P_0(z,y)$ are symmetric matrices.

3. STATEMENT OF THE ESTIMATION PROBLEM

The problem considered here is posed as follows. Given a record of measured data $Y(k)=\{Y(m),\ m=0,1,2,\dots,k\}$ and the description of the system (1)-(3) with unknown parameter θ find an optimal estimate $\hat{X}(k,z|l)$ of the state $X(k,z)$ in the mean-squared error sesse

$$E\left[\|X(k,z)-\hat{X}(k,z|l)\|^2\right]=\min \quad (6)$$

where $\|\cdot\|$ represents an Euclidean norm. For various values of discrete time k and l we hawe three estimation problems, such as:

a) prediction, if $(k>l)$
b) filtering, if $(k=l)$
c) smoothing, if $(k<l)$.

4. PROBLEM SOLUTION

In this work only the filtering and prediction problems are considered. Owing to the fact that system (1)-(3) is quasi-stationary, the problem solution is divided into two steps. The parameter identification and state estimation procedures are used alternately in the proposed adaptive algorithm.

4.1. Suboptimal filtering

State estimation problem for discrete-time DPS with known parameters was solved by many authors [2,3,8]. However, practical realization of these algorithms requires much computations and storage of digital system and the quality of the estimation process depends on the accuracy of a priori information about the system. Here, the suboptimal filtering algorithm will be considered, where the error filtering covariance matrix is estimated approximately. This algorithm is described by:

<u>filtering estimate</u> $\hat{X}(k,z|k)$

$$\hat{X}(k+1,z|k)=\mathcal{L}_z(\hat{\theta})\hat{X}(k,z|k)+B(k,z)U(k,z) \quad (7)$$

$$\beta_z(\hat{\theta})\hat{X}(k+1,z|k)=0,\quad z\in\partial\Omega \quad (8)$$

$$\hat{X}(k,z|k)\big|_{k=0}=\hat{X}_0(z),\quad z\in\Omega$$

$$\hat{X}(k+1,z|k+1)=\hat{X}(k+1,z|k)+K_N(k+1,z)\times[Y(k+1)-H(k+1)\hat{X}_N(k+1|k)] \quad (9)$$

<u>one-step ahead prediction error</u> $\delta X(k+1,z|k)$ and <u>filtering error</u> $\delta X(k+1,z|k+1)$

$$\delta X(k+1,z|k)=\mathcal{L}_z(\hat{\theta})\delta X(k,z|k)+C(k,z)\times W(k,z)+\delta(\psi(z))C_b(k,z)W_b(k,z) \quad (10)$$

$$\beta_z(\hat{\theta})\delta X(k+1,z|k)=0,\quad z\in\partial\Omega \quad (11)$$

$$\delta X(k,z|k)\big|_{k=0}=\delta X_0(z),\quad z\in\Omega$$

$$\delta X(k+1,z|k+1)=\delta X(k+1,z|k)-K_N(k+1,z)\times[Y(k+1)-H(k+1)\hat{X}_N(k+1|k)] \quad (12)$$

<u>filter gain</u> $K_N(k+1,z)$

$$K_N(k+1,z)=\hat{P}_N(k+1,z|k)H^T(k+1)\left[H(k+1)\times\hat{P}_{NN}(k+1|k)H^T(k+1)+R(k+1)\right]^{-1} \quad (13)$$

where $\hat{P}_N(k+1|k)=\left[\hat{P}(k+1,z,y^1|k)\ \dots\ \dots\ \hat{P}(k+1,z,y^N|k)\right]$

$$\hat{P}_{NN}(k+1|k)=\begin{bmatrix}\hat{P}(k+1,z^1,y^1|k)\ \dots\ \dots\ \hat{P}(k+1,z^1,y^N|k)\\ \\ \hat{P}(k+1,z^N,y^1|k)\ \dots\ \dots\ \hat{P}(k+1,z^N,y^N|k)\end{bmatrix}$$

estimate of one-step ahead prediction error covariance matrix $\hat{P}(k+1,z,y|k)$

$$\hat{P}(k+1,z,y|k)=\hat{P}(k,z,y|k-1)+\gamma(k)\left\{\frac{1}{S_z S_y}\times\int_{\Omega_z}\int_{\Omega_y}\delta X(k+1,z'|k)\delta X^T(k+1,y'|k)dz'dy'-\hat{P}(k,z,y|k-1)\right\},\quad z,y\in\Omega \tag{14}$$

$$\hat{P}(k,z,y|k-1)\big|_{k=1}=P_0(z,y) \tag{15}$$

where Ω_z, $\Omega_y\subset\Omega$ are underspaces of the space Ω for computing the sliding of stochastic field, S_z and S_y are the constant parameter which are greater than interval correlation of $\delta X(k+1,z|k)$ by the spatial coordinate z and y, $\delta(\cdot)$ is a generalized function concentrated on $\partial\Omega$, $\psi(z)$ is a scalar function, which describes the boundary $\partial\Omega$, $\psi(z)\geqslant 0\ \forall z\in\Omega$, $\gamma(k)$ is the correction coefficient. If the following conditions are satisfied

$$\gamma(k)>0,\ \sum_{k=1}^{\infty}\gamma(k)=\infty,\ \sum_{k=1}^{\infty}\gamma^2(k)<\infty \tag{16}$$

then the series (14) for k=0,1,2,... is convergent to real value P(k+1,z,y|k) with probability 1 [10]. It is a robust algorithm which is characterized by the less sensitivity to the incomplete a priori system data, than the optimal does. The presented distributed filter (7)-(15) is a direct extension of the results in the continuous-time DPS [9] to discrete-time.

4.2. Suboptimal prediction

The suboptimal prediction algorithm will be considered in the sense of suboptimal filtering. In this case the l-step ahead prediction algorithm is described by:

prediction estimate $\hat{X}(k+l,z|k)$

$$\hat{X}(k+l,z|k)=\mathcal{L}_z(\hat{\theta})\hat{X}(k+l-1,z|k)+B(k+l,z)\times U(k+l,z),\quad z\in\Omega \tag{17}$$

$$\beta_z(\hat{\theta})\hat{X}(k+l,z|k)=0,\quad z\in\partial\Omega$$
$$\hat{X}(k+l,z|k)\big|_{k=0}=\hat{X}_o(l-1,z),\quad z\in\Omega \tag{18}$$

prediction error $\delta X(k+l,z|k)$

$$\delta X(k+l,z|k)=\mathcal{L}_z(\hat{\theta})\delta X(k+l-1,z|k)+C(k+l,z)W(k+l,z)+\delta(\psi(z))C_b(k+l,z)\times W_b(k+l,z) \tag{19}$$

$$\beta_z(\hat{\theta})\delta X(k+l,z|k)=0,\quad z\in\partial\Omega$$

$$\delta X(k+l,z|k)\big|_{k=0}=\delta X_o(l-1,z) \tag{20}$$

estimate of prediction error covariance matrix $\hat{P}(k+l,z,y|k)$

$$\hat{P}(k+l,z,y|k)=\hat{P}(k+l-1,z,y|k)+\gamma(k)\left\{\frac{1}{S_z S_y}\times\int_{\Omega_z}\int_{\Omega_y}\delta X(k+l,z|k)\delta X^T(k+l,y|k)dzdy-\hat{P}(k+l-1,z,y|k)\right\},\quad z,y\in\Omega \tag{21}$$

$$\hat{P}(k+l,z,y|k)\big|_{k=0}=\hat{P}_o(l-1,z,y) \tag{22}$$

4.3. Parameter identification

The identification problem of unknown vector-parameter $\theta(k,z)$ based on using the optimal control methods for DPS is solved. Given the observation data $Y(k)=\{Y(m),\ m=0,1,\dots,k\}$ and the state estimate $\hat{X}(k,z|k)$ find the optimal parameter estimate $\hat{\theta}(k,z)$ in the squared error sense:

$$J(\hat{\theta})=\sum_{k=0}^{K}\|Y(k)-H(k)\hat{X}_N(k|k,\hat{\theta})\|^2 \tag{23}$$

The identification algorithm is derived by minimizing of the identification squared error at the measurement points. Minimization of the Lagrange function using the conjugate gradient method is made. In our case the Hamiltonian function $\mathcal{H}(\theta)$ has the form:

$$\mathcal{H}(\hat{\theta})=J(\hat{\theta})+\hat{F}^T(k+1,z|k+1)\hat{X}(k+1,z|k+1)+\delta F^T(k+1,z|k+1)\delta X(k+1,z|k+1)+\mathrm{tr}\{\hat{P}(k+1,z,z|k)\hat{S}^T(k+1,z,z|k)\} \tag{24}$$

where $\hat{F}(k+1,z|k+1)$, $\delta F(k+1,z|k+1)$ and $\hat{S}(k+1,z,y|k)$ are conjugate functions with respect to $\hat{X}(k+1,z|k+1)$, $\delta X(k+1,z|k+1)$ and $\hat{P}(k+1,z,y|k)$, $\mathrm{tr}\{\cdot\}$ denotes the trace operator of the matrix $\{\cdot\}$. The conjugate functions may be obtained from the following conditions:

$$\hat{F}(k,z|k)=\frac{\partial\mathcal{H}(\hat{\theta})}{\partial\hat{X}(k,z|k)};\qquad \delta F(k,z|k)=\frac{\partial\mathcal{H}(\hat{\theta})}{\partial\delta X(k,z|k)}$$

$$\hat{S}(k+1,z,y|k)=\frac{\partial\mathcal{H}(\hat{\theta})}{\partial\hat{P}(k+1,z,y|k)} \tag{25}$$

The minimization of the Hamiltonian function (24) with respect to $\hat{\theta}(k,z)$ was made by using the gradient method [11], where

$$\hat{\theta}^{i+1}(k,z)=\hat{\theta}^{i}(k,z)+\lambda_i\frac{\partial\mathcal{H}(\hat{\theta})}{\partial\hat{\theta}^{i}(k,z)} \qquad (26)$$

and i is the iteration number and $\lambda_i > 0$ is a parameter of the convergence procedure.

In general, the identification algorithm is defined by the solution of the sequence of direct filter equations (7)-(15), conjugate (25) and calculation of the parameter estimate (26). This self-tuning procedure is stoped, if the following conditions are satisfied:

$$\sum_{k=0}^{K}\left\|\hat{\theta}^{i}(k,z)-\hat{\theta}^{i+1}(k,z)\right\|^2 \leqslant \varepsilon(z) \qquad (27)$$

5. PREDICTION OF AIR POLLUTANT

To illustrate the application of the proposed adaptive distributed estimator, the prediction of the air pollutant concentrations in city zone will be considered. Mathematical models describing the diffusion and transport of air pollution from emission sources into the atmosphere are generally expressed by advection-diffusion equation [5, 13,14]:

$$\frac{\partial q(t,z)}{\partial t}=-(u,\operatorname{grad}q(t,z))+\alpha q(t,z)+\sum_{i=1}^{3}\frac{\partial}{\partial z_i}\left(p_i(t,z)\frac{\partial q(t,z)}{\partial z_i}\right)+f(t,z)+w(t,z) \qquad (28)$$

with boundary conditions:

$$q(t,z)=0, \quad z\in\partial\Omega_b$$

$$\frac{\partial q(t,z)}{\partial z_3}=\alpha_b q(t,z), \quad z\in\partial\Omega_o \qquad (29)$$

$$\frac{\partial q(t,z)}{\partial z_3}=0, \quad z\in\partial\Omega_H$$

and initial condition $q(t,z)\big|_{t=0}=q_o(z)$, where t is the time, Ω is the cylinder, which consists of the boundary surface $\partial\Omega_b$, the upper base $\partial\Omega_H$ ($z_3=H$) and the lower base $\partial\Omega_o$ ($z_3=0$), $q(t,z)$ is the pollutant concentration, $u(t,z)=(u_1, u_2, u_3)$ is the vector of mean wind velocities, $\operatorname{div}u(t,z)=0$, $p(p_1, p_2, p_3)$ is the vector of turbulent diffusivities, α and α_b are some constant parameters, $f(t,z)$ is a function of emission source rate and $w(t,z)$ is the stochastic input of process. Usually, the parameters $p(t,z)$ and $u(t,z)$ are unknown.

To solve the prediction problem of the pollutant concentration based on the measurements data at the selected points the approach of the first part of this work was applied. The discrete-time model of the process was obtained by using the finite difference scheme [5]. For this model the suboptimal adaptive filter (7)-(15) and predictor (17)-(22) were defined by taking into consideration the identification algorithm (25), (26). Numerical solution of these equations was obtained by using the time-splitting technique [5] and the Galerkin finite element method [14, 15].

Air pollution process was simulated in restricted two-dimensional space $\Omega=(Z_{1F},Z_{2F})=(100,100)$ for two cases: with one emission source located at $(z_1,z_2)=(20,80)$ and with two $(z_1,z_2)=(40,80)$ and $(20,60)$. In two cases the sampling interval Δt, the coordinates division Δz, the dispersion of noise process σ_w^2, parameters α and α_b, the wind velocity $u=(u_1,u_2)$ and the turbulent diffusivities $p=(p_1,p_2)$ are equal to: $\Delta t=0.1$, $\Delta z_1=\Delta z_2=5$, $\sigma_w^2=0.04$, $\alpha=\alpha_b=0$, $u=(5,-5)$, $p=(20,20)$.

Some results of the air pollution process modeling in the steady state are shown in Fig.1-3.

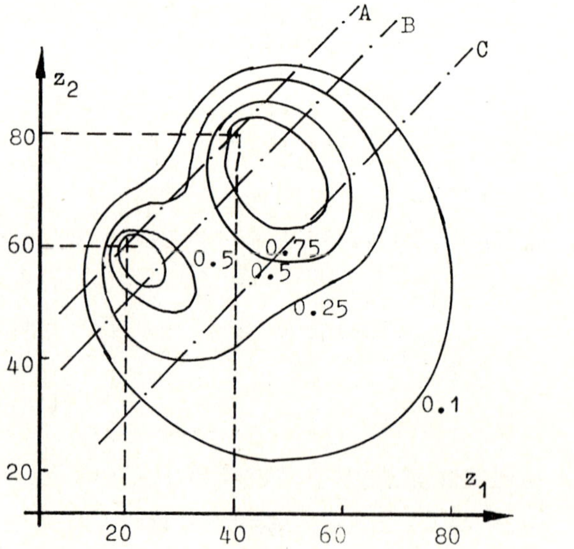

Fig.1 Lines of the concentration levels.

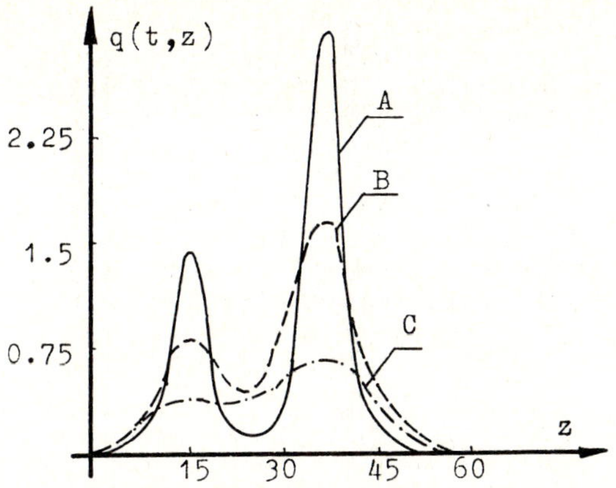

Fig.2 The distribution of the concentrations levels at different points.

In Fig.1 and Fig.2 the concentration lines and the distribution of the concentration levels for two emission sources (f=(1,0.5)) are represented. The case with one emission source (f=1.) is shown in Fig.3.

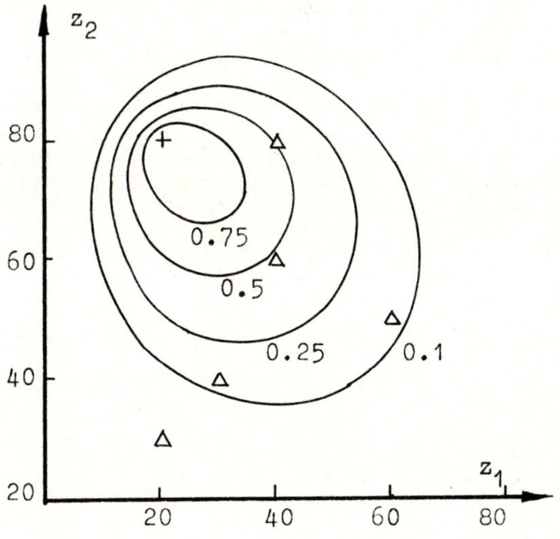

Fig.3 Lines of the concentration levels for the "object" model.

To illustrate the proposed identification algorithm the estimation of the unknown wind velocity parameter was considered. The air pollutant concentration data as the result of the process modeling with one emission source were taken as measurement data. This model was called "object" model. The measurement stations (points) coordinates are (20,30), (30,40), (40,60), (40,80), (60,50). In Fig.3 the "object" model emission source point and measurement points are labeled by "+" and "Δ", respectively.

The initial value of the wind velocity parameter for "trial" model was chosen as (4,-4). The other parameters are the same as for the "object" model. The velocity changing is shown in Fig.4.

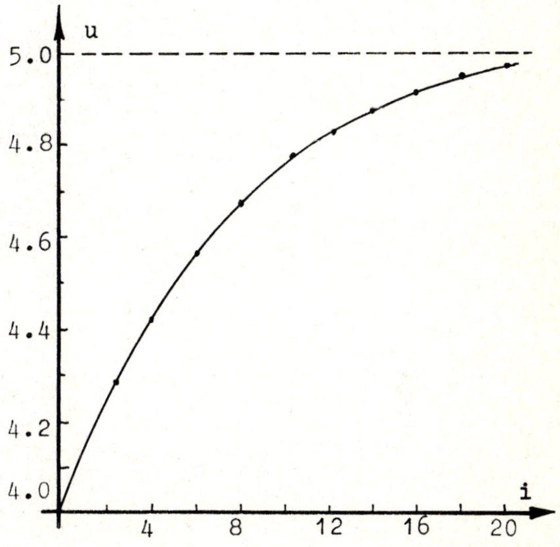

Fig.4 Wind velocity changing during identification process.

The value u ($u=u_1=-u_2$) increased from 4.0 to 4.98 during 20 iterations, and the value of square error criterion decreased from 6.87×10^{-9} to 0.66×10^{-9} at the same interval (see Fig.5).

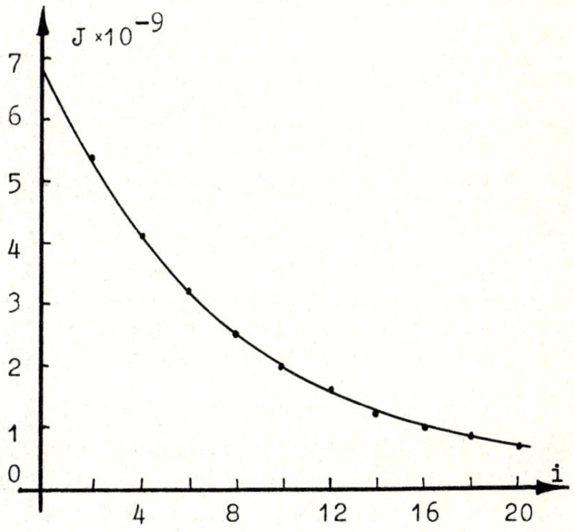

Fig.5 Square error criterion decreasing.

REFERENCES

[1] Tzafestas, S.G. Baysian approach to distributed parameter filtering and smoothing. - Int. J. Control, 1972, v.15, No.2, pp.273-296.

[2] Omatu, S., Seinfeld, J.H. A unified approach to discrete-time distributed parameter estimation by the least-squares method. - Int. J. Syst. Sci., 1981, v.12, No.6, pp.665-686.

[3] Watanabe, K., Yoshimura. T., Soeda, T. A discrete-time adaptive filter for stochastic distributed parameter systems. - Trans. ASME, J. Dynam. Syst. Meas. and Control, v. 103, pp.267-278.

[4] Korbicz, J. Adaptive algorithm of the Kalman filter for the distributed parameter systems. - Proc. Inter. Conf. "Systems Engineering", Coventry, England, 1980, pp.266-274.

[5] Marchuk, G.I. Mathematical modeling in the environment protection problem. - Moskow, Nauka, 1982, (in Russian).

[6] Chavent, G. Identification of distributed parameter. - 3rd IFAC Symp. on Identification and System Parameter Estimation, Hague, 1973, pp.649-660.

[7] Zgurovsky, M.Z., Novikov, A.N., Selin, A.N. Parameter identification of oil recovering processes.- Chem. Technol., 1984, No.3, pp.44-51, (in Russian).

[8] Omatu, S., Seinfeld, J.H. Filtering and smoothing for linear discrete-time distributed parameter systems based on Wiener-Hopf theory with application to estimation of air pollution. - IEEE Trans. Syst. Man and Cybernetics, 1981, v.SMC-11, No.12, pp.785-801.

[9] Kraskevitch, V.E., Korbicz, J. Suboptimal Kalman-filter for the distributed parameter systems. - Systems Science, 1980, v.6, No.3, pp.225-234.

[10] Korbicz, J Suboptimal state estimation algorithm for the nonlinear discrete-time distributed systems. - Int. J. Syst. Sci., 1986, v.17 (in print).

[11] Tzypkin, J.Z. Adaptive and Self-Learning in Control Systems.- Moscow, Nauka, 1968 (in Russian).

[12] Sage, A.P., White, Ch.C.111 Optimum System Control. - Printice-Hall, Inc, Englewood Cliffs, New Jersey, 1980.

[13] Zgurovsky, M.Z., Korbicz, J., Selin, A.N. Simulation and identification of air pollution. - Chem. Technol., 1984, No.5, pp.61-65 (in Russian).

[14] Yoshimura, T., Nishino, K. Estimation of air pollutant concentrations by the Galerkin finite element method. - Int. J. Syst. Sci., 1983, v.14, No.6, pp.661-672.

[15] Zienkiewicz, O.C. The Finite Element Method in Engineering Science. - London, McGraw-Hill, 1971.

Complex and Distributed Systems: Analysis, Simulation and Control
S.G. Tzafestas and P. Borne (editors)
Elsevier Science Publishers B.V. (North-Holland)

COMPUTATIONAL METHODS FOR AN OPTIMAL CONTROL PROBLEM INVOLVING A CLASS OF HYPERBOLIC PARTIAL DIFFERENTIAL EQUATIONS

K.L. TEO

Department of Industrial and Systems Engineering
National University of Singapore
Kent Ridge, Singapore 0511
Republic of Singapore

In this paper, we consider an optimal control problem involving a class of second-order hyperbolic partial differential equations with Darboux-type boundary conditions. Our aim is to present a brief survey of some recent results on the computational methods for solving this class of optimal control problem. These computaional methods are based on either the conditional gradient technique or strong variational techniques.

1. INTRODUCTION

An optimal control problem involving a class of second-order hyperbolic partial differential equations with Darboux-type boundary conditions is considered in this paper. This class of optimal control problem is initiated by Cesari in [1]. A necessary condition for optimality is derived in [2] and [3], while an existence theorem for optimal controls is proved in [4]. Computational methods are the subject matter of [5], [6], [7] and [8]. The aim of this paper is to present a brief survey of the results reported in [5] - [8].

2. THE GENERAL PROBLEM STATEMENT

For any positive integer n, we denote by E^n the n dimensional Euclidean space. For

$$x = (x_1,\dots,x_n) \in E^n \text{ , let } |x| = \sum_{i=1}^{n} |x_i| .$$

For any matrix

$$Y = (Y_{ij}) \text{ , } \qquad \text{let } |Y| = \sum_{i,j} |Y_{ij}| .$$

A^T denotes the transpose of A with A being a vector or a matrix. For any measurable subset Q of E^s, let $\bar{Q}$ denote the closure of Q. For a function $Z \in L_p(Q)$, $1 \leq p \leq \infty$, we denote by $||Z||_p$ the usual L_p-norm; in particular, $||Z||_\infty = \text{ess}_Q\sup|Z|$. Let $L_p(Q,E^n)$ denote the space of all n-vector valued functions $Z = (Z_1,\dots,Z_n)$ with $Z_i \in L_p(Q)$, $i = 1,\dots,n$. Then $L_p(Q,E^n)$ is a Banach space under the norm defined by

$$||Z||_p = \sum_{i=1}^{n} ||Z_i||_p .$$

We consider the following system of second-order nonlinear hyperbolic partial differential equations:

$$Z_{xy} = f(x,y,Z(x,y),Z_x(x,y),Z_y(x,y),u(x,y))$$

$$(x,y) \in G = [a, a + \ell_1] \times [b, b + \ell_2] \text{ ,} \qquad (1a)$$

$$Z(x,b) = \phi(x) \text{ , } x \in [a, a + \ell_1] \text{ ;}$$

$$Z(a,y) = \psi(y) \text{ , } y \in [b, b + \ell_2] \text{ .} \qquad (1b)$$

Let U be a fixed compact and convex subset of E^m, and let $\mathcal{U}$ be defined by

$$\mathcal{U} = \{u \in L_\infty(G,E^m) : u(x,y) \in U \text{ for all } (x,y) \in G\} .$$

Then any function in $\mathcal{U}$ is called an admissible control.

The following conditions are assumed throughout the paper:

A(i). $Z = (Z_1,\dots,Z_n)$ is an n-vector valued function defined on $G = [a,a+\ell_1] \times [b,b+\ell_2]$.

A(ii). $u \in L_\infty(G,E^m)$;

A(iii). $\phi(x) = (\phi_1(x),\dots,\phi_n(x))$ and $\psi(y) = (\psi_1(y),\dots,\psi_n(y))$ are given n-vector valued functions which are absolutely continuous in $a \leq x \leq a+\ell_1$ and $b \leq y \leq b+\ell_2$, respectively, and whose derivatives ϕ_x and ψ_y belong to L_p class for some $p \in [1,\infty]$;

A(iv). The function $f=(f_1,\dots,f_n)=f(x,y,Z^1, Z^2,Z^3,V)$ is defined on $G \times E^{3n} \times E^m$ such that

(a) For each $Z^1=(Z^1_1,\dots,Z^1_n)$, $Z^2=(Z^2_1,\dots,Z^2_n)$, $Z^3=(Z^3_1,\dots,Z^3_n)$, and $v=(v_1,\dots,v_n)$, $f(\cdot,\cdot,Z^1,Z^2,Z^3,v)$ is measurable in G.

(b) For each $u \in L_\infty(G,E^m)$, $f(\cdot,\cdot,0,0,0,u(\cdot,\cdot)) \in L_p(G,E^n)$ with p as in A(iii).

(c) There exist constants K_1 and K_2 such that

$$|f_\beta(x,y,z^1,z^2,z^3,v)| \leq K_1 \tag{2}$$

$$|f_\beta(x,y,z^1,z^2,z^3,v) - f_\beta(x,y,\tilde{z}^1,\tilde{z}^2,\tilde{z}^3,\tilde{v}) \leq K_2\{|z^1-\tilde{z}^1|+|z^2-\tilde{z}^2|+|z^3-\tilde{z}^3|+|v-\tilde{v}|\} \tag{3}$$

for all $(x,y) \in G$, z^k, $\tilde{z}^k \in E^n$, $(k = 1,2,3)$, and v, $\tilde{v} \in U$, where f_β stands for any of the derivative matrices

$$f_{z^k} = \left(\frac{\partial f_j}{\partial z_i^k}\right) \begin{matrix} i=1,2,\dots,n \\ j=1,2,\dots,n \end{matrix} \quad (k = 1,2,3) \text{ and}$$

$$f_v = \left(\frac{\partial f_j}{\partial v_i}\right) \begin{matrix} i=1,2,\dots,m. \\ j=1,2,\dots,n \end{matrix}$$

For each $u \in L_\infty(G,E^m)$, the system (1) admits a unique solution in the space which consists of all functions $\{Z\}$ such that Z together with its generalized partial derivatives Z_x, Z_y, and Z_{xy} all belong to $L_p(G,E^n)$. This result is reported in [3].

The general control problem considered in this paper is stated as follows: Subject to system (1), find a control $u \in \mathcal{U}$ that minimizes the cost functional:

$$J(u) = \rho \cdot Z(u)(a + \ell_1, b + \ell_2) = \sum_{i=1}^{n} \rho_i Z_i(a + \ell_1, b + \ell_2) , \tag{4}$$

where $\rho = (\rho_1,\dots,\rho_n) \in E^n$ is a given constant vector and Z(u) is the solution of system (1) corresponding to the control $u \in \mathcal{U}$.

For convenience, this control problem will be referred to as problem (P).

Remark 2.1. (i) The cost functional J (for definition, see Eq. (4)) is also well defined in $L_\infty(G,E^m)$.

(ii) In view of inequality (3.3) of [3, p. 132], we note that the cost functional J is bounded on $\mathcal{U}$.

Following the approach used in [2], system (1) will be written in the form proposed by Cesari with state equation of Dieudonné-Rashevsky type [1]. For this, we introduce the following notation:

$$z^1 = Z , \ z^2 = Z_x , \ z^3 = Z_y . \tag{5}$$

Then system (1) can be converted into the following equivalent form:

$$\begin{aligned} z_x^1(x,y) &= z^2(x,y), \\ z_x^3(x,y) &= f(x,y,z^1(x,y),z^2(x,y),z^3(x,y),u(x,y)), \\ z_y^1(x,y) &= z^3(x,y), \\ z_y^2(x,y) &= f(x,y,z^1(x,y),z^2(x,y),z^3(x,y),u(x,y)), \end{aligned} \tag{6a}$$

with boundary conditions

$$z^1(x,b) = \phi(x) , \ z^1(a,y) = \psi(y) ,$$

$$z^2(x,b) = \phi_x(x), \ z^3(a,y) = \psi_y(y). \tag{6b}$$

Cost functional (4) is equivalent to

$$\tilde{J}(u) = \int_a^{a+\ell_1} \{\tfrac{1}{2}\rho \cdot z^2(u)(x, b + \ell_2)\}dx + \int_b^{b+\ell_2} \{\tfrac{1}{2}\rho \cdot z^3(u)(a + \ell_1, y)\}dy . \tag{7}$$

The Hamiltonian function is defined by

$$H(x,y,\hat{Z},\lambda,\mu,u) = \lambda^1 \cdot z^2 + \mu^1 \cdot z^3 + (\lambda^3 + \mu^2) \cdot f(x,y,z^1,z^2,z^3,u) , \tag{8}$$

where $\hat{Z} = (z^1,z^2,z^3)$, $\lambda = (\lambda^1,\lambda^3)$, and $\mu = (\mu^1,\mu^2)$, while λ^1, λ^3, μ^1, $\mu^2 \in E^n$.

The multipliers λ and μ are required to satisfy the following system of differential equations (conjugate problem; see [3]).

$$\lambda_{ix}^1 + \mu_{iy}^1 = -\partial H/\partial z_i^1 , \ \mu_{iy}^2 = -\partial H/\partial z_i^2 ,$$

$$\lambda_{ix}^3 = -\partial H/\partial z_i^3 , \quad i = 1,\dots,n , \tag{9a}$$

with boundary conditions

$$\lambda^1(a + \ell_1, y) = \mu^1(x, b + \ell_2) = 0 ,$$

$$\mu^2(x, b + \ell_2) = \lambda^3(a + \ell_1, y) = \rho/2 . \tag{9b}$$

In [3], it has been shown that this conjugate problem (Eqs. (9)) has a family of solutions belonging to $L_\infty(G,E^{4n})$ and that $W = \lambda^3 + \mu^2$ is uniquely determined by the integral equation $W = TW$, where TW is defined by

$$\begin{aligned} (TW)_i(x,y) = \rho_i &+ \int_x^{a+\ell_1}\int_y^{b+\ell_2} W(\eta,\tau) \cdot \left\{\frac{\partial f(\eta,\tau,\hat{Z}(\eta,\tau),u(\eta,\tau))}{\partial z_i^1}\right\} d\eta d\tau \\ &+ \int_y^{b+\ell_2} \left\{W(x,\tau) \cdot \frac{\partial f(x,\tau,\hat{Z}(x,\tau),u(x,\tau))}{\partial z_i^2}\right\} d\tau \\ &+ \int_x^{a+\ell_1} \left\{W(\eta,y) \cdot \frac{\partial f(\eta,y,\hat{Z}(\eta,y),u(\eta,y))}{\partial z_i^3}\right\} d\eta \end{aligned} \tag{10}$$

(for details, see [3, pp. 134-135]).

Clearly, the multipliers λ, μ, and $W = \lambda^3 + \mu^2$ depend on the choice of $u \in L_\infty(G,E^m)$, and hence will be written as $\lambda(u)$, $\mu(u)$, and $W(u)$, respectively.

3. A CONDITIONAL GRADIENT ALGORITHM

The aim of this section is to devise a conditional gradient algorithm for solving the general optimal control problem (P). For this, we need the expression of the Fréchet derivative of the cost functional J obtained in Theorem 2.1 of [5]. It is quoted in the following:

Theorem 3.1. The cost functional $J:L_\infty(G,E^m)\to E^1$ given in (4) is Fréchet differentiable everywhere on $L_\infty(G,E^m)$. Further, the derivative of J at $u^o \in L_\infty(G,E^m)$ is given by

$$J'(u^o)(u) = \iint_G (W(u^o)(x,y))^T f_v(x,y,\hat{Z}(u^o)(x,y),u^o(x,y))u(x,y)dxdy . \qquad (11)$$

On the basis of the formula for the Fréchet derivative of the cost functional, we are now able to devise a conditional gradient algorithm to solve the problem (P). This algorithm is given below:

Algorithm (A).

Step 1. Choose constants $\alpha, \beta \in (0,1)$, and initial control $u^o \in \mathcal{U}$. Set $i = 0$.

Step 2. Calculate $J'(u^k)$ by using formula (11).

Step 3. Find a control $u^{k,*} \in \mathcal{U}$ such that

$$(W(u^k)(x,y))^T f_v(x,y,\hat{Z}(u^k)(x,y),u^k(x,y))u^{k,*}(x,y)$$
$$\leq (W(u^k)(x,y))^T f_v(x,y,\hat{Z}(u^k)(x,y),u^k(x,y))u(x,y)$$

for all $u \in \mathcal{U}$ and for almost all $(x,y) \in G$.

Step 4. If $J'(u^k)(u^{k,*} - u^k) = 0$, then set $u^{k+\ell} = u^k$ for all positive integers ℓ and stop; otherwise, go to next step.

Step 5. Choose σ^k to be the first element in the sequence $1,\alpha,\alpha^2,\ldots,$ such that

$$J(u^k+\sigma^k(u^{k,*}-u^k))-J(u^k)\leq\sigma^k\beta J'(u^k)(u^{k,*}-u^k) .$$

Step 6. Set $u^{k+1} = u^k + \sigma^k(u^{k,*} - u^k)$. Go to Step 2 with $k = k + 1$.

From Section 4 of [5], we note that Algorithm (A) is well defined.

Theorem 4.3 of [5] presents a convergence result for the algorithm. This convergence result is quoted in the following.

Theorem 3.2. Let $\{u^k\}$ be a sequence of admissible controls generated by the conditional gradient algorithm. If $\tilde{u}$ is an accumulation point of $\{u^k\}$, then

$$J'(\tilde{u})(\tilde{u}^* - \tilde{u}) = 0 \qquad (12)$$

where $\tilde{u}^*$ belongs to $\mathcal{U}$ and minimizes $J'(\tilde{u})(\cdot)$ over U.

Remark 3.1. From Theorem 4.2 of [5], we see that if $\tilde{u} \in \mathcal{U}$ is an optimal control, then it satisfies the condition (12) which is a necessary condition for optimality.

4. A SPECIAL CASE - LINEAR SYSTEM AND CONVEX COST FUNCTIONAL

In this section, we shall consider the problem (P) under certain linearity assumptions. More precisely, system (1) will take the following special form:

$$Z_{xy} = A(x,y)Z+B(x,y)Z_x+C(x,y)Z_y+D(x,y)u ,$$
$$(x,y) \in G = [a, a + \ell_1] \times [b, b + \ell_2] , \qquad (13a)$$

with fixed Darboux boundary conditions

$$Z(x,b) = \phi(x) , x \in [a, a + \ell_1] ,$$
$$Z(a,y) = \psi(y) , y \in [b, b + \ell_2] . \qquad (13b)$$

To ensure the fulfilment of the conditions stated in Section 2, the matrix valued functions A, B, C and D are assumed to satisfy the following conditions.

A(v). $A = (A_{ij})$, $B = (B_{ij})$, $C = (C_{ij})$ are $n \times n$ matrix valued functions with each of their components belonging to $L_\infty(G)$.

A(vi). $D = (D_{ij})$ is an $n \times m$ matrix valued function with each of its components belonging to $L_p(G)$, where p is as defined in A(iii).

Let $h_1(z)$, $h_2(x,y,z)$, $h_3(x,y,u)$ be real-valued continuous functions defined on E^n, $G \times E^n$, $G \times E^m$, respectively. Assume further that these functions are twice continuously differentiable and convex with respect to the variables z and u. We denote by ∇h_1 and $\nabla^2 h_1$ respectively, the gradient and Hessian of h_1 with respect to z. The gradients and Hessians of h_2 and h_3 with respect to z and u will be given the similar notation.

Let Z(u) denote the solution of the system (13) corresponding to the admissible control u. The control problem considered in this section is stated as follows: Subject to the system (13), find a control $u \in \mathcal{U}$ that minimizes the cost functional:

$$J(u) = h_1(Z(u)(a+\ell_1, b+\ell_2))$$

$$+\int_G\int[h_2(x,y,Z(u)(x,y))+h_3(x,y,u(x,y))]dxdy. \quad (14)$$

For convenience, this control problem will be referred to as problem (P1).

From Theorem 4.1 of [6], we note the optimal control problem (P1) has a solution.

The necessary and sufficient condition for optimality is obtained in Theorem 5.2 of [6]. It is quoted in the following.

Theorem 4.1. Consider the problem (P1). Then, $u^* \in \mathcal{U}$ is an optimal control if and only if u^* is such that

$$\min_{v\in U}\{\bar{w}(x,y)\cdot(D(x,y)v)+h_3(x,y,v)\}$$

$$= \bar{w}(x,y)\cdot(D(x,y)u^*(x,y))+h_3(x,y,u^*(x,y)) \quad (15)$$

for almost all $(x,y) \in G$, where the vector valued function

$$\bar{w}(x,y) = (\bar{w}_1(x,y),\ldots,\bar{w}_n(x,y))$$

satisfies the relation

$$\bar{w}_i(x,y) = \frac{\partial h_1(Z(u^*)(a+\ell_1, b+\ell_2))}{\partial Z_i}$$

$$+\int_x^{a+\ell_1}\int_y^{b+\ell_2}[\sum_{j=1}^n \bar{w}_j(\alpha,\beta)A_{ji}(\alpha,\beta)$$

$$+\frac{\partial h_2(\alpha,\beta, Z(u^*)(\alpha,\beta))}{\partial Z_i}]d\alpha d\beta$$

$$+\int_y^{b+\ell_2}[\sum_{j=1}^n \bar{w}_j(x,\beta)B_{ji}(x,\beta)]d\beta$$

$$+\int_x^{a+\ell_1}[\sum_{j=1}^n \bar{w}_j(\alpha,y)C_{ji}(\alpha,y)]d\alpha\ , \quad (16)$$

The algorithm (A) can certainly be used to solve the problem (P1). However, we can, in fact, do more as the problem (P1) is only a special case of the general problem (P). Let us begin by introducing some notation. Define Ψ be the product space:

$$\Psi = E^n \times L_\infty(G,E^n) \times L_\infty(G,E^m)\ ,$$

with the inner product of two elements:

$$\Phi^1 = (e^1,z^1,u^1)\ ,\ \Phi^2 = (e^2,z^2,u^2)\ ,$$

given by

$$<\Phi^1,\Phi^2> = e^1\cdot e^2$$

$$+\int_G\int[z^1(x,y)\cdot z^2(x,y)+u^1(x,y)\cdot u^2(x,y)]dxdy\ .$$

For any $u \in \mathcal{U}$, let $\Phi(u)$ denote the element:

$$(Z(u)(a+\ell_1, b+\ell_2),\ Z(u),\ u) \in \Psi\ ;$$

and, for any $u, v \in \mathcal{U}$, let $\nabla J(u)$ and $\nabla^2 J(u)\Phi(v)$ denote the following elements in Ψ:

$$\nabla J(u) = (\nabla h_1(Z(u)(a+\ell_1, b+\ell_2))\ ,$$

$$\nabla h_2(\cdot,\cdot,Z(u))\ ,\ \nabla h_3(\cdot,\cdot,u))\ ,$$

$$\nabla^2 J(u)\Phi(v) =$$

$$(\nabla^2 h_1(Z(u)(a+\ell_1, b+\ell_2))Z(v)(a+\ell_1, b+\ell_2)\ ,$$

$$\nabla^2 h_2(\cdot,\cdot,Z(u))Z(v)\ ,\ \nabla^2 h_3(\cdot,\cdot,u)v)\ .$$

A minimizing sequence of admissible controls $\{u^k\}$ can be generated by the following algorithm.

Algorithm (A1).

Step 1. Choose an initial control $u^1 \in \mathcal{U}$, and set $k = 1$.

Step 2. Choose $\tilde{u}^k \in \mathcal{U}$ by the requirement that

$$\min_{u\in\mathcal{U}}<\nabla J(u^k),\Phi(u)>$$

$$= \min_{u\in\mathcal{U}}\{\nabla h_1(Z(u^k)(a+\ell_1,b+\ell_2)\cdot Z(u)(a+\ell_1,b+\ell_2))$$

$$+\int_Q\int[\nabla h_2(x,y,Z(u_k)(x,y))\cdot Z(u)(x,y)$$

$$+\nabla h_3(x,y,u^k(x,y))\cdot u(x,y)]dxdy\}$$

$$= <\nabla J(u^k),\Phi(\tilde{u}^k)>\ .$$

Step 3. Let $u^{k+1} = u^k + \alpha^k(\tilde{u}^k - u^k)$, where α^k is determined by the requirement that

$$J(u^k + \alpha^k(\tilde{u}^k - u^k)) = \inf_{0\leq\alpha\leq 1} J(u^k + \alpha(\tilde{u}^k - u^k))\ .$$

Step 4. Go to Step 2, with $k = k + 1$.

All steps, except Step 2, are readily implementable. We now describe the procedure for determining $\tilde{u}^k$ appearing in Step 2. Let

$$\tilde{Z} = (Z,\ Z_{n+1})$$

be an (n+1)-vector valued function. Consider the following system:

$$Z_{xy} = A(x,y)Z+B(x,y)Z_x+C(x,y)Z_y+D(x,y)u\ , \quad (17a)$$

$$(z_{n+1})_{xy} = \nabla h_2(x,y,Z(u^k)(x,y))\cdot Z$$

$$+\nabla h_3(x,y,u^k(x,y))\cdot u\ , \quad (17b)$$

with boundary conditions

$$\tilde{Z}(x,b) = (\psi(x),0)\ ,\ x \in [a,\ a+\ell_1]\ , \quad (17c)$$

$$\tilde{Z}(a,y) = (\phi(y),0)\ ,\ y \in [b,\ b+\ell_2]\ . \quad (17d)$$

Define a cost functional as:

$$\tilde{J}(u) = \nabla h_1(Z(u^k)(a+\ell_1,b+\ell_2) \cdot Z(u)(a+\ell_1,b+\ell_2) + Z_{n+1}(u)(a+\ell_1,b+\ell_2)) \,, \tag{18}$$

where

$$\tilde{Z}(u) = (Z(u), Z_{n+1}(u))$$

is the solution of the system (17) corresponding to the control $u \in \mathcal{U}$.

Step 2 of the algorithm is exactly the problem of minimizing the cost functional (18), subject to the system (17). This problem is in a form of the problem considered in Section 3 of [6]. Thus, by virtue of the method given in that section, the control $\tilde{u}^k$ can always be constructed in one iteration. For details, see Section 3 of [6]. Note that this approach is motivated by that of [9].

To close this section, the convergence results reported in Theorems 6.1, 7.1 and 7.2 of [6] are quoted, respectively, in the following three theorems.

Theorem 4.2. The sequence $\{u^k\}$ generated by the algorithm (A1) is a minimizing sequence. That is to say

$$\lim_{k\to\infty} J(u^k) = J(u^*) \,,$$

where u^* is an optimal control for Problem (P1).

Theorem 4.3. The sequence $\{u^k\}$, generated by the algorithm (A1), has a subsequence which converges to an optimal control in the weak * topology of $L_\infty(G,E^m)$. Furthermore, if $u^* \in \mathcal{U}$ is an accumulation point of the sequence $\{u^k\}$, with respect to the weak * topology, then it is an optimal control.

Theorem 4.4. Suppose that, for almost all $(x,y) \in G$, $h_3(x,y,\cdot)$ is strictly convex in U. Then, the sequence $\{u^k\}$, generated by the algorithm (A1), converges to the optimal control in the weak * topology.

5. A STRONG VARIATION ALGORITHM

The aim of this section is to use the strong variational technique to produce a computational algorithm. However, this technique does not appear possible for the general problem (P). Thus, we shall impose some restrictions. More precisely, we consider the following linear version of the system (1).

$$\begin{aligned} Z_{xy} &= f(x,y,Z,Z_x,Z_y,u) \\ &= A(x,y,u(x,y))Z+B(x,y)Z_x+C(x,y)Z_y \\ &\quad + g(x,y,u(x,y)) \,, \quad (x,y) \in G, \end{aligned} \tag{19a}$$

with Darboux boundary conditions

$$\begin{aligned} Z(x,b) &= \phi(x) \,, \quad x \in [a, a+\ell_1] \\ Z(a,y) &= \psi(y) \,, \quad y \in [b, b+\ell_2] \end{aligned} \tag{19b}$$

This system is more general than the one considered in Section 4. More precisely, the present system allows the control variable to appear not only in the forcing term but also in the zero order coefficients.

To ensure the fulfilment of the conditions stated in Section 2, we need to assume that the matrix valued functions A, B, C and g satisfy the following conditions.

A(vii). B, C satisfy the assumption A(v).

A(viii). The matrix valued function $A(\cdot,\cdot,\cdot)$ and the vector valued function $g(\cdot,\cdot,\cdot)$ are uniformly continuous (and hence bounded) on $G \times U$.

The optimal control problem considered in this section is now stated as follows:

Subject to system (19), find a control $u \in \mathcal{U}$ that minimizes the cost functional

$$\begin{aligned} J(u) &= \rho \cdot Z(u)(a+\ell_1, b+\ell_2) \\ &= \sum_{i=1}^{n} \rho_i Z_i(u)(a+\ell_1, b+\ell_2) \end{aligned} \tag{20}$$

where $\rho = (\rho_1,\ldots,\rho_n) \in E^n$ is a constant vector and $Z(u)$ is the solution of the system (19) corresponding to the control u.

For convenience, this optimal control problem will be referred to as problem (P2).

The corresponding Hamiltonian function and the conjugate problem may be easily deduced from (8) and (9), by taking into account of the special structure of the problem (P2). More precisely, the Hamiltonian function H is defined as:

$$\begin{aligned} H(u)(x,y,v) &= \lambda^1 \cdot Z_x(u) + \mu^1 \cdot Z_y(u) \\ &\quad + (\lambda^3 + \mu^2) \cdot f(x,y,Z(u),Z_x(u),Z_y(u),v) \\ &= \lambda^1 \cdot Z_x(u) + \mu^1 \cdot Z_y(u) \\ &\quad + (\lambda^3 + \mu^2) \cdot A(x,y,v)Z(u) \\ &\quad + BZ_x(u) + CZ_y(u) + g(x,y,v) \,, \end{aligned} \tag{21}$$

and the conjugate problem takes the form:

$$\frac{\partial \lambda_i^1}{\partial x} + \frac{\partial \mu_i^1}{\partial y} = -\sum_{j=1}^{n} (\lambda_j^3 + \mu_j^2) A_{ji}(x,y,u(x,y)) \,,$$

$$\frac{\partial \mu_i^2}{\partial y} = - \lambda_i^1 - \sum_{j=1}^{n} (\lambda_j^3 + \mu_j^2) B_{ji} ,$$

$$\frac{\partial \lambda_i^2}{\partial x} = - \mu_i^1 - \sum_{j=1}^{n} (\lambda_j^2 + \mu_j^2) C_{ji} , \quad (22a)$$

with boundary conditions

$$\lambda^1(a + \ell_1, y) = \mu^1(x, b + \ell_2) = 0$$

$$\lambda^3(a + \ell_1, y) = \mu^2(x, b + \ell_2) = \rho/2 , \quad (22b)$$

where ρ is as given in (20).

From Theorem 4.1 of [3], it follows that, for each $u \in \mathcal{U}$, the conjugate problem (22) has a family of solutions belonging to $L_\infty(G,E^{4n})$. However, if λ_x^1 is chosen to be zero, then (22) has only a unique solution which can be expressed as:

$$\lambda_i^i(x,y) = 0 .$$

$$\mu_i^1(x,y) = \int_y^{b+\ell} [\sum_{j=1}^{n} W_j(u)(x,\tau) A_{ji}(x,\tau,u(x,\tau))] d\tau ,$$

$$\lambda_i^3(x,y) = \rho_i/2$$

$$+ \int_x^{a+\ell_1}\int_y^{b+\ell_2} [\sum_{j=1}^{n} W_j(u)(\eta,\tau) A_{ji}(\eta,\tau,u(\eta,\tau))] d\eta d\tau$$

$$+ \int_x^{a+\ell_1} [\sum_{j=1}^{n} W_j(u)(\eta,y) C_{ji}(\eta,y)] d\eta , \quad (23)$$

$$\mu_i^2(x,y) = \rho_i/2$$

$$+ \int_y^{b+\ell_2} [\sum_{j=1}^{n} W_j(u)(x,\tau) B_{ji}(x,\tau)] d\tau , \quad i = 1,\dots,n .$$

Where $W(u)$ stands for $\lambda^3(u) + \mu^2(u)$ and satisfies the integral equation (uniquely in the sense of $L_\infty(G,E^n)$):

$$W_i(x,y) = \rho_i$$

$$+ \int_x^{a+\ell_1}\int_y^{b+\ell_2} [\sum_{j=1}^{n} W_j(\eta,\tau) A_{ji}(\eta,\tau,u(\eta,\tau))] d\eta d\tau$$

$$+ \int_y^{b+\ell_2} [\sum_{j=1}^{n} W_j(x,\tau) B_{ji}(x,\tau)] d\tau$$

$$+ \int_x^{a+\ell_1} [\sum_{j=1}^{n} W_j(\eta,y) B_{ji}(\eta,y)] d\eta . \quad (24)$$

5.1 Successive Controls

The first order differential operator Θ is a mapping from $\mathcal{U}$ into E^1 defined by:

$$\Theta(u) = \frac{1}{\ell_1 \ell_2} \iint_G \{H(u)(x,y,\omega(u)(x,y))$$

$$- H(u)(x,y,u(x,y))\} dxdy , \quad (25)$$

where $\omega(u)$ is an admissible control such that

$$H(u)(x,y,\omega(u)(x,y)) \leq H(u)(x,y,v)$$

for all $v \in U$ and for almost all $(x,y) \in G$. Note that the existence of such a $\omega(u)$ is guaranteed by Lemma 5.1 of [8].

Definition 5.1. A control $u \in \mathcal{U}$ is said to be an extremal control if

$$\Theta(u) = 0 . \quad (26)$$

Throughout the rest of this subsection, u will be used to denote a fixed, nonextremal admissible control, that is, $u \in \mathcal{U}$ and $\Theta(u) < 0$.

Consider the set

$$\Omega(u) = \{(x,y) \in \bar{G} : H(x,y,\omega(u)(x,y))$$

$$- H(u)(x,y,u(x,y)) \leq \Theta(u)\} . \quad (27)$$

It is clear that, for all $u \in \mathcal{U}$, $\Theta(u) \leq 0$ and $|\Omega(u)| > 0$, where $|\Omega(u)|$ denotes the Lebesgue measure of the set $\Omega(u)$.

In what follows, let $\{G_\alpha, \alpha \in [0,|G|]\}$ be a class of sets which possesses the following properties.

$$\begin{cases} 1) & G_\alpha \subset \bar{G} \\ 2) & |G_\alpha| = \alpha \\ 3) & G_\alpha \subset G_{\alpha'} \quad \text{if} \quad \alpha \leq \alpha' \\ 4) & G_\alpha = \Omega(u) \quad \text{if} \quad \alpha = |\Omega(u)| . \end{cases} \quad (28)$$

Let

$$u^\alpha(x,y) = \begin{cases} \Omega(u)(x,y) , & (x,y) \in G_\alpha \\ u(x,y) , & \text{otherwise} \end{cases} \quad (29)$$

Then, u^α is an admissible control (i.e. $u^\alpha \in \mathcal{U}$) which will be referred to as the perturbed control of u of size α.

Theorem 5.6 of [8] is the heart of the strong varitional algorithm to be presented in the next subsection, and is recalled in the following.

Theorem 5.1. For an arbitrary but fixed $u \in \mathcal{U}$, let

$$\alpha^* = \sup\{\alpha \in [0,|G|] : J(u^\alpha) - J(u) \leq \Theta(u) \frac{\alpha}{2}\} .$$

Then, there exists a positive constant K_3, independent of α^* and u, such that

$$J(u^{\alpha^*}) - J(u) \leq - K_3 |\Theta(u)|^3 \quad (30)$$

Note that the admissible control $u^{\alpha*}$ is called the *successive control* of the control $u \in \mathcal{U}$. Furthermore, by virtue of Theorem 5.1, it is easy to see that the condition (26) is *a necessary condition for optimality*.

5.2 The Algorithm

Consider the problem (P2). If u^o is a given non-extremal admissible control, then it follows from Theorem 5.1 that the successive control $u^{o,\alpha*} \in \mathcal{U}$ can be constructed. This successive control improves the value of the cost functional by at least $\alpha*\Theta(u)/2$. Thus, by repeating this process, we obtain a sequence of controls $\{u^k\} \subset \mathcal{U}$ such that u^{k+1} is the successive control of the control u^k, for $k = 0,1,2,\ldots$. This sequence of controls is called the sequence of successive controls corresponding to $u^o \in \mathcal{U}$. The details of the construction are given in the following algorithm.

Algorithm (A2).

Step 1. Let $u^o \in \mathcal{U}$ be any non-extremal control and set $k = 0$.

Step 2. For u^k, solve the system (19) and then evaluate $J(u^k)$ using (20).

Step 3. Solve the conjugate problem using the equations (23) and (24).

Step 4. Find the function $\omega(u^k)$ that minimizes the Hamiltonian function $H(u^k)(x,y,\cdot)$ for almost all $(x,y) \in G$.

Step 5. Calculate the number $\Theta(u^k)$ given by (25) and determine the set $\Omega(u^k)$ defined by the equation (27).

Step 6. By some rule, specify a sequence of sets G_α for $\alpha \in [0,|G|]$ that satisfies (28).

Step 7. Find, for each $\alpha \in [0,|G|]$, the control u^α defined by (29) and then evaluate $J(u^\alpha)$.

Step 8. Find the supremum $\alpha*$ of the set

$$\{\alpha \in [0,|G|] : J(u^\alpha) - J(u^k) \leqslant \Theta(u^k)\alpha/2\}$$

Step 9. Set $u^{k+1} = u^{\alpha*}$, and go to Step 2 with k replaced by $k + 1$.

The next theorem, which contains a convergence result of the algorithm (A2), is quoted from Theorem 5.7 of [8].

Theorem 5.2. Let $\{u^k\}$ be a sequence of non-extremal controls such that u^{k+1} is a successive control of u^k for all k. If $u*$ is an accumulation point of $\{u^k\}$ in the strong topology of $L_\infty(G,E^m)$, then $u*$ is an extremal control.

6. DISCUSSION

In this paper, we considered optimal control problems involving hyperbolic partial differential equations with Darboux boundary conditions. The aim was to present a brief survey of some of the main results obtained in [5], [6], [7] and [8].

A conditional gradient algorithm was the subject matter of Section 3, where the main reference was [6]. The convergence property of the algorithm was reported in Theorem 3.1. It is of the same type as that given in [10]. In Section 4, a special class of optimal control problems was considered. To be more precise, the system is linear and the cost functional is convex. For this special case, an algorithm, which is basically a conditional gradient algorithm, was devised. This algorithm possesses very strong convergence properties. For details, see Theorems 4.2, 4.3 and 4.4.

The extension of the approach of Section 4 to the general control problem (P) does not seem possible. In fact, it does not appear possible even for the optimal control problem (P2) in which the system is again linear. However, it allows the control variable to appear not only in the forcing term but also in the zero order coefficient. For the problem (P2), a strong variational technique was used to produce a computational algorithm. Naturally, we could only obtain the same type of convergence property as that reported in [10], (i.e., as that reported in Theorem 3.2).

The convergence results reported in Theorem 3.2 and Theorem 5.2 are not satisfactory in the sense that a sequence $\{u^k\}$ of controls generated by either algorithm does not necessarily possesses L_∞ - accumulation points, because $\mathcal{U}$ is not sequentially compact in the strong topology of L_∞.

The convergence result of Theorem 5.2 can be improved by using a topology arising in the study of the relaxed controls. Although algorithm (A2) generates a sequence of ordinary controls, it is possible to associate with each ordinary control an equivalent relaxed control. The topology on the space of relaxed controls is weak enough to ensure that any infinite sequence of L_∞ - bounded relaxed controls has accumulation points. It is yet strong enough to ensure that its accumulation points satisfy a necessary condition for optimality for the corresponding relaxed optimal control problem. We refer the interested reader to [11].

For other types of distributed parameter systems, we refer the reader to [12] - [15] and the relevant references cited therein.

REFERENCES

[1] Cesari, L., Optimization with Partial Differential Equations in Dieudonné-Rashevsky Form and Conjugate Problems, Archive for Rational Mechanics and Analysis, 33 (1969), pp. 339-357.

[2] Suryanarayana, M.B., Optimization Problems with Hyperbolic Partial Differential Equations, Ph.D. Thesis, Department of Mathemtics, The University of Michigan, 1969.

[3] Suryanarayana, M.B., Necessary Conditions for Optimization Problems with Hyperbolic Partial Differential Equations, SIAM Journal on Control, (1973), pp. 130-147.

[4] Suryanarayana, M.B., Existence Theorems for Optimization Problems Concerning Linear, Hyperbolic Partial Differential Equations Without Convexity Conditions, Journal of Optimization Theory and Applications, 19 (1976), pp. 47-61.

[5] Wu, Z.S. and Teo, K.L., A Conditional Gradient Method for an Optimal Control Problem Involving a Class of Nonlinear Second-Order Hyperbolic Partial Differential Equations, Journal of Mathematical Analysis and Applications, 91 (1983), pp. 376-393.

[6] Wu, Z.S. and Teo, K.L., A Convex Optimal Control Problem Involving a Class of Linear Hyperbolic Systems, Journal of Optimization Theory and Applications, 3 (1983), pp. 541-560.

[7] Wu, Z.S. and Teo, K.L., First-Order Strong Variation Algorithm for Optimal Control Problem Involving Hyperbolic Systems, Journal of Optimization Theory and Applications 3 (1983), pp. 561-587.

[8] Wu, Z.S., On the Optimal Control of a Class of Hyperbolic Systems, Ph.D. Thesis, School of Mathematics, University of New South Wales, 1981.

[9] Barnes, E.R., An Extension of Gilbert's Algorithm for Computing Optimal Controls, Journal of Optimization Theory and Applications, 7 (1971), pp. 420-443.

[10] Mayne, D.Q. and Polak, E., First-Order Strong Variation Algorithms for Optimal Control, Journal of Optimization Theory and Applications, 16 (1975), pp. 277-301.

[11] Teo, K.L., Clements, D.J., Wu, Z.S. and Choo, K.G., Convergence of a Strong Variational Algorithm for Relaxed Controls Involving a Class of Hyperbolic Systems, Journal of Optimization Theory and Applications, 46 (1985), pp. 295-317.

[12] Teo, K.L. and Wu, Z.S., Computational Methods for Optimizing Distributed Systems (Academic Press, Orlando, 1984).

[13] Butkovskiy, A.G., Distributed Control Systems (Elsevier, New York, 1969).

[14] Ahmed, N.U. and Teo, K.L., Optimcal Control of Distributed Parameter Systems, (North Holland, New York, 1981).

[15] Lions, J.L., Optimal Control of Systems Governed by Partial Differential Equations, (Springer-Verlag, Berlin, 1971).

Complex and Distributed Systems: Analysis, Simulation and Control
S.G. Tzafestas and P. Borne (editors)
Elsevier Science Publishers B.V. (North-Holland)

OPTIMAL BOUNDARY CONTROL OF A SYSTEM DESCRIBED BY THE NON-LINEAR HEAT EQUATION

M. EL BAGDOURI and J. BURGER

Laboratoire d'Automatique de Nantes - UA CNRS 04/823
1 rue de la Noë 44072 NANTES Cédex France

Abstract: The optimal boundary control of a system discribed by a non-linear heat equation is studied, with possible phase change. Two kinds of criterion are considered according to the dependance of the desired temperature profile on time or not. The discrete gradient is calculated in each case and the conjugate gradient method is implemented. Numerical results are given and discussed.

1. INTRODUCTION

We consider the non-linear heat equation

$$(S_1)\begin{cases} C(\theta)\dfrac{\partial\theta}{\partial t} = \dfrac{\partial}{\partial x}\left(\lambda(\theta)\dfrac{\partial\theta}{\partial x}\right) + f(x,t) \\ \lambda(\theta(0,t))\dfrac{\partial\theta}{\partial x}(0,t) = a(t).\theta(0,t) - u(t) \\ \theta(1,t) = \theta_r(t) \qquad x \in \Omega =]0,1[\\ \theta(x,0) = \theta_o(x) \qquad t \in]0,T[\end{cases}$$

with $0 < \lambda_1 \leqslant \lambda(\theta) \leqslant \lambda_2$, $0 < C_1 \leqslant C(\theta) \leqslant C_2$; f, a, θ_r and θ_0 are given functions with $a(t) \geqslant 0$; u is the control term ; C may be a discontinuous function of the temperature taking into account a possible phase change. In this case we must take $\theta = \theta_1$ in (S_1) for one phase and $\theta = \theta_2$ for the other phase and then add to (S_1) the following conditions for the interphase (free boundary) :

$$\theta_1(x_F,t) = \theta_2(x_F,t) = \theta_F$$

$$L\frac{dx_F}{dt} = \lambda(\theta_F^-).\frac{\partial\theta_1}{\partial x}(x_F,t) - \lambda(\theta_F^+).\frac{\partial\theta_2}{x}(x_F,t)$$

where θ_F is the phase-change temperature, L is the latent heat and x_F is the absciss of the interphase.

Our purpose is to solve numerically the two following boundary control problems :

θ being solution of (S_1),

(P^1) : minimize $J_1(u) =$

$$\int_0^T\int_\Omega (\theta(x,t;u)-\theta_d(x,t))^2dxdt + \sigma\int_0^T u^2(t)dt\ .$$

(P^2) : minimize $J_2(u) =$

$$\int_\Omega (\theta(x,T,u)-\theta_d(x))^2dx + \sigma\int_0^T u^2(t)dt\ .$$

where θ_d is a desired temperature profile, and σ a positive real number.

This type of problem occurs in transformation industry where it is necessary to heat or cool a material respecting constraints like space and time profiles in the material, assuring its good quality.

2. EXISTENCE OF A SOLUTION OF (S_1)

To solve the system (S_1), we use the so-called Kirchoff and enthalpy transformations which are defined by :

$$y = \beta(\theta) = \int_0^\theta \lambda(\tau)\,d\tau$$

$$\frac{dH(\theta)}{dt} = C(\theta)\ ;\ H(0) = 0\ ;\ H(\theta_F^+) - H(\theta_F^-) = L.$$

We note $G(y) = H(\theta) = H(\beta^{-1}(y))$; that is $G = H \circ B^{-1}$. G is a maximal monotone operator which is multivocal in the case of a phase change.

With these transformations, the system (S_1) can be written under the form

$$(S_2)\begin{cases} \dfrac{\partial v}{\partial t} - \dfrac{\partial^2 y}{\partial x^2} = f(x,t) \\ v \in G(y) \\ \dfrac{\partial y}{\partial x}(0,t) = a(t).\beta^{-1}(y(0,t)) - u(t) \\ y(1,t) = \beta(\theta_r(t)) = y_r(t) \\ y(x,0) = \beta(\theta_o(x)) = y_o(x) \\ v(x,0) = v_o(x) \text{ with } v_o \in G(y_o). \end{cases}$$

Multiplying the first equation of (S_2) by $\phi \in H^1(\Omega)$ with $\phi(1) = 0$ and integrating by parts, we obtain the variational formulation of the direct problem. We discretize this formulation in time by dividing]0,T[in M intervals of the same length τ. Using an implicit scheme, we obtain the following semi-discretized system :

$$(S_3)\begin{cases} \left(\frac{v^{n+1}-v^n}{\tau},\phi\right)+\left(\frac{\partial y^{n+1}}{\partial x},\frac{\partial\phi}{\partial x}\right)+\left(a^{n+1}.\beta^{-1}(y^{n+1}(0))\right. \\ \left. \qquad - u^{n+1}\right).\phi(0) = (f^{n+1},\phi) \\ v^{n+1} \in G(y^{n+1}) \\ y^{n+1}(1) = y_r^{n+1} \\ y^{\circ}(x) = y_o(x) \\ v^{\circ}(x) = v_o(x) \quad \text{with} \quad v_o \in G(y_o). \end{cases}$$

Applying the method used by GRANGE and MIGNOT in [1] and taking into account the proofs given in [2], we can prove that system (S_3) admits a solution $\{y^n,v^n, n=1,\ldots,M\}$ and that this solution is continuous with respect to $\{u^n, n=1,\ldots,M\}$. Taking the limit, when τ tends towards zero, we prove the existence of a weak solution for the direct problem.

3. CONTROL PROBLEMS

3.1. Existence of a solution

We consider now the semi-discretized problem (S_3), and we approximate the two control problem (P^1) and (P^2) by :

$\{y^n, n=1,\ldots,M\}$ being solution of (S_3)

(P^1_τ) : minimize $J_1(u) =$

$$\sum_{n=1}^{M}\int_\Omega [y^n(x;u)-\beta(\theta_d^n(x))]^2dx + \sigma\sum_{n=1}^{M}(u^n)^2$$

(P^2_τ) : minimize $J_2(u) =$

$$\int_\Omega [y^M(x;u)-\beta(\theta_d(x))]^2dx + \sigma\sum_{n=1}^{M}(u^n)^2.$$

The continuous dependance of $\{y^n, n=1,\ldots,M\}$ with respect to $\{u^n, n=1,\ldots,M\}$ implies the existence of at least one solution for each problem (P^1_τ) and (P^2_τ).

3.2. Regularization

In the general situation with phase change, the operator G is not differentiable. This operator is regularized by introducing an operator G_ε defined by :

$$G_\varepsilon(y) = \int_0^y g_\varepsilon(\tau)\,d\tau$$

with $g_\varepsilon(y) = G'(y)$ if $y < y_F = \beta(\theta_F)$ or $y \geq (y_F+\varepsilon)$

$$g_\varepsilon(y) = q_\varepsilon(y) \text{ if } y_F \leq y \leq (y_F+\varepsilon)$$

where q_ε is the unique second degree polynomial defined on this interval by :

$$q_\varepsilon(y_F) = G'(y_F^-),\ q_\varepsilon(y_F+\varepsilon) = G'(y_F+\varepsilon),$$

and $$\int_{y_F}^{y_F+\varepsilon} q_\varepsilon(y)\,dy = L.$$

It is proved in [3] that for $0 < \varepsilon < 1$, the operator G_ε is a continuously differentiable Lipschitz operator.

The regularized system (S_3) is written as

$$(S_4)\begin{cases} \left(\frac{v_\varepsilon^{n+1}-v_\varepsilon^n}{\tau},\ \right)+\left(\frac{\partial y^{n+1}}{\partial x},\frac{\partial\phi}{\partial x}\right)+\left(a^{n+1}\beta^{-1}(y^{n+1}(0))\right. \\ \left. \qquad - u^{n+1}\right).\phi(0) = (f^{n+1},\phi) \\ v_\varepsilon^{n+1} = G_\varepsilon(y_\varepsilon^{n+1}) \\ y_\varepsilon^{n+1}(1) = y_r^{n+1} \\ y_\varepsilon^{\circ}(x) = y_o(x) \\ v_\varepsilon^{\circ}(x) = v_o(x) \text{ with } v_o = G_\varepsilon(y_o). \end{cases}$$

The system (S_4) which is a particular case of (S_3), admits a solution $\{y_\varepsilon^n,v_\varepsilon^n, n=1,\ldots,M\}$ depending continuously on $\{u^n, n=1,\ldots,M\}$. It is proved in [4] that when ε tends towards zero, this solution converges towards $\{y^n,v^n,n=1,\ldots M\}$ solution of (S_3) at least in the following sense

$y_\varepsilon^n \to y^n$ weakly in $H^1(\Omega)$

$v_\varepsilon^n \to v^n$ weakly in $L^2(\Omega)$.

3.3. Differentiability of y_ε^{n+1} with respect to u^{n+1}

Omitting the notations with n and (n+1) and fixing the time t, the first equation of (S_4) can be written as

$$\frac{1}{\tau}(G_\varepsilon(y_\varepsilon),\phi) + \left(\frac{\partial y_\varepsilon}{\partial x},\frac{\partial\phi}{\partial x}\right) + $$
$$+ (a.\beta^{-1}(y_\varepsilon(0)) - u).\phi(0) = (f_\varepsilon,\phi)$$

with $f_\varepsilon = f^{n+1} + \frac{1}{\tau}v_\varepsilon^n$.

We note $G_\varepsilon^\omega(y_\varepsilon) = G_\varepsilon(y_\varepsilon) - \omega y_\varepsilon$, with $0 \leq \omega < \inf\left(\frac{C(\theta)}{\lambda(\theta)}\right)$ such that the operator G_ε^ω remains monotonous and we obtain

$$\frac{1}{\tau}(G_\varepsilon^\omega(y_\varepsilon),\phi) + \frac{\omega}{\tau}(y_\varepsilon,\phi) + $$
$$+ \left(\frac{\partial y_\varepsilon}{\partial x},\frac{\partial\phi}{\partial x}\right) + (a.\beta^{-1}(y_\varepsilon(0)) - u).\phi(0) = (f_\varepsilon,\phi).$$

If we note $z_\varepsilon^\xi = \frac{1}{\xi}(y_\varepsilon(u+\xi w) - y_\varepsilon(u)) = \frac{1}{\xi}(y_\varepsilon^\xi - y_\varepsilon)$, then z_ε^ξ satisfies

$$\frac{1}{\tau}.\frac{1}{\xi}(G_\varepsilon^\omega(y_\varepsilon^\xi)-G_\varepsilon^\omega(y_\varepsilon),\phi) + \frac{\omega}{\tau}(z_\varepsilon^\xi,\phi) + \left(\frac{\partial z_\varepsilon^\xi}{\partial x},\frac{\partial\phi}{\partial x}\right) + $$
$$\left(a\,\frac{\beta^{-1}(y_\varepsilon^\xi(0)) - \beta^{-1}(y_\varepsilon(0))}{\xi} - w\right).\phi(0) = 0.$$

Putting $\phi = z_\varepsilon^\xi$ in this expression and using the fact that G_ε^ω and β^{-1} are monotonous, we obtain

$$\frac{\omega}{\tau}|z_\varepsilon^\xi|^2 + \left|\frac{\partial z_\varepsilon^\xi}{\partial x}\right|^2 - w.z_\varepsilon^\xi(0) \leq 0\ ,$$

Hence

$$\eta\,||z_\varepsilon^\xi||^2 \leq \frac{1}{2\alpha}\,w^2 + \frac{\alpha}{2}\,(z_\varepsilon^\xi(0))^2,\ \forall\alpha > 0$$

with $\eta = \inf(1, \frac{\omega}{\tau})$.

A trace theorem allows us to write

$(z_\varepsilon^\xi(0))^2 \leq \gamma\,||z_\varepsilon^\xi||^2$, γ positive constant,

hence

$$(\eta - \frac{\alpha\gamma}{2})\,||z_\varepsilon^\xi||^2 \leq \frac{1}{2\alpha}\,w^2.$$

An adequate choice of α leads to

$||z_\varepsilon^\xi||^2 \leq C$, C constant not depending on ξ.

z_ε^ξ is then bounded and we can extract a subsequence, still denoted z_ε^ξ, such that

$z_\varepsilon^\xi \to z_\varepsilon$ weakly in $H^1(\Omega)$ when $\xi \to 0$.

G_ε being a continuously differentiable operator with a bounded second derivative, it follows

$$\frac{1}{\xi}(G_\varepsilon^\omega(y_\varepsilon^\xi) - G_\varepsilon^\omega(y_\varepsilon),\phi) \to (G_\varepsilon'(y_\varepsilon).z_\varepsilon - \omega z_\varepsilon,\phi).$$

Finally, y_ε is differentiable with respect to u. If z_ε is its derivative in the direction w, it satisfies

$$\frac{1}{\tau}\,(G_\varepsilon'(y_\varepsilon).z_\varepsilon,\phi) + (\frac{\partial z_\varepsilon}{\partial x},\frac{\partial\phi}{\partial x}) +$$
$$+ (a.(\beta^{-1})'(y_\varepsilon(0)).z_\varepsilon(0) - w).\phi(0) = 0.$$

3.4. Adjoint systems and gradients

Taking into account the regularization of the direct problem, the two control problems become:

$(P^1_{\tau,\varepsilon})$: minimize $J_1(u) =$

$$\sum_{n=1}^{M}\int_\Omega (y_\varepsilon^n(x;u) - \beta(\theta_d^n(x)))^2 dx + \sigma\sum_{n=1}^{M}(u^n)^2.$$

$(P^2_{\tau,\varepsilon})$: minimize $J_2(u) =$

$$\int_\Omega (y_\varepsilon^M(x;u) - \beta(\theta_d(x)))^2 dx + \sigma\sum_{n=1}^{M}(u^n)^2.$$

We consider an adjoint state $\{p^n, n=1,\dots,M\}$ belonging to $H^1(\Omega)$ and we introduce the Lagrangian L_i (i=1,2) associated to each problem.

$$L_i(y,p;u) = J_i(u) + \sum_{n=0}^{M-1}[(\frac{G_\varepsilon(y_\varepsilon^{n+1}) - G_\varepsilon(y_\varepsilon^n)}{\tau},p^n) + (\frac{\partial y_\varepsilon^{n+1}}{\partial x},\frac{\partial p^n}{\partial x})$$

$$+ (a^{n+1}.\beta^{-1}(y_\varepsilon^{n+1}(0)) - u^{n+1}).p^n(0) - (f^{n+1},p^n)]$$

$$= J_i(u)\sum_{n=1}^{M}[(G_\varepsilon(y_\varepsilon^n),\frac{p^{n-1} - p^n}{\tau}) + (\frac{\partial y_\varepsilon^n}{\partial x},\frac{\partial p^{n-1}}{\partial x})$$

$$+ (a^n.\beta^{-1}(y_\varepsilon^n(0)) - u^n).p^{n-1}(0) - (f^n,p^{n-1})]$$

$$+ \frac{1}{\tau}\,(G_\varepsilon(y_\varepsilon^M),p^M) - \frac{1}{\tau}\,(G_\varepsilon(y_\varepsilon^\circ),p^\circ).$$

The adjoint-systems are obtained by cancelling the derivative of L_i (i=1,2), with respect to y in all directions $\phi \in H^1(\Omega)$ with $\phi(1) = 0$. We obtain then for $(P^1_{\tau,\varepsilon})$:

$$(S_5^1)\begin{cases}(\frac{p^{n-1} - p^n}{\tau},G_\varepsilon'(y_\varepsilon^n).\phi) + (\frac{\partial p^{n-1}}{\partial x},\frac{\partial\phi}{\partial x}) + \\ \quad + a^n.p^{n-1}(0).(\beta^{-1})'(y_\varepsilon^n(0)).\phi(0) \\ \quad = -2(y_\varepsilon^n - \beta(\theta_d^n),\phi) \\ p^{n-1}(1) = 0 \\ p^M(x) = 0,\end{cases}$$

and for $(P^2_{\tau,\varepsilon})$:

$$(S_5^2)\begin{cases}(\frac{p^{n-1} - p^n}{\tau},G_\varepsilon'(y_\varepsilon^n).\phi) + (\frac{\partial p^{n-1}}{\partial x},\frac{\partial\phi}{\partial x}) + \\ \quad + a^n.p^{n-1}(0).(\beta^{-1})'(y_\varepsilon^n(0)).\phi(0) = 0 \\ p^{n-1}(1) = 0 \\ \frac{1}{\tau}\,(p^M,G_\varepsilon'(y_\varepsilon^M).\phi) = -2(y_\varepsilon^M - \beta(\theta_d),\phi).\end{cases}$$

The systems (S_5^1) and (S_5^2) are linear and it can be proved that each of them admits a unique solution belonging to $H^1(\Omega)$.

When y_ε and p are respectively solutions of the direct and adjoint systems, we have

$$\nabla J_i(u).w = \frac{\partial L_i}{\partial u}.w \qquad \forall w$$

$$= 2\sigma\sum_{n=1}^{M}u^n.w^n - \sum_{n=1}^{M}p^{n-1}(0)\,w^n.$$

The gradient of the two criteria have the same expression :

$$\nabla J_i^n = 2\sigma\,u^n - p^{n-1}(0) \qquad n=1,\dots,M.$$

$i = 1,2$

3.5. Limit when $\varepsilon \to 0$

Let $(u_\varepsilon^*, y_\varepsilon^*)$ be an optimal solution of the regularized problem $(P^i_{\tau,\varepsilon})$. We prove that this solution converges, in some sense, towards an optimal solution of the non regularized problem (P^i_ε) : by definition of an optimal solution we have

$$J_i(u_\varepsilon^*,y_\varepsilon^*) \leq J_i(u,y_\varepsilon) \qquad \forall u.$$

We know that $y_\varepsilon \to y$ weakly in $H^1(\Omega)$ when $\varepsilon \to 0$. Hence $J_i(u,y_\varepsilon)$ and $J_i(u_\varepsilon^*,y_\varepsilon^*)$ converge.

If $J_i(\tilde{u}^*,\tilde{y}^*)$ is the limit, then

$$J_i(\tilde{u}^*, \tilde{y}^*) \leq J_i(u,y) \qquad \forall u,$$

hence $(\tilde{u}^*, \tilde{y}^*)$ is an optimal solution to the non-regularized problem.

3.6. Numerical implementation

Numerical resolution of (S_4) is performed after we introduce the Yosida approximate of the operator G_ε [5][6]. This procedure is very interesting because it transforms the resolution of a non-linear system into the resolution of a linear one followed by the resolution of a non-linear equation in each point of the space discretization.

The space discretization is done with finite elements of first order. The integrals are calculated by means of the Simpson formula which is exact if the integrated function is a polynomial of degree less or equal to two. We thus obtain a discrete tridiagonal system solved by the Thomas algorithm [6].

The optimal control is computed with a descent method :

$$u^{n,k+1} = u^{n,k} - \rho w^{n,k}, \qquad n = 1,\dots,M$$

where $\{w^{n,k}, n=1,\dots,M\}$ is the descent direction and ρ is a positive scalar calculated at each step k with the parabolic extrapolation method. The descent direction w^k is that of the conjugate gradient :

$$w^k = \nabla J^k + \frac{(\nabla J^k, \nabla J^k - \nabla J^{k-1})}{(\nabla J^{k-1}, \nabla J^{k-1})} \cdot w^{k-1}$$

4. APPLICATION EXAMPLE

We consider the following example with phase change

$$\frac{\partial \theta_1}{\partial t} = \frac{\partial^2 \theta_1}{\partial x^2} + 2(e^{-2t} - 1)$$

$$4. \quad \frac{\partial \theta_2}{\partial t} = \frac{\partial^2 \theta_2}{\partial x^2} + 4(4e^{-2t} - 1)$$

$$\frac{\partial \theta_2}{\partial x}(0,t) = \theta_2(0,t) - u(t)$$

$$\theta_1(1,t) = -e^{-2t}$$

$$\theta_1(x,0) = x(x-2)$$

with the interphase conditions

$$\theta_1(x_F,t) = \theta_2(x_F,t) = 0$$

$$2\frac{dx_F}{dt} = \frac{\partial \theta_1}{\partial x}(x_F,t) - \frac{\partial \theta_2}{\partial x}(x_F,t).$$

Let $u(t) = \tilde{u}(t) = 2(3 - e^{-2t})$; the analytic solution is written as follows :

$$\theta_1(x,t) = x^2 - 2x + 1 - e^{-2t}$$

$$\theta_2(x,t) = 2(x^2 - 2x + 1 - e^{-2t})$$

$$x_F(t) = 1 - e^{-t}$$

The operator G is regularized with $\varepsilon = 0.0625$ and the direct problem is solved for $u(t) = \tilde{u}(t)$ with a space step $h = 0.125$ and a time step $\tau = 0.0625$.

When the result of this simulation is taken as desired state for the control problems, then the optimal control we obtain coincides exactly with $\tilde{u}(t)$ for criterion J_1 as well as criterion J_2. Table 1 gives the evolution of criterion J_1 in this case during iterations taking $\sigma = 0$ and initializing the control by $u(t) = 4$.

On the other hand, when we take as desired state the exact solution to the direct problem, the obtained optimal control, when compared with $\tilde{u}(t)$, is given on figure 1 for criterion J_1 and on figure 2 for criterion J_2.

We must note that with criterion J_2, we try to determine $\{u^n, n=1,\dots,M\}$ only knowing the desired final state $\{\theta_d(ih), i=1,\dots,N\}$, therefore, in this case, the number M of time steps must be at most equal to the number N of space steps.

Number of iterations	Values of criterion J_1
1	0.72595 E-1
2	0.58298 E-2
3	0.42244 E-3
4	0.25857 E-3
5	0.52476 E-4
6	0.10627 E-4
7	0.63411 E-5

Table 1

The non linearity of the system yields that the two criteria are not quadratic with respect to u. As a consequence, the conjugate gradient method that is used converges after a number of steps which is greater than M. Theoretically, that number of steps is equal to the control dimension for quadratic criteria. We note also that, in our case, that number of steps may depend on the control initial value.

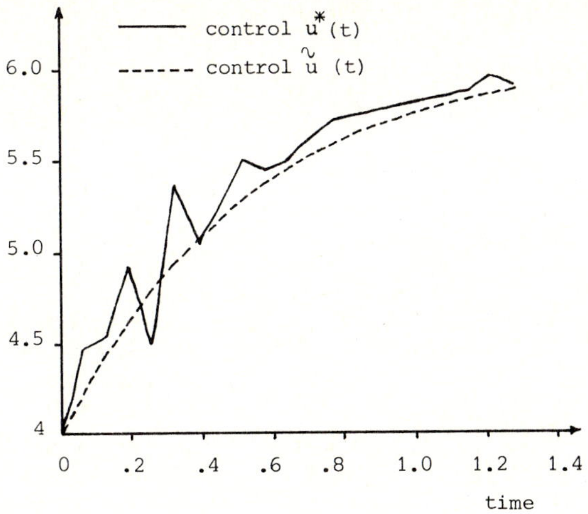

Figure 1

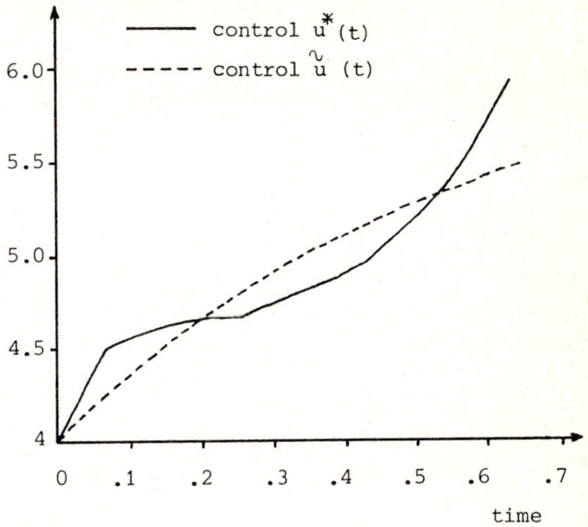

Figure 2

REFERENCES

[1] Grange, O. and Mignot, F., Sur la résolution d'une équation et d'une inéquation parabolique non linéaires, Journal of Functional Analysis, vol. 11, 1972, pp. 77-92.

[2] Saguez, C., Contrôle optimal de systèmes à frontière libre, Thèse d'Etat, Compiègne, 1980 (France).

[3] Jérome J., Existence and Approximation of weak solutions of nonlinear Dirichlet problems with discontinuous coefficients, SIAM J. Math. Anal. vol. 9, 1978, pp. 730-742.

[4] Jérome, J. and Rose, M., Error estimates for the multidimensional two-phase Stefan problem, Math. of Comp., vol. 39, 1982, pp. 377-414.

[5] Bermudez, A. and Moreno, C., Duality methods for solving variational inequalities, Comp. and Maths with Appls., vol. 7, 1981, pp. 43-58.

[6] El Bagdouri, M. and Burger, J., Résolution numérique de l'équation non linéaire de la chaleur, Comptes-rendus des Journées d'Etudes 84, ENSMA Poitiers (France).

Complex and Distributed Systems: Analysis, Simulation and Control
S.G. Tzafestas and P. Borne (editors)
Elsevier Science Publishers B.V. (North-Holland)

THE STABILITY OF A FLUID POWER SPEED CONTROL SYSTEM WITH DISTRIBUTED PARAMETER SIGNAL TRANSMISSION

J. Watton

Department of Mechanical Engineering and Energy Studies,
University College, Newport Road, Cardiff, U.K.

The dynamic characteristics of an electrohydraulic speed control system is studied for the case when servovalve and motor are connected by long transmission lines. A small signal analysis is first pursued in the frequency domain and it is shown how this fails to predict oscillations in the open-loop system. The conditions necessary for the stability of the open-loop are then established from a time domain simulation using the method of characteristics. Finally the stability of the closed-loop system is investigated for a range of motor operating speeds.

1. NOMENCLATURE

C_o	velocity of sound in the line
D_m	motor displacement
$G_v(s)$	servovalve transfer function
i_s, i_{ss}	servovalve current and its steady-state value
I_s	effective current input for dynamic model
J	motor and load inertia
k_a	servo amplifier gain
k_o, k_f	servovalve flow constants
k_v	open-loop gain $k_a k_i k_t / D_m$
k_i	servovalve linearised flow gain
k_t	tachogenerator gain
ℓ	line length
L_m	motor and load inductance J/D_m^2
n	number of line sections
N_e	motor friction non-linearity Describing Function
P_1, P_2	servovalve pressures
P_a, P_b	motor pressures
P_s	supply pressure
P_t	total motor friction pressure
P_c, P_{st}	motor coulomb and stiction pressures
Q_1, Q_2	servovalve flows
Q_a, Q_b	motor flows
Q_{ref}	simulation reference flow
R	line resistance per unit length
R_i, R_m	motor cross port and external leakage flow constants
R_v	servovalve linearised resistance
R_{ref}	simulation reference resistance
$T_1(s), T_2(s)$	transmission line functions
x	distance down line
X, Y	line simulation parameters
Y_t	transmission line shunt admittance
Z_c, Z_{ca}	transmission line characteristic impedance and its lossless line value
Z_t	transmission line series impedance
ε	friction non-linearity exponent
ξ	servovalve damping ratio
Γ	transmission line propogation constant
ω	angular frequency
ω_m	motor speed
ω_n	servovalve undamped natural frequency

2. SYSTEM MODELLING

2.1. Steady State

Figure 1 shows the system which consists of a servovalve of the force-feedback type coupled via transmission lines to an axial piston ball motor.

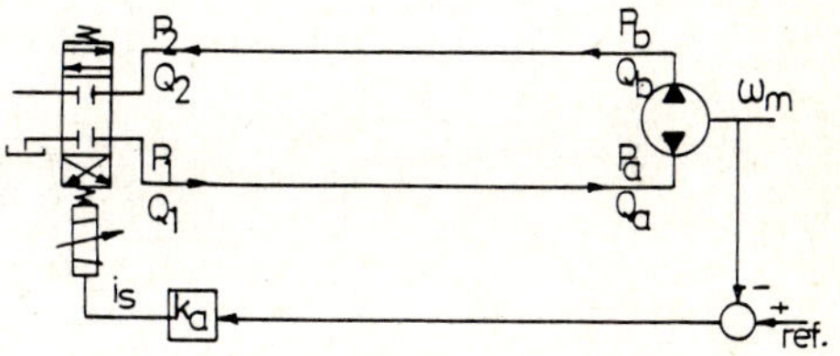

Figure 1 Speed control system

The dynamic simulation of such control systems has been extensively covered particularly for situations where line dynamics are negligible [1], and the mathematical models of electrohydraulic components tend to follow well-understood and predictable forms. Pressure/flow characteristics for the servovalve are shown in Figure 2:-

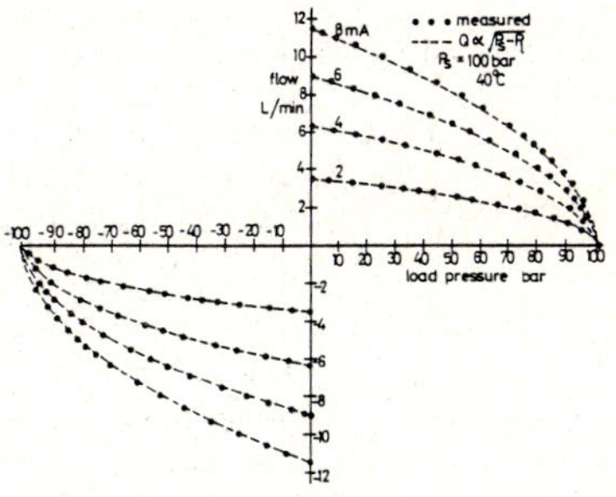

Figure 2 Servovalve flow characteristics

Spool valve underlap is present for current i_s < 0.335 mA. Consequently the flow equations for the servovalve may be written:-

$$Q_1 = (k_o + k_t i_s)\sqrt{P_s - P_1} \quad i_s > 0.335 \text{ mA} \tag{1}$$

$$Q_2 = (k_o + k_t i_s)\sqrt{P_2} \tag{2}$$

The motor no-load pressure/flow characteristics are shown in Figure 3:-

These characteristics are particularly important since it will be shown later that the stiction level is the major cause of open-loop instability. It will be deduced from the flow characteristics shown in Figure 3 that motor leakage is of minor importance although it has been included for completeness. Consequently

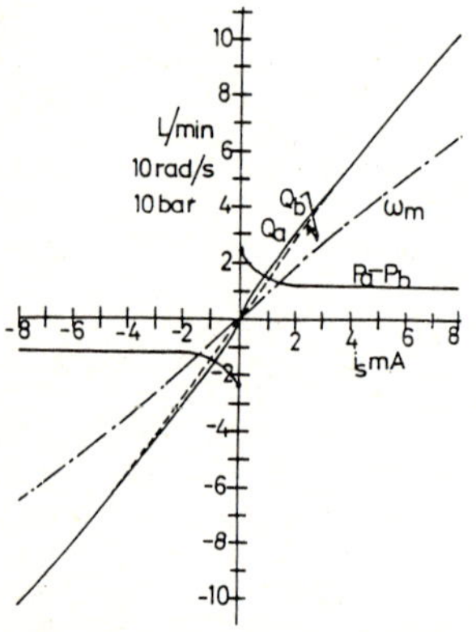

Figure 3 Motor pressure/flow characteristics

the motorflow equations are written in the linear form:-

$$Q_a = D_m \omega_m + \frac{P_a}{R_m} + \frac{(P_a - P_b)}{R_i} \tag{3}$$

$$Q_b = D_m \omega_m - \frac{P_b}{R_m} + \frac{(P_a - P_b)}{R_i} \tag{4}$$

The motor pressure drop is non-linear as shown in Figure 3 and changes from the stiction level P_{st} at zero speed to the coulomb friction level P_c for larger speeds. This characteristic is therefore modelled by the equation:-

$$P_f = |P_c + (P_{st} - P_c)e^{-\varepsilon \omega_m}|\,\text{sign}(\omega_m) \tag{5}$$

The flow in the transmission lines is laminar and the experimentally determined pressure drop equations are simply written:-

$$P_1 - P_a = R\ell Q_1 \quad (Q_1 = Q_a) \tag{6}$$

$$P_b - P_2 = R\ell Q_2 \quad (Q_2 = Q_b) \tag{7}$$

where the resistance R is defined per unit length of line.

2.2 Dynamic Equations

It is necessary to include servovalve dynamics since measured system resonant frequencies up to 25 Hz exist, this frequency being significant when compared with a first order definition of servovalve bandwidth (about 53 Hz in this case). The dynamic equation needed is that which links spool movement to applied current. This is taken from Manufacturer's data and previous studies and may be written:-

$$\frac{1}{\omega_n^2}\frac{d^2 i_s}{dt^2} + \frac{2\xi}{\omega_n}\frac{d i_s}{dt} + i_s = I_s \tag{8}$$

Equation (8) should be understood to be symbolic only.

The transmission line equations are written in the generalised form [2,3]:-

$$\frac{\partial P}{\partial x} = -Z_t Q \tag{9}$$

$$\frac{\partial Q}{\partial x} = -Y_t P \tag{10}$$

and the inclusion of motor and load inertia gives the final dynamic equation:-

$$P_a - P_b = P_f + \frac{J}{D_m}\frac{d\omega_m}{dt} \tag{11}$$

where the friction term P_f is as defined in equation (5).

3. SYSTEM STABILITY

3.1 Linearised Solution in the Frequency Domain

If the appropriate dynamic equations are linearised about the operating point, and the frequency domain solution to the transmission line equations is adopted, then the system block diagram is as shown in Figure 4:-
where the transmission line functions are:-

$$T_1(s) = \cosh\Gamma\ell + \frac{Z_c}{R_v}\sinh\Gamma\ell \tag{12}$$

$$T_2(s) = \cosh\Gamma\ell + \frac{R_v}{Z_c}\sinh\Gamma\ell \tag{13}$$

and the servovalve transfer function is:-

$$G_v(s) = \frac{1}{\left(1 + \frac{2\xi}{\omega_n} + \frac{s^2}{\omega_n^2}\right)} \tag{14}$$

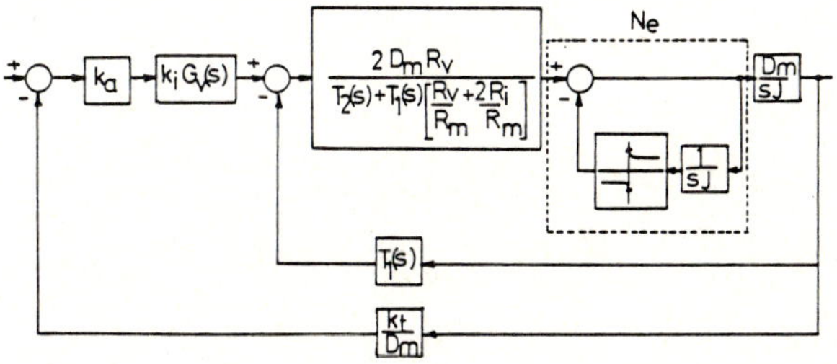

Figure 4 Linearised system block diagram

It will be noted from Figure 4 that the friction non-linearity has been arranged into a form which may be compared to a previously derived Describing Function Analysis [4]. Although these analyses assumed pure stiction and coulomb friction, the Describing Functions developed are useful if the extreme boundary values are considered. The appropriate Describing Function is shown in Figure 5.

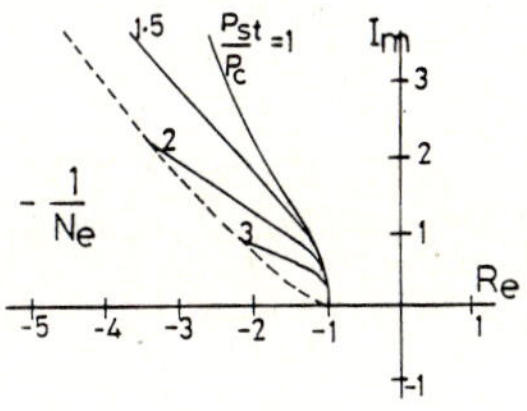

Figure 5 Describing Function for motor friction [4]

By re-arranging the block diagram of Figure 4 it is then possible to derive "open-loop" transfer functions for the minor loop and the major loop as follows:-

$$G_i(s) = \frac{T_1(s)}{\frac{sL_m}{2R_v}\left|T_2(s)+T_1(s)\left(\frac{R_v}{R_m}+\frac{2R_v}{R_i}\right)\right|} \tag{15}$$

(Minor loop)

$$G_o(s) = \frac{k_v G_v(s) + T_1(s)}{\frac{sL_m}{2R_v}\left|T_2(s)+T_1(s)\left(\frac{R_v}{R_m}+\frac{2R_v}{R_i}\right)\right|} \tag{16}$$

(Major loop)

3.2 Stability of the minor loop

A Nyquist plot of $G_i(s)$ and the Describing Function $-1/N_e$ is shown in Figure 6. Also shown in Figure 6 is a plot of the major loop transfer function $G_o(s)$ to be discussed later.

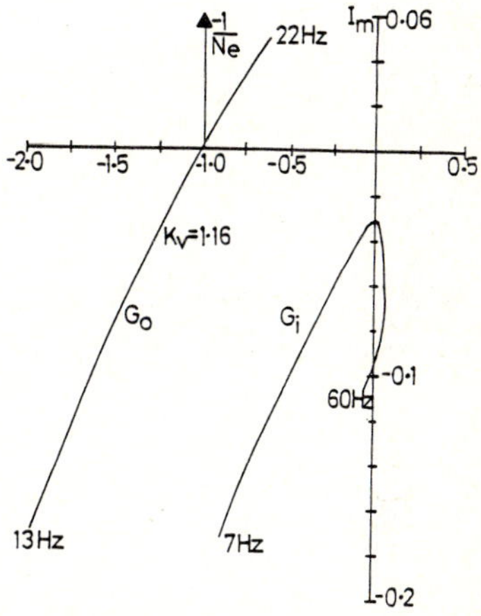

Figure 6 Nyquist plots for minor loop and major loop

These calculations apply to servovalve current of 1.5 mA and a 13 mm diameter line 10.73 m long. It is clear from Figure 6 that stability is predicted although this is not the practical case as the motor speed is suddenly lowered from a previously higher value. Severe stick-slip oscillations occur at the motor and are not prediction by a frequency response analysis. Consequently it is appropriate to pursue a time domain analysis.

3.3 Method of characteristics in the time domain

A variety of analytical techniques are available for time domain analysis of non-linear systems with distributed parameter effects [5,6]. In this study the method of characteristics is used to represent transmission line dynamics. Using this technique the partial differential equations (9) and (10) are transformed to a total differential equation which applies along either a positive or negative characteristic as follows:-

$$Z_{ca}\frac{dQ}{dt} \pm \frac{dP}{dt} + C_o RQ = \frac{C_o R}{2}\Phi \tag{17}$$

$$C_o = \pm\frac{dx}{dt}$$

where Φ represents the effect of the so called time dependent friction [2,7]. This effect has been pursued in some detail by the writer, but it will not be included here since it does not significantly modify the dynamic performance. Consider then an average friction model and a notation as shown in Figure 7 which represents one line lump:-

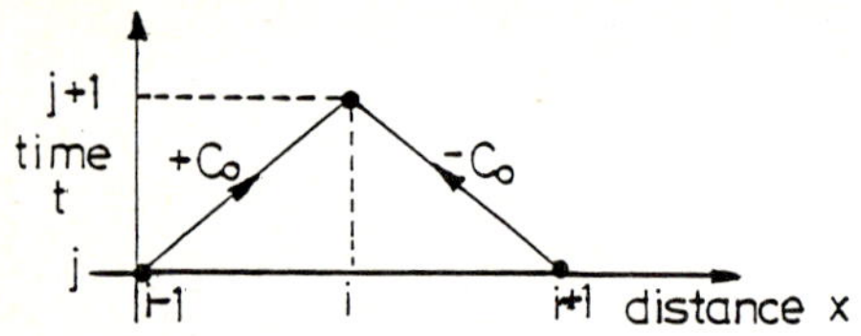

Figure 7 Characteristic grid for a single line element

The appropriate finite difference approximations for equation (17) are then:-

$$\bar{P}_i^{j+1} = \bar{P}_{i-1}^{j} - X\bar{Q}_i^{j+1} + Y\bar{Q}_{i-1}^{j} \qquad (18)$$

$$\bar{P}_i^{j+1} = \bar{P}_{i+1}^{j} + X\bar{Q}_i^{j+1} - Y\bar{Q}_{i+1}^{j} \qquad (19)$$

$$X = \frac{Z_{ca}}{R_{ref}} + \frac{R\ell}{4n\,R_{ref}} \qquad (20)$$

$$Y = \frac{Z_{ca}}{R_{ref}} - \frac{R\ell}{4n\,R_{ref}} \qquad (21)$$

$$R_{ref} = P_s/Q_{ref} \qquad (22)$$

$$Q_{ref} = (k_o + k_f i_{ss})\sqrt{\frac{P_s}{2}} \qquad (23)$$

Boundary conditions are solved directly by incorporating the servovalve equations (1) and (2) which produce a pair of quadratic equations. If time dependent friction is introduced as briefly mentioned earlier, then the solutions become more time and computer memory consuming and also requires an iterative solution for the boundary conditions.

3.4 Open-loop stability in the time domain

The solution to the problem using the method of characteristics is discussed in [8], and considering the case when servovalve current is changed from 6 mA to 1.5 mA gives a series of phase plane solutions as shown in Figure 8.

Figure 8 shows that as the line length is increased beyond a value of about 4 m then stick-slip oscillations are predicted. Hence the phenomenon measured for the system with 10.73 m lines is indeed predictable and some further comparisons for this configuration are shown in Figures 9 and 10.

The simulation predictions then indicate that these oscillations may also be removed if the line diameter is reduced, in this case to a value of less than 8 mm. Therefore 7 mm diameter lines were selected and indeed a stable transient solution was achieved as shown by the results presented in Figure 11.

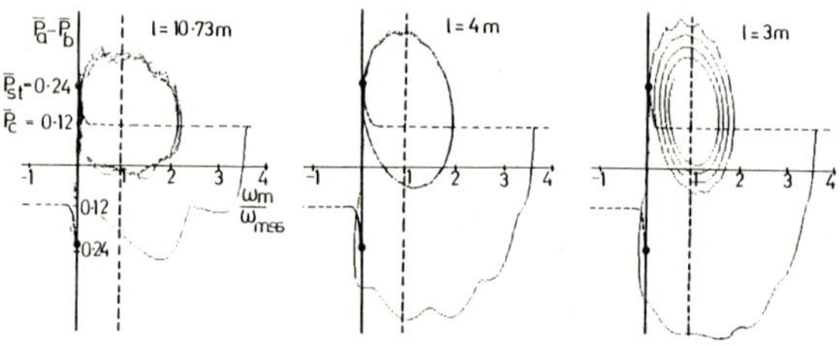

Figure 8 Phase plane solutions for the open-loop system

3.5 Closed-loop stability

Having now established a stable open-loop system by reducing line diameter, it is a relatively easy matter to predict the conditions necessary for closed-loop stability. It has been found that the small signal frequency response approach is applicable. Recalling the Nyquist plot shown as Figure 6 establishes a simple criterion for closed-loop which may be stated:-

$$G_o(s) = -\frac{1}{N_e} \rightarrow -1 \qquad s = j\omega \qquad (24)$$

Consequently the solution is given by:-

$$\tan\frac{\omega\ell}{C_o} = -\frac{\omega L_m}{2Z_{ca}}\frac{\left[\dfrac{4\xi R_v}{\omega_n L_m} + \left(1 + \dfrac{R_v}{R_m}\right)\left(1 - \dfrac{\omega^2}{\omega_n^2}\right)\right]}{\left[1 - \dfrac{\omega^2}{\omega_n^2} - \dfrac{\omega^2 L_m \xi}{\omega_n Z_{ca}}\left(\dfrac{R_v}{Z_{ca}} + \dfrac{Z_{ca}}{R_m}\right)\right]} \qquad (25)$$

$$k_v = \frac{2\xi\omega}{\omega_n}\left|\frac{Z_{ca}}{R_v}\sin\frac{\omega\ell}{C_o} + \frac{\omega L_m}{2R_v}\left(1 + \frac{R_v}{R_m}\right)\cos\frac{\omega\ell}{C_o}\right| - \left(1 - \frac{\omega^2}{\omega_n}\right)\left|\cos\frac{\omega\ell}{C_o} - \frac{\omega L_m}{2R_v}\left(\frac{R_v}{Z_{ca}} + \frac{Z_{ca}}{R_m}\right)\sin\frac{\omega\ell}{C_o}\right| \qquad (26)$$

where equation (24) has been evaluated using lossless line theory.

Results are finally presented in Figure 12 for the closed-loop speed control system operating over a wide range of steady-state conditions.

Error between measurement and theory become more evident at lower value currents due to the lossless line theory. Since servovalve resistance is drastically increased as the current is lowered, the major system damping will then be dominated by the small line resistance.

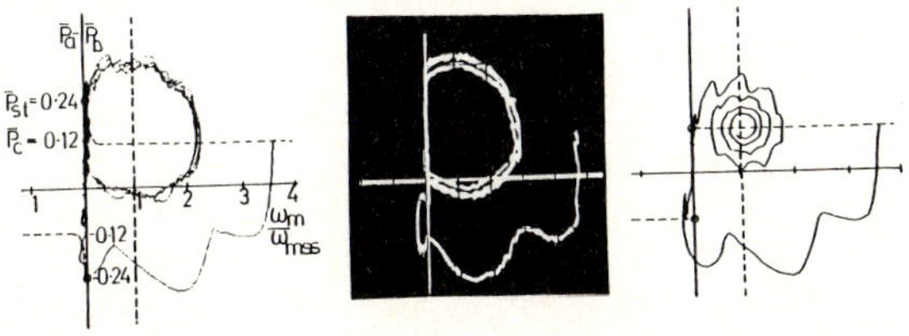

a) computed with stiction b) measured c) computed without stiction

Figure 9 Phase plane solutions for the experimental system

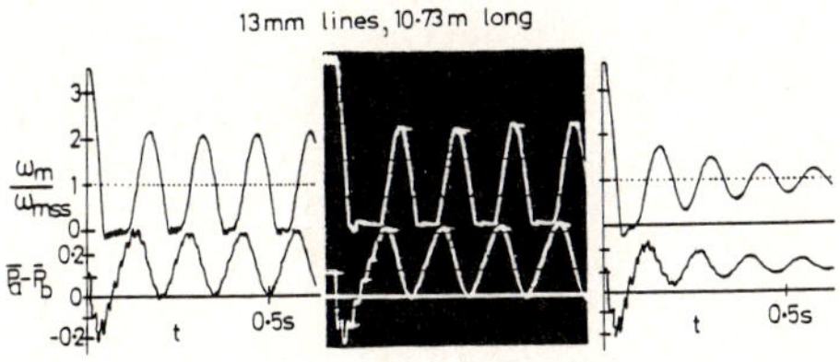

a) computed with stiction b) measured c) computed without stiction

Figure 10 Transient responses for the experimental system

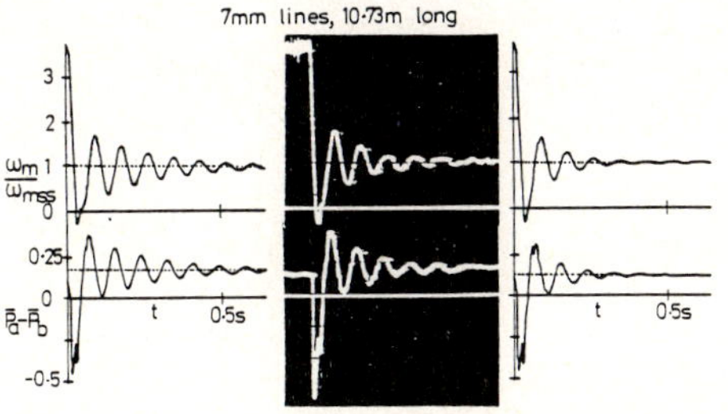

a) computed with stiction b) measured c) computed without stiction

Figure 11 Transient responses for 7 mm diameter line system

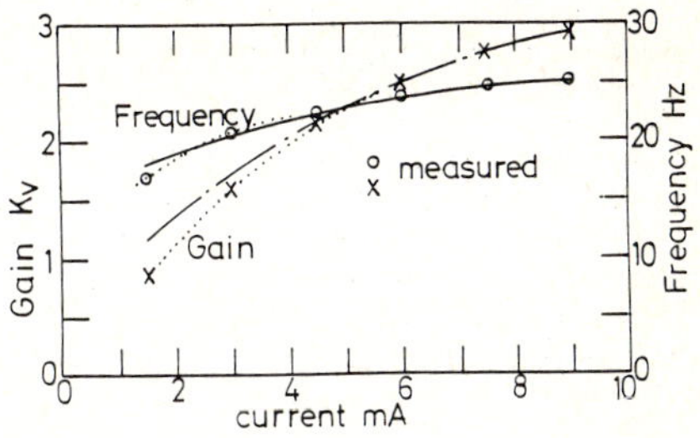

Figure 12 Stability limits for closed-loop system

CONCLUSIONS

The method of characteristics has been found to be a useful tool in predicting dynamic performance of electrohydraulic control systems. The boundary conditions are more easily handled when compared to other simulation techniques that require the derivatives of the system equations.

Motor stiction was found to have a very significant affect on the stability of the open loop system, although the frequency dependent portion of the characteristic was found to be unimportant. Simulation predictions showed that the removal of this stiction would remove the possibility of stick-slip oscillations in the open loop. However it is also possible to remove these oscillations by careful choice of either line length or diameter. A stable closed-loop system can then be achieved as verified both experimentally and theoretically.

REFERENCES

1. Watton, J., "The Generalised Response of a Servovalve-controlled, Single-rod, Linear Actuators and the Influence of Transmission Line Dynamics", Journal of Dynamic Systems, Measurement and Control, June 1984, pp 157-162.
2. Brown, F.T., "A Quasi Method of Characteristics with Application to Fluid Lines with Frequency Dependent Wall Shear and Heat Transfer", Trans. of ASME Journal of Basic Engineering, June 1969, pp 217-227.
3. Kirshner, J.M., and Katz, S., Design Theory of Fluid Components, published by Academic Press.
4. Tou, J. and Schulthesis, P.M., "Static and Sliding Friction in Feedback Systems", Journal of Applied Physics, Vol. 21, No. 9, Sept. 1953, pp 1210-1217.
5. Shearer, J.L., "Digital Simulation of a Coulomb-damped Hydraulic Servosystem", Journal of Dynamic System, Measurement and Control, Vol. 105, 1983, pp 215-221.

6. Watton, J., "The Simulation of Transmission Line Dynamics in Non-linear Electrohydraulic Systems", 1983 Winter Annual Meeting, ASME, and published in ASME Book H00278, pp 98-112.
7. Zielke, W., "Frequency-dependent Friction in Transient Pipe Flow", ASME Journal of Basic Engineering, March, 1968, pp 109115.
8. Watton, J., "An Efficient Simulation Technique for the Transient Analysis of Servovalve-controlled Motors with Transmission Line-effects", ASME 1984 WAM.

Section V

CONTROL OF COMPLEX SYSTEMS

Complex and Distributed Systems: Analysis, Simulation and Control
S.G. Tzafestas and P. Borne (editors)
Elsevier Science Publishers B.V. (North-Holland)

USE OF A HOMOGRAPHIC TRANSFORMATION FOR THE DECOUPLING OF DISCRETE MULTI-TIME SCALE SYSTEMS

A. EL MOUDNI, G. DAUPHIN-TANGUY, P. BORNE

Laboratoire d'Automatique et d'Informatique Industrielle (L. A. I. I.)
Institut Industriel du Nord (I. D. N.) - B.P. 48
59651 Villeneuve d'Ascq cédex - France

ABSTRACT : The study deals with singularly perturbed discrete systems. The proposed modelling of multi-time scale discrete systems is quite the dual of the continuous case. This duality appears on the one hand in the retained initial model, and on the other hand with the formal representation of the calculated disconnected subsystems.
When studying the slow dynamics of a discrete system, the singular perturbations method does not give enough accuracy. In order to avoid this, we propose a new approach of such systems, using a matricial "Homographic transformation" noted H. The proposed method is composed of three stages :
- starting from an initial discrete system, the transformation H leads to an equivalent continuous model, without modification of the dynamical behaviour,
- the singular perturbations method is applied to this continuous model, and leads to the slow and fast continuous disconnected parts,
- using the inverse transformation H^{-1}, slow and fast discrete decoupled parts are then derivated.
The slow part obtained by this way gives a very good modelling of the initial slow variables of the discrete system. The fast dynamics are directly approximated by usual singular perturbations method.
Theses results can be applied to both transfert functions and matrical models. It is available for linear systems, but can be extended to nonlinear ones.
The method is implemented on the decoupling of a discrete model of a steam power system and the results are compared by simulation with the initial system and with results obtained by usual singular perturbations technique.

KEYWORDS : Discrete multi-time scale systems, Singular perturbations method, Homographic transformation, Order reduction.

1. INTRODUCTION

Many physical systems are composed of parts with different dynamical behaviours. It is often useful to disconnect dynamics in order to reduce dimensionnality, to simplify calculation of optimal control, to suppress imprecisions due to very different magnitude order terms.

The singular perturbations (SP) method applied to discrete two-time-scale systems allows to obtain a reduced fast subsystem which takes care of the influence of the slow neglected part ; it gives a good accuracy in opposition with the results concerning the slow reduced part. It is important to notice that a duality appears with the well-known results on continuous two-time-scale systems decoupled by SP-method, for which the precision concerns the slow reduced part. So we proposed to use a homographic transformation H which leads from the discrete system to an equivalent continuous one. A new decoupling approach, called (SP + H) method, gives then the same accuracy on the slow and the fast disconnected part.

2. SOME WELL-KNOWN RESULTS ON THE STUDY OF TWO-TIME-SCALE SYSTEMS

2.1. Continuous case |1| |2|

Let us consider the two-time-scale continuous system modelled through a singularly perturbed form as :

$$(1)\quad \left|\begin{array}{l} \begin{bmatrix} \overset{\circ}{x} \\ \mu\, \overset{\circ}{z} \end{bmatrix} = \begin{bmatrix} A_{11} & A_{12} \\ A_{21} & A_{22} \end{bmatrix} \begin{bmatrix} x \\ z \end{bmatrix} + \begin{bmatrix} B_1 \\ B_2 \end{bmatrix} u \\ y = (C_1 \quad C_2) \begin{bmatrix} x \\ z \end{bmatrix} \\ x(t_o),\ z(t_o) \text{ initial values} \\ t_o \text{ initial moment} \end{array}\right.$$

where $x \in \mathbb{R}^{q_1}$ and $z \in \mathbb{R}^{q_2}$ denote respectively the slow and the fast parts of the state vector ; $u \in \mathbb{R}^m$ and $y \in \mathbb{R}^p$ are the input and output vectors.

The small parameter $\mu \in]0, 1]$ allows to compare the different time scales.

By setting μ equal to zero (iff A_{22} non singular), we obtain the slow reduced system as :

$$(2)\quad \begin{cases} \mathring{x}_s = A_s x_s + B_s u_s \\ y_s = C_s x_s + D_s u_s \\ z_s = - A_{22}^{-1} (A_{21} x_s + B_2 u_s) \\ \text{with : } A_s = A_{11} - A_{12} A_{22}^{-1} A_{21} \\ \qquad B_s = B_1 - A_{12} A_{22}^{-1} B_2 \\ \qquad C_s = C_1 - C_2 A_{22}^{-1} A_{21} \\ \qquad D_s = - C_2 A_{22}^{-1} B_2 \end{cases}$$

The initial values are conserved on $x_s(t_o)$ $(x_s(t_o) = x(t_o))$. The vector z_s is the "quasi-steady state" of z at $\mu = 0$, and in most cases $z_s(t_o)$ is different from $z(t_o)$.

By introducing the fast time scale :

$$\tau = \frac{t - t_o}{\mu} \qquad (\tau = 0 \text{ at } t = t_o)$$

and by defining a new variable

$$z_f = z - z_s$$

we obtain the boundary layer system as :

$$(3)\quad \begin{cases} \dfrac{dz_f}{d\tau} = A_{22} z_f + B_2 u_f \\ y_f = C_2 z_f \\ z_f(t_o) = z(t_o) + A_{22}^{-1} A_{21} x(t_o) \end{cases}$$

Then a two time-scale approximation of the state of (1) is :

$$(4)\quad \begin{cases} x(t) = x_s(t) + 0(\mu) \\ z(t) = z_s(t) + z_f(\tau) + 0(\mu) \end{cases}$$

The slow decoupled subsystem takes into account the fast neglected part through the state quadruplet (A_s, B_s, C_s, D_s). In opposition the fast variables z_f are not obtained with the same accuracy, the slow influence just appears as static in the initial values $z_f(t_o)$.

2.2. Discrete case

It does not exist an unic model to characterize a discrete two-time-scale system as in continuous case (1).

Let us consider the following model (5), where the small parameter μ appears in the right side of the state equation |3| as :

$$(5)\quad \begin{cases} \begin{bmatrix} x_{n+1} \\ z_{n+1} \end{bmatrix} = \begin{bmatrix} A_{11} & A_{12} \\ A_{21} & A_{22} \end{bmatrix} \begin{bmatrix} x_n \\ \mu z_n \end{bmatrix} + \begin{bmatrix} B_1 \\ B_2 \end{bmatrix} u_n \\ y_n = (C_1 \quad C_2) \begin{bmatrix} x_n \\ \mu z_n \end{bmatrix} \\ x_{n_o}, z_{n_o} \text{ initial values} \end{cases}$$

where $x_n \in \mathbb{R}^{q1}$ and $z_n \in \mathbb{R}^{q2}$ denote respectively the slow and the fast parts of the global state vector. The small parameter μ normalizes here the magnitude orders of the matrix terms.

The same scheme as in continuous case is then used to obtain the slow and the fast disconnected subsystems, by setting μ equal to zero.

The slow reduced part is defined by :

$$(6)\quad \begin{cases} x_{s_{n+1}} = A_{11} x_{s_n} + B_1 u_{s_n} \\ y_{s_n} = C_1 x_{s_n} \\ z_{s_n} = A_{21} A_{11}^{-1} x_{s_n} + (B_2 - A_{21} A_{11}^{-1} B_1) u_{s_n} \\ \text{with } x_{s_{n_o}} = x_{n_o} \end{cases}$$

The boundary-layer equation, obtained by defining the fast decoupled variables z_{f_n} as :

$$z_{f_n} = z_n - z_{s_n}$$

is :

$$(7)\quad \begin{cases} z_{f_{n+1}} = A_f z_{f_n} + B_f u_{f_n} \\ y_{f_n} = C_f z_{f_n} + D_f u_{f_n} \\ \text{with : } A_f = (A_{22} - A_{21} A_{11}^{-1} A_{12}) \mu \\ \qquad B_f = B_2 - A_{21} A_{11}^{-1} B_1 \\ \qquad C_f = (C_2 - C_1 A_{11}^{-1} A_{12}) \mu \\ \qquad D_f = - C_1 A_{11}^{-1} B_1 \\ z_{f_{n_o}} = z_{n_o} - A_{21} A_{11}^{-1} x_{n_o} \end{cases}$$

A two-time-scale approximation of the state of (5) is :

$$(8)\quad \begin{cases} x_n = x_{s_n} + O(\mu) \\ z_n = z_{s_n} + z_{f_n} + O(\mu) \end{cases}$$

REMARK

The duality continuous/discrete points out on the one hand in the introduction of the small parameter μ (in the state derivate vector/in the state vector), and on the other hand in the formal representation of the reduced decoupled subsystems. One of the dynamics is obtained with a better accuracy, the slow part in continuous case and the fast one in discrete case. The loose of accuracy on the disconnected continuous fast part has been compensated by introducing the reciprocal transformation concept |2|.

We present here a similar approach in discrete case, using simultaneously the notion of homographic transformation and the singular perturbations technique, in order to increase the precision on the slow reduced discrete part. The homographic transformation H, which will be extensively developped in the following section, gives a continuous model equivalent to the initial discrete system.

2.3. Definition of the class of studied systems

We study here discrete systems supposed to be obtained by sampling of a two-time-scale continuous process. The real part of the eigenvalues of the state matrix will be always positive.

The Gershgorin theorem |6| defines the location domain of eigenvalues of matrix A as a set of circles (C_i) centered in diagonal terms a_{ii} and with radius R_i equal to :

$$R_i = \sum_{i=1}^{q} |a_{ij}| \qquad \text{(q dimension of A)}$$

When applied to a two-dynamic-system, it gives two disconnected sets of circles, as in the following Figure 1.

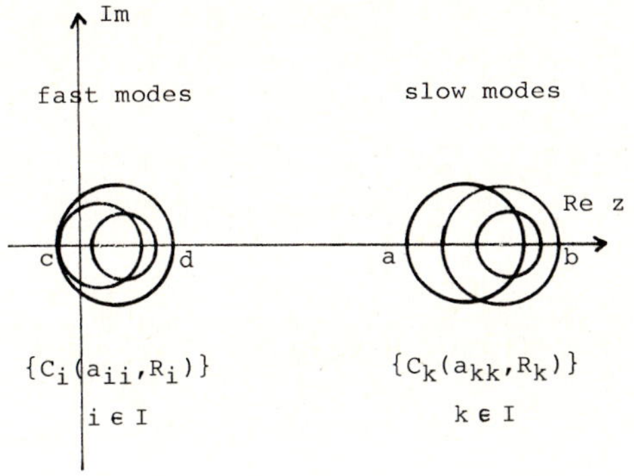

FIGURE 1

This approach allows to identify the slow and the fast state variables, which leads to model (5). The parameter μ can then be chosen as |2|.

$$\mu = \max_{\substack{i \in I \\ k \in K}} \frac{R_i + R_k}{|a_{ii} - a_{kk}|}$$

3. HOMOGRAPHIC TRANSFORMATION

3.1. Definitions

3.1.1. In state space :

This notion was introduced by |5| in order to solve optimal control problems.

1° - If a discrete system is modelled by the equation :

$$(9)\quad \begin{cases} x_{n+1} = A\, x_n + B\, u_n \\ y_n = C\, x_n + D\, u_n \end{cases} \qquad x_n \in \mathbb{R}^q$$

then the equivalent continuous system is modelled by the relations :

$$(10)\quad \begin{cases} \overset{o}{x} = M\, x + F\, u \\ y = C\, x + D\, u \end{cases} \qquad x \in \mathbb{R}^q$$

with

$$(11)\quad \begin{cases} M = (A - I) \left[\frac{A + I}{2}\right]^{-1} \\ F = \left[\frac{A + I}{2}\right]^{-1} B \end{cases}$$

(I denote the identity matrix of right dimension)

The matrices M and F are obtained by expressing the term $(x_{n+1} - x_n)$ as a function of :

$$\frac{x_{n+1} + x_n}{2}.$$

2° - If a continuous system is modelled by equations (10), then the equivalent discrete system is modelled by (9) with :

$$(12)\quad \begin{cases} A = (I + \frac{M}{2})\,(I - \frac{M}{2})^{-1} \\ B = (I - \frac{M}{2})^{-1}\, F \end{cases}$$

The matrices A and B are obtained by expressing the term $(x + \overset{o}{x}/2)$ as a function of $(x - \overset{o}{x}/2)$.

3° - The homographic transformation H is defined as shown on the following scheme :

discrete system		continuous system
A	$\xrightarrow{H}$	$M = H(A)$
$A = H^{-1}(M)$	$\xleftarrow{H^{-1}}$	M

with : $H(A) = (A - I)\left[\frac{A+I}{2}\right]^{-1}$

$$H^{-1}(M) = (I + \frac{M}{2})\left[I - \frac{M}{2}\right]^{-1}$$

3.1.2. By use of transfert function :

If z and s denote symbolic operators associated respectively to z-transformation and Laplace transformation, then symbolic transfert functions can be deduced from (9) and (10) as :

(13-a) $W_D(z) = C\,(zI - A)^{-1}\,B$

and

(13-b) $W_C(s) = C\,(sI - M)^{-1}\,F$

(D supposed equal to zero)

1° - The continuous transfert function $\tilde{W}_C(s)$ is deduced from the discrete one $W_D(z)$ by the relation :

$$(14)\quad \tilde{W}_C(s) = \frac{1}{1 - s/2}\,W_D\left(\frac{1 + s/2}{1 - s/2}\right)$$

this equation is easily obtained by changing in the initial discrete function $W_D(z)$ the triplet (A,B,C) by its equivalent continuous :

$$(H(A)\,,\,(\frac{A+I}{2})^{-1}\,B\,,\,C).$$

2° - The discrete transfert function $\tilde{W}_D(z)$ is deduced from the continuous one $W_C(s)$ by the relation :

$$(15)\quad \tilde{W}_D(z) = \frac{2}{z+1}\,W_C\,(2\,\frac{z-1}{z+1})$$

It is obtained by changing in the continuous function W(s) the triplet (M,F,C) by its equivalent discrete :

$$(H^{-1}(M)\,,\,(I - \frac{M}{2})^{-1}\,F,\,C)$$

3.2. Properties

The continuous/discrete equivalent system obtained by H/H^{-1}-transformation reflects the properties of controllability, observability and stability of the initial discrete/continuous process |5| |4|. The aim of the study is here the separation of dynamics ; so we consider only the problem of the modification of initial values and dynamical behaviour due to homographic transformation.

3.2.1. Initial values :

The equivalent continuous state vector x obtained by H-transformation has been defined in section 3.1.1 as an average between two successive values of the discrete state vector as :

$$x(t) = \frac{x_{n+1} + x_n}{2}$$

where t corresponds to the middle point of time interval [n , n+1].

In order to have an equivalent continuous system which has the same static properties than the discrete initial one, we extend by continuity to origin the value $x(t = n_0 + \frac{1}{2})$ to $x(t = n_0)$.

So in the following study, we shall set as an hypothesis that :

$$x(t_0) = x_{n_0}.$$

The same approach can be made in the transformation H^{-1} (continuous → discrete), which can be written as :

$$x_{n_0} = x(t_0)$$

3.2.2. Dynamical behaviour :

In section 2.3, we shown that the location domain of eigenvalues of a matrix A is a set of circles.

ASSUMPTION

The homographic transformation H keeps circles. So the eigenvalues of the equivalent continuous system belong to circles which are the transformed by H of the initial circles.

This property is presented geometrically on Figure 2, for particular extremal real points of Gershgorin circles.

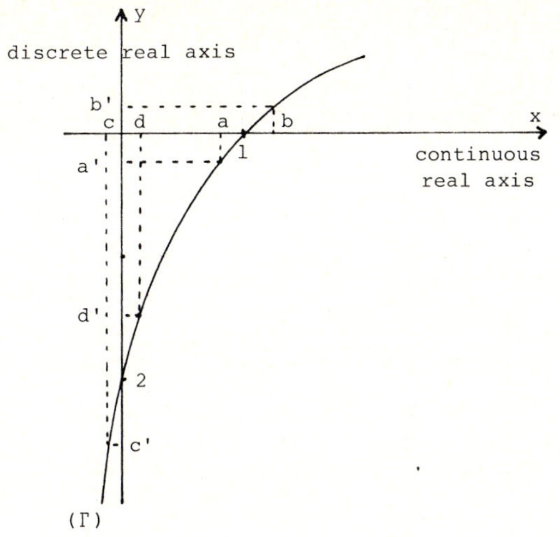

FIGURE 2

$$(\Gamma) : \{x/y = \frac{2(x-1)}{x+1}\}$$

represents the bond between points localized on discrete and equivalent continuous real axis. The points (a,b)/(c,d) are the extremal real values of discrete location domain of slow/fast eigenvalues. (a',b')/(c',d') obtained by projection on (Γ) are then the extremal real values of equivalent continuous location domain of slow/fast eigenvalues.

If a and d are well separated, it will be the same for a' and d' as shown Figure 3.

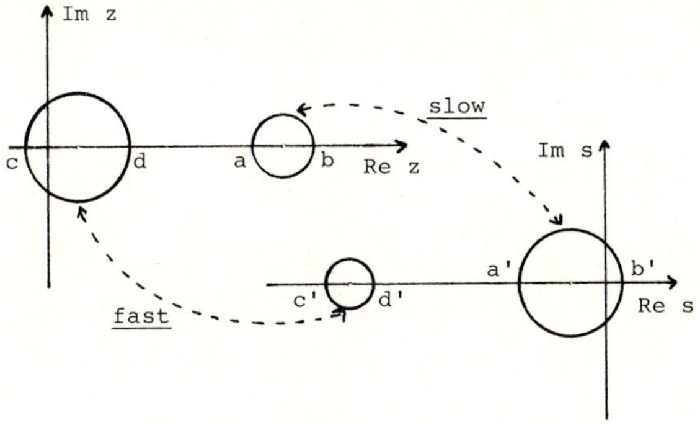

FIGURE 3

The homographic transformation H does not modify the dynamical characteristics of state variables.

4. APPLICATION OF H-TRANSFORMATION TO STUDY SLOW REDUCED DISCRETE PART

The proposed approach is the following :

$$\begin{array}{ccc} (\Sigma_D) & \xrightarrow{H} & (\tilde{\Sigma}_C) \\ \downarrow & & \downarrow \text{SP} \\ (\Sigma_D)_{\tilde s} + (\Sigma_D)_{\tilde f} & \xleftarrow{H^{-1}} & (\tilde{\Sigma}_C)_s + (\tilde{\Sigma}_C)_f \end{array}$$

FIGURE 4

which can also be expressed in terms of matricial quadruplets :

$$\begin{array}{ccc} (A,B,C)_D & \xrightarrow{H} & (\tilde{A}_C,\tilde{B}_C,C) \\ \downarrow & & \downarrow \text{SP} \\ (A_{\tilde s},B_{\tilde s},C_{\tilde s},D_{\tilde s}) + & \xleftarrow{H^{-1}} & (\tilde{A}_C,\tilde{B}_C,C,D)_s + \\ +(A_{\tilde f},B_{\tilde f},C_{\tilde f}) & & +(\tilde{A}_C,\tilde{B}_C,C)_f \end{array}$$

From the initial model (5), we obtain then, by (SP+H) method for the boundary layer domain :

$$(16) \quad \left| \begin{array}{l} z_{\tilde f_{n+1}} = A_{\tilde f}\, z_{\tilde f_n} + B_{\tilde f}\, u_{f_n} \\ y_{\tilde f_n} = C_{\tilde f}\, z_{\tilde f_n} \\ \text{with :} \\ A_{\tilde f} = (A_{22} - A_{21}(A_{11}+I)^{-1}A_{12})\,\mu \\ B_{\tilde f} = B_2 - A_{21}\,(A_{11}+I)^{-1}\,B_1 \\ C_{\tilde f} = C_2\,\mu \end{array} \right.$$

and for the slow reduced part :

$$\left| \begin{array}{l} x_{\tilde s_{n+1}} = A_{\tilde s}\, x_{\tilde s_n} + B_{\tilde s}\, u_{s_n} \\ y_{\tilde s_n} = C_{\tilde s}\, x_{\tilde s_n} + D_{\tilde s}\, u_{s_n} \end{array} \right.$$

$$(17) \quad \left| \begin{array}{l} A_{\tilde s} = \Big[\big[A_{11} - (A_{11}+I)^{-1}A_{12}(A_{\tilde f}-I)^{-1}A_{21}\big] \\ \qquad \big[I + (A_{11}+I)^{-1}A_{12}(A_{\tilde f}-I)^{-1}A_{21}\big]^{-1}\Big] \\ B_{\tilde s} = \Big[\big[I + A_{12}(A_{\tilde f}-I)^{-1}A_{21}(A_{11}+I)^{-1}\big]^{1} \\ \qquad \big[B_1 - A_{12}(A_{\tilde f}-I)^{-1}B_{\tilde f}\big]\Big] \\ C_{\tilde s} = \big[C_1 - 2\,C_2\,(A_{\tilde f}-I)^{-1}A_{21}(A_{11}+I)^{-1}\big] \\ D_{\tilde s} = -\,C_2\,(A_{\tilde f}-I)^{-1}B_{\tilde f} \end{array} \right.$$

with the quasi-steady state $z_{\tilde{s}_n}$ given by :

(18) $z_{\tilde{s}_n} = 2\,(A_{\tilde{f}} - I)^{-1} A_{21} (A_{11} + I)^{-1} x_{\tilde{s}_n}$

INTERPRETATION

The approximation of the slow reduced part in (17) takes into account the fast influence, with in addition the introduction of a direct transmission of input by $D_{\tilde{s}}$. These results are better than in (6) and present a form similar to continuous case in (2).

So we propose, in the following scheme, different approachs for decoupling according to dynamical or frequential domain of the study.

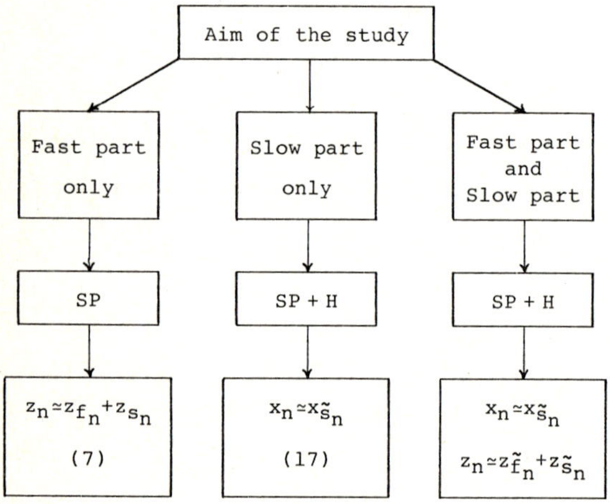

FIGURE 5

5. IMPLEMENTATION ON AN EXAMPLE

The proposed example concerns a steam generator |7| which is modelled in the form (5) as :

$$\begin{bmatrix} x^1_{n+1} \\ x^2_{n+1} \\ \hline z^1_{n+1} \\ z^2_{n+1} \\ z^3_{n+1} \end{bmatrix} = \begin{bmatrix} & & \\ & A & \\ & & \end{bmatrix} \begin{bmatrix} x^1_n \\ x^2_n \\ \hline z^1_n \\ z^2_n \\ z^3_n \end{bmatrix}$$

$$A = \left[\begin{array}{cc:ccc} 0.9014 & 0.1179 & 0.2625 & 0.0835 & 0.1052 \\ -0.0196 & 0.8743 & 0 & 0.125 & 0.1467 \\ \hdashline -0.0071 & 0.7342 & 1.00875 & 0.065 & 1.05335 \\ -0.75 & -0.0557 & -0.16 & 0.96587 & -0.0703 \\ -0.306 & -0.01614 & -0.0055 & 0.7139 & 0.066085 \end{array}\right]$$

with : $x^1_{n_o} = 1$; $x^2_{n_o} = -0.8$; $\mu = 0.2$

We want just to compare results obtained on the slow decoupled part by SP and (SP+H) methods.

By SP method, the reduced equations deduced from (6) are :

$$\begin{bmatrix} x^1_{n+1} \\ x^2_{n+1} \end{bmatrix}_s = \begin{bmatrix} 0.9014 & 0.1179 \\ -0.0196 & 0.8743 \end{bmatrix} \begin{bmatrix} x^1_n \\ x^2_n \end{bmatrix}_s$$

with :

$$\begin{bmatrix} x^1_{n_o} \\ x^2_{n_o} \end{bmatrix}_s = \begin{bmatrix} 1 \\ -0.8 \end{bmatrix}$$

and by (SP+H) approach, from (17) :

$$\begin{bmatrix} x^1_{n+1} \\ x^2_{n+1} \end{bmatrix}_{\tilde{s}} = \begin{bmatrix} 0.874834 & 0.163395 \\ -0.055618 & 0.870434 \end{bmatrix} \begin{bmatrix} x^1_n \\ x^2_n \end{bmatrix}_{\tilde{s}}$$

with :

$$\begin{bmatrix} x^1_{n_o} \\ x^2_{n_o} \end{bmatrix}_{\tilde{s}} = \begin{bmatrix} 1 \\ -0.8 \end{bmatrix}$$

The simulation results presented Figure 6 show the advantage of the new (SP+H) technique.

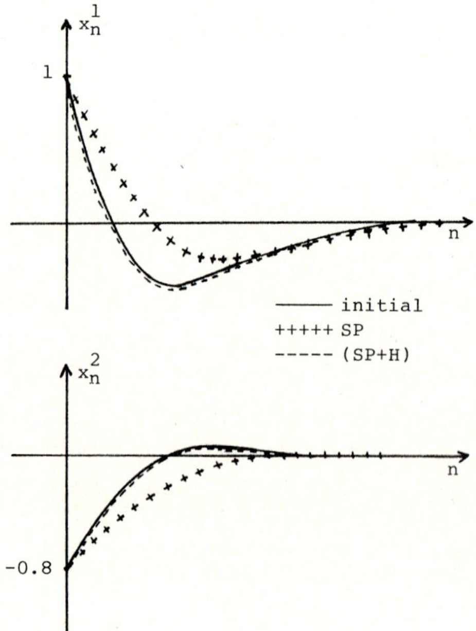

FIGURE 6

6. CONCLUSION

The (SP+H) method presented here brings more precision for study of the slow part of discrete two-time-scale systems. In order to have the best disconnected process, it is possible to keep for the fast part the SP reduced fast subsystem and for the slow part the (SP+H) reduced slow one. This method can be implemented very easily, and gives very interesting results for simulation and optimal control. It can be extended to non linear systems admitting a state representation.

REFERENCES

|1| Kokotovic, P.V., O'Malley, R.E. and Sannuti, P. "Singular perturbations and order reduction on control theory. An overview", Automatica (1976), Vol. 12, pp. 123-132.

|2| Dauphin-Tanguy, G. "Sur la représentation multi-modèle des systèmes singulièrement perturbés", Doctorat d'Etat (Lille, 1983).

|3| El Moudni, A., Dauphin-Tanguy, G. and Borne, P. "Sur une nouvelle méthode de séparation des dynamiques d'un système discret à double échelle de temps", Congrès A.M.S. (Nice, 1984), pp. 92-98.

|4| Gentina, J.C. "Sur la notion de modèle dans la description d'un processus", Acte du Colloque organisé par l'ECAM et l'ENSTA sous l'étude de l'ANRT, (Paris, 1975).

|5| Borne, P. and Gentina, J.C. "Definition of a matrical transformation permitting the discretization of a set of differential equations", IFIP Congress, (Stocholm, 1974), pp. 453-455.

|6| Bauer, F.L. "Fields of values and Gershgorin disk", Numerische Mathematik, n° 12, pp. 91-95.

|7| Phillips, R.G. "Reduced order modelling and control of two time-scale discrete systems", Int. J. Control (1980), Vol. 31, n° 4, pp. 765-780.

Complex and Distributed Systems: Analysis, Simulation and Control
S.G. Tzafestas and P. Borne (editors)
Elsevier Science Publishers B.V. (North-Holland)

CONTROL SYSTEMS IN PHASE SPACE

Zdzislaw JACYNO

Department of Physics
University of Quebec at Montreal
Canada

Designer of automatic control systems must conform to specifications imposed on steady-state and transient response to applied input control signals. This is often achieved through a trial and error approach. Design problems may be greatly simplified, providing existence of some guidelines on effects of each parameter on overall dynamic characteristics. Classification of parameters according to their contribution to total energy of the system looks promising in this respect. Identification of dissipative, restoring and inertial elements and their relative energetic importance, provide qualitative indications ways of changing dynamic response. Furthermore, this classification points to a quantitative effects as well and indicates control strategies of the feedback controller required. An approach, proposed in this paper, is based on generalized phase space and related energetic decomposition of dynamic systems, introduced earlier. Control systems, given in state space form, are first transferred into a phase space, through an appropriate linear transformation. Initial state space representation is replaced by a representation with respect to generalized position and velocity modes. A new coefficient-matrix of the system, broken into dissipative and storing elements, indicates structural parameters responsible for damping and oscillations.
A few examples are given. They show, that for even order systems, parameters from the state space on the main diagonal have dissipative characteristics; the same parameters for odd order systems tend to increase oscillations. It is also shown, that certain parameters, such as those on the secondary diagonal, affect directly control action.

1. INTRODUCTION

Consider an open-loop, strongly internally connected system in the state space, given by

$$\dot{X} = AX + DU, \ X \in R^n, U \in R^m, \ m \le n. \tag{1}$$

The matrix

$$A = [a_{ij}], \ i,j = 1,\ldots,n \tag{2}$$

has almost all elements non null. To evaluate the response of the system to a particular input signal U, it is necessary to get estimation of the eigenvalues of A, solve the equation (1)

$$X(t) = e^{At}X_0 + e^{At}\int e^{-At}DUdt, \tag{3}$$

or simulate the system. Either approach will provide a satisfactory response. However, results will not be easily forthcoming, when the design problem arises: modify the structure of the system and/or values of its parameters, in order to achieve required transient and static performances. It will be even more difficult to suggest a suitable strategy for closed-loop operation. Choice of a right controller becomes a matter of art rather then of an enlightened craft, especially in less conventional and more complicated situations. A rule: put the PID controller, may not suit a multivariable, hard to decouple system.

These problems are the result of a still unsatisfactory understanding of internal mechanisms, responsible for systems behaviour. In many instances we act as outside observers, taking notes about damped or oscillatory responses, or instability, and look for available remedies. Very rarely the question is asked, what is behind a particular behaviour. And it is evident, that those hidden mechanisms stem from fundamental laws of physics. There exist even a contradiction in our approach to automatic control problems. First, a mathematical model is developed, based on the knowledge of the system and applicable laws. Then, at the stage of analysis and design, those laws are entirely forgotten and very formal procedures applied. Physical reality of the system is lost out of sight. Even when talking about the state of the system, the reference is made to a particular mathematical formalism used.

It may be profitable to take an another look at the system dynamics and automatic control problems. The dynamic behaviour will be seen as an external manifestation of some internal mechanisms, defined by energetic relations of a system. The energy of the system may be exchanged with its environment or stored. The exchange proceeds either inward or outward, i.e. the energy is either absorbed or dissipated. For linear systems, the first case means the instability and the latter one - the asymptotic stability. The limit case, where there is no energy exchange, corresponds typically to sustained oscillations.

The energetic concepts have been introduced in our several previously published papers [1-12]. It is necessary, at this point, to take a complete and thorough review of a proposed approach and its possible applications to the design of automatic control systems. This is the main purpose of this paper.

2. PHASE SPACE

In order to apply general principles, discussed above, let transfer the system (1) from the natural state space X into an another state space Y. At first, the system (1) under zero-input conditions

$$U = 0 \tag{4}$$

is considered. A new state vector is defined as follows

$$Y = PX, \quad \det P \neq 0, \tag{5}$$

with

$$Y = [Y_1^T \; Y_2^T]^T \tag{6}$$

$$\dot{Y}_1 = CY_2, \tag{7}$$

where

C - intermodal coupling matrix.

In general

$$\dim Y_2 \neq \dim Y_1 \tag{8}$$

and the coupling matrix C assures a compatibility between the subvectors Y_1 and Y_2, particularly important for odd-order systems. The partitionning of Y will be assumed in the sequel to provide

$$\dim Y_2 = \dim Y_1 \tag{9}$$

and

$$C = I, \tag{10}$$

I - identity matrix,

for even-order systems, and

$$\dim Y_2 = 1 + \dim Y_1, \tag{11}$$

with C - an almost identity matrix, for odd-order systems. Under these conditions, equation (1) becomes

$$\dot{Y} = BY, \quad Y \in R^n \tag{12}$$

and a new coefficient - matrix B takes an imposed form

$$B = \begin{bmatrix} 0 & C \\ -G & -H \end{bmatrix}. \tag{13}$$

Taking into account, the nonsingularity of P, (5), the matrices A and B are both similar to the same diagonal (or, in general, Jordan bloc) matrix

$$\Lambda = [S(A)]^{-1} AS(A) = [S(B)]^{-1} B\, S(B), \tag{14}$$

S(A), S(B) - modal matrices of A and B, and obviously they are similar one to another through the transformation P

$$B = PAP^{-1} \tag{15}$$

From (14) and (15), the required transformation P can be found

$$P = S(B)\, [S(A)]^{-1} \tag{16}$$

The matrix S(A) is known from a given matrix A. The modal matrix for B seems, at first glance, to be unspecified. However, the form imposed on B, (13), implies a following structure of its eigenvectors

$$E_i = \begin{bmatrix} E_{i1} \\ E_{i2} \end{bmatrix} = \begin{bmatrix} \lambda_i^{-1} CE_{i2} \\ E_{i2} \end{bmatrix}, \quad i = 1,\ldots,n, \tag{17}$$

where the subvectors E_{i1}, E_{i2} are compatible with Y in (6-7). This, in turn, indicates the structure of S(B)

$$S(B) = \begin{bmatrix} \lambda_1^{-1} CE_{12} \ldots & \lambda_i^{-1} CE_{i2} & \ldots & \lambda_n^{-1} CE_{n2} \\ E_{12} \ldots & E_{i2} & \ldots & E_{n2} \end{bmatrix}, \tag{18}$$

i.e. that of the Vandermonde generalized matrix. The eigenvalues in (18) are those for the matrix A, due to the similarity between A and B. As a result, the transformation P in (16) is expressed in terms of known eigenvalues of A and its elements a_{ij}, (2). In practice, though, P is given by a_{ij} and algebraic combination of products and sums of the eigenvalues, so that the direct computation of the eigenvalues is not required. Instead, the combinations of the eigenvalues are substituted by appropriate subdeterminants of A. This gives the transformation P specified solely by the entries of A.

In the new state space, system (12) may be described equivalently be a set of two differential equations of the first-order

$$\left.\begin{aligned}\dot{Y}_1 &= CY_2 \\ \dot{Y}_2 &= GY_1 - HY_2\end{aligned}\right\} \qquad (19)$$

or by one equation of the second-order

$$\ddot{Y}_2 + H\dot{Y}_2 + GCY_2 = 0. \qquad (20)$$

An analogy between the space Y and the phase plane is apparent. In fact, the space Y represents an extension of the phase plane, applicably to second-order systems only, onto systems of an arbitrary order. It is called then phase space. The phase space is spanned by two hypercoordinates: Y_1 - generalized position and Y_2 - generalized velocity, the latter in accordance with (7).

3. ASSOCIATED SYSTEMS AND ENERGY

Set of two first-order differential equations (19) may be put into a parametric form. To do that, let premultiply the second equation by C, then transpose it and postmultiply by the first equation. As a result, the time, appearing explicitly in (19), is eliminated. The following bilinear differential form is obtained

$$(CY_2)^T d(CY_2) + Y_1^T(CG)^T dY_1 = -(CHY_2)^T dY_1. \qquad (21)$$

It is the differential of a trajectory of the system (12) in the phase space. A particular trajectory originates from a point specified by initial conditions and proceeds according to (21). It may be graphically drawn using a discretized form of (21).

The left-hand side of (21) represents a differential of some scalar Q(Y), i.e. it is a quadratic differential form of the phase space vector Y. In general

$$CG \neq (CG)^T, \qquad (22)$$

and, to find Q(Y), a direct integration is not possible. Let though consider a system with

$$H = 0, \qquad (23)$$

which has the following quadratic differential form

$$(CY_2)^T d(CY_2) + Y_1^T(CG)^T dY_1 = 0. \qquad (24)$$

This system will be called associated system with the system (12). It is also given by a set of differential equations

$$\left.\begin{aligned}\dot{Y}_1 &= CY_2 \\ \dot{Y}_2 &= GY_1\end{aligned}\right\} \qquad (25)$$

and by one second-order equation

$$\ddot{Y}_2 + GCY_2 = 0, \qquad (26)$$

obtained from (19) and (20) under the condition (23).

In order to obtain a decoupled, directly integrable form of (24), a new set of phase coordinates is introduced

$$Z_1 = [S(CG)]^{-1}Y_1. \qquad (27)$$

A transformation matrix S(CG) is the matrix diagonalizing CG

$$\Lambda(CG) = [S(CG)]^{-1}CGS(CG), \qquad (28)$$

$\Lambda(CG)$ - diagonal matrix, similar to CG.

As a result, system (25) takes a decoupled form

$$\left.\begin{aligned}\dot{Z}_1 &= Z_2 \\ \dot{Z}_2 &= -\Lambda(CG)Z_1\end{aligned}\right\} \qquad (29)$$

with

$$Z_2 = [S(CG)]^{-1}CY_2, \qquad (30)$$

and its second-order representation (26)

$$\ddot{Z}_2 + \Lambda(CG)Z_2 = 0, \qquad (31)$$

respectively. Similarly, the quadratic differential form (24) becomes now

$$Z_2^T dZ_2 + Z_1^T \Lambda(CG) dZ_1 = 0 \qquad (32)$$

or

$$\tfrac{1}{2}d[Z_2^T Z_2 + Z_1^T\Lambda(CG)Z_1] = 0. \qquad (33)$$

The quadratic form for the associated system

$$Q(Z) = \tfrac{1}{2}Z_2^T Z_2 + \tfrac{1}{2}Z_1^T\Lambda(CG)Z_1 = \text{const.} \qquad (34)$$

provides also its first integral. It describes a trajectory in the phase space Z, originating from a point, defined by initial conditions of the system. If all eigenvalues in $\Lambda(CG)$ are positive, i.e. if the matrix CG is positive-definite

$$CG > 0, \qquad (35)$$

relation (34) represents a set of closed, convex shells, centered at the origins of the phase space. These hypersurfaces are topologically equivalent to a family of spheres with center at Z = 0. In fact, they are the hyperellipsoids. Any trajectory originating on the surface of a particular hyperellipsoid remains permanently confined to it.

The transformation from Y to Z, under the condition (35), with all eigenvalues of CG distinct, is nonsingular, so the topology of the space

Y is similar to that for Z-space. The associated system in Y-space will have its first integral given by a quadratic form

$$Q(Y) = \tfrac{1}{2}(CY_2)^T[S(CG)S^T(CG)]^{-1}CY_2 + \tfrac{1}{2}Y_1^T[S(CG)\Lambda^{-1}(CG)S^T(CG)]^{-1}Y_1 = \text{const.} \quad (36)$$

Its trajectories will also remain bounded to closed, convex surfaces of hyperellipsoids centered at Y = 0, Fig. 1.

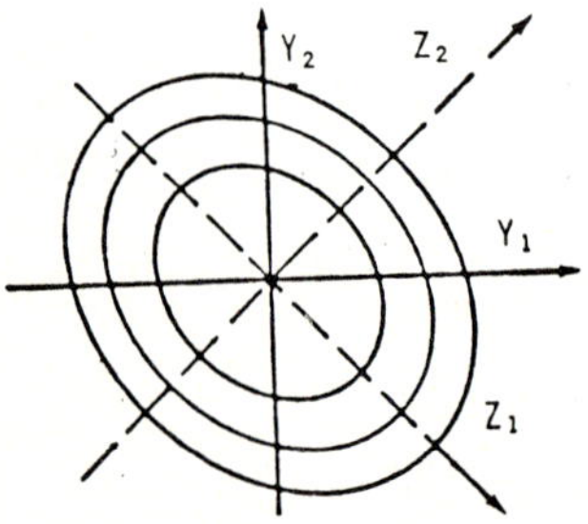

Fig. 1. Trajectories of associated systems in the space.

A characteristic feature of systems associated with nondegenerated systems (1),

$$\det A \neq 0, \quad (37)$$

is their continuous oscillating nature. This may be shown from the second-order representation (31). It has the solution of exponential type

$$Z_2 = (\exp R_1 t)C_1 + (\exp R_2 t)C_2, \quad (38)$$

where

C_1, C_2 - constant column-matrices, given by initial conditions,

R_1, R_2 - root-matrices of the characteristic equation for (31).

The characteristic equation for (31) is

$$R^2 + \Lambda(CG) = 0 \quad (39)$$

and, under the assumption (35), its roots are all purely imaginary

$$R_{1,2} = \pm j\sqrt{\Lambda(CG)}. \quad (40)$$

The same holds for the Y-space. The second-order representation (26) with respect to the subvector CY_2 has the characteristic equation

$$[R(CY_2)]^2 + CG = 0, \quad (41)$$

with two root-matrices

$$R(CY_2)_{1,2} = \pm j\sqrt{CG}. \quad (42)$$

But now, because of (22), they cannot be computed directly. To do that, the matrix CG in (41) is first diagonalized, using (28),

$$R^2 + S\Lambda S^{-1} = 0, \quad (43)$$

and then, the auxiliary roots found

$$(S^{-1}RS)_{1,2} = \pm j\sqrt{\Lambda}, \quad (43a)$$

leading finally to

$$R(CY_2)_{1,2} = \pm jS(CG)\sqrt{\Lambda(CG)}[S(CG)]^{-1}. \quad (44)$$

It is worth to notice, that the approach, used to find the square roots of a given matrix, may be extended to computations of roots of any order of square matrices, symmetrical or not, real or complexe-valued.

The quadratic form (34) and (36) for the associated systems contain two terms. First is a quadratic form of the phase subvector, which is the time derivative of the subvector appearing in the second term. Consequently, the first term may be interpreted as a quadratic form for generalized velocity and the second term - as a quadratic form with respect to a generalized position in the phase space. The similarity between the quadratic forms and expressions for kinetic and potential energies are apparent. In fact, it will be shown, that equations (34) and (36) are indeed the equations for the energy, stored within associated systems For zero-input control systems, this energy comes from the outside of systems, in the form of initial conditions.

Let consider a conservative system in the phase space. The increment of the state vector Y for such system is, in general,

$$dY = BYdt, \quad (45)$$

given by (12), with the matrix B appropriately reflecting conservative nature of the system. Furthermore, consider a real-valued function over Y

$$Q = \tfrac{1}{2}Y^TNY, \quad (46)$$

defined by a positive definite matrix

$$N > 0. \quad (47)$$

The increment vector (45) and the gradient of Q, denoted by

$$F = \text{grad}Q, \quad (48)$$

form in the space Y two fields: increment - vector field and associated gradient field. A line integral, taken along the piecewise smooth closed path in Y, vanishes in accordance to Green's theorem,

$$\oint F^T dY = 0. \quad (49)$$

This condition equals the requirement, that

$$\frac{d}{dY}(\text{grad}Q) = N \tag{50}$$

and

$$N = N^T. \tag{51}$$

Consider next the action integral in Y-space

$$K = \int_{t_1}^{t_2} L(Y)dt, \tag{52}$$

with the Lagrangian

$$L(Y) = T - U, \tag{53}$$

T,U - kinetic and potential energies.

The energies in the Lagrangian are arbitrarily defined from (36), as

$$T = \tfrac{1}{2}Y_2^T C^T [S(CG)S^T(CG)]^{-1} CY_2, \tag{54}$$

$$U = \tfrac{1}{2}Y_1^T [S(CG)\Lambda^{-1}(CG)S^T(CG)]^{-1} Y_1. \tag{55}$$

According to the principle of the least effort, the action integral must be minimized. This requires the nullification of the first variation of (52)

$$\frac{d}{dt}\frac{\partial K}{\partial Y_2} - \frac{\partial K}{\partial Y_1} = 0 \tag{56}$$

and a positive definite second variation

$$\frac{\partial^2 K}{\partial Y_1^2} > 0. \tag{57}$$

With the Lagrangian defined as above, the principle of the least effort imposes on the associated system two conditions

$$\det[S(CG)]^{-T}\Lambda(CG)[S(CG)]^{-1} \neq 0, \tag{58}$$

$$[S(CG)]^{-T}\Lambda(CG)[S(CG)]^{-1} > 0. \tag{59}$$

In other words, it is required, that

$$\det CG \neq 0 \tag{60}$$

and, in addition, that the matrix CG is positive definite, both conditions met under (35). Consequently, the quadratic form (36) is the Hamiltonian of the associated system and formulae (54) and (55) define the kinetic and potential energies of the system. The quadratic form (36) is then the total energy accumulated in the system. Associated systems are conservative.

The extremum point for the total energy is situated at the origins of the phase space and condition (59) indicates that it is a minimum. Trajectories for conservative associated systems may be then seen as projections of fixed energetic equipotential levels on the phase space, Fig. 2.

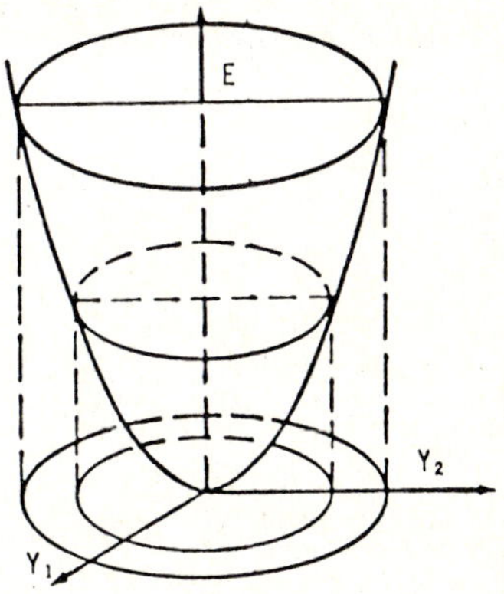

Fig. 2. Trajectories of associated systems seen as projections of equienergetic levels on the phase space.

Taking into account the way associated systems have been introduced, through the condition (23), applied to any dynamic system, they represent, in fact, a conservative state of a given system. At the same time, they point to the structural components of the system, which are responsible for oscillating tendencies and which establish values of such dynamic specifications as natural frequencies. Those parameters are gathered in the submatrix G.

4. EXAMPLES OF ASSOCIATED SYSTEMS

4.1. Second-order companion system.

First, the second-order system with a companion matrix in the state space X

$$A = \begin{bmatrix} 0 & 1 \\ -a_0 & -a_1 \end{bmatrix}, \tag{61}$$

is considered. The transformation to phase space Y,

$$P = I, \tag{62}$$

leads to the following parameters for a new system

$$B = A, \tag{63}$$

with the submatrices in (13), defining B, given by

$$G = a_0, \tag{64}$$

$$H = a_1. \tag{65}$$

The state and the phase spaces coincide. In

fact, the phase space is, for this system, the well known phase plane.

For the considered system, its associated system conforms to the requirement (23), implying

$$a_1 = 0. \tag{66}$$

The associated system will be conservative under (35):

$$a_0 > 0. \tag{67}$$

Conservative trajectories in the phase space, which also describe the equienergetic levels (36)

$$\tfrac{1}{2}(\dot{x})^2 + \tfrac{1}{2}a_0 x^2 = \text{const.} \tag{68}$$

take the form of ellipsis with principal axes along the coordinates of the phase plane. Terms in (68) refer evidently to kinetic and potential energies of the associated system.

This analysis points to a conservative character of a_0 in the original system (61) and, through (42), to its role in establishing value of natural frequency.

4.2. General second-order system.

In state space, this system has a coefficient-matrix (2)

$$A = \begin{bmatrix} a_{11} & a_{12} \\ a_{21} & a_{22} \end{bmatrix}. \tag{69}$$

It requires a transformation

$$P = \begin{bmatrix} 1 & 0 \\ a_{11} & a_{12} \end{bmatrix}, \tag{70}$$

(16), to the phase space. A new coefficient-matrix B, (13), is specified by the following submatrices

$$C = 1, \tag{71}$$

$$G = \det A = a_{11}a_{22} - a_{12}a_{21}, \tag{72}$$

$$H = \mathrm{Tr}A = -a_{11} - a_{22}. \tag{73}$$

An associated conservative system in Y has the matrix B as above, but subject to conditions (23),

$$\mathrm{Tr}A = 0, \tag{74}$$

and (35),

$$\det A > 0, \tag{75}$$

or

$$a_{11}^2 + a_{12}a_{21} < 0, \tag{76}$$

respectively. Equienergetic curves for this system are, (36),

$$\tfrac{1}{2}(\dot{y})^2 + \tfrac{1}{2}(\det A)y^2 = \text{const.} \tag{77}$$

Analysis of obtained results shows, that in order to change the undamped natural frequency without affecting other dynamic parameters, for example damping, the adjustment in values of a_{12} and/or a_{21} is required.

4.3. Third-order companion system.

This system, described in state space by a state vector

$$X = [x_1\ x_2\ x_3]^T = [x\ \dot{x}\ \ddot{x}]^T \tag{78}$$

and by a companion matrix

$$A = \begin{bmatrix} 0 & 1 & 0 \\ 0 & 0 & 1 \\ -a_0 & -a_1 & -a_2 \end{bmatrix}, \tag{79}$$

is transferred to the phase space via the transformation (16)

$$P = \begin{bmatrix} 0 & 1 & 0 \\ 1 & 0 & 0 \\ 0 & 0 & 1 \end{bmatrix}. \tag{80}$$

The system is characterized now by a phase vector Y with two subvectors, according to (6) and (7),

$$Y_1 = [x_2] = [\dot{x}], \tag{81}$$

$$Y_2 = [x_1\ x_3]^T = [x\ \ddot{x}]^T, \tag{82}$$

and by a coefficient-matrix B, specified in terms of its submatrices in (13)

$$C = [0\ \ 1], \tag{83}$$

$$G = \begin{bmatrix} -1 \\ a_1 \end{bmatrix}, \tag{84}$$

$$H = \begin{bmatrix} 0 & 0 \\ a_0 & a_2 \end{bmatrix}. \tag{85}$$

A corresponding associated conservative system must fulfill the condition (23), i.e.

$$a_0 = a_2 = 0 \tag{86}$$

and (35)

$$a_1 > 0. \tag{87}$$

It has the following expression for equienergetic levels in phase space

$$\tfrac{1}{2}x_3^2 + \tfrac{1}{2}a_1 x_2^2 = \text{const.} \tag{88}$$

These results are confirmed through the simulation, Fig. 3.

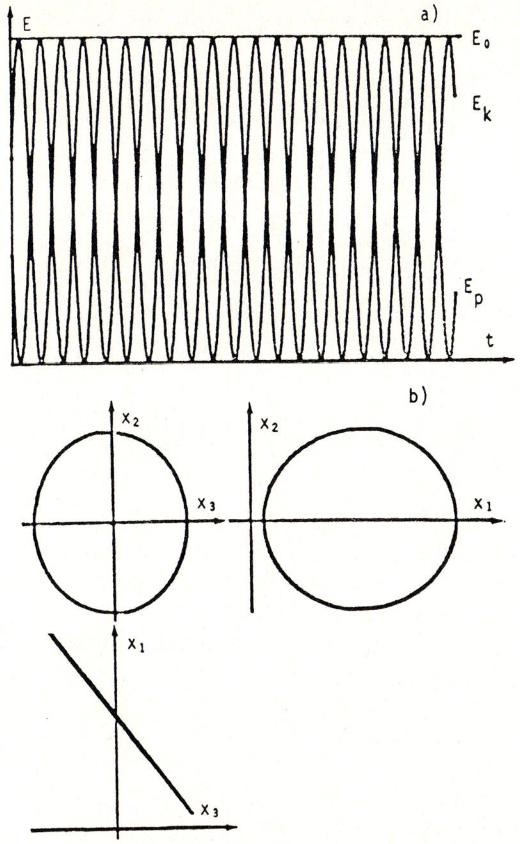

Fig. 3. Third-order associated conservative system simulated on an analog computer: a) total energy and its kinetic and potential components, b) trajectories in phase space.

It may be concluded, that natural frequency of third-order companion-type systems is determined by the coefficient a_1 only, i.e. by a coefficient related to the first-order derivative of x.

4.4. General third-order system.

It will be given in the state space X by a matrix A, (2), with all entries nonnul, and by a state vector

$$X = [x_1 \; x_2 \; x_3]^T. \tag{89}$$

It requires the transformation to the phase space

$$P = \begin{bmatrix} a_{11} & 1 & a_{11}^2 + a_{12}a_{21} + a_{13}a_{31} \\ a_{12} & 0 & a_{11}a_{12} + a_{12}a_{22} + a_{13}a_{32} \\ a_{13} & 0 & a_{11}a_{13} + a_{12}a_{23} + a_{13}a_{33} \end{bmatrix}, \tag{90}$$

where the phase vector, defined in (6) and (7), is

$$Y_1 = [\dot{y}], \tag{91}$$

$$Y_2 = [y \; \ddot{y}]^T, \tag{92}$$

and the matrix B, (13),

$$C = [0 \;\; 1], \tag{93}$$

$$G = \begin{bmatrix} -1 \\ g_{12} \end{bmatrix}, \tag{94}$$

$$H = \begin{bmatrix} 0 & 0 \\ h_{21} & h_{22} \end{bmatrix}, \tag{95}$$

accordingly. Elements of the submatrices G and H are

$$g_{12} = \Sigma A_{ij}, \tag{96}$$

A_{ij} - principal 2 x 2-minors of A,

$$h_{21} = -\det A, \tag{97}$$

$$h_{22} = -\mathrm{Tr}A. \tag{98}$$

For the associated conservative system, following conditions hold:

$$\det A = 0, \tag{99}$$

$$\mathrm{Tr}A = 0, \tag{100}$$

and

$$CG = g_{12} > 0. \tag{101}$$

Condition (99) means, that third-order systems may have oscillating behaviour only in the second-order phase subspace (see Fig. 3). This statement may be generalized to higher-order systems as well.

Odd-order oscillating conservative systems are not possible. Oscillations may occur in an even-order phase subspace only. The coefficient-matrix for an odd-order oscillating system must be singular, (99) and quasi skew-symmetrical, i.e. it is similar to some skew-symmetrical matrix.

Equienergetic curves for the third-order system are given by (36):

$$\left.\begin{aligned} \tfrac{1}{2}(\ddot{y})^2 + \tfrac{1}{2}g_{12}(\dot{y})^2 &= \text{const.}, \\ y &= \text{const.} \end{aligned}\right\} \tag{102}$$

The undamped natural frequency of the initial system depends upon g_{12} and corresponding elements of A, when, in addition, relations (99), (100) are taken into account.

4.5. Fourth-order companion system.

The state vector

$$X = [x_1 \; x_2 \; x_3 \; x_4]^T = [x \; x^{(1)} \; x^{(2)} \; x^{(3)}], \tag{103}$$

$x^{(i)}$ - i-th time derivative of x,

and the matrix

$$A = \begin{bmatrix} 0 & 1 & 0 & 0 \\ 0 & 0 & 1 & 0 \\ 0 & 0 & 0 & 1 \\ -a_0 & -a_1 & -a_2 & -a_3 \end{bmatrix}, \tag{104}$$

are transferred to the phase space using a linear unitary operator

$$P = \begin{bmatrix} 1 & 0 & 0 & 0 \\ 0 & 0 & 1 & 0 \\ 0 & 1 & 0 & 0 \\ 0 & 0 & 0 & 1 \end{bmatrix}. \tag{105}$$

There, the system is described by vector Y with subvectors

$$Y_1 = [x_1 \; x_3]^T = [x \; x^{(2)}]^T, \tag{106}$$

$$Y_2 = [x_2 \; x_4]^T = [x^{(1)} \; x^{(2)}]^T \tag{107}$$

and a new coefficient matrix B, given by submatrices

$$C = I_2, \tag{108}$$

I_2 - 2 x 2 identity matrix,

$$G = \begin{bmatrix} 0 & -1 \\ a_0 & a_2 \end{bmatrix}, \tag{109}$$

$$H = \begin{bmatrix} 0 & 0 \\ a_1 & a_3 \end{bmatrix}. \tag{110}$$

This system has an associated conservative system, if

$$a_1 = a_3 = 0, \tag{111}$$

(23), and if, according to (35),

$$\lambda_{1,2}(CG) = \lambda_{1,2} = \tfrac{1}{2}a_2 \pm [(\tfrac{1}{2}a_2)^2 - a_0]^{\frac{1}{2}} > 0, \tag{112}$$

$\lambda_i(CG)$ - eigenvalues of CG,

equivalent to

$$a_0 > 0, \quad a_2 > 0. \tag{113}$$

Matrix CG is not symmetrical, so quadratic differential form (24) is not directly integrable. The passage through the phase space Z is required, via the modal matrix of CG, (27),

$$S(CG) = \begin{bmatrix} 1 & 1 \\ -\lambda_1 & -\lambda_2 \end{bmatrix}. \tag{114}$$

From there, integration of (24) may be carried out, providing equienergetic curves (36) and kinetic, (54),

$$T = \tfrac{1}{2} \begin{bmatrix} x_2 \\ x_4 \end{bmatrix}^T \begin{bmatrix} a_2^2 - 2a_0 & a_2 \\ a_2 & 2 \end{bmatrix} \begin{bmatrix} x_2 \\ x_4 \end{bmatrix} \tag{115}$$

and potential, (55)

$$U = \tfrac{1}{2} \begin{bmatrix} x_1 \\ x_3 \end{bmatrix}^T \begin{bmatrix} a_0 a_2 & 2a_0 \\ 2a_0 & a_2 \end{bmatrix} \begin{bmatrix} x_1 \\ x_3 \end{bmatrix} \tag{116}$$

energies for the associated system.

Verification of a conservative nature of the associated system has been done through the simulation, Fig. 4

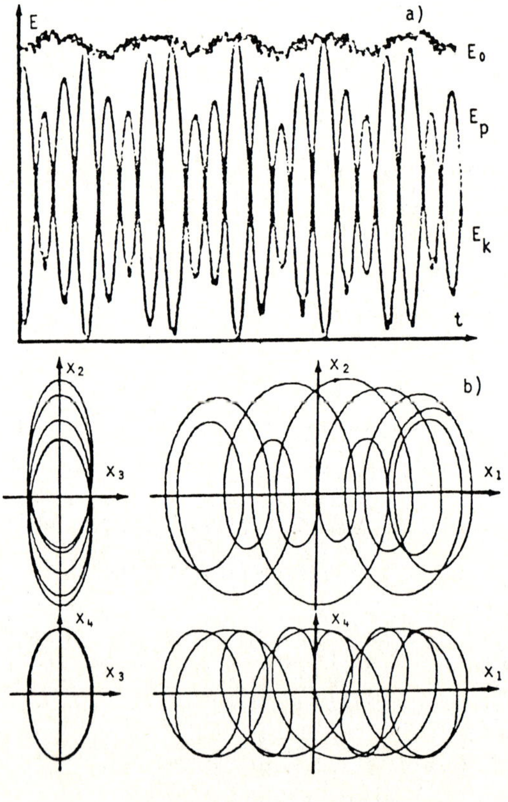

Fig. 4. Fourth-order associated conservative system simulated on an analog computer: a) total energy and its kinetic and potential components, b) trajectories in the phase space.

and through the direct solution

$$X(t) = \frac{1}{\omega_1^2 - \omega_2^2} \left\{ \begin{bmatrix} \omega_1^2 - a_2 & 0 & -1 & 0 \\ 0 & \omega_1^2 - a_2 & 0 & -1 \\ a_0 & 0 & \omega_1^2 & 0 \\ 0 & a_0 & 0 & \omega_1^2 \end{bmatrix} \cos\omega_1 t - \right.$$

$$- \begin{bmatrix} \omega_2^2-a_2 & 0 & -1 & 0 \\ 0 & \omega_2^2-a_2 & 0 & -1 \\ a_0 & 0 & \omega_2^2 & 0 \\ 0 & a_0 & 0 & \omega_2^2 \end{bmatrix} \cos\omega_2 t \; +$$

$$+ \begin{bmatrix} 0 & \omega_1-\frac{a_2}{\omega_1} & 0 & -\frac{1}{\omega_1} \\ \frac{a_0}{\omega_1} & 0 & \omega_1 & 0 \\ 0 & \frac{a_0}{\omega_1} & 0 & \omega_1 \\ -a_0\omega_1 & 0 & \frac{a_0}{\omega_1}-a_2\omega_1 & 0 \end{bmatrix} \sin\omega_1 t \; -$$

$$\left. - \begin{bmatrix} 0 & \omega_2-\frac{a_2}{\omega_2} & 0 & -\frac{1}{\omega_2} \\ \frac{a_0}{\omega_2} & 0 & \omega_2 & 0 \\ 0 & \frac{a_0}{\omega_2} & 0 & \omega_2 \\ -a_0\omega_1 & 0 & \frac{a_0}{\omega_2}-a_2\omega_2 & 0 \end{bmatrix} \sin\omega_2 t \right\} X_0 \quad (117)$$

The results indicate the existence of two natural frequencies

$$\omega_1 = \{\tfrac{1}{2}a_2 + [(\tfrac{1}{2}a_2)^2 - a_0]^{\frac{1}{2}}\}^{\frac{1}{2}}, \quad (118)$$

$$\omega_2 = \{\tfrac{1}{2}a_2 - [(\tfrac{1}{2}a_2)^2 - a_0]^{\frac{1}{2}}\}^{\frac{1}{2}} \quad (119)$$

for the system (104) and points on parameters a_0 and a_2 as responsible for oscillating tendencies of the system.

5. CONTROL SYSTEMS

Control systems (1) can be transferred to phase space, using the transformation P, (5), an new state vector Y, (6-7),

$$\dot{Y} = BY - V, \quad (120)$$

and a new control vector

$$V = PDU = [V_1^T \; V_2^T]^T, \quad (121)$$

compatible with Y

$$\dim V_1 = \dim Y_1, \quad (122)$$

$$\dim V_2 = \dim Y_2. \quad (123)$$

Equation (120) has an equivalent representation by a set of two differential equations of the first-order

$$\left. \begin{aligned} \dot{Y}_1 &= CY_2 - V_1 \\ \dot{Y}_2 &= -GY_1 - HY_2 - V_2 \end{aligned} \right\} \quad (124)$$

or by one second-order non-homogeneous equation

$$\ddot{Y}_2 - H\dot{Y}_2 - GCY_2 = V_2 - GV_1. \quad (125)$$

A parametric differential form (21) for trajectories in phase space may be extended to systems under control signals

$$(CY_2)^T d(CY_2) + Y_1^T(CG)^T dY_1 = -(CHY_2)^T dY_1 + (CV_2)^T dY_1 - V_1^T d(CY_2). \quad (126)$$

The left-hand side differential quadratic form is integrable upon the diagonalization of CG, through the passage by Z-phase space, (27-30):

$$\tfrac{1}{2}d[(CY_2)^T\Phi(CY_2) + Y_1^T\Psi Y_1] = [(CV_2)^T - Y_2^T(CH)^T]\Phi dY_1 - Y_1^T\Phi d(CY_2), \quad (127)$$

where

$$\Phi = \Phi^T = [S(CG)S^T(CG)]^{-1}, \quad (128)$$

$$\Psi = \Psi^T = [S(CG)\Lambda^{-1}(CG)S^T(CG)]^{-1}. \quad (129)$$

Relation (127) contains a quadratic form on the left-hand and bilinear forms on the right-hand side. The quadratic form with respect to position and velocity subvectors in the phase space, represents total energy accumulated within the system in potential and kinetic forms, as it has been shown previously, (54), (55). These energies were obtained for systems in conservative adiabatic state, with no exchange of energy to outside world. Consequently, the right-hand side must also represent some form of energy. The bilinear forms appearing there are given in terms of two subvectors of the state vector Y and in terms of input control signals, acting on position and velocity modes. It becomes evident, that (127) describes the balance of energy for a system interacting with its environment. Interactions take place through the control action (terms with components of V) and through some other mechanism (term with components of Y), dependent upon parameters of the system itself. The parameters involved in the exchange of energy are given by a submatrix H, (13). We shall show now, that the submatrix H is indeed responsible for dissipative or absorptive characteristics of the system.

Let consider a zero-input system (12). It has a parametric differential trajectory in the phase space in the form (126) or (127) under V = 0. The system may be also solved directly in the time domain

$$Y = e^{Bt}Y_0, \quad (130)$$

Y_0 - initial conditions.

The exponential matrix can be expanded into

series

$$e^{Bt} = I + Bt + \frac{1}{2!}(Bt)^2 + \ldots \quad (131)$$

For system with distinct eigenvalues, a similar diagonal matrix Λ exists, (14), so that (131) is also given by

$$e^{Bt} = S(B)S^{-1}(B) - S(B)\Lambda S^{-1}(B) +$$

$$+ \frac{1}{2!}S(B)\Lambda^2 S^{-1}(B)t^2 + \ldots = S(B)\, e^{\Lambda t} S^{-1}(B), \quad (132)$$

where

$$\Lambda = \lambda_{ij}, \quad i,j = 1,\ldots,n, \quad (133)$$

$$\lambda_{ij} = \begin{cases} \lambda_i(B), & \forall i = j \\ 0, & \forall i \neq j \end{cases} \quad (134)$$

with

$$\lambda_i(B) = \alpha_i + j\beta_i. \quad (135)$$

The norm for (132) is

$$||e^{Bt}|| \le ||S(B)||\; ||S^{-1}(B)||\; ||e^{\Lambda t}|| =$$

$$= ||e^{\Lambda t}||. \quad (136)$$

Similarly, for the associated system to (12) under (23)

$$||e^{B_c t}|| \le ||e^{\Lambda_c t}||, \quad (137)$$

B_c - matrix (13) with $H = 0$,

where

$$\Lambda_c = [\lambda_{cij}], \quad (138)$$

$$\lambda_{cij} = \begin{cases} \lambda_{ci}, & \forall i = j \\ 0, & \forall i \neq j \end{cases} \quad (139)$$

with

$$\lambda_{ci} = j\gamma_i. \quad (140)$$

Let define the norm

$$||e^{\Lambda_c t}|| = \sum_i^n |e^{j\gamma_i}|; \quad (141)$$

then the following estimation for the associated system holds

$$||e^{\Lambda_c t}|| \le n. \quad (142)$$

For the system (136), taking into account (133-135), the norm is given by

$$||e^{\Lambda t}|| = \sum_i^n e^{\alpha_i t}|e^{j\beta_i t}|. \quad (143)$$

Suppose, that real parts of all eigenvalues of B are negative

$$\alpha_i < 0, \forall i \quad (144)$$

and

$$\alpha = \inf(|\alpha_i|), \quad (145)$$

then the norm (143) can be estimated as follows

$$||e^{\Lambda t}|| \le e^{-\alpha t}\sum_i^n |e^{j\beta_i t}| \le ne^{-\alpha t}. \quad (146)$$

Comparing (146) with (142) it may be concluded, that

$$||e^{\Lambda t}|| \le ||e^{\Lambda c t}||, \quad (147)$$

i.e. for a stable system, its trajectory in the phase space is bounded within the trajectory for corresponding associated conservative system, Fig. 5. It tends toward the equilibrium point at the origins of the phase space. The

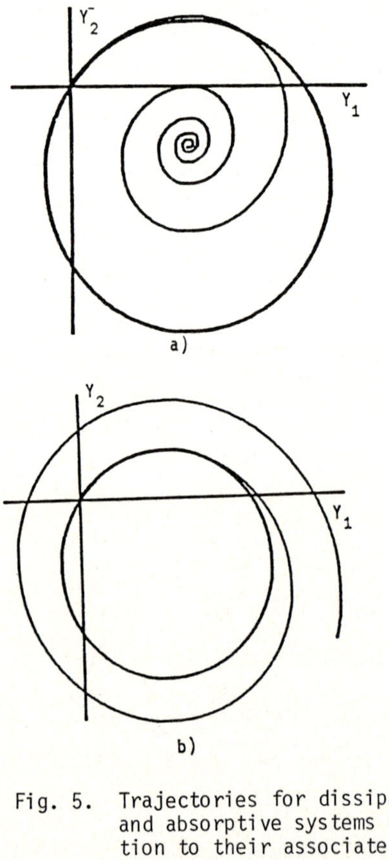

Fig. 5. Trajectories for dissipative and absorptive systems in relation to their associated systems in the phase space.

stability is achieved because the accumulated energy, injected by initial conditions, has been dissipated to outside. The instability, on the other hand, results from absorptive nature of the system, causing continuous increase of its internal energy. Parameters, responsible for the exchange of energy are those given by the submatrix H. The submatrix H determines the stability and damping characteristics of transients.

The phase space representation (124) unveils internal mechanisms of dynamic behaviour of system under input control action or in a free evolution, under initial conditions, Fig. 6.

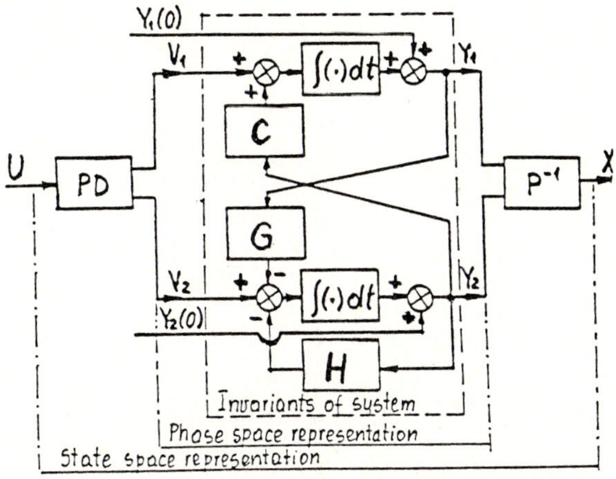

Fig. 6. Internal structure of dynamic systems.

It can be seen, that oscillating tendencies are due to the existence of internal positive feedback, between the velocity and position modes, through the intermodal coupling matrix C, balanced by the negative feedback through G. In associated conservative systems, an equilibrium state between both feedbacks exists. Tendencies to increase the distance from equilibrium point in phase space (position mode) are returned and act toward slowing the transients (velocity mode). Then, dicreasing velocity will cause an increase in the distance from origins.

For full systems, an additional, velocity type feedback through the matrix H is added. In stable systems, this feedback is negative and in unstable ones - positive (due to negative matrix H). The existence of negative feedback with respect to the velocity mode provides dissipation of internal energy. Loose or tigh negative feedback will result in weakly or strongly damped transients.

6. CONCLUDING REMARKS

Energetic decomposition of systems in phase space, showing internal mechanisms, may be used in establishing strategies of feedback controls, Return signals in a closed loop are taken from certain outputs and then transformed according to algorithm of the controller. Together with set points, they form a total input signal. The controller may be decomposed into elements, which act on position and velocity modes. It will either reinforce or weaken dissipation and/or accumulation of internal energy. These considerations allow a choice of structure of control system in order to obtain required performance in transient and steady states.

REFERENCES

[1] JACYNO, Z., An Energetic Approach to the Dynamics of Linear Systems; "Proceedings of the 20th Midwest Symposium on Circuits and Systems", Part 2, pp. 514-519; Texas Tech University, Lubbock, Texas, August 15-17, 1977.

[2] JACYNO, Z., A Hyperphase Space; "Proceedings of the 21st Midwest Symposium on Circuits and Systems", pp. 336-340; Iowa State University, Ames, Iowa, August 14-15, 1978.

[3] JACYNO, Z., Analysis and Design of Control Systems in Generalized Phase Space; "Proceedings of the International Symposia Measurement and Control MECO'79 and Computer Applications in Industry", pp. 132-136; June 13-15, 1979, Grenoble, France; Editor ACTA Press.

[4] JACYNO, Z., An Energetic Decomposition of Dynamic Systems; "1979 International Symposium on Circuits and Systems Proceedings" - sponsored by the IECE Circuits and Systems Society and the IEEE Circuits and Systems Society, pp. 154-157; July 17-19, 1979, Tokyo, Japan.

[5] JACYNO, Z., Automatic Controls: A Hyperphase Space Approach; "Actas del 4° Congreso Informatica y Automatica" organized by Asociacion Espanola de Informatica y Automatica, vol. I, pp. 14-22; October 16-19, 1979, Madrid, Spain.

[6] JACYNO, Z., Control Problems in General Phase Space; "IV Congreso Asociacion Chilena de Control Automatico", August 4-9, 1980, vol. I, pp. 18-26; Santiago, Chile.

[7] JACYNO, Z., LEMAIRE, B., Automatic Control of interconnected Dynamical Systems; V Congreso de Informatica y Automatica organizado por Asociacion Espanola de Informatica y Automatica, vol. I, pp. 17-20; May 4-6, 1982, Madrid, Spain.

[8] JACYNO, Z., Energetic Relations in General Dynamic Systems; "1982 IEEE International Symposium on Circuits and Systems Proceedings", vol. 2, pp. 389-392; May 10-12, 1982, Rome, Italy.

Complex and Distributed Systems: Analysis, Simulation and Control
S.G. Tzafestas and P. Borne (editors)
Elsevier Science Publishers B.V. (North-Holland)

INVESTIGATIONS OF METHODS FOR THE DIRECT ASSESSMENT OF PARAMETER SENSITIVITY IN LINEAR CLOSED-LOOP CONTROL SYSTEMS

D.J. MURRAY-SMITH

DEPARTMENT OF ELECTRONICS AND ELECTRICAL ENGINEERING, UNIVERSITY OF GLASGOW, GLASGOW G12 8QQ, SCOTLAND.

1. SUMMARY

The paper describes direct techniques for determination of the parameter sensitivity functions of linear closed-loop systems. These techniques, which may be applied both in the time-domain and in the frequency-domain, can yield sensitivity functions of all of the output variables with respect to all of the adjustable parameters of the closed-loop system using minimal experimentation and without the need for explicit identification of the controlled system.

2. INTRODUCTION

Conventional techniques for the assessment of parameter sensitivity in closed-loop control systems usually involve either a parameter perturbation approach or the use of a sensitivity cosystem [1]. Perturbation methods are based upon calculation of the difference between system responses before and after a change of each parameter and thus require p+1 separate tests to obtain sensitivity information for any p parameters of the system. Since this approach involves the determination of small differences between similar response curves, the accuracy of the results is often limited by problems of noise. Although use of the sensitivity cosystem method can provide accurate simultaneous estimates of all of the parameter sensitivity functions for a linear system, this method does require precise knowledge of the structure and parameters of the equations defining the system under consideration. Such information is seldom completely available in studies involving practical control systems.

An alternative, and direct, approach to parameter sensitivity analysis in closed-loop systems, which has been considered previously only in the context of electrical power systems applications [2,3], is based upon a special case of the three points method of parameter sensitivity analysis [4]. This is a technique which does not require identification of the controlled system and may be applied readily in experimental studies associated with on-line parametric optimisation. The method has been used successfully in research associated with synchronous-generator excitation control [3] and provides a basis for other applications involving single-input, single-output multivariable control systems. The most important feature of this direct method is that, for most purposes, it can allow investigation of closed-loop system sensitivity to variations of controller parameters without identification of the controlled system and allows simultaneous assessment of all the sensitivity functions of interest from a single set of time-domain or frequency-domain measurements.

3. MULTIVARIABLE SYSTEMS

The block diagram of Figure 1 shows a multivariable feedback system involving a plant transfer function matrix $\underline{G}(s)$ together with a controller transfer function matrix $\underline{C}(s)$ and a transfer function matrix $\underline{H}(s)$ in the feedback path.

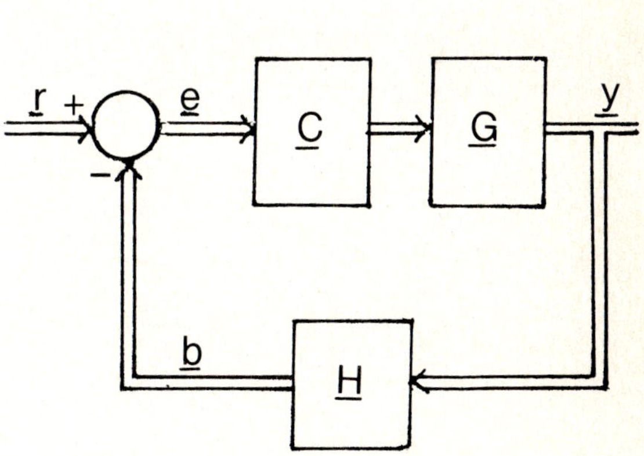

Figure 1. Multivariable system block diagram.

For the system of Figure 1

$$\underline{b}(s) = \underline{H}(s)\ \underline{y}(s) \tag{1}$$

$$\underline{e}(s) = \underline{r}(s) - \underline{b}(s) \tag{2}$$

and $$\underline{y}(s) = \underline{G}(s)\ \underline{C}(s)\ \underline{e}(s) \tag{3}$$

giving

$$\underline{y}(s) = \left[\underline{I} + \underline{G}(s)\,\underline{C}(s)\,\underline{H}(s)\right]^{-1} . \underline{G}(s)\,\underline{C}(s)\,\underline{r}(s) \quad (4)$$

$$= \underline{W}(s)\,\underline{r}(s) \quad (5)$$

where $\underline{W}(s) = \left[\underline{I} + \underline{G}(s)\,\underline{C}(s)\,\underline{H}(s)\right]^{-1} . \underline{G}(s)\,\underline{C}(s)$ is the transfer function matrix of the closed-loop system.

If the elements $\underline{C}(s)$ and $\underline{H}(s)$ within this system are dependent upon a set of parameters $\underline{a}$ (but the plant transfer function matrix $\underline{G}(s)$ is independent of $\underline{a}$) then, for a given parameter a_i, we have from equation (1),

$$\frac{\partial \underline{b}(s)}{\partial a_i} = \frac{\partial \underline{H}(s)}{\partial a_i}\,\underline{y}(s) + \underline{H}(s)\,\frac{\partial \underline{y}(s)}{a_i} \quad (6)$$

Similarly, from equation (2)

$$\frac{\partial \underline{e}(s)}{\partial a_i} = -\frac{\partial \underline{b}(s)}{\partial a_i} \quad (7)$$

and thus, in terms of equation (3), it may be shown that

$$\frac{\partial \underline{y}(s)}{\partial a_i} = \underline{G}(s)\,\frac{\partial \underline{C}(s)}{\partial a_i}\,\underline{e}(s) - \underline{G}(s)\,\underline{C}(s)\,\underline{H}(s)\,\frac{\partial \underline{y}(s)}{\partial a_i} - \underline{G}(s)\,\underline{C}(s)\,\frac{\partial \underline{H}(s)}{\partial a_i}\,\underline{y}(s) \quad (8)$$

For cases where $\frac{\partial \underline{H}(s)}{\partial a_i} = 0$ it therefore follows that the parameter sensitivity function for the output $\underline{y}(s)$ is

$$\frac{\partial \underline{y}(s)}{\partial a_i} = \left[\underline{I} + \underline{G}(s)\,\underline{C}(s)\,\underline{H}(s)\right]^{-1} . \underline{G}(s)\,\frac{\partial \underline{C}(s)}{\partial a_i}\,\underline{e}(s) \quad (9)$$

i.e. $$\frac{\partial \underline{y}(s)}{\partial a_i} = \underline{W}(s)\,\underline{C}(s)^{-1}\,\frac{\partial \underline{C}(s)}{\partial a_i}\,\underline{e}(s) \quad (10)$$

For cases where $\frac{\partial \underline{C}(s)}{\partial a_i} = 0$

$$\frac{\partial \underline{y}(s)}{\partial a_i} = -\left[\underline{I} + \underline{G}(s)\,\underline{C}(s)\,\underline{H}(s)\right]^{-1} . \underline{G}(s)\,\underline{C}(s)\,\frac{\partial \underline{H}(s)}{\partial a_i}\,\underline{y}(s) \quad (11)$$

$$= -\underline{W}(s)\,\frac{\partial \underline{H}(s)}{\partial a_i}\,\underline{y}(s)$$

4. SINGLE-INPUT SINGLE-OUTPUT SYSTEMS

In the single-input single-output closed-loop system of Figure 2 the transfer function C(s) represents a controller or cascade compensator which incorporates p parameters $a_1, a_2, a_3 \ldots a_p$ forming a parameter set a. The transfer function H(s), in the feedback path, incorporates q additional parameters $b_1, b_2, b_3, \ldots b_q$ forming a parameter set $\underline{b}$. The transfer function G(s) represents the system which is being controlled.

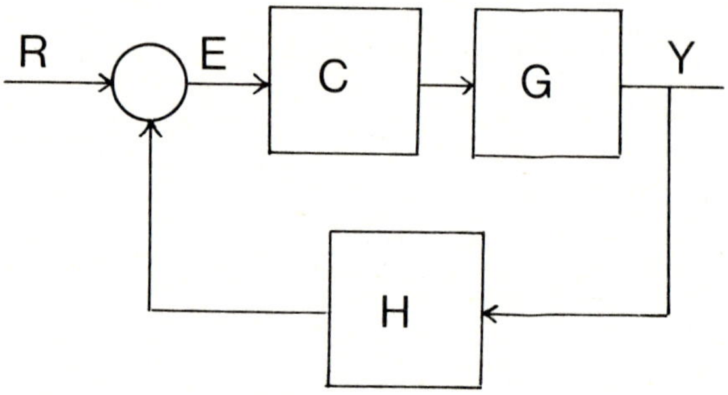

Figure 2. Single-input, single-output system block diagram.

For this closed-loop system the output Y(s) is related to the reference input R(s) by the equation

$$Y(s) = \frac{G(s)\,C(s)}{1 + G(s)\,C(s)\,H(s)} . R(s) \quad (13)$$

Differentiating with respect to parameter a_i gives the parameter sensitivity function

$$\frac{\partial Y(s)}{\partial a_i} = \frac{Y(s)}{R(s)}\,\frac{1}{C(s)}\,\frac{\partial C(s)}{\partial a_i}\,E(s) \quad (14)$$

Similarly, for parameter b_j

$$\frac{\partial Y(s)}{\partial b_j} = -\frac{Y(s)}{R(s)} . \frac{\partial H(s)}{\partial b_j}\,Y(s) \quad (15)$$

Equations (14) and (15) may be re-written as

$$\frac{\partial Y(s)}{\partial a_i} = \frac{Y(s)}{R(s)} \cdot F_{a_i}(s).E(s) \qquad (16)$$

and
$$\frac{\partial Y(s)}{\partial b_j} = \frac{Y(s)}{R(s)} \cdot F_{b_j}(s).Y(s) \qquad (17)$$

where
$$F_{a_i}(s) = \frac{1}{C(s)} \frac{\partial C(s)}{\partial a_i} \qquad (18)$$

and
$$F_{b_j}(s) = -\frac{\partial H(s)}{\partial b_j} \qquad (19)$$

5. TIME-DOMAIN ANALYSIS

The sensitivity functions of equations (16) and (17) may be evaluated readily in the time-domain for the special cases of step function or impulsive test signals applied at the reference input of the closed-loop system.

When the test signal is a unit step function equation (16) becomes

$$\frac{\partial Y(s)}{\partial a_i} = s\,Y(s)\,Z_{a_i} \qquad (20)$$

where
$$Z_{a_i}(s) = F_{a_i}\,E(s) \qquad (21)$$

Hence, in the time-domain, the sensitivity function for parameter a_i is given by the convolution integral

$$\frac{\partial y(t)}{\partial a_i} = \int_0^t \frac{dy}{d\tau}\,Z_{a_i}(t-\tau)\,d\tau \qquad (22)$$

Similarly, for parameter b_j

$$\frac{\partial Y(s)}{\partial b_j} = s\,Y(s)\,Z_{b_j}(s) \qquad (23)$$

where
$$Z_{b_j}(s) = F_{b_j}(s)\,Y(s) \qquad (24)$$

Thus

$$\frac{\partial y(t)}{\partial b_j} = \int_0^t \frac{dy}{d\tau}\,Z_{b_j}(t-\tau)\,d\tau \qquad (25)$$

The corresponding results for a unit impulse test signal are as follows:-

$$\frac{\partial y(t)}{\partial a_i} = \int_0^t y(\tau)\,Z_{a_i}(t-\tau)\,d\tau \qquad (26)$$

and

$$\frac{\partial y(t)}{\partial b_j} = \int_0^t y(\tau)\,Z_{b_j}(t-\tau)\,d\tau \qquad (27)$$

In both cases, therefore, the sensitivity functions may be evaluated using a convolution integral involving only the quantities $Z_{a_i}(t)$ or $Z_{b_j}(t)$ and the system output, y(t), or its time derivatives.

For both types of test signal the quantity $Z_{a_i}(t)$ may be obtained by passing the error signal e(t) through a filter having the transfer function $F_{a_i}(s)$ defined in equation (18). Similarly $Z_{b_j}(t)$ may be obtained by filtering of the output y(t) by the transfer function $F_{b_j}(s)$ given in equation (19). Since $F_{a_i}(s)$ and $F_{b_j}(s)$ are independent of the plant transfer function, G(s), detailed knowledge of the plant characteristics is not required in order to use equations (26) and (27) to calculate the sensitivity of the system output y(t) to variation of the controller parameters $\underline{a}$ and $\underline{b}$. The most significant constraints on the method are those due to the assumptions of linearity of the system which are implicit in the above analysis.

6. FREQUENCY-DOMAIN ANALYSIS

From equation (13) the closed-loop system transfer function for the system shown in the block diagram of Figure 2 is given by

$$W_c(s) = \frac{G(s)\,C(s)}{1 + G(s)\,C(s)\,H(s)} \qquad (28)$$

The sensitivity of this closed-loop transfer function to variation of parameter a_i is therefore given by

$$\frac{\partial W_c(s)}{\partial a_i} = \frac{1}{C(s)} \frac{\partial C(s)}{\partial a_i} \frac{E(s)}{R(s)} \cdot W_c(s) \qquad (29)$$

$$= P(s)\,W_c(s) \qquad (30)$$

$$\text{where } P(s) = \frac{1}{C(s)} \frac{\partial C(s)}{\partial a_i} \frac{E(s)}{R(s)} \quad (31)$$

Similarly for variations of parameter b_j the parameter sensitivity function is given by

$$\frac{\partial W_c(s)}{\partial b_j} = - \frac{\partial H(s)}{\partial b_j} \frac{Y(s)}{R(s)} \cdot W_c(s) \quad (32)$$

$$= Q(s)\, W_c(s) \quad (33)$$

$$\text{where } Q(s) = - \frac{\partial H(s)}{\partial b_j} \frac{Y(s)}{R(s)} \quad (34)$$

In general, in terms of the harmonic response, the transfer function of the closed-loop system may be expressed in polar form as

$$W_c(j\omega) = M(\omega)\, e^{j\alpha(\omega)} \quad (35)$$

Similarly, the functions $P(j\omega)$ and $Q(j\omega)$ may be expressed as

$$P(j\omega) = U(\omega)\, e^{j\beta(\omega)} \quad (36)$$

$$\text{and } Q(j\omega) = V(\omega)\, e^{j\gamma(\omega)} \quad (37)$$

Hence, from equation (30),

$$\frac{\partial W_c(j\omega)}{\partial a_i} = U(\omega)\, M(\omega)\, e^{j(\alpha+\beta)} \quad (38)$$

$$\text{But } W_c(j\omega, \underline{a} + \Delta a_i) = W_c(j\omega, \underline{a}) + \frac{\partial W_c(j\omega,\underline{a})}{\partial a_i} \Delta a_i \quad (39)$$

$$= M(\omega) \left(e^{j\alpha} + U(\omega)\, e^{j\beta}\, \Delta a_i \right) \quad (40)$$

Hence it may be shown that

$$|W_c(j\omega, \underline{a} + \Delta a_i)| = \left(M(\omega)^2 + 2M(\omega)^2\, U(\omega) \cdot \cos(\alpha-\beta)\, \Delta a_i \right)^{1/2} \quad (41)$$

which may be reduced to

$$|W_c(j\omega, \underline{a} + \Delta a_i)| = M(\omega) \left(1 + U(\omega) \cos\beta \Delta a_i \right) \quad (42)$$

$$\text{or } \frac{\Delta M(\omega)}{\Delta a_i} = U(\omega)\, M(\omega) \cos\beta \quad (43)$$

$$\text{Similarly } \frac{\Delta M(\omega)}{\Delta b_j} = V(\omega)\, M(\omega) \cos\gamma \quad (44)$$

The products $U(\omega).M(\omega).\cos\beta$ and $V(\omega).M(\omega)\cos\gamma$ are therefore the parameter sensitivity functions for the closed loop system transfer function. In the context of sinusoidal or random signal testing of the closed loop system these quantities may usually be determined without difficulty from measurements.

7. EXAMPLE

Figure 3 shows a block diagram representing an aircraft flight control system in the altitude hold mode of operation [5]. This mode is generally used during cruise conditions for which the aircraft's flight path angle, and thus its altitude, is controlled by the elevators. When this mode is operational airspeed is controlled, either manually or automatically, by use of the throttle. The mathematical model is of an aircraft cruising at an altitude of 12300 m at an airspeed of 185 m s^{-1}.

The aircraft transfer function in Figure 4 relating the altitude, $N(s)$, to the variable $E_A(s)$ is given by [5]:

$$G(s) = - \frac{1.8765(s - 4.89)(s + 4.89)}{s^2(s + 0.8)(s^2 + 10s + 29)} \quad (45)$$

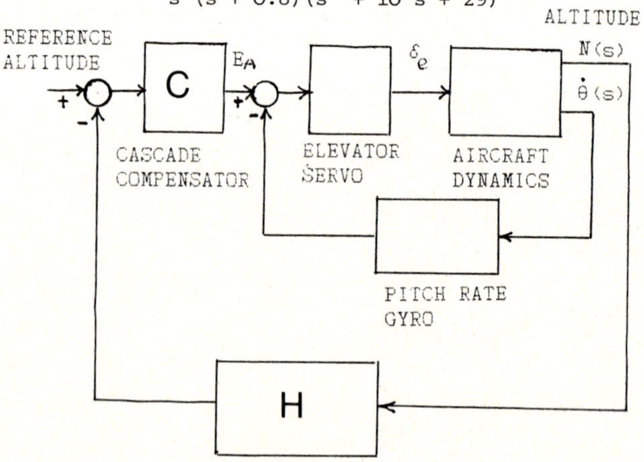

Figure 3. Block diagram of altitude hold mode flight control system.

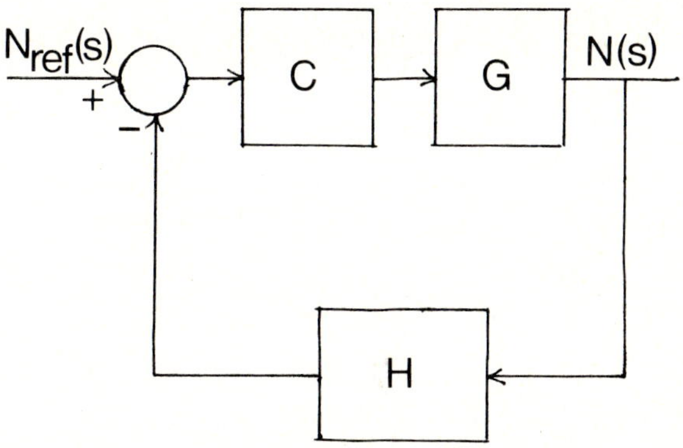

Figure 4. Reduced block diagram of altitude hold system.

The cascade compensator, C(s), has the form

$$C(s) = \frac{a_1 (s + a_2)}{s + a_3} \quad (46)$$

where a_1, a_2 and a_3 have typical values of 0.5, 0.8 and 8.0 respectively. The feedback compensator transfer function is

$$H(s) = b_1(s + b_2) \quad (47)$$

where $b_1 = 10$ and $b_2 = 0.1$, typically.

For this system the filter transfer functions needed for the generation of the sensitivity functions for each of the parameters are as follows:-

$$F_{a_1}(s) = \frac{1}{C(s)} \cdot \frac{\partial C(s)}{\partial a_1} = \frac{1}{a_1} \quad (48)$$

$$F_{a_2}(s) = \frac{1}{C(s)} \cdot \frac{\partial C(s)}{\partial a_2} = \frac{1}{s + a_2} \quad (49)$$

$$F_{a_3}(s) = \frac{1}{C(s)} \cdot \frac{\partial C(s)}{\partial a_3} = - \frac{1}{s + a_3} \quad (50)$$

$$F_{b_1}(s) = -(s + b_2) \quad (51)$$

$$F_{b_2}(s) = -b_1 \quad (52)$$

These transfer functions may be implemented without difficulty in either analogue or digital form and may be applied to the system as shown in Figure 5. The outputs of these filters form the auxiliary signals needed for both the time-domain and frequency-domain methods of analysis.

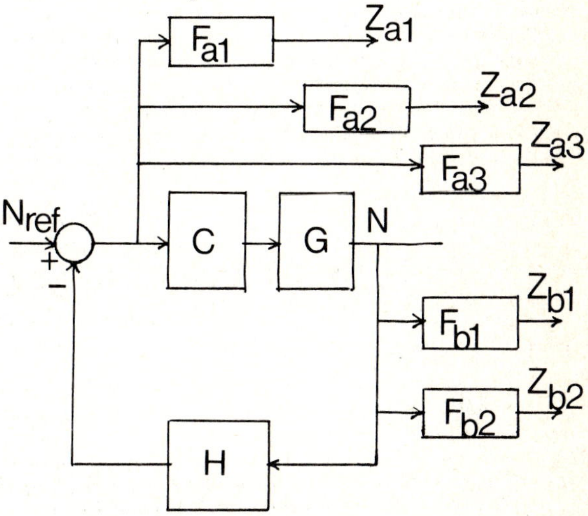

Figure 5. Block diagram of system with auxiliary filters for sensitivity analysis by direct method.

8. DISCUSSION

Most methods for the simultaneous assessment of parameter sensitivity functions of a closed-loop system require accurate knowledge of both the structure and parameters of the plant. The direct method outlined in this paper avoids the problems which arise in many practical control systems applications due to lack of detailed structural and parametric information about the system being controlled. This direct sensitivity technique can be applied in the frequency domain or in the time-domain using simple auxiliary filters which can be implemented using computer simulation methods.

The technique has been applied successfully, in the time domain, to the determination of sensitivity functions of an excitation control system of a synchronous machine [2,3]. That application was concerned principally with the use of parameter sensitivity information for the on-line optimisation of controller parameters. Tests carried out on a micro-machine system showed clearly that the method can give useful results even in the presence of measurement noise [2,3].

For on-line applications involving continuous systems the technique can be implemented using a small digital computer with analogue to digital conversion equipment and a real-time clock. The technique is also applicable to linear digital control systems.

Applications in which this approach to the sensitivity analysis of closed-loop systems appears especially appropriate include systems which are unstable under open-loop conditions and multivariable systems which show significant cross-coupling effects and provide difficulties in precise plant modelling. Such problems are encountered in the helicopter industry where augmentation system gains predicted in design studies often have to be altered significantly at the development flight test stage [6].

9. REFERENCES

[1] Tomovic, R., Sensitivity Analysis of Dynamic Systems (McGraw-Hill, New York, 1964).

[2] El-Shirbeeny, E.H.T., Murray-Smith, D.J. and Winning, D.J., Technique for direct evaluation of parameter sensitivity functions, Electronics Letters 10 (1974) 530-531.

[3] Winning, D.J., El-Shirbeeny, E.H.T., Thompson, E.C. and Murray-Smith, D.J., Sensitivity method for online optimisation of a synchronous-generator excitation controller, Proc. I.E.E. 124 (1977) 631-637.

[4] Rutman, R.S., The method of Three Points in Sensitivity Theory, in: Cruz, J.B. (ed.), System Sensitivity Analysis (Dowden, Hutchison and Ross, 1973) pp. 57-67.

[5] Blakelock, J.H., Automatic Control of Aircraft and Missiles (Wiley, New York, 1965).

[6] Chen, R.T.N. and Hindson, W.S., Analytical and flight investigation of the influence of rotor and other high-order dynamics on helicopter flight-control system bandwidth, Paper presented at the 1st Annual Forum of the International Conference on Basic Rotorcraft Research, Research Triangle Park, North Carolina, Feb. 19-21, 1985.

Complex and Distributed Systems: Analysis, Simulation and Control
S.G. Tzafestas and P. Borne (editors)
Elsevier Science Publishers B.V. (North-Holland)

CONTROL SYSTEM TREATMENT BY PROGRAM PACKAGE ANA

M.ŠEGA, S.STRMČNIK, R.KARBA*, D.MATKO*

University Edvard Kardelj
Institut "Jožef Stefan", Jamova 39
*Faculty of Electrical Engineering, Tržaška 25
61000 Ljubljana, Yugoslavia

The work contains a brief description of an interactive program package ANA which can be used as a tool for analysis and design of system control. It is designed for univariable or multivariable, continuous or discrete systems. It can be of great help in all steps of control design (transformations, analysis, synthesis). For more complex and possibly nonlinear systems another concept was developed. It enables simulation of control loop which can include also optimization and state estimation. Nonlinearities are introduced in the scheme through special FORTRAN block which has to be written by user for each special case. The usefulness of the package is shown especially through the great interactiveness of programming which enables fast and simple work for different users which are familiar with system theory. For communication the simplest dialogues are used (question and answer dialogue, menu-selection dialogue and dialogue using mnemonics) but they are not tedious and time consuming for experienced users because indirect running (BATCH) of the package is possible to repeat the whole design procedure with changed parameters or to repeat a part of the procedure and then proceed in direct way. The features of the package are illustrated through the example - modelling and control of semibatch distillation column.

1. INTRODUCTION

Over the last decade a large amount of work was devoted to the development of computer-aided-design tools because it is obvious that the variety of control methods and algorithms is so wide that an engineer could not master both areas i.e. the control theory and the necessary programming /1,2,3,4,5/. Nowadays so called interactive program package for computer aided design (CAD) represent the highest level of digital computer usage. So we developed interactive program package ANA for analysis and design of system control /6,7/. It is designed for univariable or multivariable, continuous or discrete systems. The fundamental principles for computer aided design software development were taken into account, namely the package modularity, flexibility, interacitveness and portability. It includes all necessary operations from control theory (transformations, analysis, synthesis) and appropriate I/O facilities. The usefulness of the package is shown especially through the great interactiveness. In our opinion the simplest man-machine dialogues in connection with BATCH processing are appropriate for differently skilled users as well as for educational purposes.

The aim of the work is to represent the usefulness of the package ANA. So the procedure of the concrete industrial plant treatment is given. For its understanding the shortest possible description of ANA capabilities is also included while more informations can be found in /8/.

2. SUMMARY OF ANA CAPABILITIES

2.1. Information structure

The information scheme is shown in Fig. 1. It consists of:
- logical input units for man-machine interaction,
- logical output units for representation of results and documentation,
- the main program for uniting all programs (it is called supervisor),
- programs for executing elementary operations (from system theory and I/O facilities),
- internal data base as virtual memory,
- protocol file as a base for BATCH processing.

2.2. Groups of programs for elementary operations

Internal data base generation. Programs

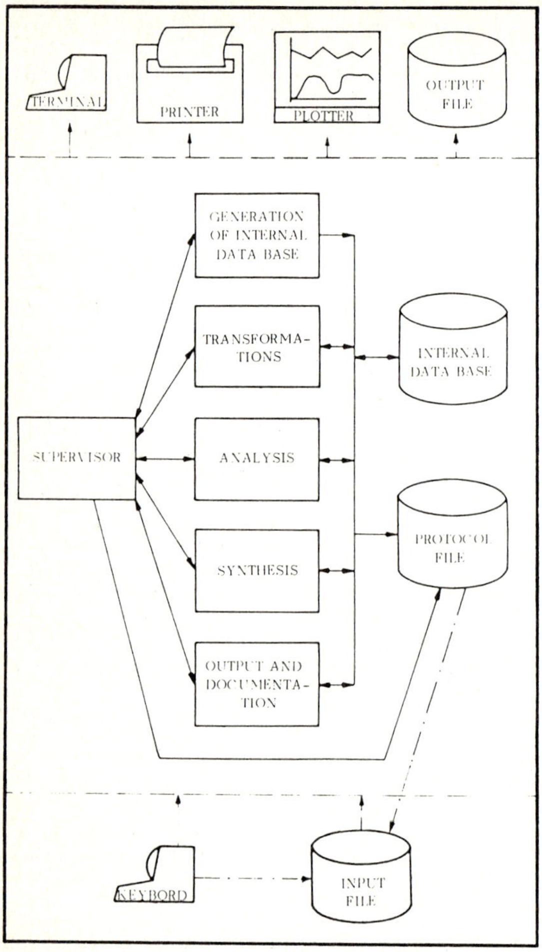

Fig.1. Information scheme of ANA

enable the creation of internal data base with initial data of treated systems. Systems can be given in one of the following forms (appropriately coded for man-machine dialogue):
- I/O data time series (TTI),
- transfer function matrix for continuous system (TGS),
- transfer function matrix for discrete system (TGZ),
- continuous system represented in state space (TMS),
- discrete system represented in state space (TMZ).

User written FORTRAN subroutine as a representation of the system (TBL), which can be also included, must be generated (edited) before the usage of the package.

Transformations. Programs enable the following conversions:
- simulation (TGZ → TTI, TMZ → TTI),
- identification (TTI → TMZ),
- discretization (TGS → TGZ, TMS → TMZ),
- from frequency domain to state space representation and vice versa (TGS ⇄ TMS, TGZ ⇄ TMZ),
- from discrete to continuous state space form (TMZ → TMS),
- modifications of forms (canonical state space forms (TMS → TMS, TMZ → TMZ), transformation between polynomial and factorized form of transfer function (TGS → TGS, TGZ → TGZ), filtering and smoothing of I/O data time series (TTI → TTI)).

Analysis. There are programs for analysis in:
- time domain (overshoot, settling time, time delay, rise time, damping factor),
- frequency domain (Bode and Nyquist plots and root locus diagram),
- state space domain (observability, controllability, ciclycity, rank and norms of matrices, eigenvalues and eigenvectors).

In the analysis block also the statistical comparison of two or more I/O data time series is included.

Synthesis. Programs in synthesis´ block enable:
- controller design ((optimal) PID, generalized deadbeat, optimal state controller),
- Kalman filter calculation,
- two ways of feedback structure treatment (the calculation of I/O state space equivalent for the linear, time invariant (sub)structure of the block diagram in Fig. 2; general feedback structure simulation with possibility of optimization (UNICUS) /9/).

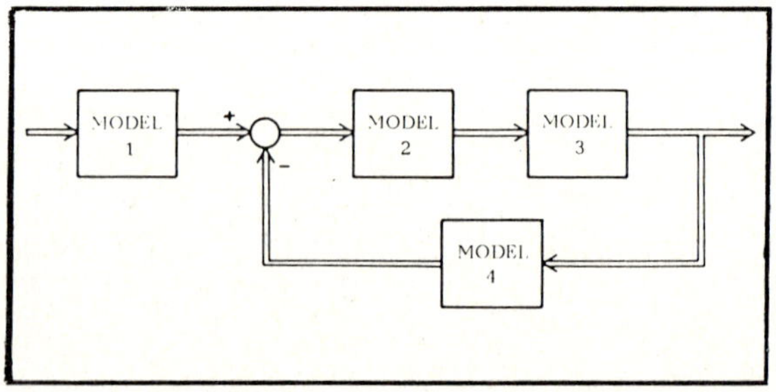

Fig.2. Block diagram of feedback structure for I/O state space equivalent calculation

Output facilities. Programs for presentation of numerical and graphical results on available output units.

2.3. Man-machine interaction

Direct. Through three kinds of dialogues (question and answer dialogue, menu selection dialogue and dialogue using mnemonics) where terminal´s keybord is logical input unit.

Batch. Indirect running of the package is possible to repeat the whole design procedure with changed parameters or to repeat a part of the procedure and then proceed in direct way. Namely by editing the protocol file we get the input file for indirect processing.

3. REAL PROBLEM TREATMENT BY ANA

For the separation of components on the basis of volatility, different kinds of distillation are used in chemical industry among which semibatch rectification is relatively poorly treated. For the description of such multivariable system the nonlinear and linearized model of semi-batch distillation for two component mixture (real pilot plant) in state space were developed which are suitable for the corresponding control design. As linearized model has relatively high order (eq.1) we tried to reduce it through the identification procedure in discrete space (black box method).

The model (eq.1) was successfully simplified to the fourth order (eq.2) what can be seen from Fig. 3 to Fig. 5.

$$A=\begin{bmatrix} -5.0E-4 & 1.8E-4 & 0.0 & 0.0 & 0.0 & 0.0 & 0.0 & 0.0 & 0.0 \\ 3.912E-2 & -6.194E-2 & 2.531E-2 & 0.0 & 0.0 & 0.0 & 0.0 & 0.0 & 0.0 \\ 0.0 & 4.156E-2 & -6.845E-2 & 3.156E-2 & 0.0 & 0.0 & 0.0 & 0.0 & 0.0 \\ 0.0 & 0.0 & 4.433E-2 & -7.8E-2 & 4.164E-2 & 0.0 & 0.0 & 0.0 & 0.0 \\ 0.0 & 0.0 & 0.0 & 4.75E-2 & -9.212E-2 & 5.614E-2 & 0.0 & 0.0 & 0.0 \\ 0.0 & 0.0 & 0.0 & 0.0 & 5.115E-2 & -0.112 & 7.38E-2 & 0.0 & 0.0 \\ 0.0 & 0.0 & 0.0 & 0.0 & 0.0 & 5.542E-2 & -0.135 & 9.224E-2 & 0.0 \\ 0.0 & 0.0 & 0.0 & 0.0 & 0.0 & 0.0 & 6.045E-2 & -0.161 & 0.0 \\ 0.0 & 0.0 & 0.0 & 0.0 & 0.0 & 0.0 & 0.0 & 5.32E-5 & -9.25E-5 \end{bmatrix} \quad B=\begin{bmatrix} 0.0 & -5.0E-6 \\ 0.131 & -8.941E-2 \\ 0.232 & -0.157 \\ 0.36 & -0.244 \\ 0.466 & -0.316 \\ 0.485 & -0.329 \\ 0.41 & -0.278 \\ 0.295 & -9.682E-2 \\ 4.96E-5 & -9.7E-4 \end{bmatrix}$$

$$C=\begin{bmatrix} 0 & 1 & 0 & 0 & 0 & 0 & 0 & 0 & 0 \\ 0 & 0 & 0 & 0 & 0 & 0 & 0 & 1 & 0 \end{bmatrix} \qquad D=\begin{bmatrix} 0 & 0 \\ 0 & 0 \end{bmatrix} \qquad (1)$$

$$A=\begin{bmatrix} -0.282 & 0.284 & 3.092E-3 & -4.482E-3 \\ -0.22 & 0.219 & -2.466E-3 & 3.675E-3 \\ 6.866E-2 & -7.465E-2 & -0.354 & 0.363 \\ -5.019E-2 & 5.465E-2 & -0.168 & 0.161 \end{bmatrix} \quad B=\begin{bmatrix} 0.131 & -8.923E-2 \\ 0.124 & -8.431E-2 \\ 0.294 & -9.771E-2 \\ 0.227 & -9.702E-2 \end{bmatrix} \qquad (2)$$

$$C=\begin{bmatrix} 1 & 0 & 0 & 0 \\ 0 & 0 & 1 & 0 \end{bmatrix} \qquad D=\begin{bmatrix} 0 & 0 \\ 0 & 0 \end{bmatrix}$$

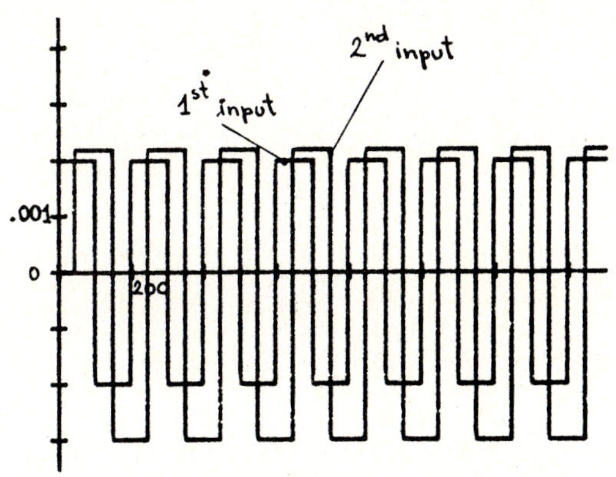

Fig.3. Input signals for the reducing procedure

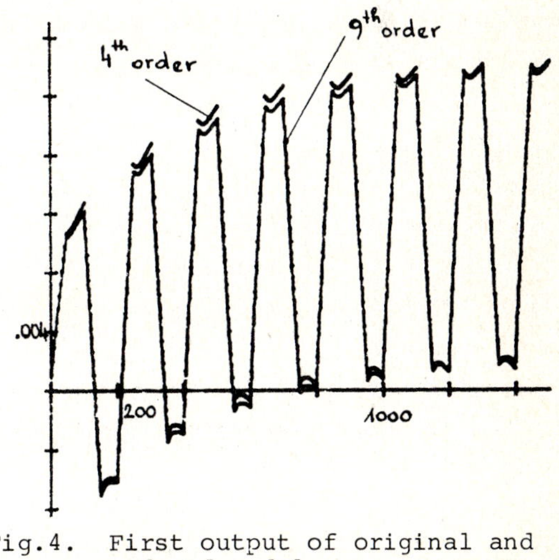

Fig.4. First output of original and reduced model

The obtained reduced model was then analysed and as an illustration a part of communication which represents a definition and results of controllability and observability test is given in Fig. 6. Though it is in slovene the main principles of man-machine communication can

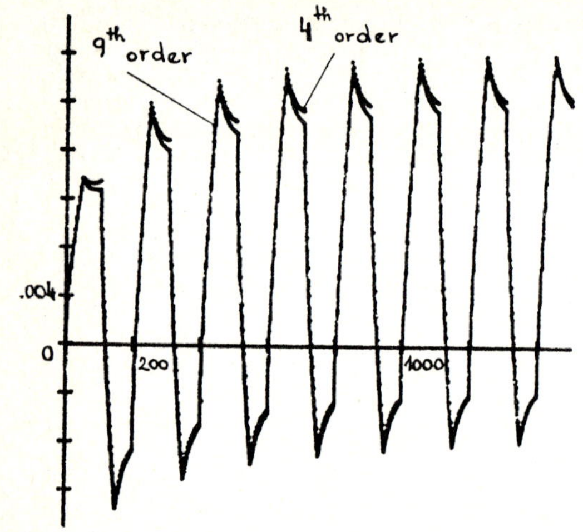

Fig.5. Second output of original and reduced model

be seen from the listing in Fig.6.

For the controller design well known and widely used pole assignment method was chosen. For the reduced model multivariable output feedback pole assignement PI controller was designed (eq.3) using program package POLASS which is also planned to be included in ANA.

P-part: I-part:

$$\begin{bmatrix} -0.9242 & 5.1189 \\ -0.2929 & -8.0053 \end{bmatrix} \qquad \begin{bmatrix} -5.8847 & 8.3357 \\ 0.3276 & -2.0862 \end{bmatrix} \qquad (3)$$

Designed controller which ensures good behaviour of closed loop with the reduced model was then tested also in closed loop with original linearized model using simulation facilities of UNICUS in ANA. Fig.7 shows interactively defined structure for the simulation where

```
ANA:   PA

ANALI: AMA

> koda zelene analize v prostoru stanj
  (ce kode ne poznas, pritisni tipko <CR>) :   AVOS

> obravnavani sistem je zvezen (Da ali Ne) :   D

> datoteka, ki jo zelis uporabiti
  (samo IME,ker je EXT default-TMS)
  (ce ne zelis nobene,pritisni tipko <CR>)

          MOD4
          KOL        : MOD4

> zelis celotno informacijo o vodljivosti in spoznavnosti
  (Da-izpis vodljivostne (Qv) in spoznavnostne (Qs) matrike ;
   Ne-le informacija o vodljivosti in spoznavnosti sistema) :   D

> pogresek za linearno ( ne ) odvisnost stolpcev
  vodljivostne oz. spoznavnostne matrike ( EPS )
  (EPS bi bil 1.E-6,ce ne bi zelel celotne informacije) :   1.0000000E-06


    SISTEM JE VODLJIV

Vodljivostni index ni vecji od :  4
Vodljivostno matriko sestavljajo vektorji :

     A ** nv(i) * b( 1)   i=0, 3


VODLJIVOSTNA MATRIKA Qv[ 4, 4]
-----------------------------

    1. stolpec        2. stolpec        3. stolpec        4. stolpec
  0.13083199     -0.18331483E-02   0.86918398E-04    0.22083581E-05
  0.12421600     -0.14812510E-02   0.86439921E-04   -0.18243487E-05
  0.29367000     -0.21563159E-01   0.30759337E-02   -0.49954106E-03
  0.22748800     -0.12491169E-01   0.16217826E-02   -0.25513687E-03

    SISTEM JE SPOZNAVEN

Spoznavnostni index ni vecji od :   3
Spoznavnostno matriko sestavljajo vektorji :

                T
     (c( 1)) * A ** ns(i)  i=0, 3


SPOZNAVNOSTNA MATRIKA Qs[ 4, 4]
-------------------------------

    1. stolpec        2. stolpec        3. stolpec        4. stolpec
   1.0000000     -0.28232199        0.17655530E-01   -0.12107365E-02
  0.00000000      0.28349999       -0.18334280E-01    0.12749634E-02
  0.00000000      0.30920999E-02   -0.19132263E-02    0.32173031E-03
  0.00000000     -0.44823997E-02    0.27090587E-02   -0.40548060E-03

ANALI: EX
```

Fig.6. The part of communication (in slovene)

the block PR is the distillation column model, block M1 contains the P-part of controller and block R1 is controller´s I-part. Block M2 which is user written subprogram (TBL) is included to obtain results of closed-loop simulation as well as model simulation (switch).

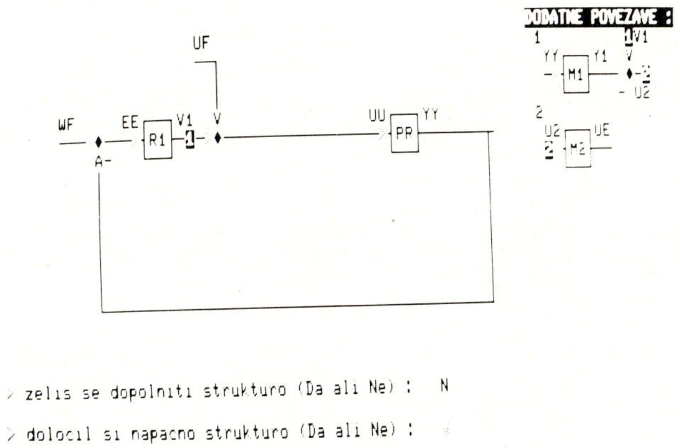

Fig.7. Interactively defined structure for the simulation (photo of terminal´s ecran)

Results of closed loop simulation are shown in Fig.8 and Fig.9. They exhibit satisfactory behaviour of the controlled loop in the case of disturbances and initial conditions on the model states.

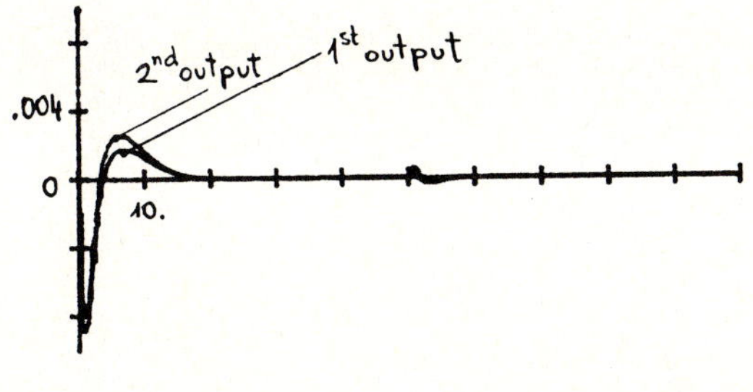

Fig.8. Output time responses of closed loop system

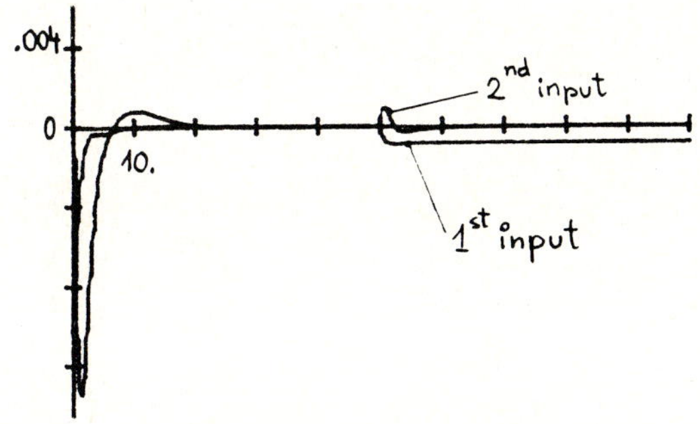

Fig.9. Control inputs

4. CONCLUSION

The experiences with the package ANA show that its concept and program organization ensure simple extensions and modifications. Batch running enables comfortable work also for experienced users in spite of the simple dialogues which are on the other hand suitable for unexperienced users.

In the near future some extensions are foreseen like introducing digital simulation language what will result in ANA´s greater capabilities in the sense of nonlinear systems treatment. We shall also include some other approaches to the controller design which have been already developed and programmed in the past (like POLASS).

Discussed distillation column was previously treated using different approaches to controller design and different program equipment (analogue simulation, digital simulation with CSSL-3 language and simulation on the hybrid computer system combined with controller design on digital computer). So it was used as testing example for program package ANA. We can establish that ANA gives expected results in significantly shorter time than using mentioned tools. Similar conclusions can be made also on the basis of other concrete problems which were studied till now.

REFERENCES

1. Federick, D.K. (1982). Software summaries on computer aided control system design software packages. Control Systems Magazine, Vol. 2., No. 4, 37-44.
2. Aström, K.J. (1983). Computer aided analysis and design of control system-a perspective. Control Systems Magazine, Vol. 3, No. 2, 4.16.
3. Emani - Naeini, A., and G.F. Franklin (1981). Interactive computer - aided design of control systems. Control Systems Magazine, Vol. 1, No. 4, 31-36.
4. Bingulac, S., and M.A. Farias (1977). L-A-S language and its use in automatic control eductation and research. Automatika, Vol. 5-6, 213-224.
5. Wieslander, J. (1979). Design principles for computer aided design software. Proc. IFAC Zürich Symp., 493-496.
6. Šega, M., S.Strmčnik and J.Čretnik (1984). Analysis and synthesis of systems with program package ANA-1. Proc. ETAN Split Yugoslavia, Vol.4, 35-42 (in slovene).

7. Šega M. (1984). Computer aided analysis and control design of dynamical systems. Faculty of Electrical Engineering Ljubljana, Yugoslavia (work for M.S. Degree in slovene).
8. Šega, M., S.Strmčnik, R.Karba and D.Matko (1985). Interactive program package ANA for system analydis and control design. Proc. 3rd IFAC Symp. CADCE´85 in Copenhagen, 145-150.
9. Matko, D., B.Zupančič and M.Šega (1985). UNICUS-A new approach to the simulation and design of control systems, this volume.

Complex and Distributed Systems: Analysis, Simulation and Control
S.G. Tzafestas and P. Borne (editors)
Elsevier Science Publishers B.V. (North-Holland)

SYNTHESIS AND SIMULATION OF OPTIMAL CONTROL SYSTEMS FOR TURNABLE TELESCOPES

E. HASENJÄGER

Institute of Mechanics and Control Engineering
University of Siegen, Paul-Bonatz-Str., D-5900 Siegen, F.R.G.

Increasing requirements on telescope driving systems demand improved control strategies. The objective of this contribution is to present the synthesis and the hybrid simulation of a multivariable control system for turnable telescopes. The proposed control scheme serves for the active damping of the vibrational modes of the plant and considers external signals as well as nonlinearities like friction and backlash. Hybrid real-time simulation is a substantial step between the design of the control system and its implementation on the real plant. The simulation results yield significant reductions of the tracking errors and considerable improvements of the closed-loop behaviour in comparison with conventional cascade controllers.

1. INTRODUCTION

The mechanical structures and the drive units of turnable radio-telescopes and of optical telescopes can be modelled as multi-body systems with elastic couplings and non-linearities like friction and backlash. Similar models are obtained by analyzing the dynamics of machine tools and industrial robots. To ensure improved position control performances for these systems, the application of state space methods, considering external signals and non-linear effects, seems to be very promising. In the following, a digitally realized state control system will be presented, which accomplishes the given control tasks optimally. These tasks are:

- active damping of the dominant vibrating modes,
- consideration of non-dominant dynamics,
- accommodation of servocommand and disturbance signals,
- compensation of effects caused by friction and backlash,
- reconstruction of non-measurable state variables.

Whereas conventional cascade controllers, in many cases, do not sufficiently meet the requirements on the control performances, state feedback control systems are more suitable to the mentioned types of driving systems. In addition, digital computer control allows more complex control schemes, like structure-variable strategies for non-linear systems. Hybrid real-time simulations of the entire control system give a good insight into the closed-loop properties and permit extensive tests of the digital control program under different operational modes. It should be mentioned that simulation studies of telescope control systems are established as essential investigation steps at the different phases of a realization project.

The first part of the paper describes the synthesis of the control system for a 20 meter radio-telescope in rough outlines. The second part deals with the digital realization on a microcomputer and presents the entire configuration for hybrid real-time simulations. These simulation results demonstrate significant improvements of the position control performance, compared with the cascade controller.

2. NON-LINEAR TELESCOPE MODELS

Detailed finite-element calculations for the telescope mechanics, followed by numerical and physical model simplifications, lead to linear vibrational systems with rotatory and translatory degrees of freedom /1/. A main result of these investigations is, that the two motional axis (azimuth and elevation) can be considered as independent. Defining the n-dimensional vector $\underline{q}(t)$ of the coordinates $q_i(t)$, $i = 1,..n$, the differential equations of motion for each of the two axis can be written as

$$J\ddot{\underline{q}}(t) + D\dot{\underline{q}}(t) + S\underline{q}(t) = E\underline{u}(t) + F\underline{v}(t) \qquad (1)$$

with the matrices of mass, damping and stiffness J, D and S. The input matrices E and F correspond to the control input signals $\underline{u}(t)$ (motor torques) and to the disturbance signals $\underline{v}(t)$ (wind torques and forces). The derived telescope models typically are described by n = 5 to 10 degrees of freedom. Figure 1 shows the sketch of the radio-telescope mentioned above with the according linear mechanical models of the two telescope axis. These linear models are added by non-linear friction models for the reflector bearings and for the drive systems. They are also added by backlash functions located in the gear trains (subsystem 1 in Fig.1).

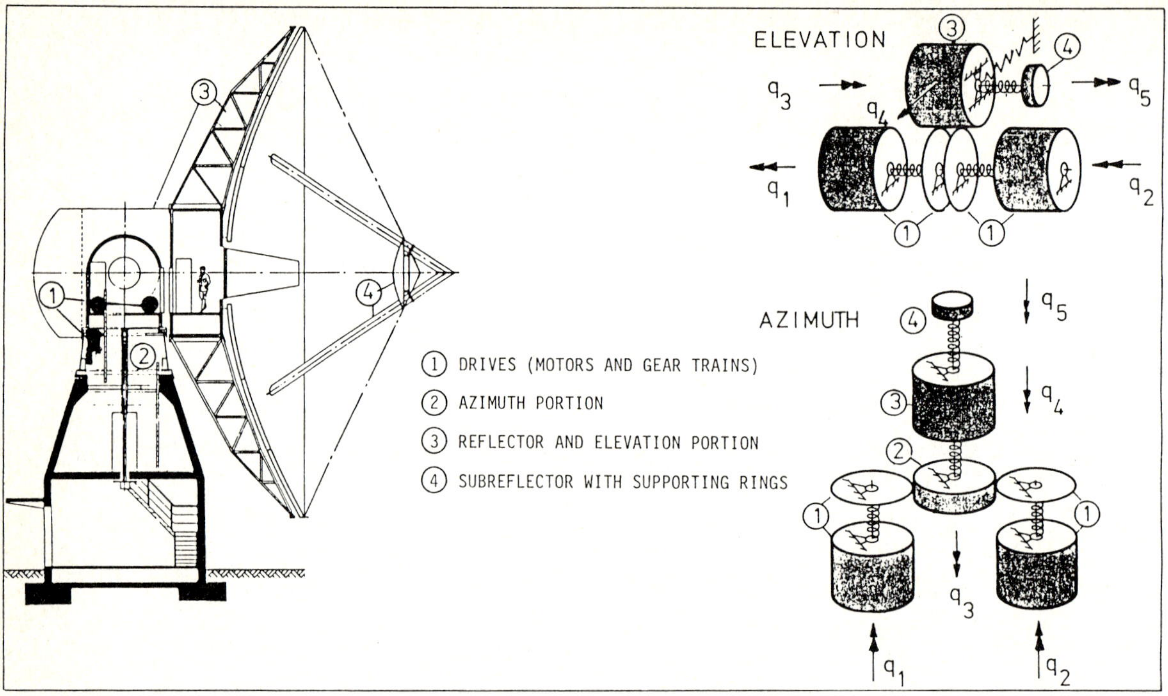

Figure 1: Sketch and mechanical models for the two axis of a 20 meter radio-telescope

The state space representation of Eq.(1) is given with

$$\dot{\underline{x}}(t) = A\underline{x}(t) + \underline{f}(\underline{x},t) + B\underline{u}(t) + B_v\underline{v}(t)$$
$$\underline{y}(t) = C\underline{x}(t) \tag{2}$$

where $\underline{x}^T = [\underline{q}^T, \dot{\underline{q}}^T]$ is the 2n-dimensional state vector and $\underline{y}$ the m-dimensional vector of output signals. Regarding Eq.(1), the system matrices in Eq.(2) are

$$A = \begin{bmatrix} 0 & I_n \\ -J^{-1}S & -J^{-1}D \end{bmatrix}, \quad B = \begin{bmatrix} 0 \\ J^{-1}E \end{bmatrix}, \quad B_v = \begin{bmatrix} 0 \\ J^{-1}F \end{bmatrix}. \tag{3}$$

The vector $\underline{f}$ is made up of all separated non-linearities, and therefore with $\underline{f} = \underline{0}$, Eq.(2) describes the linear system behaviour. The non-zero elements of $\underline{f}$ are for both of the two axis:

$$f_{n+i} = \frac{1}{J_i}\,[c_i \mathrm{sat}_{bi}(\Delta q_i) - m_i \mathrm{sgn}(\dot{q}_i)]$$
$$f_{n+3} = -\frac{1}{J_3}\{(\Sigma i\; c_i \mathrm{sat}_{bi}(\Delta q_i)) + m_3 \mathrm{sgn}(\dot{q}_3)\} \tag{4}$$

with $\Delta q_i = q_i - q_3$ for $i = 1,2$.

In Eq.(4) the parameters J_i, c_i, b_i and m_i are the moments of inertia, spring rigidities, backlash and Coulomb friction rates. Beside the modelling of the mechanical dynamics, the time lag T_S of the servo amplifiers and the time delay T_0 of the digital controller have to be taken into account.

Assuming the controller output $\underline{u}_S$, the actual input signal $\underline{u}$ for the telescope can be modelled with

$$\dot{\underline{u}}(t) = \frac{1}{T_s}[\underline{u}_s(t-T_0) - \underline{u}(t)] \ . \tag{5}$$

Furthermore, the disturbance signals $\underline{v}(t)$ as well as the servocommand signals $\underline{w}(t)$ are modelled as deterministic state equations. The disturbance signal model is given by

$$\dot{\underline{z}}(t) = M_z\underline{z}(t) \ , \quad \underline{v}(t) = H_z\underline{z}(t) \tag{6}$$

and the servocommand signal model is given by

$$\dot{\underline{r}}(t) = M_r\underline{r}(t) \ , \quad \underline{w}(t) = H_r\underline{r}(t) \ . \tag{7}$$

The linear telescope models of Eq.(1) describe n rotatory and translatory degrees of freedom with different relevance to the global system behaviour. Therefore, in order to restrict the controller design to the active damping of the dominant (and controllable) vibrating modes, modal reduction procedures are applied to the linear systems (Eq.(2) with $\underline{f} = \underline{0}$) /2/. Analyzing the dominance of modes, n_c essential modes and therewith a corresponding number of essential state variables can be defined:

$$\underline{x}_c(t) = C_c\underline{x}(t) \ . \tag{8}$$

The state vector $\underline{x}_c$ consists of relevant and measurable process signals and of those state variables which determine the non-linear function $\underline{f}$ in Eq.(4).

The reduced order system then can be formulated as

$$\dot{\underline{x}}_c(t) = \tilde{A}\underline{x}_c(t) + \tilde{\underline{f}}(\underline{x}_c,t) + \tilde{B}\underline{u}(t) + \tilde{B}_v\underline{v}(t) \; . \qquad (9)$$

Hence, the synthesis of the suggested control system, outlined in the following, bases on the Eqs. (5),(6),(7) and (9).

3. SYNTHESIS OF THE CONTROL SYSTEM

Refering to the controller tasks (stated in chapter 1), the first step of the design procedure is to determine the optimal feedback law

$$\underline{u}_s(t) = -K\underline{x}_c(t) \; . \qquad (10)$$

The feedback matrix K is obtained by quadratic optimization techniques, here using a modal weighting matrix Q_1 for the integral performance index

$$I = \int_0^\infty (\underline{x}_c^T Q_1 \underline{x}_c + \underline{u}_s^T Q_2 \underline{u}_s)dt \rightarrow \min! \qquad (11)$$

In a further step, the criterion (11) is used in direct (digital or hybrid) optimization runs in order to adapt matrix K to the time constants T_S and T_0 in Eq.(5). The resulting matrix K^* also avoids spillover effects: It must be guaranteed that the control input $\underline{u}$ does not destabilize the closed loop by excitation of the $n-n_c$ non-dominant (and in Eq.(9) neglected) modes /3/. In order to consider external signals $\underline{v}(t)$ and $\underline{w}(t)$ Eq.(10) has to be completed by feeding forward the states $\underline{z}(t)$ and $\underline{r}(t)$:

$$\underline{u}_s(t) = -K^*\underline{x}_c(t) + K_z\underline{z}(t) + K_r\underline{r}(t) \; . \qquad (12)$$

If we define an error state vector

$$\underline{e}_R(t) = \underline{x}_R(t) - \underline{x}_c(t) \qquad (13)$$

which describes the difference between the (dominant) system states and the rigid body reference states

$$\underline{x}_R(t) = R\underline{r}(t) \; , \qquad (14)$$

analytical calculations, given in /4/,/5/, lead to a modification of Eq.(12):

$$\underline{u}_s(t) = K^*[\underline{x}_R(t) + \underline{x}_L(t) - \underline{x}_c(t)] + \underline{u}_L(t) \; . \qquad (15)$$

In Eq.(15), two reference states are established: the rigid body reference $\underline{x}_R(t)$ and the load-depending reference deformation of the mechanical structure $\underline{x}_L = \underline{x}_L(\underline{z},\underline{r})$. Wind loads and inertial torques are compensated by feeding forward the signals $\underline{u}_L = \underline{u}_L(\underline{z},\underline{r})$. These signals, together with the reference states in Eq.(15), ensure the exact positioning of the reflector. Introducing reference states is also very useful in order to accommodate backlash effects.

Since the rate of the backlash b is of the same size as the admissable pointing errors or even higher, the closed linkage of the pinions has to be realized by counter-distortion of the motor shafts. Whereas conventional controllers use distortion torques, in the approach of this paper appropriate motor position references are generated. The backlash is compensated by the reference state $\underline{x}_B$, and additional distortion torques $\underline{t}_D$ (for maintaining the closed linkage) can be realized by an additional reference $\underline{x}_D = \underline{x}_D(\underline{t}_D, K^*)$. Thus, finally the resulting control law for the non-linear system can be formulated as

$$\underline{u}_s(t) = K^*[\underline{x}_{REF}(\underline{z},\underline{r},\underline{f}) - \underline{x}_c(t)] + \underline{u}_L(\underline{z},\underline{r}) \qquad (16)$$

with the non-linear reference state

$$\underline{x}_{REF}(\underline{z},\underline{r},\underline{f}) = \underline{x}_R(\underline{r}) + \underline{x}_L(\underline{z},\underline{r}) + \underline{x}_B(\underline{f}) + \underline{x}_D(\underline{f},K^*) \qquad (17)$$

- $\underline{x}_D$: distortion
- $\underline{x}_B$: backlash compensation
- $\underline{x}_L$: load-depending deformation reference
- $\underline{x}_R$: rigid-body motion reference

In cases of high load torque demands, the concept of the reference state generation, together with a structure-variable feedback matrix K^*, provides state-controlled and load-depending changes of pinion meshes. Finally, it should be mentioned that the unknown disturbance state vector $\underline{z}$ and the non-measurable elements of $\underline{x}_c$ are reconstructed by a combined system state and disturbance state observer of reduced order. As unknown non-linear friction torques are interpreted as external disturbances, they can be reconstructed by the mentioned observer. A servo-command signal observer estimates non-measurable components of $\underline{r}(t)$, if, for example at auto-tracking mode, the reference speeds and acceleration are not available directly. The entire dynamic system of all mentioned observers shall be formulated in the following form:

$$\dot{\hat{\underline{\psi}}}(t) = F\hat{\underline{\psi}}(t) + G\underline{y}(t) + B\underline{u}_s(t) + H\underline{w}(t) \; . \qquad (18)$$

The application of the linear observer theory to the given non-linear system can be realized by introducing a transformed state $\bar{\underline{x}}(t)$, considering the elements of the nonlinear reference state $\underline{x}_{REF}$ in Eq.(17). Summarizing the resulting non-linear state controller system, it consists of the following components:

- state and disturbance signal observer,
- servocommand signal observer,
- non-linear reference state generation,
- load torque feedforward,
- variable-structure feedback.

4. HYBRID REAL-TIME SIMULATIONS

Regarding the structure of the proposed control scheme, it is clear that only digital computers can be suggested for the realization. Transparent and userfriendly programming assumed, digital control programs are more flexible than analog circuits and they are of high advantage for the implementation on varying types of telescopes. For feasibility studies and as a preliminary step toward the implementation of the state control system on real plants, hybrid real-time simulations of the entire control systems are indispensable. As the given systems and evaluated controllers are non-linear, time-discretization by z-transforms are not applicable. Therefore, refering to Eq.(16), a quasi-continuous form of the control law has been chosen here:

$$\underline{u}_s^{k+1} = K*[\underline{x}_{REF}^k(\underline{z}^k,\underline{r}^k,\underline{f}^k)-\underline{x}_c^k] + \underline{u}_L^k(\underline{z}^k,\underline{r}^k) \ . \qquad (19)$$

Eq.(19) consists of time series at the discrete times $t_k = t_0 + kT$, $k = 0,1,2,..$, where T is the sampling period. The observer equations (18) are integrated by numerical integrations methods. As a good compromise between effort of computation time and numerical stability the ADAMS-BASHFORTH algorithm of 2nd order has been chosen:

$$\hat{\underline{\psi}}^{k+1} = \hat{\underline{\psi}}^k + \frac{T}{2}[3\underline{g}^k - \underline{g}^{k-1}] \qquad (20)$$

with $\underline{g}^k = F\hat{\underline{\psi}}^k + G\underline{y}^k + B\underline{u}_s^k + H\underline{w}^k$.

The entire state controller and observer system is realized as a modular assembler program, using process-oriented macros on a PDP 11/23 microcomputer. Different operational modes are provided in the program for a step-wise implementation and for the testing of the control program. The program requires 24 kByte memory and a sampling time of 5 ms for each of the two telescope axis. The digital computer is interfaced with an analog computer EAI 2000, on which the non-linear telescope models are patched. Furthermore the servo amplifier dynamics and the encoder and tacho characteristics are programmed on the analog computer as well as disturbance signal and servocommand signal generators. Figure 2 illustrates the hybrid simulation configuration. As analog computers operate within the signal range of +/- 10 volts, it is advisable to simulate only the error state $\underline{e}_R(t) = \underline{x}_R(t) - \underline{x}(t)$ similar to Eq.(13). Regarding Eqs.(2),(7),(14), the simulated error system results as:

$$\dot{\underline{e}}_R(t) = A\underline{e}_R(t) - \underline{f}(\underline{e}_R) - B\underline{u}(t) - B_v\underline{v}(t) + (RM_r - AR)\underline{r}(t) . \quad (21)$$

Figure 3 shows two examples of hybrid real-time simulations. They base on data of the mentioned 20 meter radio-telescope. This plant, at the present, is selected for the implementation of the controller software. On the left side of the figure a 30 arcsec step in the reference position is simulated. The diagram shows the test signal, the actual reflector position, the motor positions and the input signal for one servo amplifier.

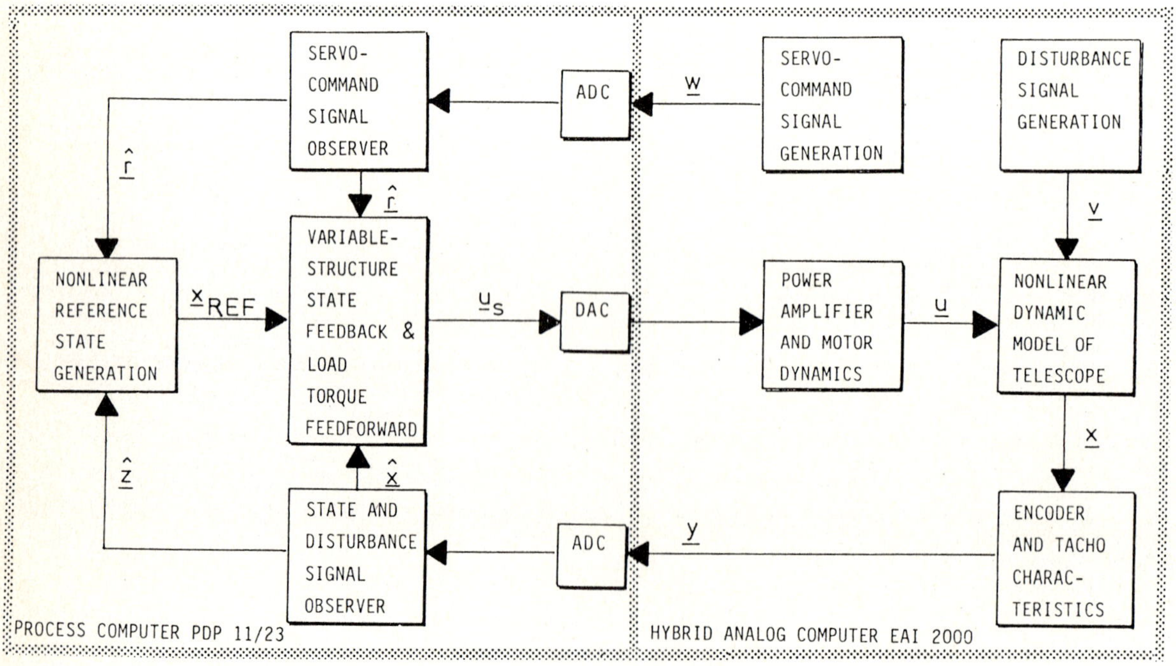

Figure 2: Signal flow diagram of hybrid real-time control system simulation

In the second simulation (right part of the figure) a transient wind load is simulated. Both simulations compare the bevaviour of the digital state controller (SC), described here, with an optimized analog cascade controller (CC). Beside the significantly improved dynamic behaviour of the closed-loop system, attention should directed to the exact reconstruction of the wind load and to the compensation of the backlash by distorting the motor shafts.

5. CONCLUSIONS

With the elaborated control system for turnable telescopes, it is possible to improve the position control performance significantly. The main reasons for that fact are (i) the feedback of essential state variables and (ii) the dynamic compensation of external disturbances and servo-command signals, using deterministic signal models and signal state observers. Linear state space methods have been extended to typical non-linearities of driving systems by generating non-linear reference states. The hybrid real-time simulation is a highly valuable and indispensable tool for the development and the implementation of new digital control schemes.

The modular structure of the presented concept permits its transferability to other vibrating drive systems. Current investgations are concerned with the application to optical telescopes and to industrial robots.

REFERENCES

/1/ Mäder,H.F., G. Henning and H.W. Kiedrowski (1981): Vibrations of turning-head-antennas around the azimuth axis (in German). Tech. Mitt. Krupp, Forsch.-Ber., Vol.36, pp.47-64.

/2/ Litz, L. (1979): Reduction of the order of linear state space models retaining the dominant modes (in German). Regelungstechnik, Vol.27, pp. 80-86.

/3/ Gülich, H. (1984): The avoidance of spill-over at the control of vibrating systems (in German). 2. Symposium Simulationstechnik, Wien. Berlin, Heidelberg, New York, pp.666.

/4/ Hasenjäger, E. (1981): Optimal state controller design for tracking antenna driving systems. Control and Computers, Vol.9, pp.30-35.

/5/ Hasenjäger, E. (1985): Digital state control of parabolic antennas with consideration of non-linearities (in German). Fortschr.-Ber. VDI Reihe 8 Nr.87, Düsseldorf.

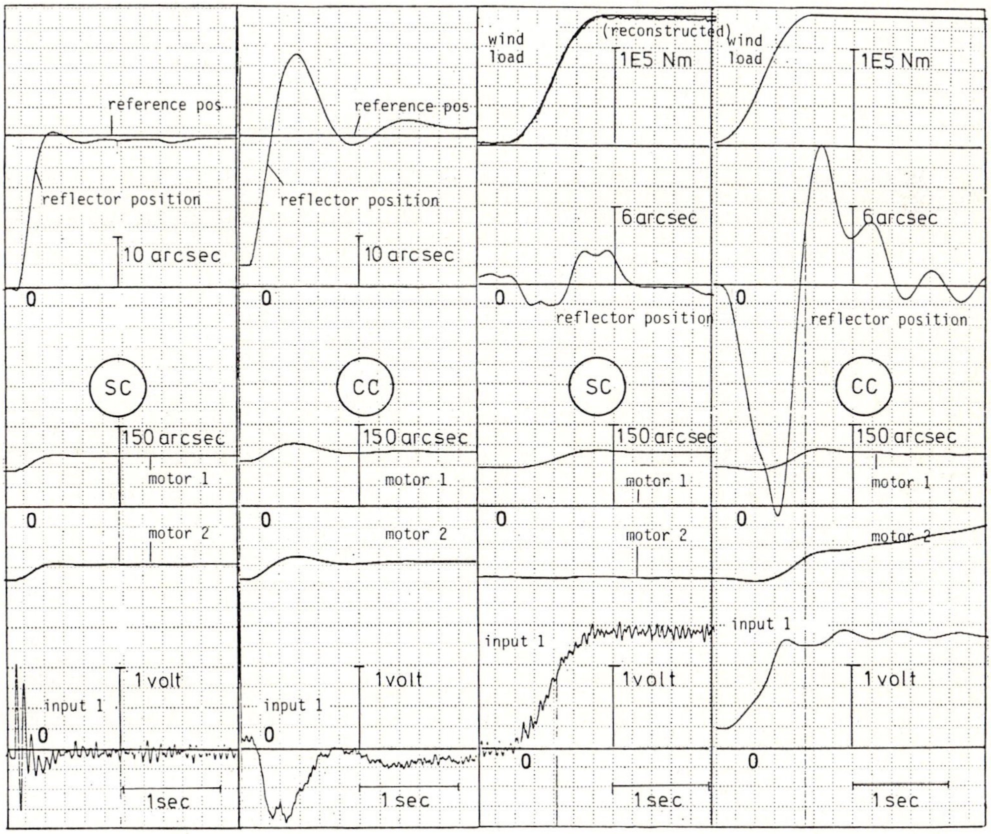

Figure 3: Hybrid real-time simulations of a 20 meter radio-telescope (azimuth axis)

Complex and Distributed Systems: Analysis, Simulation and Control
S.G. Tzafestas and P. Borne (editors)
Elsevier Science Publishers B.V. (North-Holland)

DESIGN OF OPTIMAL CONSTRAINED CONTROLLERS FOR LINEAR SYSTEMS

Salem A.K. Al-Assadi*, Rashad M. Al-Ansari**, Ausama A. Al-Ani**

*Electrical Engineering Department, University of Baghdad, Iraq
**Computer department, Scientific Documentation Centre, Baghdad, Iraq, P.O.B. 2441

A computer-aided method to design optimal constrained controllers for single-variable time invariant linear control systems is presented in this paper.
The design procedure was implemented on a digital computer using an automatic method based on time domain optimization technique utilized with constraints. This method uses certain search technique to locate optimum values for the constrained parameters of different types of controllers, in order to improve the performance of the closed loop control system and to meet certain required specifications. The measure of goodness of optimality is taken as the minimum value of the Integral Squared Error (ISE) formulated in the time domain.
Examples are taken to illustrate the usefulness and efficiency of the mechanism of this method using off-line computer control design. Moreover, this paper includes a comparison of the results obtained by this method with those obtained by other methods in literature.

1. INTRODUCTION

The problem of designing optimal controllers has been stuided by several researchers (1,2,3), some of them have utilized the classical methods based on root locus and frequency response techniques(1). The main disadvantage of these methods is the tedius trail and error procedure, specially for complicated systems and when there are some phisical limitations on the controller parameters. Other researchers have made use of the numerical methods to solve this problem such as the method of inequalities used by Zakian and Al-Naib(2), which satisfies certain constraints on the controller parameters with some limitations on the dynamic system performance. Moreover this method is unpreferable due to its complexity. Some other designers have applied certain optimization techniques, such as Bach(3) who aimed to satisfy certain system performance by obtaining the values of the controller parameters which are not optimal through the minimization of the Integral Time Squared Error (ITSE) criterion. The proposed method is to use a time domain optimization technique(4) with certain search strategy to satisfy the predefined constraints on the controller parameters and to give at the same time an optimal system performance.

The Integral Squared Error (ISE) criterion is minimized to achieve the optimal system performance, taking into consideration the reduction of the dynamic specifications such as steady-state error, rise time, settling time, overshoot,....,etc. to the minimum possible values. To illustrate the efficiency of this technique, specific control problems in single-variable systems, are considered and designs obtained by the proposed method are compared with those obtained by other methods.

2. FORMULATION OF THE PROBLEM

Consider a block diagram for the standard form of a control system shown in figure (1).

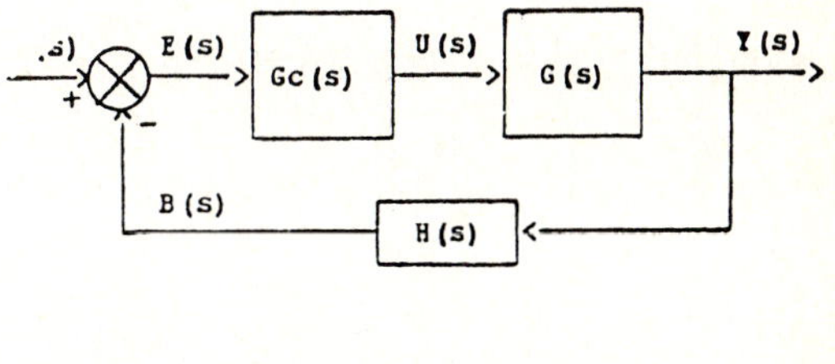

Figure (1) control system block diagram.

where,

- $Gc(s)$ represent the controller transfer function with certain structure.
- $G(s)$ represent the plant transfer function.
- $H(s)$ represent the feedback transfer function.

For control systems such as electrical networks and other analogous systems the design of the controller is subjected to some physical limitations. The design limitations imposed on controller parameters may appear as a set of inequalities as follows:

$$L_i \leqslant P_i \leqslant U_i \quad ; \quad i=1,2,\ldots n \qquad (1)$$

where,
P is the controller parameters vector.
L and U are the vectors of the lower and upper limits of the constraints on P.

It is required to determine optimal values for the elements of the vector P in order to achieve an optimal system response $y(t)$, satisfying the dynamic specifications formulated in the time domain such as steady-state error, rise time, settling time, and overshoot which can be represented as follows (3):

$$D_i(y) \leqslant W_i \quad ; \quad i=1,2\ldots m \qquad (2)$$

where,
D is the dynamic specification vector.
W is the constraints vector on D.
For unit step input i.e. ($R(s) = 1/s$), the steady-state error $D1(y)$ is defined by :

$$D1(y) = 1 - y(\infty) \qquad (3)$$

$y(\infty)$ can be computed by the well known relation :

$$y(\infty) = \lim_{s \to 0} sY(s) \qquad (4)$$

The rise time $D2(y)$ is the least value of t such that $y(t)=0.9y(\infty)$.
The settling time $D3(y)$ is the least value of t1 such that :

$$|y(\infty) - y(t)| < 0.02|y(\infty)| \qquad (5)$$
for all $t \geqslant t1$

The overshoot $D4(y)$ is defined by :

$$D4(y) = (\hat{y} - |y(\infty)|) / |y(\infty)| \quad \text{if } \hat{y} > |y(\infty)| \qquad (6)$$
$$= 0 \quad \text{if } \hat{y} \leqslant |y(\infty)|$$

where,

$$\hat{y} = \max_{0 \leqslant t \leqslant 2D3(y)} |y(t)| \qquad (7)$$

A time domain optimization technique is utilized to find the optimal controller parameters vector P such that equation (1) is satisfied and the dynamic specifications given in equation (2) have to be reduced to give an optimal system performance. This is achieved by minimizing the value of the Integral Squared Error (ISE) which is formulated in the time domain as follows:

$$Q = \int_0^{tf} [r(t) - y(t)]^2 \cdot dt \qquad (8)$$

where,
tf is the final time required by the system to reach steady- state value.

During this time domain searching technique, in each success cycle the constrains vector W is replaced by the new dynamic specifications vector D which is less than the old one for a success path. This will continue until the optimal values are reached.
The minimization procedure for the technique adopted utilizes both Rosenbrock's automatic hill-climbing method which optimzes the direction of search using Gram-Schmidt procedure, and Fibanacci search method for step size optimization. The reader can refer to references (4,5,6,7) for more details of the algorithm and different applications of this technique to control problems.

3. ILLUSTRATIVE EXAMPLES

3.1 Example (1)

Consider the control system (2) shown in figure (1) with :

$$G(s) = 10/s(s+1)(s+5) \qquad (9)$$
$$R(s) = 1/s \qquad (10)$$
$$H(s) = 1 \qquad (11)$$
$$Gc(s) = p1(1+p2s)/(1+p2p3s) \qquad (12)$$

The upper and lower limits on the controller parameters Pi for (i=1,2,3), are given as follows:

$$\begin{aligned} U1 &= p1 \leqslant 100 \\ L1 &= -p1 \leqslant -0.01 \\ U2 &= p2 \leqslant 20 \\ L2 &= -p2 \leqslant 0 \\ U3 &= p3 \leqslant 10 \\ L3 &= -p3 \leqslant -.01 \end{aligned} \qquad (13)$$

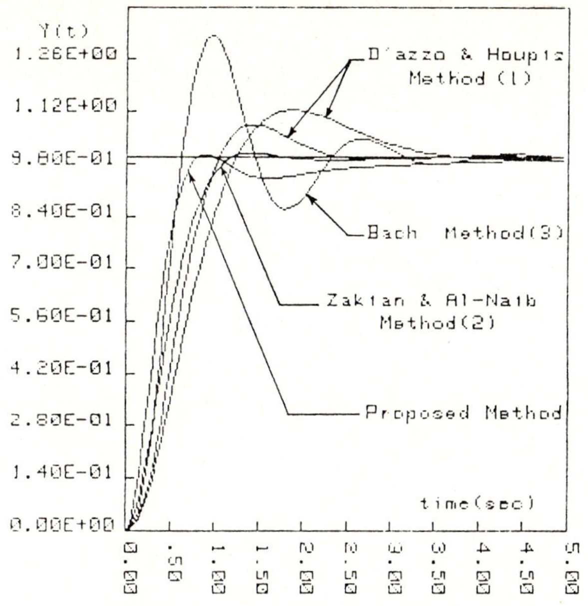

Fig.(2) Comparison of unit step responses for example(1).

The dynamic specifications have to be minimized and they are subjected to the following constraints:

$$D2(y) \leq 1$$
$$D4(y) \leq 0.01 \qquad (14)$$

For the purpose of comparison,table(1) illustrate different desings obtained by different methods,with the corresponding unit step responses shown in fig(2).It is clear from this figure and the table that the best controller was computed by the proposed method in terms of minimum values of (ISE) and more improvement in the required dynamic specifications of the controlled system.

3.2 Example(2)

The block diagram shown in figure (3) describes the components of a sun-seeker control system (8).The system may be mounted on a space vehicl so that it will track the sun with high accuracy.The variable θr represent the reference angle of the solar ray,and θo denotes the vehicle axis.The parameters of the system are given as follows:

Rf=10000 , Kb=0.0125 V/rad/sec
Ki=0.0125 N-m/A , Ra=6.25
$J=10^{-6}$ $Kg\text{-}m^{2}$, Ks=0.1 A/rad
B=0 , n=800 , K=1

The series controller Gc(s) of a phase-lead type is considered to be designed with optimum parameters values represented by the following transfer function:

$$Gc(s) = \frac{p1(s + 1/p1p2)}{(s + 1/p2)} \qquad (15)$$

The upper and lower limits on the controller parameters Pi for (i=1,2) are given as follows:

$$\begin{aligned} U1 &= p1 \leq 15 \\ L1 &= p1 \geq 1.0 \\ U2 &= p2 < 0.133 \\ L2 &= p2 \geq 0.0 \end{aligned} \qquad (16)$$

Table (1) Results for example (1)

Method	The controller parameters			The dynamic specifications		Integral Squared Error
	p1	p2	p3	D2	D4	
Dazzo & Houpis (1) (Classical Method)	1.0	1.0	0.1	0.9	0.0849	4.480×10^{-1}
	1.0	0.666	0.1	1.07	0.1238	5.184×10^{-1}
Bach (3)	1.7	1.2	0.12	0.59	0.3230	3.907×10^{-1}
Zakian & Al-Naib (2)	0.9970	1.1190	0.0125	0.92	0.0091	3.635×10^{-1}
Proposed	1.0526	1.7383	0.0010	0.61	0.0049	2.719×10^{-1}

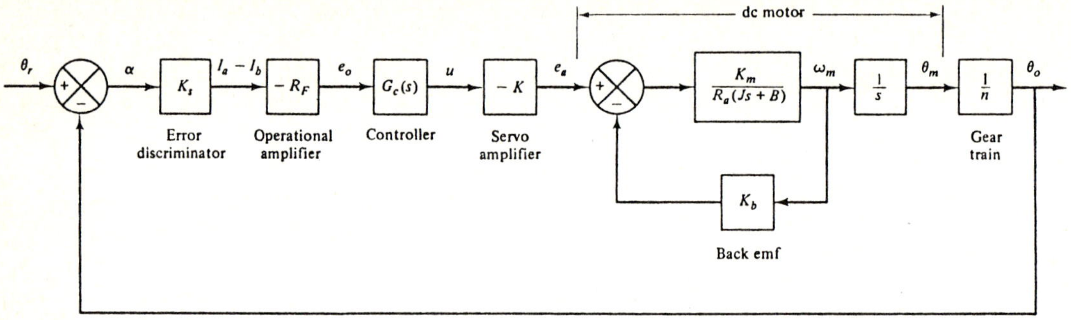

Figure (3) Block diagram of a sun-seeker control system.

The open loop transfer function of the system is :

$$\frac{\theta o(s)}{\alpha(s)} = \frac{Ks\ Rf\ p1(s + 1/p1p2)K\ Ki/n}{(Ra\ J\ s + Ki\ Kb\ s)(s + 1/p2)} \quad (17)$$

substituting the numerical values of the system parameters, Eq. (17) gives :

$$\frac{\theta o(s)}{\alpha(s)} = \frac{2500p1(s + 1/p1p2)}{s(s + 25)(s + 1/p2)} \quad (18)$$

The objective of the sun-seeker control system is to miantain the steady-state error D1(θo) and the peak overshoot D4(θo) to be minimum and should not exceed the following limits :

$$D1(\theta o) \leqslant 0.01$$
$$D4(\theta o) \leqslant 0.1 \quad (19)$$

In order to make a comparison between the classical methods based on frequency response and root locus techniques used by Kuo (8) with the proposed method, different controllers are designed as illustrated in table (2), while fig. (4) shows a similar comparison between the time responses for the controlled system using different controllers designed by different methods, together with the desired response of unity step change in the refrence angle of the solar ray θr(t). The optimal controller designed by the proposed method gives more improvement in the transient behaviour with the optimal values of the required specifications and the minimum value of the integral squared error.

Table (2) Results for example (2)

Method	The controller parameters		The dynamic specifications		Integral Squared Error
	p1	p2	D1	D4	
Frequency Response (8)	5.828	0.00588	0.0	0.0774	9.727×10^{-3}
Root Locus (8)	12.0	0.00200	0.0	0.0669	9.517×10^{-3}
Proposed	14.0	0.00222	0.0	0.0288	8.250×10^{-3}

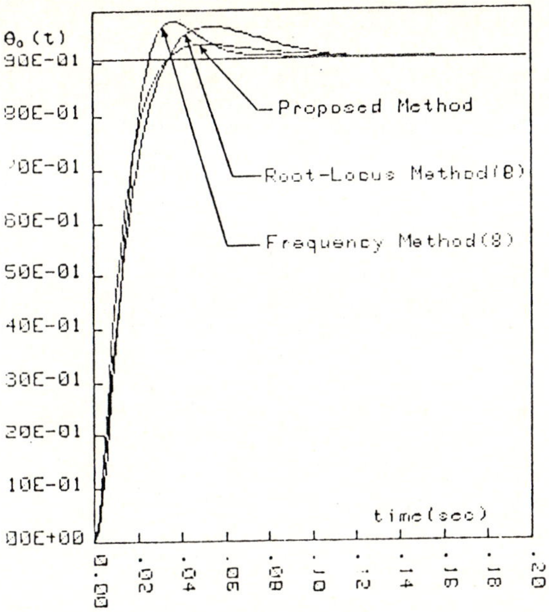

Fig.(4) Comparison of unit step responses for example(2).

4. CONCLUSION

The results obtained by this method are superior to those obtained by other methods in terms of the improvement in the required dynamic specifications of the controlled system and the minimum value of the Integral Squared Error. The constraints updating during each success cycle leads to the best dynamic specifications zone which can not be achieved by any other design technique for the same controller structure.More over the program developed for this method can be applied very straigt forword and efficiently to design such type of controllers. For further application of this method multivariable systems can be considered and this is a continuation work for this paper which is developing now by the first author.

REFERENCES

1) D'AZZO J.J. and HOUPIS C.H.,"Feed back control system analysis and synthesis", Mcgraw-Hill,1966,2nd edn.

2) ZAKIAN V. and Al-NAIB U., "Design of dynamical and control system by the method of inequalities",procc.,IEE, vol.12 ,No.11,p.1421,1973.

3) BACH R.E.Jun.,"A practical approach to control system optimization", procc.,IFAC Tokyo symposium on systems engineering for control system design, Japan,pp.129-135,1965.

4) AL-ASSADI S.A.K,"Time domain optimization technique for the simplification of a control system ", procc. of National Aerospace and Electronic Conference,(NAECON'83),Dayton,Ohio, 17-19 may,1983.

5) AL-ASSADI S.A.K. and HAMEED K.H.A., "Simplification of non-linear control system",procc. of National Aerospace and Electronic Conference,(NAECON'84), Dayton,Ohio,21-25 may,1984.

6) AL-ASSADI S.A.K. and AL-ANSARI R.M., "An iterative method for model reduction of multivariable system ",Advance in Modelling and Simulation,(AMSE) press, Vol.4,No.3, p.11-36, 1985.

7) AL-ASSADI S.A.K. and ALLOS J.E.,"Computation of optimal time-varient gain controller via optimization technique", procc. of SCS Multi-Conference on Modeling and Simulation on Microcomputer, P.86,San Diego,california,2-4 feb,1984.

8) KUO B.C.,"Automatic control system", Prentic-Hall,1982,4th edn.

Complex and Distributed Systems: Analysis, Simulation and Control
S.G. Tzafestas and P. Borne (editors)
Elsevier Science Publishers B.V. (North-Holland)

ON THE NONLINEAR BOUNDARY VALUE PROBLEM FOR ORDINARY DIFFERENTIAL EQUATIONS IN THE STATICS OF LONG TETHERED SATELLITES

Nicola BELLOMO, Riccardo RIGANTI and Maria Teresa VACCA

Dipartimento di Matematica, Politecnico di Torino
Corso Duca degli Abruzzi, 24
I - 10129 Torino, Italy

SUMMARY. The equations of the relative equilibrium of a long wire in a system of tethered satellites,subject to the earth gravitational field and random molecular fluid dynamic drag,are considered. A general modelling of the random fluctuations of the air density leads to a stochastic, nonlinear boundary value problem whose approximate analytical solution is derived on the basis of the Adomian decomposition method. Quantitative results on the first-order statistics of the equilibrium configuration are then presented in a pertinent application.

1. INTRODUCTION

It is known that recent technological projects in astronautics have emploied tethered satellites suspended, by a long wire, from a spacecraft. One of the mathematical problems arising in the analysis of these systems consists in the study of the equations of the static equilibrium, in the frame-work of linear elasticity, of the wire under the action of the fluid-dynamical drag and of the gravitational field. More in details, the mathematical problem is defined by a set of nonlinear coupled ordinary differential equations with random parameters and boundary values at the extremities of the long tether.

The randomness of the parameters arises whenever the fluid-dynamical drag, which is certainly non-negligible because of the lenght of the wire, which can reach a lenght over 100 Km, is stochastically modelled. The drag is, in fact, proportional to the air density, which, as accurately discussed in [1] or [2],undergoes consistent random fluctuations due to the day-to—day effects, solar activity and semi-annual variations in a fashion that the afore mentioned fluctuations are predictable only on the basis of probabilistic descriptions. See also the references cited in [1, 2] as well as in ref.[3].

More in details on the contents of this paper, the second section supplies a description of the mathematical system. The third section is an analysis of the nonlinear stochastic boundary value problem for ordinary differential equation on the basis of the Adomian decomposition method [4], which is here stressed in order to supply approximate analytical solution of the mathematical problem considered in this paper.

The method has shown to be flexible and accurate enough to supply sufficiently precise solution as documented in the final section of the paper where quantitative results and applications are proposed.

2. THE MATHEMATICAL SYSTEM

The equations of the relative equilibrium of a long wire in the (x, y) - plane, with reference to fig. 1, in a tethered satellite system in the earth gravitational field and under the action of the molecular fluid dynamic drag have been already derived elsewhere, see ref. [5] and related bibliography. According to [4] as well as to the mathematical models proposed in [2] referring to the fluid-dynamic drag, these equations can be written, in the framework of linear elasticity, in the following form:

$$\text{(1a)}\quad \frac{d}{ds}\left(\tau(s)\frac{dx}{ds}\right) + 3\delta(g(r_B)/r_B)\,x = 0$$

$$\text{(1b)}\quad \frac{d}{ds}\left(\tau(s)\frac{dy}{ds}\right) + \tfrac{1}{2}\,\rho(\omega,x,y)AC_D^{(T)}\dot{\phi}^2(r_B-x)^2\frac{dx}{ds} = 0$$

$$\text{(1c)}\quad (dx/ds)^2 + (dy/ds)^2 = 1$$

subject to the following boundary conditions:

$$x\ (s=0) = y\ (s=0) = 0$$

$$(2)\quad f_1(\tau,x,dx/ds;s=\ell) = -\,\tau(\ell)\,(dx/ds)_{s=\ell} + 3\, m\cdot g_B\cdot x(s=\ell)/r_B = 0$$

$$f_2(\tau,x,dy/ds,dx/ds,s=\ell) = -\tau(\ell)\cdot(dy/ds)_{s=\ell} + \tfrac{1}{2}\,\rho(\omega,x(s=\ell),y(s=\ell)C_D^{(A)}A^*\,\dot{\phi}^2(r_B-x(s=\ell))^2\cdot (dx/ds)_{s=\ell} = 0$$

where in the above equations:

s is the curvilinear coordinate of the wire with lenght ℓ.
$x=x(s)$ and $y=y(s)$ are the coordinates of the point of the wire at the coordinate s.
$\tau=\tau(s)$ is the elastic tension in the wire.
δ is the mass per unit lenght of the wire.
$g(r_B)$ is the earth's gravitational acceleration at the point B indicated in the figure.
A and A* are the cross section, per unit lenght of the satellite and of the wire, respectively; $C_D^{(A)}$ and $C_D^{(T)}$ being their drag coefficients assumed to be experimentally known.
$\dot{\phi}$ is the rotational mean motion of the outer body B.
ρ is the air mass density.

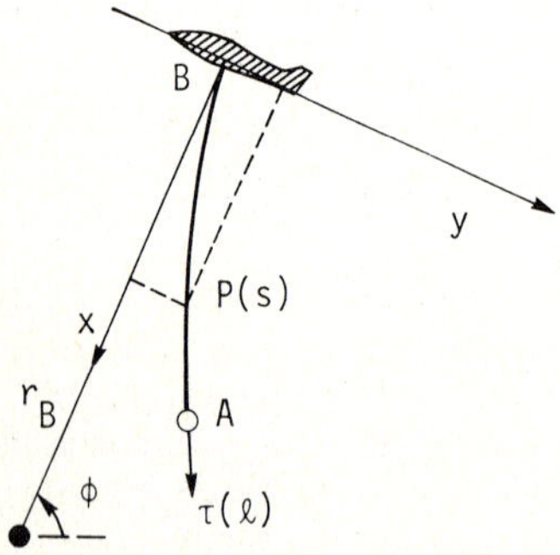

Fig. 1 - Geometry of the system

In particular, the air density can be, following [2], modelled by the equation:

$$(3)\quad \rho(\omega;x,y) = \rho_0 \exp\left\{-\frac{((r_B-x)^2+y^2)^{\frac{1}{2}}-R_e}{H}\right\}(1+\frac{\beta_1(\omega)}{2})\cdot(1+\frac{\beta_2(\omega)}{5})$$

where R_e is the Earth radius and $\beta_{1,2}$ are constant random variables modelled by a beta-type probability density defined in the domain $[-\frac{1}{2},\frac{1}{2}]$ with zero mean value and a variance $V(\beta_{1,2})$ to be experimentally determined.

However, the mathematical analysis, developed in the next sections, is realized in a general way and the results can be interpreted for the most general modelling of the random fluctuations of the air density.

3. ANALYSIS

The objective of the analysis realized in this section consists in finding, for real physical situations, the solutions of eqs. (1,2). A suitable re-writing of the afore mentioned equations is then useful towards the said objective. Therefore, the following dimensionless vector state-variable can be defined:

$$(4)\quad \underset{\sim}{z}=\{z_1=x/\ell,\, z_2=y/\ell,\, z_3=\tau/\tau_B,\, z_4=dz_1/d\sigma\}\in D_z \subset \mathbb{R}^4$$

where $\sigma=s/\ell \in [0,1]$ can be regarded as a new independent variable and $\tau_B = \tau(\sigma=0)$.

After the position (4), eqs. (1,2) can be rewritten in normal vector form:

$$(5)\quad i = 1,\dots,4: \qquad dz_i/d\sigma = \psi_i(\underset{\sim}{z};\, \underset{\sim}{\beta}(\omega))$$

where $\underset{\sim}{\beta} = \{\beta_1, \beta_2\}$ and:

$$(6a)\quad \psi_1 = z_4$$

$$(6b)\quad \psi_2 = (1 - z_4^2)^{\frac{1}{2}}$$

$$(6c)\quad \psi_3 = ((\ell/\tau_B)\cdot(z_4/(1-2z_4^2))\cdot\{a\ell z_1 - b(\underset{\sim}{z};\underset{\sim}{\beta}(\omega))\cdot(1-z_4^2)^{\frac{1}{2}}\}$$

$$(6d)\quad \psi_4 = ((\ell/\tau_B)\cdot(z_3(1 - 2z_4^2)))\cdot \{b(\underset{\sim}{z};\underset{\sim}{\beta}(\omega))z_4^2(1-z_4^2)^{\frac{1}{2}} - a\ell z_1(1-z_4^2)\}$$

where

$$(7a)\quad a = 3\,\delta g(r_B)/r_B$$

$$(7b)\quad b = \tfrac{1}{2}\rho(z_1,z_2;\,\underset{\sim}{\beta}(\omega))A\,\dot{\phi}^2 C_D^{(T)}(r_B - \ell z_1)^2$$

and where eqs. (5, 6) are subject to the following boundary conditions:

$$(8a)\quad z_1(0) = z_2(0) = 0$$

$$(8b)\quad f_1(\underset{\sim}{z};\,\sigma=1)=dz_1(1)-\tau_B z_3(1)z_4(1) = 0$$

(8c) $f_2(\underset{\sim}{z};\ \sigma=1)=b(\underset{\sim}{z}(1);\underset{\sim}{\beta}(\omega))z_4(1)\ -$

$$\tau_B z_3(1)(1-z_4^2(1))^{\frac{1}{2}} = 0$$

where $d = am\ell/\delta$ and where clearly the mathematical model is consistent if: $(\sqrt{2}/2) < z_4 \le 1$.

At this end, considering that $z_3(0) = 1$, by definition, the mathematical problem can now be stated as follows:

- *The "mathematical problem" consists in finding a solution $\underset{\sim}{z} = \underset{\sim}{z}(\sigma;\ \underset{\sim}{z}_0)$ of eqs.(5,6) and two constants τ_B and $z_4(\sigma = 0) = z_{40}$ such that eqs. (5, 6) are satisfied together with the boundary conditions:*

$$f_1(1) = f_2(1) = 0;\ z_1(0) = z_2(0) = 0;\ z_3(0) = 1$$

Such a problem can be solved after suitable application of the Adomian decomposition method [4]. In fact, if the initial conditions of eq. (5) are assumed to be given, this equation can be re-written in integral form:

(9) $$z_i(\omega,\sigma;\underset{\sim}{z}_0)=z_{io}+\lambda\int_0^\sigma \psi_i(\underset{\sim}{z}(\omega,u;\underset{\sim}{z}_0,\underset{\sim}{\beta}(\omega))du, \qquad \lambda=1$$

where both z_i and ψ_i can be decomposed in terms of powers of the parameter λ as follows:

(10) $$z_i = \sum_j \lambda^j z_i^{(j)}\ ,\ \psi_i = \sum_j \lambda^j \psi_i^{(j)}$$

The arguments of the functions appearing in eq. (10) have been omitted for simplicity.

Equaling the terms with the same power of the decomposition parameter "λ" supplies the following sequence of quadratures:

(11a) $$z_i^{(0)} = z_i^{(0)}(\underset{\sim}{\beta}(\omega)) = z_{i0}$$

(11b) $$z_i^{(1)} = z_i^{(1)}(\sigma;\ \underset{\sim}{\beta}(\omega)) = \int_0^\sigma \psi_i^{(0)}(\underset{\sim}{z}^{(0)}(..))du$$

.....

(11c) $$z_i^{(j)} = z_i^{(j)}(\sigma;\ \underset{\sim}{\beta}(\omega)) = \int_0^\sigma \psi_i^{(j-1)}(\underset{\sim}{z}^{(0)}..),\dots,\ \underset{\sim}{z}^{(j-1)}(u,..))\ du$$

Details of the above calculations are supplied in the Appendix with regards to the first terms. Continuation is a routine.

Once the solution has been formally obtained, the unknown constants "τ_B" and "z_{40}", which in the solution appear as free parameters, can be determined by the boundary conditions (8) to be satisfied in the mean:

(12) $$i = 1,2:\quad \int f_i(\tau_B, z_{40}; \sigma=1, \underset{\sim}{\beta}(\omega))P(\underset{\sim}{\beta})d\underset{\sim}{\beta} = 0$$

The application considered in the last section clearly indicates how the said constants can be efficiently computed by a suitable application of the univariate search [6] to the system of nonlinear algebraic equations defined in (12).

After the determination of the said constants, the solution is fully defined in the form:

(13) $$z_i \cong z_{id}(\sigma;\ \omega) = \sum_{j=0}^{n} z_i^{(j)}(\sigma;\ \underset{\sim}{\beta}(\omega))$$

and the pertinent statistics on the solution can be realized:

(14) $$e_{ij}^{mn}(\sigma)=E(z_{id}^m \cdot z_{jd}^n) = \int z_{id}^m(\sigma;\underset{\sim}{\beta}(\omega))z_{jd}^n(\sigma;\underset{\sim}{\beta}(\omega))P(\underset{\sim}{\beta})d\underset{\sim}{\beta}$$

for $m, n = 0, 1, 2, \dots,$.

4. APPLICATION AND DISCUSSION

The application considered in this section refers to detailed calculations realized in order to test the mathematical method proposed in the preceding section. Of course, these calculations utilized the results concerning the decomposition which are summarized in the Appendix.

In particular, the application has been realized for the following values of the parameters:

$\delta = 6.5 \cdot 10^{-4}$ Kg/m; $A = 1\cdot 10^{-3}$ m^2; $A^* = .7$ m^2;

$C_D^A = 2.2$; $C_D^T = 3$; $\rho_o = .125$ Kg$\cdot$sec^2/m^4;

$r_B = 6.585\cdot 10^6$ m; $R_e = 6.35\cdot 10^6$ m; $H=1.3\cdot 10^4$ m

$V(\beta_1) = V(\beta_2) = .1$;

and for two values of the rotational mean motion of the outer body: $\dot{\phi} = 7.29\cdot 10^{-5}$ and $2\cdot 10^{-4}$ sec^{-1}. The computed values of the constants τ_B, z_{40} were, respectively:

$\tau_B = 9.58$; $z_{40} = .9995$ for $\dot{\phi} = 7.29\cdot 10^{-5}$;

$\tau_B = 9.45$; $z_{40} = .9922$ for $\dot{\phi} = 2.10^{-4}$.

The main results, for these particular cases, are visualized in Fig.s 2 and 3 which indicate, respectively, $E\{z_1\}$ versus $E\{z_2\}$ and $E\{z_3\}$ versus σ, namely the deformation in the mean and the behaviour of the internal tension along the wire. Analogous calculations can be realized for the second order statistics, or for different values of the parameters caracterizing the model. This result can be easily recovered following the solution procedure indicated in the preceding section according to the decomposition terms listed in the Appendix.

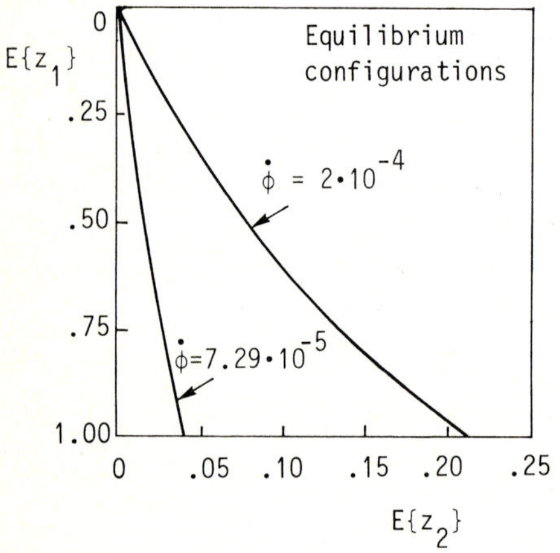

Fig. 2 - $E\{z_1\}$ versus $E\{z_2\}$ with a four-terms solution (n = 3).

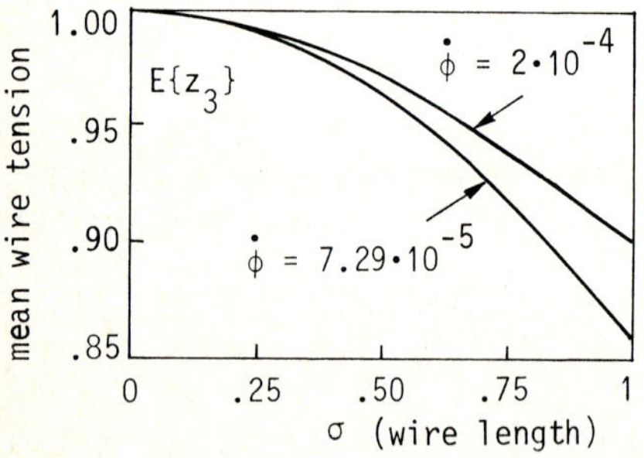

Fig. 3 - $E\{z_3\}$ versus σ with a four-terms solution, (n = 3).

The method has shown to be very efficienty in comparison with the direct numerical treatment of the boundary value problem with univariate search of the unknown boundary values. Addition of further terms in the decomposition does not supply significant improvement. Let us finally comment how a further stressing of the solution procedure can hopefully bring a solution of the relatively more difficult problem consisting in the analysis of the dynamics of the tether. The relatively easier problem consisting in the analysis of the statics of the system has been studied in this paper.

ACKNOWLEDGEMENTS. This research has been partially supported by the Italian Minister for Education, M.P.I..

REFERENCES

1. D. King-Hele, Research results from the analysis of satellite orbits, in: Space Craft Dynamics, (ESA, Paris, 3-11, 1981);
2. N. Bellomo, R. Riganti and M.T. Vacca, On the dynamics of dumbell satellites in two gravitational fields, in: Mathematical modelling in Science and Technology (Avula, Kalman, Liapis & Rodin, Eds.), Pergamon Press, New York (1984) p. 320-325.
3. COSPAR International Reference Atmosphere 1972 (Akademie Verlag, Berlin, 1972).
4. G. Adomian, Stochastic Systems (Academic Press, New York, 1983).
5. G. Colombo, E.M. Gaposchkin, M. Grossi and G. Weiffenbach, The "Skyhook":a Shuttle-borne tool for low orbital altitude research (Meccanica, J. of the Italian Ass. Theor. Appl. Mech., 1, 1975, p. 3-20).
6. L. Cooper and D. Steinberg, Introduction to Methods of Optimization (Saunders Co., Philadelphia, 1970).

APPENDIX

The terms $\psi_i^{(j)}$ in the decomposition (10) of ψ_i are defined as:

$$\psi_i^{(o)} = \psi(\underset{\sim}{z}^{(o)};\ \beta(\omega)),\ i = 1,..,4$$

$$\psi_i^{(j)} = \frac{1}{j!}\left(\frac{d^j\psi_i}{d\lambda^j}\right)_{\lambda=0},\ i = 1,..,4;\ j = 1,2,..$$

Explicit calculation of the first three ones yields:

$$\psi_1^{(o)} = z_4^{(o)} = z_{40}$$

$$\psi_1^{(1)} = z_4^{(1)}$$

$$\psi_1^{(2)} = z_4^{(2)}$$

$$\psi_2^{(o)} = (1 - z_{40}^2)^{\frac{1}{2}}$$

$$\psi_2^{(1)} = - z_{40} z_4^{(1)} (1-z_{40}^2)^{-\frac{1}{2}}$$

$$\psi_2^{(2)}=-\{z_{40}z_4^{(2)}+(z_4^{(1)})^2/[2(1-z_{40}^2)]\}(1-z_{40}^2)^{-\frac{1}{2}}$$

$$\psi_3^{(o)}=-\ell z_{40}b_o(1-z_{40}^2)^{\frac{1}{2}}/\{\tau_B(1-2z_{40}^2)\}$$

$$\psi_3^{(1)}=\ell b_o\{z_{40}z_1^{(1)}(a\ell/b_o-(1-z_{40}^2)^{\frac{1}{2}}h_1)-$$
$$-z_4^{(1)}(1-z_{40}^2)^{-\frac{1}{2}}(1-2z_{40}^2)^{-1}\}/\{\tau_B(1-2z_{40}^2)\}$$

$$\psi_3^{(2)}=-\ell b_o\Big[z_{40}(1-z_{40}^2)^{\frac{1}{2}}\big[h_2(z_1^{(1)})^2-$$
$$-\ell^2(z_2^{(1)})^2/(Hr_B)+(z_4^{(1)})^2(9-10z_{40}^2)\cdot(1-$$
$$-z_{40}^2)^{-2}\cdot(1-2z_{40}^2)^{-2}+2h_1z_1^{(2)}\big]+$$
$$+2z_4^{(2)}\cdot(1-z_{40}^2)^{-\frac{1}{2}}(1-2z_{40}^2)^{-1}-$$
$$-2z_{40}z_1^{(2)}a\ell/b_o\Big]/\{2\tau_B(1-2z_{40}^2)\}\,.$$

$$\psi_4^{(o)}=\ell b_oz_{40}^2(1-z_{40}^2)^{\frac{1}{2}}/\{\tau_B(1-2z_{40}^2)\}$$

$$\psi_4^{(1)}=\ell b_o(1-z_{40}^2)^{\frac{1}{2}}\Big[z_1^{(1)}\big[h_1z_{40}^2-a\ell(1-z_{40}^2)^{\frac{1}{2}}/b_o\big]-$$
$$-z_3^{(1)}z_{40}^2+z_4^{(1)}z_{40}\{2(1-2z_{40}^2)^{-1}-$$
$$-z_{40}^2(1-z_{40}^2)^{-1}\}\Big]/\{\tau_B(1-2z_{40}^2)\}$$

$$\psi_4^{(2)}=\ell b_o(1-z_{40}^2)^{\frac{1}{2}}\Big[(z_1^{(1)})^2z_{40}^2h_2-$$
$$-(z_2^{(1)})^2z_{40}^2\ell^2/(Hr_B)+$$
$$+(z_4^{(1)})^2(2+3z_{40}^2(1-2z_{40}^2))\cdot(1-$$
$$-2z_{40}^2)^{-2}\cdot(1-z_{40}^2)^{-2}+2(z_3^{(1)})^2z_{40}^2+$$
$$+2z_1^{(2)}\big[h_1z_{40}^2-a\ell(1-z_{40}^2)^{\frac{1}{2}}/b_o\big]-2z_3^{(2)}z_{40}^2+$$
$$+2z_4^{(2)}z_{40}\big[2(1-2z_{40}^2)^{-1}-z_{40}^2(1-z_{40}^2)^{-1}\big]\Big].$$
$$\cdot\{2\tau_B(1-2z_{40}^2)\}^{-1}$$

where:

$$h_1=\ell/H-2\ell/r_B$$

$$h_2=2\ell^2/r_B^2-4\ell^2/(Hr_B)+\ell^2/H^2;$$

$$b_o=\tfrac{1}{2}\rho_o\exp\{-(r_B-R_e)/H\}A\,\dot{\phi}^2\,C_D^T\,r_B^2\,(1+$$
$$+\beta_1(\omega)/2)\cdot(1+\beta_2(\omega)/5).$$

Inserting the above terms into Eqs.(11) and integrating, it is obtained:

$$z_1^{(o)}=0$$

$$z_1^{(1)}=\sigma\cdot z_{40}$$

$$z_1^{(2)}=\ell b_oz_{40}^2(1-z_{40}^2)^{\frac{1}{2}}\sigma^2/\{2\tau_B(1-2z_{40}^2)\}$$

$$z_1^{(3)}=\sigma^3\chi_3(z_{40},\tau_B,\omega)/6$$

$$z_2^{(o)}=0$$

$$z_2^{(1)}=\sigma(1-z_{40}^2)^{\frac{1}{2}}$$

$$z_2^{(2)}=-\ell b_oz_{40}^3\sigma^2/\{2\tau_B(1-2z_{40}^2)\}$$

$$z_2^{(3)}=-\sigma^3\{z_{40}\chi_3(z_{40},\tau_B,\omega)+\chi_1^2(z_{40},\tau_B,\omega)\cdot(1-$$
$$-z_{40}^2)^{-1}\}/\{6(1-z_{40}^2)^{\frac{1}{2}}\}$$

$$z_3^{(o)}=1$$

$$z_3^{(1)}=-\ell b_oz_{40}(1-z_{40}^2)^{\frac{1}{2}}\sigma/\{\tau_B(1-2z_{40}^2)\}$$

$$z_3^{(2)}=\ell b_oz_{40}^2\sigma^2\{a\ell/b_o-h_1(1-z_{40}^2)^{\frac{1}{2}}-$$
$$-\ell b_o(1-2z_{40}^2)^{-2}/\tau_B\}/\{2\tau_B(1-2z_{40}^2)\}$$

$$z_3^{(3)} = -\chi_1(z_{40},\tau_B,\omega)\sigma^3\{h_2 z_{40}^2 - \ell^2(1-z_{40}^2)/(Hr_B) + (9-10z_{40}^2)\chi_1^2(z_{40},\tau_B,\omega)(1-2z_{40}^2)^{-2}(1-z_{40}^2)^{-2} + (h_1 - a\ell(1-z_{40}^2)^{-\frac{1}{2}}/b_o)\chi_1(z_{40},\tau_B,\omega) + \chi_3(z_{40},\tau_B,\omega)\cdot(1-z_{40}^2)^{-1}(1-2z_{40}^2)^{-1}/z_{40}\}/(6z_{40})$$

$$z_4^{(o)} = z_{40}$$

$$z_4^{(1)} = \ell b_o z_{40}^2(1-z_{40}^2)^{\frac{1}{2}}\sigma/\tau_B(1-2z_{40}^2)\}$$

$$z_4^{(2)} = \ell b_o z_{40}\sigma^2(1-z_{40}^2)^{\frac{1}{2}}\{h_1 z_{40}^2 - a\ell(1-z_{40}^2)^{\frac{1}{2}}/b_o + \ell b_o z_{40}^2(1-z_{40}^2)^{\frac{1}{2}}[1+2(1-2z_{40}^2)^{-1} - z_{40}^2(1-z_{40}^2)^{-1}]/[\tau_B(1-2z_{40}^2)]\}/\{2\tau_B(1-2z_{40}^2)\}$$

$$z_4^{(3)} = \frac{\sigma^3}{6}\chi_1(z_{40},\tau_B,\omega)\Big[h_2 z_{40}^2 + \big[(2+3z_{40}^2(1-2z_{40}^2))/\{z_{40}^2(1-z_{40}^2)^2(1-2z_{40}^2)^2\} + 2/z_{40}^2\big]\chi_1^2(z_{40},\tau_B,\omega) - \ell^2(1-z_{40}^2)/(Hr_B) - \chi_2(z_{40},\tau_B,\omega) + \big[h_1 - a\ell(1-z_{40}^2)^{\frac{1}{2}}/(b_o z_{40}^2)\big]\cdot\chi_1(z_{40},\tau_B,\omega) + \big[2/\{z_{40}(1-2z_{40}^2)\} - z_{40}(1-z_{40}^2)^{-1}\big]\chi_3(z_{40},\tau_B,\omega)\Big]$$

where:

$$\chi_1(z_{40},\tau_B,\omega) = \ell b_o z_{40}^2(1-z_{40}^2)^{\frac{1}{2}}/\{\tau_B(1-2z_{40}^2)\}$$

$$\chi_2(z_{40},\tau_B,\omega) = \chi_1(z_{40},\tau_B,\omega)(1-z_{40}^2)^{-\frac{1}{2}}\{a\ell/b_o - h_1(1-z_{40}^2)^{\frac{1}{2}} - \ell b_o/\{\tau_B(1-2z_{40}^2)^2\}\}$$

$$\chi_3(z_{40},\tau_B,\omega) = \chi_1(z_{40},\tau_B,\omega)\{h_1 z_{40}^2 - a\ell(1-z_{40}^2)^{\frac{1}{2}}/b_o + [1+2(1-2z_{40}^2)^{-1} - z_{40}^2(1-z_{40}^2)^{-1}]\cdot\chi_1(z_{40},\tau_B,\omega)\}/z_{40}.$$

Complex and Distributed Systems: Analysis, Simulation and Control
S.G. Tzafestas and P. Borne (editors)
Elsevier Science Publishers B.V. (North-Holland)

REMARKS ON VARIATIONAL PROBLEMS DESCRIBING DYNAMICAL SYSTEMS AND NETWORKS

Valery Yakovlevitch ZUIKOV

Institute of Industrial Electronics, Kiev Polytechnical Institute
Brest-Litovsky Road 39, 252056 Kiev, USSR

Mirosław KOWALSKI

Institute of Electronics, Technical University of Łódź
Gdańska 176, 90-924 Łódź, Poland

The sufficient conditions for the existence of extremum at a stationary curve of the action functional of a dissipative network are formulated. Linear time-invariant, time-varying and nonlinear networks are considered in the paper. It has been proved that the action functional attains a strong global minimum in the set X, defined in (2) below, at the solution of a linear system of state equations having a constant and diagonalizable state matrix possessing real eigenvalues.

1. INTRODUCTION

In solving many problems of variational calculus it is significant to investigate the extreme properties of functionals in a given set of functions. The dynamical systems and networks provide an important and interesting basis for such investigation. The variational principle most frequently used for the description of dynamical systems and networks is Hamilton's variational principle [2]. In the classical formulation this principle requires that the action functional

$$Q \ni q \longrightarrow J[q] := \int_{t_o}^{t_f} \mathcal{L}(q,\dot{q},t)dt \qquad (1)$$

where Q is a given set of functions (possible trajectories), $q = [q_1 \dots q_n]^T \in R^n$ is the vector of generalized coordinates, $\dot{q} = [\dot{q}_1 \dots \dot{q}_n]^T \in R^n$ is the vector of generalized velocities, $\mathcal{L} : R^n \times R^n \times [t_o,t_f] \longrightarrow R$ is the corresponding Lagrangian function and $t_o, t_f \in R$ are the initial and final moments of the considered time interval $[t_o,t_f]$, is stationary at the solution of the system equations of motion. However, this principle does not reveal the conditions under which the functional $J[q]$ has an extremum. In this paper with the aid of well known theorems of variational calculus we formulate the sufficient conditions for a strong local extremum of the newly found functional $J[q]$ for a class of linear and nonlinear dissipative networks [3]. The results obtained in the work may be applied when finding out approximate solutions of variational problems describing dynamical systems and networks with the aid of direct variational methods [1].

2. LINEAR TIME-INVARIANT NETWORKS

Consider a linear time-invariant network containing two-terminal R, L, C, E, J elements and linear multi-ports whose mathematical description on the interval $[t_o,t_f]$ is given by the n-th order state equation system

$$\dot{x} = Ax + Bv \qquad (2)$$

where $x \in X$, $v \in C^o([t_o,t_f] ; R^m)$, $X := \{x | \; x \in C^1([t_o,t_f] ; R^n), x(t_o) = x_o, x(t_f) = x_f\}$, $A \in R^{nxn}$, $B \in R^{nxm}$ are constant matrices, $x_f = \phi(t_f;x_o,t_o)$ is the solution at $t = t_f$ of system (2) with the initial condition $x_o = \phi(t_o; x_o,t_o)$ and $x^s(t) = \phi(t;x_o,t_o)$ is the transition function. Differentiate the system of equations (2) with respect to t (or introduce new variables $X_i := \int_{t_o}^{t} x_i dt + X_i(t_o)$, $i = 1,2,\dots,n$ as it was proposed in [1]) and introduce new variables $q_1,\dots,q_n$ by using a linear transformation W^{-1}: $R^n \ni x \to q \in R^n$ with nonsingular matrix $W^{-1}(t) = e^{-0.5At}$ $\forall t \in [t_o, t_f]$. From Liouville's formula it it follows that $\det W(t) \neq \infty$ $\forall t \in [t_o, t_f]$ if $-\infty < t_o < t_f < \infty$. If $t_f = \infty$, invertibility of the matrix $W^{-1}(\infty)$ requires $\mathrm{tr}A \leqslant 0$. This condition is obviously satisfied if the network is stable. Thus the system (2) takes the form

$$\ddot{q} + S(t)q = e(t) \qquad (3)$$

where $S = S(\cdot) = -\frac{1}{4}A^2$ and $e = e(\cdot) = W^{-1}(\cdot)B\dot{v}$. Assume that the matrix S is a symmetric one, e.g. $S = S^T$. Since the system (3) is self-adjoint it may be considered as the Euler-Lagrange system of equations [2], associated with the action functional (1). The set Q in this case is obtained from X by performing the linear transformation W^{-1}, e.g. $Q = \{q \mid q \in C^1([t_o,t_f]; R^n),\ q(t_o) = W^{-1}(t_o)x_o = q_o,\ q(t_f) = W^{-1}(t_f)x_f = q_f\}$. The Lagrangian function of system (3) may be expressed in the form

$$\mathcal{L} = \frac{1}{2}\dot{q}^T\dot{q} + \frac{1}{8}q^TA^2q + q^Te \qquad (4)$$

or

$$\mathcal{L} = \frac{1}{2}\dot{q}^T\dot{q} - \frac{1}{4}tq^TA^2q + q^Te. \qquad (5)$$

Observe that in the case when $A^2 = (A^2)^T$ we don't have to use the linear transformation of variables to obtain the self-adjoint form of the network equations. Indeed, the system (2) after its differentiating with respect to t and after excluding $\dot{x}$ takes the form (3), where $q = x$, $S = -A^2$ and $e = e' = ABv + B\dot{v}$. The Lagrangian function in this case may have the form

$$\mathcal{L} = \frac{1}{2}\dot{q}^T\dot{q} + \frac{1}{2}q^TA^2q + q^Te' \qquad (6)$$

or

$$\mathcal{L} = \frac{1}{2}\dot{q}^T\dot{q} - t\,\dot{q}^TA^2q + q^Te' \qquad (7)$$

and corresponds to the motion of the system in the state space. Assume that the minimal polynomial of the matrix A is a product of distinct linear factors corresponding to the real eigenvalues of A. In this case $A = H\Lambda H^{-1}$, where $H \in R^{n\times n}$ is nonsingular matrix, $\Lambda = \mathrm{diag}(\lambda_1,\ldots,\lambda_n)$, $\lambda_i \in \mathrm{Spect}(A) \subset R$, $i = 1,2,\ldots,n$. Spect(A) denotes the set of the eigenvalues of the matrix A. Differentiating the system (2) with respect to t and then multiplying it from the left by nonsingular symmetric matrix $D = D(t) = D_oe^{-At}$, where

$$D_o = \begin{cases} I_n & \text{for } A = A^T \\ (H^{-1})^TH^{-1} & \text{for } A \neq A^T \end{cases}$$

$I_n \in R^{n\times n}$ denotes the unit matrix, one can obtain the self-adjoint form of the network equations. If $-\infty < t_o < t_f = \infty$, then we assume that $\mathrm{tr}A \leqslant 0$. From a physical point of view, the requirement for the self-adjoint form of the network equations expresses the fact that the Lagrangian function of the dissipative network must be explicitly time-dependent regardless of whether the network is excited or not. The function $\mathcal{L}$ may be found in the form

$$\mathcal{L} = \frac{1}{2}\dot{q}^TD\dot{q} + q^Te'' \qquad (8)$$

where $q = x$ and $e'' = e''(t) = D(t)B\dot{v}$

<u>Theorem 1.</u> The action functional $J[q]$ with the Lagrangian function given by (8) and with $t_f < \infty$ attains a strong global minimum in the set X at the solution $q^s(t) = x^s(t)$ of system (2).

Proof. As is known from the theory of strong extremum [2], in the case when $\frac{\partial^2\mathcal{L}}{\partial\dot{q}\partial q} = \left(\frac{\partial^2\mathcal{L}}{\partial\dot{q}\partial q}\right)^T$ the curve $q^s(\cdot)$ is the argument of the strong local minimum of $J[q]$ if the following conditions are satisfied:

1. The curve $q^s(\cdot)$ is a stationary curve of $J[q]$.
2. The matrix $\frac{\partial^2\mathcal{L}}{\partial\dot{q}^2}(q,z,t)$ is a positive definite symmetric matrix in some region containing $q^s(\cdot)$ for all $z = [z_1 \ldots z_n]^T \in R^n$ such that $|z_i| < \infty$, $i = 1,2,\ldots,n$.
3. The quadratic functional $\delta^2J[q^s][h] = \int_{t_o}^{t_f}(\dot{h}^TP_{\mathcal{L}}\dot{h} + h^TQ_{\mathcal{L}}h)dt$ is positive definite for all $h \in C^1([t_o,t_f]; R^n)$ such that $h(t_o) = h(t_f) = \theta_n$, where $\theta_n \in R^n$ denotes the zero column vector,

$$P_{\mathcal{L}} := \frac{1}{2}\frac{\partial^2\mathcal{L}}{\partial\dot{q}^2}\Bigg|_{\substack{q=q^s(t)\\ \dot{q}=\dot{q}^s(t)}}$$

$$Q_{\mathcal{L}} := \frac{1}{2}\left[\frac{\partial^2\mathcal{L}}{\partial q^2} - \frac{d}{dt}\left(\frac{\partial^2\mathcal{L}}{\partial\dot{q}\partial q}\right)\right]\Bigg|_{\substack{q=q^s(t)\\ \dot{q}=\dot{q}^s(t)}}.$$

In the case of functional $J[q]$ with the Lagrangian function in form (8) conditions 1) - 3) are satisfied. This results from the fact that $\mathcal{L}$ is a strictly convex function with respect to $\dot{q}$ and is convex with respect to $[\dot{q}^Tq^T]^T$

in the domain $R^n \times R^n \times [t_o, t_f]$. Indeed, the Hessian matrix $\frac{\partial^2 \mathcal{L}}{\partial \dot{q}^2} = (H^{-1})^T e^{-\Lambda t} H^{-1}$ is positive definite since it is congruent with the positive definite matrix for all $t \in [t_o, t_f]$. If $t_f = \infty$, then positive definiteness of the matrix $e^{-\Lambda t}$ requires $\lambda_i \leq 0$, $i = 1,2,\dots,n$, e.g. the network stability. Moreover, $Q_{\mathcal{L}} = \mathbf{0}_n$, where $\mathbf{0}_n \in R^{n \times n}$ denotes the zero matrix. From here in particular it follows that condition 3) is satisfied. Since $J[q]$ is a quadratic functional, it has a strong global minimum in X at $x^s(\cdot)$. Q.E.D.

Observe that in the case of functional $J[q]$ with the Lagrangian function in form (4), (5) or (6), (7), the matrix $Q_{\mathcal{L}} = -\frac{1}{2} S$. Thus the following remarks are valid

Remarks

1. If the matrix $-S$ of the linear network with the Lagrangian function in form (4), (5) or (6), (7), is a positive semi-definite symmetric matrix on the interval $[t_o, t_f]$, then $J[q]$ has a strong global minimum in the sets Q or X at the solution $q^s(\cdot)$ of system (3).
2. If the functional $J[q]$ of a linear time-invariant (time-varying) network attains an extremum at its stationary curve, then this extremum is a strong global minimum (a strong global minimum in place, e.g. the strong global minimum is attained for a certain sufficiently small interval $[t_1, t_2] \subset [t_o, t_f]$) in the sets Q or X.

For linear networks with time-varying parameters the form of the matrix S and the Lagrangian functions (9) and (10) or (12) and (13) shall be derived in the following section.

Assume that the matrix S is negative semi-definite, e.g. $-S \geq 0$. Thus the functional $J[q]$ with the Lagrangian function in form (4) or (6) and (9) or (12) is identical with the energetic method functional

$$J'[q] := \frac{1}{2} (Fq, q) + (q, e)$$

where $(\cdot,\cdot)$ denotes a standard scalar product in the Hilbert space $L_2^n(t_o, t_f)$ of n-dimensional vector-valued functions possessing the second powers of coordinates integrable on the interval $[t_o, t_f]$. The functional $J'[q]$ was derived in Reference [5] for a symmetric and positive definite operator

$$F: Q \ni q \longrightarrow Fq \in L_2^n(t_o, t_f)$$

of the two-point boundary value problem

$$Fq = -\frac{d}{dt}\left(\frac{dq}{dt}\right) - Sq = -e, \quad q_o = q_f = \Theta_n.$$

It is evident that in the case of time-invariant networks $S = S^T$ if $A = \pm A^T$, e.g. if the network state matrix is a symmetric or skew-symmetric matrix.

Observe that if $A = A^T$, then $J[q]$ with the Lagrangian function (4), (5) ((6), (7)) has a strong global minimum in Q (X) at $q^s(\cdot)$ $(x^s(\cdot))$. If the Lagrangian function has the form (8), then the action functional is identical with $J'[q]$ which in this case is accompanied by a positive definite operator of the two-point boundary value problem

$$Fq = -\frac{d}{dt}\left(D\frac{dq}{dt}\right) = -e, \quad q_o = q_f = \Theta_n$$

with the input $e = e'' = D\dot{B}v$. When solving these particular problems with the aid of direct variational methods, the well known properties of the Ritz method in the energetic space may be used (e.g. the stable approximate solution, the finite value of a spectral conditioning factor of the Ritz matrix, etc.). These properties result from the adequate choice of the system of basis functions [5].

3. LINEAR TIME-VARYING NETWORKS

Consider the system (2). Assume that the elements of the matrices A and B are real C^1-time functions. If the matrix A is a symmetric matrix, then the self-adjoint form of the network equations may be obtained by differentiating the system (2) with respect to t and then excluding $\dot{x}$. Thus the system (2) takes the form (3), where $q = x$, $S = -[A^2(t) + \dot{A}(t)]$ and $e = [A(t)B(t) + \dot{B}(t)]v + B(t)\dot{v}$. The Lagrangian function may be found in the form

$$\mathcal{L} = \frac{1}{2}\dot{q}^T\dot{q} + \frac{1}{2} q^T(A^2 + \dot{A})q + q^T e \quad (9)$$

or

$$\mathcal{L} = \frac{1}{2}\dot{q}^T\dot{q} - \dot{q}^T\left(\int A^2 dt + A\right)q + q^T e \quad (10)$$

and corresponds to the motion of the network in the state space. The Lagrangian function may also be found in the transformed configuration space. Differentiating the system (2) with respect to t and performing a linear transformation of variables the system (2) takes the form (3). The inverse W of the transformation matrix may be found as the solution of the homogeneous system of equations

$$\dot{W} = \frac{1}{2} A(t)W \quad (11)$$

with time-varying coefficients and the initial condition $W(0) = I_n$. In some special cases the system (11) can be integrated. For example, this situation occurs when the elements of matrix A are T-periodic time functions and the system (11) is equivalent to a certain linear time-invariant system. Let $A(t) = A(t+T)$ $\forall t \in R$, where $T \in R$. If the matrix

$$S = W_T^{-1} S_T \Big\{ \ddot{(S_T^{-1})} + S_T^{-1} B_T (2I_n + B_T) - A\big[\dot{(S_T^{-1})} + S_T^{-1} B_T\big] + \dot{A} S_T^{-1} \Big\} W_T$$

where $W_T = e^{B_T t}$, $B_T = \frac{1}{T} \ln W(T)$, $S_T = W_T W^{-1}$, is a symmetric one, then the Lagrangian function may be found in the form

$$\mathcal{L} = \frac{1}{2} \dot{q}^T \dot{q} - \frac{1}{2} q^T S q + q^T e \qquad (12)$$

or

$$\mathcal{L} = \frac{1}{2} \dot{q}^T \dot{q} + \dot{q}^T (S^T dt) q + q^T e \qquad (13)$$

where $e = W^{-1}(\dot{B}v + B\dot{v})$.

The case when $A(t) = A(t+T)$ and $B(t) = B(t+T)$ $\forall t \in R$ is of special interest. This results from the possibility of the determination of the periodic solutions of linear time-varying networks with the aid of direct variational methods if the period of oscillations is a priori known. In this approach, a solution would be sought by using optimization techniques on the action functional [1].

4. EXAMPLE OF LINEAR NETWORK

As an example of the linear network described by the system (2) for n = 1, consider a simple time-invariant network of Figure 1.

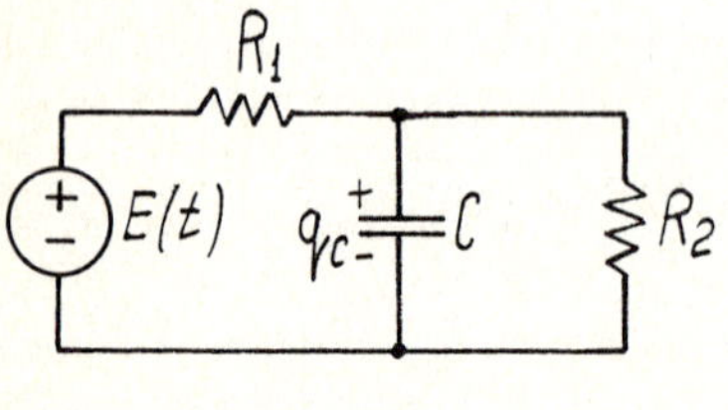

Figure 1

The system (2) takes the form

$$x = - \frac{R_1 + R_2}{R_1 R_2 C} x + \frac{1}{R_1} v \qquad (14)$$

where $x = [q_c]$ is the charge of the capacitor and $v = [E(t)]$ is the input voltage. The Lagrangian function of the network may be found in the form (4), (5) or (6), (7), (8), where

$$W^{-1} = e^{\frac{R_1 + R_2}{2R_1R_2C} t}, \quad A^2 = \frac{(R_1 + R_2)^2}{R_1^2 R_2^2 C^2},$$

$$D = e^{\frac{R_1 + R_2}{R_1R_2C} t}, \quad e = \frac{v}{R_1} e^{\frac{R_1 + R_2}{2R_1R_2C} t},$$

$$e' = \frac{\dot{v}}{R_1} - \frac{R_1 + R_2}{R_1^2 R_2 C} v, \quad e'' = \frac{\dot{v}}{R_1} e^{\frac{R_1 + R_2}{R_1R_2C} t}.$$

From Theorem 1 and Remark 1 it follows that in the considered case the action functional attains a strong global minimum in the set Q (X), at the curve $q^s(t) = W^{-1} x^s(t)$ $(x^s(t))$, where $x^s(\cdot)$ denotes the solution of equation (14). If the network of Figure 1 is time-varying, e.g. $R_1 = R_1(t)$, $R_2 = R_2(t)$ and $C = C(t)$ are C^1-time functions, then the Lagrangian function of the network may be found in the form (9) or (10), where

$$S = -(A^2 + \dot{A}) = - \frac{(R_1 + R_2)^2 + R_1^2 \dot{R}_2 C}{R_1^2 R_2^2 C^2} - \frac{\dot{R}_1 R_2^2 C + (R_1 R_2^2 + R_1^2 R_2)\dot{C}}{R_1^2 R_2^2 C^2} \quad \text{and}$$

$$e = \frac{\dot{v}}{R_1} - \frac{v}{R_1^2}\Big(\dot{R}_1 + \frac{R_1 + R_2}{R_2 C}\Big).$$

From Remark 1 it follows that if $(R_1 + R_2)^2 + (R_1^2 \dot{R}_2 + \dot{R}_1 R_2^2)C + (R_1 R_2^2 + R_1^2 R_2)\dot{C} \geqslant 0$, then the action functional with the Lagrangian function in the form (9) or (10) attains a strong global minimum in the set X at the solution of equation (14).

5. NONLINEAR NETWORKS

Consider a class of nonlinear networks described by the 2n-th order system of differential equations

$$L\ddot{q} + (K^{cv} + R)q + F(q,t) = E(t) \qquad (15)$$

where $q \in R^n$, $L, K^{cv}, R \in R^{n \times n}$, $F(\cdot) \in C^1([t_0, t_f]; R^n)$, $E(\cdot) \in L_1^n(t_0, t_f)$ and $L_1^k(t_0, t_f)$, $k \in N$ denotes the totality of k-dimensional vector-valued functions having absolute values of coordinates integrable on the interval $[t_o, t_f]$. The system (15) may be derived, for instance, with the aid of loop or node method. In the case when q is the vector of loop charges, the matrices entering into equation (15) are the respective loop matrices, e.g.: L is a symmetric positive semi-definite loop inductance matrix, R is a symmetric positive semi-definite loop resistance matrix, K^{cv} is loop matrix of the coefficients of linear current controlled voltage sources, $F(\cdot)$ is a nonlinear vector-valued function whose coordinates are loop capacitance voltages and $E(\cdot)$ is loop source voltages vector.

When constructing the Lagrangian function of the considered class of networks two special cases occur:

1. The matrix $\tilde{A} := (K^{cv} + R)L^{-1}$ is diagonalizable (in particular, the matrix $\tilde{A}$ possesses distinct eigenvalues).
2. The matrix $\tilde{A}$ is not diagonalizable.

Note that if L is a symmetric positive definite matrix and $(K^{cv} + R)$ is a symmetric matrix, then the matrix $\tilde{A}$ is diagonalizable. Indeed, in this case there exists a transformation which diagonalizes matrices L and $(K^{cv} + R)$. The transformations are written as follows

$$TLT^T = I_n \quad \text{and} \quad T(K^{cv} + R)T^T = \tilde{\Lambda}$$

where $\tilde{\Lambda} = \mathrm{diag}(\tilde{\lambda}_1, \dots, \tilde{\lambda}_n)$, $\tilde{\lambda}_i \in \mathrm{Spect}(\tilde{A})$, $i = 1,2,\dots,n$, $T \in R^{n\times n}$. Hence, $\tilde{A} = \tilde{H}\tilde{\Lambda}\tilde{H}^{-1}$ is the diagonalizable matrix, and $\tilde{H} = T^{-1}$ is the corresponding modal matrix. Assume that the matrix $\tilde{A}$ is diagonalizable $(\tilde{A} = \tilde{H}\tilde{\Lambda}\tilde{H}^{-1})$ and

$$F(q,t) = \int_{\Theta_n}^{q} S(q)dq, \text{ where } S(\cdot) =$$

$= [s_{ij}(\cdot)] \in R^{n\times n}$, $s_{ii}(\cdot) \in C^0(R; R)$, $s_{ij}(\cdot) = \mathrm{const}\ \forall i \neq j$, $i,j = 1,2,\dots,n$.

In the case when q is the vector of loop charges, the matrix $S(\cdot)$ is the loop elastance matrix. The restrictions imposed on the elements of this matrix are equivalent to the requirement that the constitutive relations of nonlinear capacitances are represented by the functions of only one variable of configuration space. The Lagrangian function may be obtained from the equality

$$\left(\frac{\partial^2 \mathcal{L}}{\partial \dot{q}^2}\right)\ddot{q} + \left(\frac{\partial^2 \mathcal{L}}{\partial \dot{q}\partial q}\right)\dot{q} + \frac{\partial}{\partial t}\left(\frac{\partial \mathcal{L}}{\partial \dot{q}}\right) - \frac{\partial \mathcal{L}}{\partial q} =$$

$$= D(t)\left[L\ddot{q} + (K^{cv} + R)\dot{q} + F(q,t) - E(t)\right] = \Theta_n \qquad (16)$$

where $D(\cdot) = [d_{ij}(\cdot)] \in R^{n\times n}$ is nonsingular matrix whose elements are C^1-time functions. Assume that $DS = (DS)^T$. In this case it is possible to construct the Lagrangian function. Indeed, by solving the system of equations $\frac{\partial^2 \mathcal{L}}{\partial \dot{q}^2} = DL$, where $DL = L^T D^T$, following from (16), we obtain

$$\mathcal{L} = \frac{1}{2}\dot{q}^T DL\dot{q} - \left[\int_{\Theta_n}^{q} dq^T\left(\int_0^t Ddt\right)S\right]\dot{q} + q^T DE - \varphi_1 \qquad (17)$$

or

$$\mathcal{L} = \frac{1}{2}\dot{q}^T DL\dot{q} + q^T DE - \varphi_2 \qquad (18)$$

where $\varphi_1 = 2\varphi_2 = 2\int_\Gamma\left(\int_{\Theta_n}^{q} dq^T DS\right)dq$. The symbol $\int_\Gamma f^T(q,t)dq$, where $f(\cdot): R^{n+1} \to R^n$, denotes the line integral in some simply connected domain of configuration space. It is independent of the path of integration between Θ_n and $q(t)$. The dissipation matrix D is obtained from the system of equations $\dot{D}L = D(K^{cv} + R)$, derived by substituting $\mathcal{L}$ from (17) or (18) into (16). By using the condition $DL = L^T D^T$, we obtain $D = L^T(\tilde{H}^{-1})^T\tilde{H}^{-1}e^{\tilde{\Lambda}t}$ or $D = (\tilde{H}^{-1})^T\tilde{H}^{-1}e^{\tilde{A}t}L^{-1}$. The formulae (17) and (18) were obtained under the assumption that $\det D(t) \neq 0$ $\forall t \in [t_o, t_f]$ and $\det L \neq 0$.

From Liouville's formula it follows that $\det D(t) \neq 0\ \forall\, t \in [t_o, t_f]$, if $-\infty < t_o < t_f < \infty$. If $t_f = \infty$, nonsingularity of the matrix $D(t_f)$ requires $\mathrm{tr}\tilde{A} \geq 0$.

If $\mathrm{tr}\tilde{A} < 0$, then the matrix $\tilde{A}$ possesses at least one eigenvalue with negative real part and the trivial solution of the homogeneous system $\ddot{q} = -L^{-1}\tilde{A}L\dot{q}$ describing the unexcited network with short-circuited capacitances (if q is the vector of loop charges) is Lyapunov unstable.

If the matrix L is not invertible some of the variables $q_{k_1}, \dots, q_{k_m}$, $k_i \in \{1,2,\dots,n\}$, $i = 1,2,\dots,m$, $m \leq n$, should be excluded from the system (15). The parasitic inductances may also be added to the independent R, C, CE, RC, RE and RCE loops. The following criterion for invertibility of the loop inductance matrix may be used.

If there exists a cotree containing n inductances on the network graph, then $\det L \neq 0$. Observe that in some cases the system (15) with invertible matrix L can be obtained even if the loop inductance matrix is not invertible. This situation, for instance occurs if the network contains linear R, C, CE, RC, RE or RCE loops. Thus the required form of the network equations may be obtained by differentiating the respective loop equations.

Note that formulae (17) and (18) are valid for the networks with the symmetric or orthogonal matrix $\tilde{A}$ with arbitrary eigenvalues. If $\tilde{A} = \tilde{A}^T$, then $D = Le^{\tilde{A}t}$.

Assume that the matrix $\tilde{A}$ is not diagonalizable. In this case the Lagrangian function can be obtained, if the motion of the network is considered in the transformed configuration space. The new generalized coordinates $\tilde{q}_1,\ldots,\tilde{q}_n$ are obtained from $q_1,\ldots,q_n$ by performing the linear transformation W^{-1}: $R^n \ni q \to \tilde{q} \in R^n$ with nonsingular matrix $W^{-1}(t) = e^{-0.5\tilde{A}t}$ $\forall t \in [t_o,t_f]$, where $\tilde{q} = [\tilde{q}_1 \ldots \tilde{q}_n]^T$. From Liouville's formula it follows that $\det W^{-1}(t) \neq 0$ $\forall t \in [t_o,t_f]$, if $-\infty < t_o < t_f < \infty$.

If $t_f = \infty$, nonsingularity of the matrix $W^{-1}(t_f)$ requires that $\mathrm{tr}\tilde{A} \geqslant 0$.

Note that in the case when q is the vector of loop charges and the linear network obtained from the original one by short-circuiting capacitances is stable, then $\mathrm{tr}\tilde{A} \geqslant 0$. The Lagrangian function can be represented in the form of the sum

$$\tilde{\mathcal{L}} = \frac{1}{2}\dot{\tilde{q}}^T\dot{\tilde{q}} + a^T(\tilde{q},t)\dot{\tilde{q}} + b(\tilde{q},t) + \tilde{q}^T e(t) \qquad (19)$$

where $e(t) = W^{-1}(t)L^{-1}E(t)$. The vector-valued function $a(\cdot)$ and the scalar function $b(\cdot)$ satisfy the condition

$$\frac{\partial a}{\partial t} - \frac{\partial b}{\partial \tilde{q}} = \tilde{S} \qquad (20)$$

where $\tilde{S} = \tilde{S}(\tilde{q},t) = -\frac{1}{4}L^{-1}\tilde{A}^2L\tilde{q} + W^{-1}(t)L^{-1}F(W(t)\tilde{q},t)$. The condition (20) can be obtained by substituting $\tilde{\mathcal{L}}$ from (19) into the Euler-Lagrange system of equations and comparing the corresponding terms with the respective terms of the network transformed equations. In order to obtain the function $\tilde{\mathcal{L}}$ in form (19) the restriction must be imposed on the vector-valued function $\tilde{S}$. This restriction is of the form $\partial\tilde{S}/\partial\tilde{q} = (\partial\tilde{S}/\partial\tilde{q})^T$ and arises due to the requirement that the motion of the network in the transformed configuration space should be described by the Euler-Lagrange system of equations. The functions $a(\cdot)$ and $b(\cdot)$ may be represented, for instance, in the form

$a(\tilde{q},t) = \int_t \tilde{S}(\tilde{q},t)dt$ and $b(\tilde{q},t) = 0$ or

$a(\tilde{q},t) = \Theta_n$ and $b(\tilde{q},t) = -\int_\Gamma \tilde{S}^T(\tilde{q},t)d\tilde{q}$.

The symbol $\int_t g(\tilde{q},t)dt$, where $g(\cdot)$: $R^{n+1} \to R^n$, denotes the indefinite integral with respect to t with $\tilde{q}_1,\ldots,\tilde{q}_n$ treated as the parameters.

Let $q^s(t) \in R^n$ be the solution of the system (15) satisfying the boundary conditions $q^s(t_o) = q_o$ and $q^s(t_f) = q_f$. Consider the sets of functions $Q := \{q \mid q \in C^1([t_o,t_f];R^n),\ q(t_o) = q_o,\ q(t_f) = q_f\}$ and $\tilde{Q} := \{\tilde{q} \mid \tilde{q} \in C^1([t_o,t_f];R^n),\ \tilde{q}(t_o) = W^{-1}(t_o)q_o,\ \tilde{q}(t_f) = W^{-1}(t_f)q_f\}$.

If $t_f < \infty$ and $\mathcal{L}, \tilde{\mathcal{L}} \in C^3(R^n x R^n x [t_o,t_f];R)$, then the following theorems hold true

<u>Theorem 2</u>. If $\mathrm{Spect}(\tilde{A}) \subset R$ and the matrix $-DS$ is positive semi-definite in some region $\Omega \subset R^n x [t_o,t_f]$, containing $q^s(\cdot)$, then $J[q]$ has a strong local minimum in Q at $q^s(\cdot)$.

Proof. Under the hypothesis of Theorem 2 $P_{\mathcal{L}} = \frac{1}{2}D(t)L$ is a positive definite matrix and $Q_{\mathcal{L}} = -\frac{1}{2}D(t)S(q^s)$, where $P_{\mathcal{L}}$ and $Q_{\mathcal{L}}$ denote matrices defined in the proof of Theorem 1. Assume that the matrix $-DS$ is positive-semi-definite $\forall [q^T\ t]^T \in \Omega$. From the proof of Theorem 1 it follows that in this case the quadratic functional $\delta^2 J[q^s][h]$ is positive definite. Hence, the functional $J[q]$ has a strong local minimum in Q at $q^s(\cdot)$. Q.E.D.

<u>Theorem 3</u>. If the matrix $-\partial\tilde{S}/\partial\tilde{q}$ is positive semi-definite in some region $\tilde{\Omega} \subset R^n x [t_o,t_f]$, containing the curve $\tilde{q}^s(t) = W^{-1}(t)q^s(t)$, then the functional $J[\tilde{q}] = \int_{t_o}^{t_f} \tilde{\mathcal{L}}(\tilde{q},\dot{\tilde{q}},t)dt$ has a strong local minimum in $\tilde{Q}$ at $\tilde{q}^s(\cdot)$. The proof of this theorem is analogous to the proof of Theorem 2 and therefore will be omitted.

For a certain class of nonlinear networks described by the second order autonomous system of state equations

$$\dot{x} = f(x) \qquad (21)$$

where $x = [x_1\ x_2]^T \in R^2$, $f(\cdot) \in C^1(R^2;R^2)$ the Lagrangian function can be obtained in a different way. In this case the Hamiltonian form of state equations is used.

Lemma [7]. Let $\bar{A} \in R^{2x2}$ be an orthogonal matrix of the form $\bar{A} = \begin{bmatrix} 0 & -1 \\ 1 & 0 \end{bmatrix}$. If there exists a function $\eta(\cdot) : R^2 \ni x \to R$ such that function $\bar{f}(\cdot) := \frac{1}{\eta(\cdot)} \bar{A} f(\cdot)$ is a C^1-state function mapping R^2 into itself, then the function $\bar{F}: R^2 \ni x \to R$ defined as the line integral $\bar{F}(x) := \int_{\Gamma} \bar{f}^T(x)dx$ (independent of the path of integration between Θ_2 and $x(t)$) is an integral of motion of the system (21).

The proof of this lemma is given in Reference [7].

Consider the network $\mathcal{N}$ shown in Figure 2.

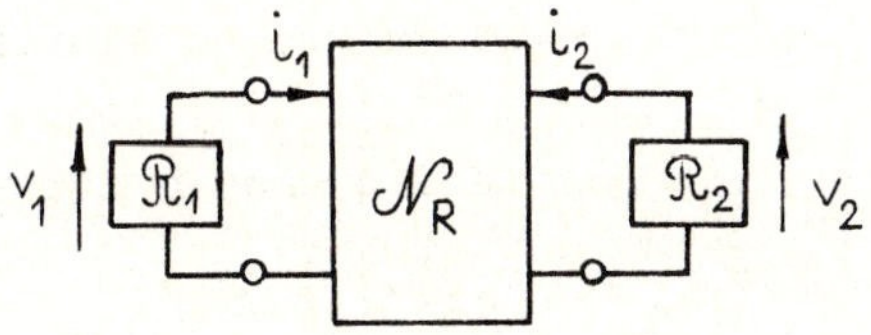

Figure 2

$\mathcal{N} = \mathcal{N}_R \cup \mathcal{R}_1 \cup \mathcal{R}_2$, where $\mathcal{R}_1$ and $\mathcal{R}_2$ denote two-terminal (generally nonlinear) reactive elements. The constitutive relation of the resistive two-port $\mathcal{N}_R$ is given by

$$y = g(z) \qquad (22)$$

where $g(\cdot) \in C^1(R^2;\ R^2)$. Substituting the component equations of reactive elements into equation (22) we obtain the system (21), where $f(x) = g(h(x))$ and $h(\cdot) \in C^1(R^2;\ R^2)$. Under the hypothesis of lemma this system can be rewritten in the form

$$\dot{x} = \eta(x) \bar{A}^T \frac{\partial \bar{F}(x)}{\partial x}.$$

Thus the following theorem holds true

Theorem 4. If there exists a function $\eta(\cdot) \in C^1(R^2;\ R)$ such that $\bar{f}(\cdot) = \frac{1}{\eta(\cdot)} \bar{A} g(h(\cdot))$ is a C^1-state function, then the Lagrangian function of the network shown in Figure 2 may be found in the form

$$\mathcal{L} = \dot{x}_2 \int_0^{x_1} \frac{1}{\eta(y,x_2)} dy + r_p(x_2)\dot{x}_2 + \bar{F}(x_1,x_2) \qquad (23)$$

where $r_p(\cdot) \in C^1(R;\ R)$.

Proof. Function (23) is the Lagrangian function of a given network if and only if the following system of the Euler-Lagrange equations

$$\frac{d}{dt}\left(\frac{\partial \mathcal{L}}{\partial \dot{x}}\right) - \frac{\partial \mathcal{L}}{\partial x} = \Theta_2$$

is satisfied. Substituting $\mathcal{L}$ from (23) into this system we obtain the system (21). Q.E.D.

Note that formula (23) contains only one of coordinates of the generalized velocity vector - $\dot{x}_2$. The dual function $\mathcal{L}$ has the form

$$\mathcal{L} = -\dot{x}_1 \int_0^{x_2} \frac{1}{\eta(x_1,y)} dy + r_d(x_1)\dot{x}_1 + \bar{F}(x_1,x_2) \qquad (24)$$

where $r_d(\cdot) \in C^1(R;\ R)$.

6. EXAMPLES OF NONLINEAR NETWORKS

As an example of the nonlinear network with a diagonalizable matrix $\tilde{A}$, consider the fifth order network of Figure 3.

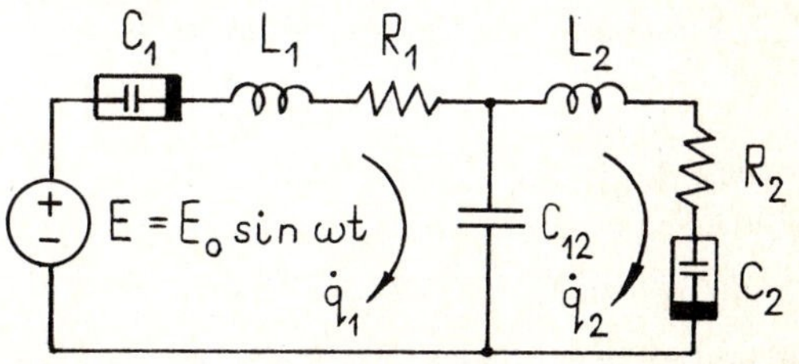

Figure 3

Assume for simplicity that $R_1 = R_2 = 1\Omega$, $L_1 = L_2 = 1H$, $C_i = (2q_i)^{-1}$, $i = 1,2$, $C_{12} = 1F$, $E_o = 1V$ and $\omega = 1s^{-1}$, where $q = [q_1\ q_2]^T$ is the vector of loop charges. Matrix $\tilde{A}$ in this case is a diagonal matrix with the eigenvalues $\lambda_1 = \lambda_2 = 1s^{-1}$. The Lagrangian function of the network of Figure 3 can be represented in the form (18) for $n = 2$

$$\mathcal{L} = e^t\left(\frac{1}{2}\dot{q}_1^2 + \frac{1}{2}\dot{q}_2^2 - \frac{1}{2}q_1^2 - \frac{1}{2}q_2^2 - \right.$$

$- \frac{1}{3} q_1^3 - \frac{1}{3} q_2^3 + q_1 q_2 + q_1 \sin t).$

Consider the circuit model shown in Figure 4.

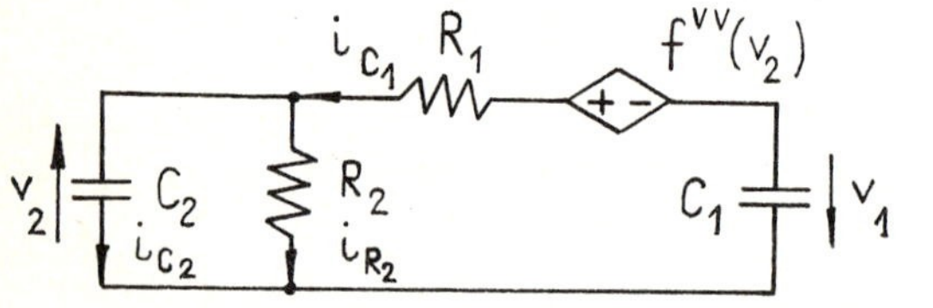

Figure 4

Its mathematical description is given by the second order autonomous system of state equations

$$\dot{v}_1 = \frac{1}{R_1C_1}\left[f^{vv}(v_2) - v_2 - v_1\right]$$
$$\dot{v}_2 = \frac{1}{R_1C_2}\left[f^{vv}(v_2) - \frac{R_1+R_2}{R_2} v_2 - v_1\right].$$

Choose $R_1 = R_2$, $C_1 = C_2$ and $f^{vv}(v_2) = 3v_2$. Let $x_1 = v_1$, $x_2 = v_2$, $\eta = 1/(R_1C_1)$. Thus the function $\overline{f}(x) = \overline{A}f(x)/\eta = \left[x_1 - x_2 \;\; 2x_2 - x_1\right]^T$ is well defined for all $x \in R^2$. Observe that $\overline{f}(x)$ is a C^1-state function and $\overline{F}(x) = \frac{1}{2} x_1^2 + x_2^2 - x_1x_2$. The Lagrangian function of the network is given by formula (23) which in this case takes the form

$$\mathcal{L} = R_1C_1x_1\dot{x}_2 + r_{pe}(x_2)\dot{x}_2 + \frac{1}{2} x_1^2 + x_2^2 - x_1x_2$$

where $r_{pe}(\cdot) \in C^1(R; R)$.

7. CONCLUDING REMARKS

The results presented in the paper may be applied in the network modeling of variational problems. Thus the solutions of these problems may be obtained by the use of widely available circuit simulation programs. There also exists a possibility of determination of the periodic solutions of nonlinear and linear time-varying networks with the aid of direct variational methods. The results obtained in the work may be effective when used for the qualitative analysis of the behaviour of dynamical systems and networks [4],[6] as well as in the optimal-control theory and network synthesis.

REFERENCES

[1] L.O. Chua and J.D. McPherson, "Explicit topological formulation of Lagrangian and Hamiltonian equations for nonlinear networks", IEEE Trans. Circuits Syst., vol. CAS-21, pp. 277-286, Mar. 1974.

[2] I.M. Gelfand and S.V. Fomin, "Calculus of Variations". Englewood Cliffs, N.J.: Prentice-Hall, 1972.

[3] T. Kacprzak and M. Kowalski, "Formulation of the Lagrangian functions for certain class of nonlinear electrical networks", Proc. 5th International Symposium on Network Theory, p. 238, (Sarajevo, Yugoslavia), 1984.

[4] T. Kacprzak and M. Kowalski, "The Hamiltonian approach to stability analysis of a certain class of linear time-varying networks", Proc. 7th European Conference on Circuit Theory and Design, pp. 232-235, (Prague, Czechoslovakia), 1985.

[5] S.G. Mikhlin and K.L. Smolitski, "Approximate Methods for Solutions of Differential and Integral Equations". New York: Elsevier, 1967.

[6] R.A. Rohrer, "Analytical mechanics and the analysis of linear time-varying networks", J. Franklin Inst., vol. 278, no. 2, pp. 84-107, Aug. 1964.

[7] A. Szatkowski, "Remarks on the existence of periodic solutions of nonlinear autonomous systems and networks", Rozprawy Elektrotechniczne, vol. 27, no. 2, pp. 327-338.

AUTHOR INDEX